Nanosatellites

Nanosatellites

Space and Ground Technologies, Operations and Economics

Edited by

Rogerio Atem de Carvalho
Reference Center for Embedded and Aerospace Systems (CRSEA)
Polo de Inovação Campos dos Goytacazes (PICG)
Instituto Federal Fluminense (IFF)
Brazil

Jaime Estela
Spectrum Aerospace Group
Germering
Germany

Martin Langer
Institute of Astronautics
Technical University of Munich
Garching
Germany

and

Orbital Oracle Technologies GmbH
Munich
Germany

The right of Rogerio Atem de Carvalho, Jaime Estela, and Martin Langer to be identified as the authors of the editorial material in this work has been asserted in accordance with law.

Registered Office
John Wiley & Sons, Inc., 111 River Street, Hoboken, NJ 07030, USA

Editorial Office
The Atrium, Southern Gate, Chichester, West Sussex, PO19 8SQ, UK

For details of our global editorial offices, customer services, and more information about Wiley products, visit us at www.wiley.com.

Wiley also publishes its books in a variety of electronic formats and by print-on-demand. Some content that appears in standard print versions of this book may not be available in other formats.

Library of Congress Cataloging-in-Publication Data

Names: Carvalho, Rogério Atem de, author. | Estela, Jaime, 1972- author. |
    Langer, Martin, 1986- author.
Title: Nanosatellites : space and ground technologies, operations and
    economics / Professor Rogerio Atem de Carvalho, University of Fluminese,
    Rio, Brazil, Jaime Estela, Spectrum Aerospace Group, Germering, Germany,
    Martin Langer, Technical University of Munich & Orbital Oracle
    Technologies GmbH, Bavaria, Germany.
Description: First edition. | Hoboken, NJ : Wiley, [2020] | Includes
    bibliographical references and index.
Identifiers: LCCN 2019049523 (print) | LCCN 2019049524 (ebook) | ISBN
    9781119042037 (hardback) | ISBN 9781119042068 (adobe pdf) | ISBN
    9781119042051 (epub)
Subjects: LCSH: Microspacecraft.
Classification: LCC TL795.4 .C37 2020 (print) | LCC TL795.4 (ebook) | DDC
    629.46–dc23
LC record available at https://lccn.loc.gov/2019049523
LC ebook record available at https://lccn.loc.gov/2019049524

Cover Design: Wiley
Cover Image: © Stocktrek Images/Getty Images

Set in 9.5/12.5pt STIXTwoText by SPi Global, Chennai, India

Printed in the UK by Bell & Bain Ltd, Glasgow

10 9 8 7 6 5 4 3 2 1

Contents

List of Contributors

Fernando Aguado-Agelet
Department of Signal Theory and
Communications
University of Vigo
EE Telecomunicación
Spain

Lucas Rodrigues Amaduro
Reference Center for Embedded and
Aerospace Systems (CRSEA)
Polo de Inovação Campos dos Goytacazes
(PICG)
Instituto Federal Fluminense (IFF)
Brazil

Kelly Antonini
GomSpace A/S
Aalborg
Denmark

Nicolas Appel
Institute of Astronautics
Technical University of Munich
Garching
Germany

Scott Armitage
Space Flight Laboratory (SFL)
UTIAS,
Toronto
Canada

Alim Rüstem Aslan
Space Systems Design and Test Lab
Department of Astronautical Engineering
Istanbul Technical University
Turkey

Andrew Barron
Broadspectrum (New Zealand) Limited
Christchurch
New Zealand

Merlin F. Barschke
Institute of Aeronautics and Astronautics
Technische Universität Berlin
Germany

Cesar Bernal
ISIS – Innovative Solutions In Space B.V.
Delft
The Netherlands

Grant Bonin
Space Flight Laboratory (SFL)
UTIAS
Toronto
Canada

Eduardo Escobar Bürger
Federal University of Santa Maria (UFSM)
Brazil

Franciele Carlesso
National Institute for Space Research
São José dos Campos
Brazil

Nicolò Carletti
GomSpace A/S
Aalborg
Denmark

Rogerio Atem de Carvalho
Reference Center for Embedded and
Aerospace Systems (CRSEA)
Polo de Inovação Campos dos Goytacazes
(PICG)
Instituto Federal Fluminense (IFF)
Brazil

Chantal Cappelletti
University of Nottingham
United Kingdom

Michele Coletti
Mars Space Ltd.
Southampton
United Kingdom

Marcos Compadre
Clyde Space Limited
Glasgow
United Kingdom

Lucas Lopes Costa
National Institute for Space Research
São José dos Campos
Brazil

Kevin Cuevas
GomSpace A/S
Aalborg
Denmark

Matteo Emanuelli
GomSpace A/S
Aalborg
Denmark

Jaime Estela
Spectrum Aerospace Group
Germering
Germany

Katharine Brumbaugh Gamble
Washington D.C.
United States of America

Ausias Garrigós
Miguel Hernández University of Elche
Spain

Anna Gregorio
Department of Physics
University of Trieste
Italy

Philipp Hager
European Space Agency
Noordwijk
The Netherlands

Lucas Ramos Hissa
Reference Center for Embedded and
Aerospace Systems (CRSEA)
Innovation Hub
Instituto Federal Fluminense (IFF)
Campos dos Goytacazes
Brazil

Siegfried W. Janson
xLab
The Aerospace Corporation
El Segundo
United States of America

Richard Joye
KCHK – Key Capital Hong Kong Limited
Hong Kong

Christopher Kebschull
Institute of Space Systems
Technical University of Braunschweig
Germany

Kaitlyn Kelley
Spaceflight Industries
Seattle
United States of America

Matthias Killian
Institute of Astronautics
Technical University of Munich
Garching
Germany

Rolf-Dieter Klein
Multimedia Studio Rolf-Dieter Klein
München
Germany

Per Koch
GomSpace A/S
Aalborg
Denmark

David Krejci
ENPULSION
Wiener Neustadt
Austria

and

Massachusetts Institute of Technology
Cambridge
United States of America

Martin Langer
Orbital Oracle Technologies GmbH
Munich
Germany

and

Institute of Astronautics
Technical University of Munich
Garching
Germany

Vaios J. Lappas
Department of Mechanical Engineering
and Aeronautics
University of Patras
Greece

Jürgen Letschnik
Institute of Astronautics
Technical University of Munich
Garching
Germany

and

Airbus
Taufkirchen/Ottobrunn
Germany

Geilson Loureiro
Laboratory of Integration and Testing (LIT)
National Institute for Space Research
(INPE)
São José dos Campos
Brazil

Jean-Francois Mayence
Belgian Federal Science Policy Office
(BELSPO)
Brussels
Belgium

Mike Miller
Sterk Solutions Corporation
Philipsburg
United States of America

Sergio Montenegro
University Würzburg
Germany

Alberto González Muíño
University of Vigo
EE Telecomunicación
Spain

Flavia Tata Nardini
Fleet Space Technologies
Beverley
Australia

Josh Newman
Space Flight Laboratory (SFL)
UTIAS
Toronto
Canada

Neta Palkovitz
ISIS – Innovative Solutions In Space B.V.
Delft
The Netherlands

and

International Institute of Air and Space
Law (IIASL)
Leiden University
The Netherlands

Laura León Pérez
GomSpace A/S
Aalborg
Denmark

Jordi Puig-Suari
Cal Poly
Aerospace Engineering Department
San Luis Obispo
United States of America

Philipp Reiss
Institute of Astronautics
Technical University of Munich
Garching
Germany

Alexander Reissner
ENPULSION
Wiener Neustadt
Austria

Ben Risi
Space Flight Laboratory (SFL)
UTIAS
Toronto
Canada

Niels Roth
Space Flight Laboratory (SFL)
UTIAS
Toronto
Canada

Sebastian Rückerl
Institute of Astronautics
Technical University of Munich
Garching
Germany

Kenan Y. Şanlıtürk
Department of Mechanical Engineering
Istanbul Technical University
Turkey

Klaus Schilling
University Würzburg

and

Zentrum für Telematik
Germany

Daniel Smith
GomSpace A/S
Aalborg
Denmark

Willem Herman Steyn
University of Stellenbosch
South Africa

Enrico Stoll
Institute of Space Systems
Technical University of Braunschweig
Germany

Andrew Strain
Clyde Space Limited
Glasgow
United Kingdom

Murat Süer
Gumush AeroSpace & Defense
Maslak
Istanbul
Turkey

Bob Twiggs
Morehead State University
United States of America

Kirk Woellert
ManTech International supporting DARPA
Arlington
United States of America

Robert E. Zee
Space Flight Laboratory (SFL)
UTIAS
Toronto
Canada

Foreword: Nanosatellite Space Experiment

Bob Twiggs

Morehead State University, Morehead, USA

The use of small satellites in general initiated the space program in 1957 with the launching of Russian Sputnik 1, and then by the United States with Vanguard 1 satellite, which was the fourth artificial Earth orbital satellite to be successfully launched (following Sputnik 1, Sputnik 2, and Explorer 1).

The concept of the CubeSat was developed by Professor Bob Twiggs at the Department of Aeronautics and Astronautics at Stanford University in Palo Alto, CA, in collaboration with Professor Jordi Puig-Suari at the Aerospace Department at the California State Polytechnic University in San Luis Obispo, CA, in late 1999. The CubeSat concept originated with the spacecraft OPAL (Orbiting Picosat Automated Launcher), a 23 kg microsatellite developed by students at Stanford University and the Aerospace Corporation in El Segundo, CA, to demonstrate the validity and functionality of picosatellites and the concept of launching picosatellites and other small satellites on-orbit from a larger satellite system. Picosatellites are defined having a weight between 0.1 and 1 kg. OPAL is shown in Figure 1, with four launcher tubes containing picosatellites. One of the picosatellites is shown being inserted into the launcher tube in Figure 2.

The satellites developed by students within university programs in 1980s and 1990s were all nanosatellites (1–10 kg size) and microsatellites (10–50 kg size). The feasibility of independently funding launch opportunities for these nanosatellites and microsatellites was limited, as the costs typically were up to $250 000—a price point well beyond the resources available to most university programs. At that time, the only available option was to collaborate with government organizations that would provide the launch. The OPAL satellite was launched in early 2000 by the US Air Force Space Test Program (STP) with sponsorship from the Defense Advanced Research Projects Agency (DARPA) for the Aerospace Corporation picosatellites.

The OPAL mission represented a significant milestone in the evolution of small satellites by proving the viability of the concept of the picosatellite and an innovative orbital deployment system. The picosatellite launcher concept used for the OPAL mission represented a major advancement that would enable the technological evolution of small satellites, setting the stage for the development of the CubeSat form factor and the Poly Picosatellite Orbital Deployer (P-POD) orbital deployer system. OPAL demonstrated a new capability

Figure 1   Picosatellite loaded into OPAL.

Figure 2   OPAL and SAPPHIRE microsatellites.

with the design of an orbital deployer that could launch numerous very small satellites contained within the launcher tube that simplified the mechanical interface to the upper stage of the launch vehicle and greatly simplified the satellite ejection system. While the OPAL mission was extremely successful and established the validity of a picosatellite orbital deployer, Professor Twiggs and Professor Puig-Suari wanted to find a lower-cost means of launching the satellites built by university students. The stage was set for the development of the CubeSat form factor and its evolution toward an engineering standard.

CubeSat Engineering Design Standard

The primary intent of the development of the CubeSat standard was to provide a standard set of dimensions for the external physical structure of picosatellites that would be compatible with a standardized launcher. Unlike the development of most modern engineering standards, there was no consulting with other universities or with the commercial satellite industry to establish this standard because most other university satellite programs and commercial ventures were concentrating on larger satellites rather than smaller satellites. There were discussions in the late 1990s within the Radio Amateur Satellite Corporation (AMSAT) community in the United Kingdom centering on building a small amateur satellite, but there were never any attempts to develop a standardized design.

The concept of a design standard for a picosatellite and associated launcher that could be used by many universities, the developers believed, would lead to many picosatellites being launched at a time. They envisioned launch vehicles accommodating several launcher tubes, each containing a few picosatellites. The final concept of the CubeSat structural standard was developed by Professor Twiggs and Professor Puig-Suari, and currently adopted by the small satellite community. The developers believed that if one organization could provide the integration of the launcher with the launch vehicle through a carefully orchestrated interface process with the launch services provider, then it seemed possible to acquire launch opportunities for university programs that would be affordable (less than $50 000 per 1 kg satellite).

Evolution History of the CubeSat Program

The first CubeSats were launched on a Russian Dnepr in 2003 through the efforts of Professor Jordi Puig-Suari at Cal Poly. Professor Puig-Suari and his students through the CubeSat integration program at Cal Poly took the initial concept design, established the standards for the 1U CubeSat, designed the P-POD deployer, and planned for the Russian launch.

Initial reaction from the aerospace industry was quite critical of the CubeSat concept. The comments were—"stupidest idea for a satellite," "would have no practical value," "academic faculty did not have the capability to design and launch a satellite." This came mostly from the amateur satellite community that had established building and launching satellites many years prior to this academic program.

Fortunately, these comments did not deter the academic community from pursuing the CubeSat program. In 2008, the National Science Foundation had a conference to explore the use of the CubeSat to do space experiments for space weather. Their initiation and funding of using CubeSats for real scientific space experiments seemed to validate that the CubeSat concept had merit in space experiments.

Today: The CubeSat Concept

As of the present, the CubeSat concept is being called a disruptive technology. It seems to have been one of the new concepts in the space industry along with new launch concepts starting with SpaceX that has brought about a new interest in space. With the commercial programs from Planet, with CubeSat space imaging, and Spire with its multisatellite constellations, there is significant investment by the venture capital community in the space industry. To date, there have been more than 900 CubeSats launched since 2003.

The CubeSat concept from the original 1U CubeSat to the 3U CubeSat in the P-POD has expanded larger to now considering 27U concepts. One of the consequences of this new interest is that, to date, there have been more than 900 CubeSats launched in near-Earth orbit as well as two MarCO CubeSats to Mars, and there are plans to launch 13 6U CubeSats on the first Space Launch System (SLS) in the next Moon mission.

The Future of the CubeSat Concept

One of the consequences of the new acceptance of the CubeSat concept is that the cost of launch from the initial cost of $40 000 for a 1U from Cal Poly has now risen to over $120 000. This has had the greatest impact of having CubeSat programs for educational training and new entrants into space experimentation. There are several small launch vehicles in development to meet this demand, but whether they can launch for lower costs is debatable. One approach to reducing the launch cost is to use the same volume as provided by the P-POD or similar deployer, but keeping launch spacecraft smaller than the 1U, thus reducing the costs of individual experiment launch.

Cornell University has the ChipSats being launched from the 3U CubeSat, as shown in Figure 3. There is also the PocketQube being promoted by Alba Orbital, shown in Figure 4.

Figure 3 Cornell University ChipSats. Source: Image credit: NASA.

Figure 4 Alba Orbital PocketQube.

In addition to the conventional means of launches for the International Space Station (ISS) and from expendable launch vehicles, the Virginia Commercial Space Flight Authority, a state economic agency of the state of Virginia, along with Northrop Grumman Corp., is providing launches from the NASA Wallops Island flight facilities on the second stage of the Antares launch vehicle that is used to launch the Cygnus resupply capsule for the ISS. This is a unique launch opportunity not used previously. Even though it releases satellites from the Planetary Systems Corporation's canisterized satellite deployer (like the P-POD) at an altitude of near 250 km, it only provides an orbital life of the satellites for a few days. This short orbital lifetime of the satellites provides an excellent opportunity regarding science, technology, engineering, and mathematics (STEM) experience to students. In addition, all spacecrafts will deorbit, leaving no debris or collision problems.

The spacecraft proposed for this program is of a sub-CubeSat size called ThinSat™, shown in Figure 5.

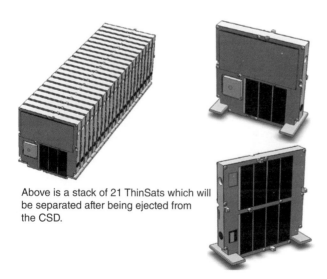

Above is a stack of 21 ThinSats which will be separated after being ejected from the CSD.

Figure 5 ThinSat sub-CubeSat satellites.

This program starting with launches in spring 2019 has the capability of launching 84 of the small satellites at one time. Also, the Antares launches every six months to resupply to the ISS. The goal for this STEM program is to provide a full year of education using real hardware to collect data, and launching a ThinSat and recovering space data for a total cost of less than $50 000.

Introduction by the Editors

Since the 1990s, Information and Communication Technologies (ICTs) have played a fundamental role in the restructuring of organizations, which have been able to horizontalize their structures and, through low-cost ICTs, to distribute and commoditize production. This impact came to be felt more strongly in the late 1990s and early 2000s in many areas of the economy. A strong tendency to "distribute" and "horizontalize" the production became commonplace.

Moreover, a need to have flexible systems to adapt to constant innovations not only in ICTs but also in electronics and materials has led production systems to move increasingly toward solutions that could be quickly prototyped, tested, implemented, and modified. ICTs have made it possible to bring production to a level of flexibility and innovation never seen before.

The space sector has not been left out of these trends, and small satellites have begun to show themselves, with all their known limitations, as technologically and economically viable platforms to test and even implement innovations. Moreover, players who until then were kept apart or operating marginally in space missions, such as universities, small businesses and research centers in countries with less space tradition, could now design and build spacecrafts almost from the bottom up.

Although small satellites have existed since the early days of spaceflight, such as Sputnik 1 itself, whose dimensions were 58.5 cm in diameter and 83.6 kg in weight, it was only with the launch of the CubeSat standard in 1999 that the growth of applications in small satellites has become exponential. In 1999, California Polytechnic State University and Stanford University developed the CubeSat specifications to promote and develop the skills necessary for the design, manufacture, and testing of small satellites intended for low Earth orbit (LEO) that perform a number of scientific research functions and explore new space technologies.

The CubeSat (U-class spacecraft) is a type of miniaturized satellite that is made up of multiple $10 \, \text{cm} \times 10 \, \text{cm} \times 10 \, \text{cm}$ cubic units. CubeSats have a mass of no more than 1.33 kg per unit, and often use commercial off-the-shelf (COTS) components for their electronics and structure. Over 1000 CubeSats have been launched as of January 2019. Over 900 have been successfully deployed in orbit and over 80 have been destroyed in launch failures.

Even before the success of CubeSats, the term "nanosatellites" was coined in 1992 and popularized. This term defines any satellite with a mass between 1 and 10 kg, including most CubeSats. Thus, seeking to treat smaller satellites more comprehensively, within a more flexible term, this book brings together a number of techniques, technologies, methodologies, and legislation applicable to nanosatellites.

Following this goal, this book is divided into three main parts:

I) *Nanosatellite Technologies:* This part presents the various technologies employed to design, test, and construct nanosatellites.

II) *Ground Segment:* Deals with the terrestrial segment and its peculiarities to deal with nanosatellites, operating alone or in networks.

III) *Policies, Legislation, and Economical Aspects:* This part deals with policies and legislation applicable to nanosatellites, and gives insight into economical aspects and future trends.

To deal with a relatively emerging and evolving subject, a team of prominent authors in the subareas described above has been brought together and, together with the editors, accepted the challenge of writing about an area of space technology that is certainly one of the most innovative and therefore in constant change.

Hence, this book seeks to treat nanosatellites through an integrated vision, focusing on technologies, methods, and techniques that form the foundation of the implementation and operation of these spacecrafts, while still employing real examples when necessary.

April 2019

Rogerio Atem de Carvalho
Jaime Estela
Martin Langer

1 I-1

A Brief History of Nanosatellites

Siegfried W. Janson

xLab, The Aerospace Corporation, Los Angeles, USA

1.1 Introduction

The term "nanosatellite" first appeared in print in a paper by the University of Surrey in 1992 [1]. Although originally defined as a spacecraft with a mass of less than 10 kg, nanosatellites are now more narrowly defined as spacecraft with a mass between 1 and 10 kg. Solid-state electronics and primitive solar cells enabled the first active nanosatellite, Vanguard 1, launched by the USA in 1958. It carried two continuous wave (CW) transmitters that enabled monitoring of spacecraft's internal temperatures and the total integrated electron density between the satellite and a ground station. This first nanosatellite is still in orbit but has been silent since 1964. Continued advancements in microelectronics/nanoelectronics, solar cells, microelectromechanical systems (MEMS), and computer-aided design (CAD) now enable visible imaging nanosatellites with 5 m ground resolution and nanosatellites that use global positioning system radio occultation measurements to measure ionospheric electron densities, tropospheric and stratospheric humidity levels, and temperatures. Much has happened in the intervening 60 years.

1.2 Historical Nanosatellite Launch Rates

Figure 1.1 shows the launch history of active nanosatellites from 1958 through 2017. Over 600 passive nanosatellites, operating basically as reflectors of radiofrequency (RF) radiation or as air drag monitors, were launched by the USA and the former Soviet Union during this period of time. These simple spacecraft did not have the typical spacecraft systems, such as power conditioning, command and control, and communications, so were therefore not included in Figure 1.1. Figure 1.1 shows historical data for active nanosatellites within the 1–10 kg mass range that were successfully launched and deployed into orbit. It includes all 1U CubeSats. The original CubeSat Specification Document mandated a mass limit of 1 kg per "U" of volume, but later versions increased the "U" mass to 1.33 kg. In addition, many early 1U CubeSats came within 10–20 g of the original 1000 g target, a mass deficiency

Nanosatellites: Space and Ground Technologies, Operations and Economics, First Edition.
Edited by Rogerio Atem de Carvalho, Jaime Estela, and Martin Langer.

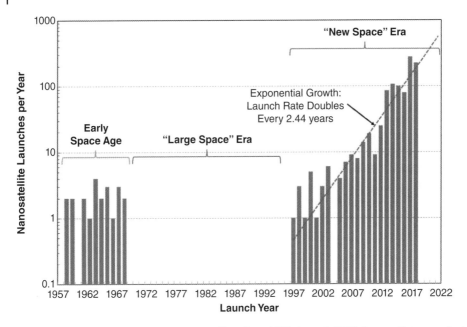

Figure 1.1 Yearly launch rates of nanosatellites from 1958 through 2017. Source: Data compiled from [2–4]. (*See color plate section for color representation of this figure*).

of 2% or less. All "sub-U" CubeSats (e.g. PocketQubes and SunCubes) are well within the picosatellite mass range and are not included in Figure 1.1. Note that the vertical scale in Figure 1.1 is logarithmic in order to show the dramatic rise in launch rates over the past two decades. This is an "integer logarithmic" plot, because the lowest launch rate has been labeled zero rather than the mathematically correct value of 0.1 for this plot. Since spacecraft come in integer values, this should be acceptable.

Figure 1.1 shows that nanosatellites were flown during the first decade of the Space Age at a rate of about two per year, and then disappeared for almost three decades. Hundreds of passive nanosatellites were flown during these three decades, along with many 11–13 kg mass, 23 cm^3 "almost" nanosatellites, but no active nanosatellites. Active nanosatellites finally reappeared in 1997, with launch rates doubling every 2.44 years since then. Launch rate predictions for the next 4 years are consistent with another 4 years of similar exponential growth. This remarkable, two-decade-long exponential growth parallels Moore's law for microelectronic/nanoelectronics—the doubling of performance every 2–3 years due to a continual reduction in transistor size. Exponential growth of nanosatellite launch rates was initiated by a combination of continuing advancements in microprocessor performance, memory storage capability, microelectromechanical sensors and actuators, image sensors, the development of CubeSats, and commercial forces.

Two nanosatellites from the Technical University of Berlin, Tubsat-N and Tubsat-N1, were inexpensively launched in 1998 from a Russian submarine as a commercial service [5]. CubeSats were launched starting in 2003 on relatively inexpensive Russian Dnepr launch vehicles, also based on converted intercontinental ballistic missiles (ICBMs). Newly minted Russian capitalists were able to compete with established Western launch providers by repurposing military hardware. Continued exponential growth of nanosatellite launch rates in the 2000s was due to government and university projects, while commercial exploitation of new space markets using CubeSats has been driving exponential growth in the 2010s.

1.3   The First Nanosatellites

Table 1.1 shows the first nanosatellites put into orbit, or beyond, from the start of the Space Age through 1968. Note that all of these spacecraft are from the USA. The former Soviet Union started the Space Age with larger launch vehicles that enabled heavier spacecraft using heavier components and systems. The 83.6 kg mass Sputnik 1, launched on October 4, 1957, became humanity's first artificial satellite. It had a polished 58.5 cm diameter sphere with a pressurized internal volume that contained 51 kg of batteries to drive a 1 W output transmitter alternating between 20.005 and 40.002 MHz. The 3.5 kg mass transmitter

Table 1.1   The first successfully launched nanosatellites from 1957 through 1968.

Name	Launch date	Mass (kg)	Perigee (km)	Apogee (km)	Inclination (degrees)	Country
Vanguard 1	17 March 1958	1.47	652	3961	34.3	USA
Pioneer 3	06 December 1958	5.88		102 300		USA
Vanguard 2	17 February 1959	10.75	558	3321	32.9	USA
Pioneer 4	03 March 1959	5.88	Lunar Flyby	Lunar Flyby		USA
Explorer 9	16 February 1961	7	636	2582	38.6	USA
OSCAR 1	12 December 1961	4.5	235	415	81.2	USA
OSCAR 2	6 February 1962	4.5	208	386	74.3	USA
Flashing Light	15 May 1963	5	161	267	32.5	USA
ERS 9	19 July 1963	1.5	3662	3731	88.4	USA
ERS 12	17 October 1963	2.1	208	103 830	36.7	USA
Explorer 19	19 December 1963	7	589	2393	78.6	USA
ERS 13	17 July 1964	2.1	193	104 400	36.7	USA
Explorer 24	21 November 1964	9	334	1551	81.4	USA
ERS 17	20 July 1965	5.4	208	112 184	35	USA
ERS 16	9 June 1966	5	180	3622	90	USA
ERS 15	19 August 1966	5	3669	3701	90.1	USA
ERS 18	28 April 1967	9	8619	111 530	32.9	USA
ERS 20	28 April 1967	8.6	8619	111 530	32.9	USA
ERS 27	28 April 1967	6	8619	111 530	32.9	USA
Explorer 39	8 August 1968	9.4	673	2533	80.6	USA
ERS 28	26 September 1968	10	175	35 724	26.3	USA

Inclination and perigee values are given for spacecraft that completed more than one Earth orbit.
Source: Data compiled from [6–8].

used vacuum tubes and required a fan to provide temperature control [9]. Sputnik 1 was designed to produce a strong radio signal that could be received by millions of common shortwave receivers all across the planet, and to be visible in small telescopes. Sputnik 2, at 508 kg mass, was launched on November 3, 1957. It contained the backup satellite to Sputnik 1, plus an additional science instrument package, as well as a pressurized capsule that held the small dog named Laika. The Soviets launched the 1327 kg mass Sputnik 3 into orbit on April 27, 1958. The Soviets were focused on putting the first human into space, and that occurred on April 12, 1961 when Yuri Gagarin took the 4730 kg mass Vostok 1 into orbit. Whether hindered by their much heavier spacecraft design, or blessed by their massive launch capability, the former Soviet Union had little need for active nanosatellites. The United Kingdom, Canada, Italy, France, Australia, West Germany, the People's Republic of China, and Japan also lofted their own satellites into orbit through 1970 using either US or indigenous launch vehicles, but these were microsatellites (10–100 kg mass) and minisatellites (100–500 kg mass) [10].

Vanguard 1 was the USA's second artificial satellite, but it was also the first nanosatellite from any country to reach orbit, the first satellite with solar cells, and is now the oldest man-made object still in space. Figure 1.2 shows a photograph of a Vanguard 1 model at the Smithsonian National Air and Space Museum's Steven F. Udvar-Hazy Center, Chantilly, Virginia, USA. Like Sputnik 1, it was spherical and shiny, but it was a lot smaller at 16.5 cm in diameter. Radio output power was 10 mW at 108 MHz for the battery-powered transmitter, and 5 mW at 108.030 MHz for the solar cell-powered transmitter. These transmitters used one transistor each, not vacuum tubes, operated in a vacuum. Transistors were first demonstrated in 1947 in the USA, and offered more efficient, reliable, and compact electronics due to the elimination of the hot, glowing filaments required in vacuum tubes. Note that both Sputnik 1 and Vanguard 1, and many early spacecraft, did not contain radio receivers; they could not be controlled from the ground. Attitude control, either active or passive, was not included. Thermal control was provided by the polished outer shell, with an internal instrument capsule thermally and electrically isolated from the outer

Figure 1.2 Photo of Vanguard 1 at the Smithsonian National Air and Space Museum's Steven F. Udvar-Hazy Center, Chantilly, Virginia. Source: Photo by author.

shell using Teflon rods. This vacuum Dewar approach effectively averages out external shell temperature extremes caused by exposure to sunlight, followed by eclipse, and return to sunlight.

The next nanosatellite put into space was Pioneer 3. This was NASA's third attempt to at least fly by the Moon, but an upper-stage malfunction prevented attainment of Earth escape velocity. It reached an altitude of 102 300 km and then. fell back to Earth, where it re-entered 38 hours after launch. Pioneer 3 was basically cone-shaped with a 58 cm high, 25 cm diameter thin fiberglass conical shell covering a squat instrument base [11]. The shell was coated in gold to provide electrical conductivity, with stripes of paint to adjust the over-all thermal balance. Pioneer 3 was spin-stabilized along the long axis, so this axis had to be a principal axis with the maximum moment of inertia to maintain spin stability. Pioneer 3 had two Geiger–Müller (GM) tubes and a 0.5 kg mass transmitter producing 0.18 W at 960.05 MHz. Pioneer 3 transmitted radiation data for 25 hours before re-entry, and these data showed the existence of a second radiation belt at higher altitudes. The inner radiation belt was discovered by Explorer 1, and both radiation belts are now named after Dr. James Van Allen, professor of physics and astronomy at the University of Iowa.

Figure 1.3 shows a photo of Pioneer 4, a copy of Pioneer 3 with one GM tube replaced by a lunar photosensor and a camera, held by Dr. Wernher von Braun, John Casani, and Dr. James Van Allen. The layout of the components, particularly the outer ring of mercury batteries, provided the needed maximum moment of inertia about the spin axis for both Pioneer 3 and Pioneer 4. Pioneer 4 flew past the Moon at an altitude of 60 000 km and became the first US spacecraft to reach heliocentric orbit [12].

Vanguard 2, which flew after Pioneer 3 but before Pioneer 4, was the first weather satellite. The spherical satellite had an aluminum-coated magnesium shell with a diameter of 50.8 cm and used two internal optical telescopes to monitor the brightness of each field-of-view. The intent was to monitor cloud cover around the Earth. Each telescope scanned the Earth, and space, as the spacecraft rotated at 50 rotations per minute, and a raster scan image could be created as the satellite moved across the Earth. Unfortunately, Vanguard 2 was deployed with a significant wobble about the intended spin axis, so the

Figure 1.3 Dr. Wernher von Braun (left), John Casani of the Jet Propulsion Laboratory (center), and Dr. James Van Allen, holding parts of the Pioneer 4 spacecraft. Source: Photo courtesy of NASA.

brightness data were hard to use. This spacecraft had a data tape recorder for recording brightness as a function of time, and a command receiver to activate playback of the tape as the spacecraft went over a ground station [13]. A 1 W output telemetry transmitter operated at 108.03 MHz, and a 10 mW, 108 MHz beacon transmitter provided a CW signal for tracking. Although both transmitters stopped working after 26 days, Vanguard 2 has been optically tracked for six decades, providing initial gravity field measurements for the Earth, and continuous atmospheric density measurements based on air drag.

The next nanosatellite launched into space was Explorer 9. It was the first satellite launched from Wallops Island, Virginia, and it was almost an order of magnitude larger than Vanguard 2. Explorers 9, 19, 24, and 39 were 3.6 m diameter balloons used for determining atmospheric density as a function of altitude. These balloons were inflated using nitrogen gas and were composed of alternating layers of aluminum foil and Mylar (polyester) film. The two hemispheres were electrically isolated so that they could function as a dipole antenna for the 15 mW output, 136 MHz tracking beacon. Power for the transmitters was supplied by solar cells and rechargeable batteries. Thermal control was supplied by the application of many white paint spots of 5.1 cm diameter to the surface. Figure 1.4 shows a photograph of Explorer 24 with a human for size comparison.

The first Orbiting Satellite Carrying Amateur Radio (OSCAR 1), a nanosatellite, was ejected into orbit in 1961. A group of amateur radio operators in Sunnyvale, California, decided to "build a continuous-wave beacon to transmit Morse code, to hook it up to batteries, place it into a metal box with an antenna coming out of the top, and have it launched into space" [14]. OSCAR 1 was the world's first private, nongovernment spacecraft; and, unlike previous nanosatellites, this one went into orbit as a secondary payload. It was launched for free as ballast on a launch vehicle of the US Air Force, thus explaining the curved shape of the $30 \times 25 \times 12\,cm^3$ box shown in Figure 1.5, as a section cut from a thick ring. Ejection off of the primary mission "was done with a very sophisticated technique:

Figure 1.4 Photo of Explorer 24. Source: Photo courtesy of NASA.

Figure 1.5 Photograph of OSCAR 1 at the Smithsonian National Air and Space Museum's Steven F. Udvar-Hazy Center, Chantilly, Virginia. Source: Photo by author.

a $1.15 (USD) spring purchased from a local hardware store. The total out-of-pocket cost (not including donated material) was only $68 (USD)" [15]. It transmitted the message "HI" in Morse code (•••• ••) at a rate proportional to the internal temperature using a two-transistor transmitter operating at 140 mW output at 144.98 MHz [16]. It transmitted for 3 weeks. Average spacecraft temperature was set by the tape pattern on the external magnesium surface, but it was warmer than expected at greater than 50°C after 60 orbits. OSCAR 2 was a copy of OSCAR 1, but with a lower-power transmitter (100 mW) for longer battery life, and modified surface treatments to lower the average spacecraft temperature to 15 ± 5°C. It started out in a lower orbit and was able to monitor the temperature increase as it started to re-enter.

The Flashing Light Unit listed in Table 1.1 was a subsatellite ejected by the Faith 7 Mercury capsule piloted by the US astronaut L. Gordon Cooper. It was a 15 cm diameter sphere that contained a pair of xenon strobe lights powered by batteries to test the visual acquisition and tracking of other spacecraft by astronauts [17]. Cooper was able to see the flashes one, two, and three orbits after ejection.

The remaining nanosatellites in Table 1.1 belong to the US Air Force Environmental Research Satellite (ERS) series. These spacecraft were used as technology test beds for spacecraft systems and components at medium Earth altitudes (greater than 1500 km apogee). They monitored radiation levels and radiation damage to solar cells. Due to their high perigees, ERS spacecraft were in sunlight most of the time, thus eliminating the need for batteries and battery charge regulators. Figure 1.6 shows a photograph of the 2 kg mass ERS 11 satellite on display at the Smithsonian National Air and Space Museum's Steven F. Udvar-Hazy Center in Chantilly, Virginia. This satellite is identical to the ERS 12 satellite flown in 1963. ERS 12 had an omnidirectional radiation detector to monitor electron fluxes for energies in excess of 0.5 and 5 MeV, and proton fluxes for energies between 10 and 20 MeV, and 50–100 MeV [8]. Note the large number of solar cells on each face, and the coiled

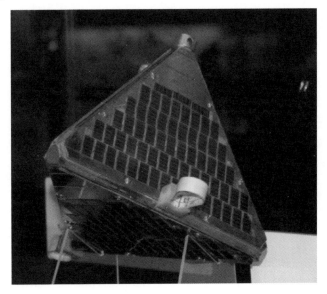

Figure 1.6 Photo of ERS 11 at the Smithsonian National Air and Space Museum's Steven F. Udvar-Hazy Center, Chantilly, Virginia. Source: Photo by author.

"carpenter tape" antennas (bottom right and left) that form a dipole when released. All of the ERS spacecraft were launched as secondary payloads.

The Early Space Age was a time of excitement and exploration fueled by the Cold War between the USA and the former Soviet Union. Both parties were learning how to exploit the high ground of space in order to monitor each other visually and electronically, and were vying to be the first to put a human being on the Moon. The world's largest rocket, the Saturn V of the USA, was developed to win the space race to the Moon. Weather satellites and geosynchronous communications satellites improved the lives of billions of people, about 3.5 billion in 1967, and precursors of today's ubiquitous global navigational satellites were flown. Average satellite mass had increased well beyond the microsatellite and nanosatellite mass range, and active nanosatellites would disappear for almost three decades.

1.4 The Large Space Era

The Large Space Era started in 1968 and ended by around 1996. No active nanosatellites were launched, but microsatellites continued at an average rate of 18.9 per year. This was a period of government-funded space programs and large commercial geosynchronous communications satellites.

At the start of the Space Age, an international telephone call required a reservation because the extremely limited total bandwidth available using cables and radio waves bounced off the ionosphere [18]. Satellites offered vastly increased bandwidth over the long term, and, realizing this, the Communications Satellite Act was passed in 1962 by the United States Congress. The first paragraph of Public Law 87-624 states: "The Congress hereby declares that it is the policy of the United States to establish, in conjunction and in cooperation with other countries, as expeditiously as practical, a commercial communications satellite system, as part of an improved global communications network, which will be responsive to public needs and national objectives, which will serve the communication needs of the United States and other countries, and which will contribute to world peace and understanding" [19]. The intent of this act was to foster competition in this new area of commerce and enable nondiscriminatory access to global communications. This was a major paradigm shift, since, at least in the USA, AT&T had almost complete control of both fixed terrestrial and undersea (cable) communications. Competition fueled continuous growth both in value of services and in spacecraft size over the next 55 years.

Hughes Aircraft, based in California, developed the first geosynchronous communications satellite, Syncom 2, launched in July 1963. Hughes invested US$2 million (about US$17 million in today's values) to develop the 35.5 kg mass, spin-stabilized microsatellite prototype, but they needed additional funding to build and launch an operational version. The US Department of Defense provided additional support, and NASA launched the spacecraft. This government–industry collaboration helped start a new industry in the early 1960s, and this model still works today.

Figure 1.7 shows the almost linear growth in geosynchronous communications satellite launch mass over time for Intelsat satellites. Beginning-of-life (BOL) mass at orbit is typically about half of the launch mass because most geosynchronous satellites are injected into a geosynchronous transfer orbit and contain a thruster and significant propellant to circularize the orbit at geosynchronous altitude. The first two generations were microsatellites, but BOL mass grew to ~800 kg during the 1970s (fourth generation), ~1000 kg during most of the 1980s (fifth generation), and ~2000 kg starting in 1989 (sixth generation). Satellite mass growth through 1997 exemplifies the "Large Space" era, and this trend has continued through today for more than half of the Intelsat spacecraft. A lower-mass branch started in 1997 with the eighth generation, leading to a series of half-mass satellites produced by Orbital Sciences debuting in 2007.

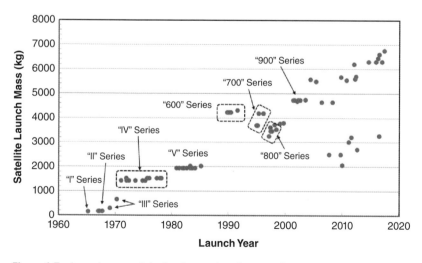

Figure 1.7 Launch mass of the Intelsat series of geosynchronous communications satellites.

Spacecraft became ever more capable during the Large Space era, but electronic and actuator technologies were still too primitive to enable active nanosatellites. A number of "almost nanosatellites," spacecraft with a mass slightly higher than 10 kg, were launched starting in 1990. Three 13 kg mass AMSAT "MicroSats" (PACSAT, OSCAR 17, and LUSAT), a 16 kg version called Webersat or OSCAR 18, and two ~46 kg mass microsatellites of the University of Surrey (UoSAT-OSCAR 14 and UoSAT-OSCAR 15) were put into orbit as secondary payloads with SPOT 2, an 1869 kg mass Earth observation satellite [20–26]. The cost to launch the secondary payloads was about US$1 million [27].

The AMSAT MicroSats were 23 cm^3, and covered with solar cells with a maximum output power of 15.7 W [28]. They used NEC V40 microprocessors, had up to 10 Mbytes of solid-state memory, and used a 15 cm long local area network to link the five electronics trays together. They operated as flying digital bulletin boards; you could upload data or have a file broadcast to all available ground stations within the satellite footprint. Figure 1.8 shows a photo of the author's ground station that was used with these and other amateur radio satellites. A personal computer with a color cathode ray tube monitor is on the left, various radio modems occupy the shelf to the right of the monitor, three transceivers are on the shelf below them, and a 50 W power amplifier for 435 MHz operation is on the extreme bottom right of the image. Data rates ranged from 300 to 9600 bits per second.

The 23 cm^3 (cube) form factor pioneered by AMSAT became popular for small satellites. Table 1.2 lists launch date, mass, orbit, and country data for 31 of these satellites. Their masses range from 10 to 13 kg, and some individual satellites have a range in mass, depending on which reference you use. Seven flew during the Large Space era, the rest during the New Space era. Note that most went into sun-synchronous orbits.

The majority of spacecraft in Table 1.2 used a permanent magnet to provide passive attitude control. This created two flips per orbit, and the antennas were painted in such a way that solar pressure would cause a small rotation rate about the magnetic axis. Instead of

Figure 1.8 Photo of author's 1990s era amateur satellite ground station. Source: Photo by author.

Table 1.2 Nanosatellites and "almost nanosatellites" with a 23–25 cm³ design.

Name	Launch date	Mass (kg)	Orbit (km)	Inclination (degrees)	Country
PACSAT (OSCAR 16)	22 January 1990	11	787 × 804	98.72	USA
Dove (OSCAR 17)	22 January 1990	11	787 × 804	98.72	USA
Webersat (OSCAR 18)	22 January 1990	13	787 × 804	98.72	USA
LUSAT (OSCAR 19)	22 January 1990	11	787 × 804	98.72	USA
ITAMSAT (OSCAR 26)	26 September 1993	11	789 × 803	98.7	Italy
EYESAT 1 (OSCAR 27)	26 September 1993	11	792 × 805	98.8	USA
UNAMSAT B	5 September 1996	11	988 × 1010	82.9	Mexico
SaudiSat 1a	26 September 2000	10	625 × 656	64.56	Saudi Arabia
SaudiSat 1b	26 September 2000	10	632 × 664	64.55	Saudi Arabia
PCSat-1	30 September 2001	10	790 × 800	67	USA
LatinSat A	19 December 2002	12	602 × 675	64.56	Saudi Arabia
LatinSat B	19 December 2002	12	613 × 710	64.56	USA
SaudiSat 1c	22 December 2002	10	633 × 690	64.56	Saudi Arabia
OSCAR 51	29 June 2004	12	697 × 817	98.2	USA
LatinSat C	29 June 2004	12	698 × 766	98.26	USA
LatinSat D	29 June 2004	12	695 × 852	98.26	USA
SaudiComsat 1	29 June 2004	12	701 × 749	98.2	Saudi Arabia
SaudiComsat 2	29 June 2004	12	700 × 783	98.2	Saudi Arabia
SaudiComsat 3	17 April 2007	12	653 × 718	98.1	Saudi Arabia
SaudiComsat 4	17 April 2007	12	650 × 751	98.1	Saudi Arabia
SaudiComsat 5	17 April 2007	12	652 × 729	98.1	Saudi Arabia
SaudiComsat 6	17 April 2007	12	649 × 762	98.1	Saudi Arabia
SaudiComsat 7	17 April 2007	12	651 × 740	98.1	Saudi Arabia
AprizeSat 3	29 July 2009	13	564 × 664	98.3	USA
AprizeSat 4	29 July 2009	13	605 × 763	98.3	USA
AprizeSat 5	17 August 2011	13	610 × 693	98.4	USA
AprizeSat 6	17 August 2011	13	627 × 694	98.4	USA
AprizeSat 7	21 November 2013	13	596 × 654	97.6	USA
AprizeSat 8	21 November 2013	13	597 × 669	97.6	USA
AprizeSat 9	19 June 2014	13	619 × 718	97.8	USA
AprizeSat 10	19 June 2014	13	620 × 737	97.8	USA

a permanent magnet, LatinSat A and LatinSat B used a ferromagnetic rod surrounded by a solenoid coil. Current pulses would set, and reset, the magnetic dipole moment of the rod in one of two antiparallel directions. This allowed these satellites to control their flip, and to thus present the best orientation to both the northern and southern hemispheres [29].

1.5 The New Space Era

Table 1.2 lists satellites from two nanosatellite-class "little LEO" communications constellations in the New Space era (1997 through the present); AprizeSat, which includes LatinSat A through LatinSat D; and SaudiComsat. The AprizeSat satellites provide store-and-forward, machine-to-machine (M2M) communications for applications, such as monitoring of oil and gas wells, mobile asset tracking, and environmental monitoring [30]. The seven SaudiComsat satellites are also store-and-forward M2M satellites for commercial applications. The first commercial M2M constellation used nadir-pointing Orbcomm 46 kg mass microsatellites, in 720 km altitude, 45° inclination orbits [31]. Launches of experimental satellites started in 1995, and this constellation provided near-continuous surface coverage between 60 °N and 60 °S latitudes using multiple ground stations. The AprizeSat system uses fewer, smaller satellites for applications that do not require continuous connectivity, but can do so over the entire planet. One specific application is the monitoring of automatic identification system (AIS) data packets transmitted by maritime vessels. These packets contain ship-identifier plus global positioning system (GPS)–based position and heading data. AIS is extremely helpful in monitoring ship traffic near coastlines and rivers, and satellite-based AIS extends this monitoring to the open oceans. AprizeSat 3 and AprizeSat 4, together, recorded 460 000 AIS transmissions from 22 000 ships each day in 2010 [32].

The first nanosatellite ejected into space in the New Space era was Sputnik 40 in 1997. Four one-third-scale Sputnik 1 spacecraft were built by Aéro-Club de France, AMSAT, and Rosaviakosmos to commemorate the 40th anniversary of the historic launch [33]. They were 23 cm in diameter, 4 kg in mass, and transmitted at 145 MHz amateur radio band [34]. They had no vacuum tubes or pressure vessels. The first was hand-deployed from the Mir space station on October 5, 1997; the second was never deployed; the third (Sputnik 41) was hand-deployed on November 10, 1998; and the fourth (Sputnik 99) was hand-deployed on April 2, 1999. The last one was turned off before deployment due to a radio licensing issue; it was going to transmit advertising for Swatch, the Swiss watchmaker, using amateur radio frequencies. New Space had some birthing issues.

Tubsat-N (8.5 kg mass) and Tubsat-N1 (3 kg mass) were launched in July 1998 from a K-407 Novomoskovsk Russian submarine from the Barents Sea. These store-and-forward communications nanosatellites became the first submarine-launched satellites. Tubsat-N had a charge-coupled device (CCD) star sensor and a reaction wheel as experiments, plus active magnetic attitude control using a magnetometer and two coils [35]. Active attitude control, a necessity for most interesting satellite missions, was being miniaturized for nanosatellites. Energy storage was supplied by decades-old nickel–cadmium battery technology.

The year 2000 was a major milestone for nanosatellites. On January 27, 2000, an Orbital Sciences Minotaur rocket, a refurbished ICBM, put JAWSAT into orbit. JAWSAT released the 22 kg Optical Calibration Sphere Experiment (OCSE, a 3.5 m diameter balloon), the 52 kg FalconSat 1 from the US Air Force Academy, the 5.9 kg ASUSat 1, and the 25 kg mass Orbiting Picosatellite Automated Launcher (OPAL). ASUSat 1 was cylindrical, with a diameter of 32 cm and a height of 25 cm. It was gravity-gradient stabilized using a 2 m long boom with a 125 g tip mass and contained a pair of 496×365 pixel cameras with a 0.5 km/pixel ground resolution, and a GPS receiver [36].

The Stanford-built OPAL microsatellite ejected three picosatellites from Santa Clara University (the 0.2 kg JAK, the 0.5 kg Thelma, and the 0.5 kg Louise), a 0.23 kg amateur radio picosatellite called StenSat, and two 0.25 kg Aerospace Corporation PicoSats [37]. Only the Aerospace PicoSats appeared to be functional. The Aerospace Corporation in El Segundo, California, helped broker a deal with the US Defense Advanced Research Projects Agency (DARPA) and the US Air Force Office of Scientific Research (AFOSR) during a US Air Force Research Laboratory (AFRL) meeting in Albuquerque, New Mexico, in 1998 to support US-built nanosatellites and to redesign the OPAL mission [38, 39]. This support was initiated by Prof. Twiggs' head-snapping comment: "I built a microsatellite (OPAL) for 50 thousand dollars!" OPAL was previously funded by the NASA's Jet Propulsion Laboratory to eject spinning picosatellies, but nonspinning spacecraft would be easier. The Aerospace Corporation assisted Stanford in developing a simplified containerized ejection mechanism for OPAL, and supplied two picosatellites for on-orbit ejection. The OPAL ejection mechanism later led to Prof. Twiggs' invention of CubeSats.

The next nanosatellites in 2000 were SaudiSat 1a and SaudiSat 1b, SNAP-1, and Munin. SNAP-1 was built by Surrey Satellite Technology Limited (SSTL) in the United Kingdom. The 6.5 kg mass satellite was designed to test technologies needed for a satellite inspection mission, and thus had three-axis attitude control, a GPS receiver, a butane warm gas thruster, four video cameras with 350×288 pixel resolution, and an ultra high frequency (UHF) intersatellite link to exchange data with the target spacecraft, the Tsinghua-1 microsatellite [40]. At the time, it was the most advanced nanosatellite ever launched. SNAP-1 and Tsinghua-1 were attached to a Nadezhda, a Russian navigation satellite, launched on June 28, 2000 into a 683×706 km, 98.13° inclination orbit. SNAP-1 and Tsinghua-1 were subsequently released by the Nadezhda and started to separate from each other. Spacecraft checkout and software fixes to remedy an anomalous spacecraft dipole moment resulted in the first thruster firing on August 15, 2000. By this time, the dispersion in initial deployment vectors and differential drag caused SNAP-1 to have a 2 km lower semi-major axis than Tsinghua-1. The 3.5 m/s delta-V from the butane thruster was insufficient to match orbits, with correct phasing, to rendezvous with Tsinghua-1. Flight results showed a lower demonstrated delta-V, most likely caused by liquid, rather than gaseous, butane ejection during initial thruster firings [41]. Ullage (empty space in the propellant tank) was increased for future thrusters, along with the addition of a sponge to keep liquid away from the nozzle.

Munin was a $21 \times 21 \times 22 \text{ cm}^3$ nanosatellite with a mass of 6 kg. It carried a miniaturized electrostatic dual-top-hat spherical analyzer, a device to measure ion and electron energy distributions with energies up to 18 keV per charge, and a 340×240 pixel CCD imager to image the aurora in visible light [42]. It was a space weather satellite and Europe's first scientific nanosatellite. It made the technological leap to lithium-ion batteries for significantly decreased battery mass per unit stored energy, but it relied on a mere 2 Mbytes of random-access memory for data storage, and passive magnetic stabilization.

The year 2001 saw one nanosatellite put into orbit, PCSat-1. Figure 1.9 shows a photograph of the 10 kg mass PCSat-1 satellite on display at the Smithsonian National Air and Space Museum's Steven F. Udvar-Hazy Center in Chantilly, Virginia. A panel, which normally holds a 144 MHz monopole antenna, has been removed to show the internal compartments. Note the simple yellow carpenter tape antennas for use at 435 MHz. This satellite received and logged amateur radio X.25 data packets [43].

The Aerospace Corporation, under funding by DARPA, created two 0.8 kg mass, $10 \times 10 \times 12.5 \, \text{cm}^3$ MEMS-Enabled PicoSatellite Inspector (MEPSI) nanosatellites that were deployed by the US Space Shuttle Endeavour on December 2, 2002, using the Aerospace Picosatellite Orbital Deployer (A-POD) [44]. They were connected by a 15 m long tether. The A-POD was designed for a man-rated launch system and flew one year before the first CubeSat Polytechnic Picosatellite Orbital Deployer (P-POD), the workhorse CubeSat deployer. The other nanosatellite launched during 2002 was the 10 kg mass SaudiSat 1c. The year 2003 marked the start of an unprecedented revolution in small satellite evolution and deployment: the CubeSat era. Pumpkin, Inc. in San Francisco started delivering CubeSat kits that could be ordered online. The founder, Dr. Andrew Kalman, was an expert in embedded microcomputer systems, and had happened to meet Prof. Twiggs at Stanford in 1998. Kalman helped Twiggs create the CubeSat specifications,

Figure 1.9 Photograph of the PCSat-1 satellite on display at the Smithsonian National Air and Space Museum's Steven F. Udvar-Hazy Center in Chantilly, Virginia. Source: Photo by the author.

and fostered the idea of using inexpensive, state-of-the-art, commercial off-the-shelf (COTS) electronics instead of expensive, tested, high-quality electronics that could be a generation or two behind current mass-market designs [45].

Five 1U CubeSats were launched on June 30, 2003: CanX-1 (Canadian Advanced Nanospace eXperiment from the University of Toronto), AAU-CubeSat (Aalborg University in Denmark), DTUSat-1 (Danmarks Tekniske Universitet Satellite), XI-IV (X-factor Investigator, University of Tokyo), and CUTE 1 (CUbical TITech Engineering Satellite from the Tokyo Institute of Technology), each about 1 kg in mass, were deployed into an ~820 km altitude, sun-synchronous orbit. CanX-1 had active magnetic stabilization, a Complementary Metal–Oxide–Semiconductor (CMOS) horizon and star tracker experiment, and a GPS receiver [46]. AAU-CubeSat had three-axis magnetic coil stabilization and a 1280 × 1024 pixel color camera with a depth of 24 bits per color pixel. Ground resolution was about 150 m, but only basic housekeeping data were downloaded due to a significant reduction in battery capacity [47]. DTUSat-1 was dead on arrival [48]. The XI-IV CubeSat used passive magnetic stabilization and had a 640 × 480 color CMOS camera as a payload [49]. CUTE 1 was unstabilized, but contained a CMOS image sensor as a sun sensor. QuakeSat 1, a 3U CubeSat of 4.5 kg mass, was launched with the previous 5 1U CubeSats. It was built in collaboration between Stanford and QuakeFinder of Palo Alto, California [50]. QuakeFinder's goal was to save human lives by developing earthquake monitoring systems. QuakeSat 1 monitored extremely low frequency (ELF) electromagnetic fields that were predicted to occur before major earthquakes. It monitored the magnetic component of these fields and used on-board magnets for passive magnetic stabilization. This first CubeSat launch put six university-class CubeSats into orbit using P-PODs mounted on the upper stage of an inexpensive Russian Rockot RS launch vehicle. Launch cost was about US$40 000 per "U" of volume. After 15 years, CUTE 1 (aka CubeSat-OSCAR 55) and XI 4 (aka CubeSat-OSCAR 57) are still active [51]. This is an amazing tribute to Japanese engineering, especially when one considers the radiation environment at ~820 km altitude. No other nanosatellites were launched that year, and no nanosatellites were launched in 2004.

The first nanosatellite of 2005 was TNS 0, a 4.5 kg mass, 25 cm long, 17 cm diameter cylinder with no solar cells, that contained a pair of Globalstar modems, batteries, and a timer [52, 53]. It was developed by the Russian Scientific Institute of Space Device Engineering (RSIDE)987+ and hand-deployed from the International Space Station (ISS) to test the use of the Globalstar network for continuous command and control and download of telemetry. UWE-1 (University of Würzburg's experimental satellite 1), XI-V (University of Tokyo), and nCube-2 (Norway) were of 1 kg mass, 1U CubeSats that flew in October 2005 on the second CubeSat mission, using a Kosmos-3M launch vehicle. UWE-1 was designed to test the use of established Internet protocols such as TCP, UDP, STCP, and HTML in a space environment [54]. It used passive magnetic stabilization, a micro-Linux operating system, and completed its mission within a few weeks [55]. The XI-V CubeSat was a backup to XI-IV, but included a few upgrades, such as thin-film copper–indium–gallium–selenide (CIGS) solar cells on one face and improvements in the camera control software. nCube-2 was designed to monitor maritime AIS transmissions using active magnetic stabilization and a gravity-gradient boom. Unfortunately, no signals were received, and it was believed that it never deployed from the host spacecraft. The year 2005 also saw the founding of Clyde Space in Glasgow,

Scotland, to provide CubeSats, CubeSat systems, and nanosatellite systems to universities, commercial companies, and government organizations.

The year 2006 had an international mix of two "official" CubeSat launches, plus three CubeSat-like launches. The first one put the CUTE 1.7 + APD (aka OSCAR 56) CubeSat into orbit using a Japanese M-5 launch vehicle. This 3.6 kg mass, 2U CubeSat was built by the Tokyo Institute of Technology and was ejected using their own CubeSat deployer. Although it was 2U in size, it did not conform to the established CubeSat design specifications. The next flight in July used a Russian Dnepr launch vehicle. Unfortunately, it had a launch failure, resulting in the complete loss of 14 CubeSats. The third flight in September used a Japanese M-5 launch vehicle to put the $12 \times 12 \times 12 \, \text{cm}^3$, 2.7 kg mass, satellite HITSAT-1 (aka OSCAR 59) of Hokkaido Institute of Technology into orbit using a custom deployer. A CubeSat launch in December used a US Minotaur 1 launch vehicle to put the 3U CubeSat Genesat-1 into orbit. GeneSat-1, shown partially assembled in Figure 1.10, was built by the NASA's Ames Research Center and multiple universities to perform genetic experiments on *E. coli* bacteria growing in microgravity and shielded ~400 km altitude radiation environment. Note the silver "tuna can" on the top side. This volume was made available by cutting a hole in the pusher plate that separates the deployment spring from the CubeSat. The tuna can fit inside the helical deployment spring.

Four additional nanosatellites were put into orbit in December on the US space shuttle Discovery. MEPSI 2A and MEPSI 2B, built by The Aerospace Corporation, were

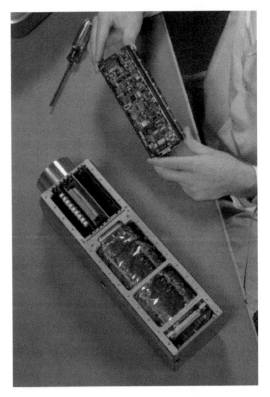

Figure 1.10 Photo of a partially assembled GeneSat-1. Source: Photo courtesy of NASA.

$10 \times 10 \times 12.5 \, \text{cm}^3$ nanosatellites connected by a 15 m long tether. The "target" had six 640×480 pixel color cameras pointing out the +X, −X, +Y, −Y, +Z, and −Z faces, while the "inspector" had three reaction wheels, two cameras, and an additively manufactured 5-thruster cold gas propulsion system. We believe that this was the first additively manufactured part in any satellite. The other nanosatellites on this mission were RAFT 1 and MARScom built by the US Naval Academy in Annapolis, Virginia. These were $12.5 \, \text{cm}^3$ deployed by an expanded-volume A-POD.

The year 2006 also saw Innovative Solutions in Space (ISIS), based in the Netherlands, open the online CubeSat shop (www.cubesatshop.com), a "one-stop webshop for CubeSats and nanosats." ISIS was a spin-off from the Delfi-C3 nanosatellite project from Delft University of Technology [56]. GomSpace, originally based in Aalborg, Denmark, followed in 2007 as another supplier of CubeSats, CubeSat systems, and services [57]. CubeSats were slowly evolving from university experiments to scientific and commercial platforms, and at least four commercial companies were selling products to the growing market on the internet.

The year 2007 saw eight nanosatellites put into orbit. Seven CubeSats (CP 3, CP 4, CAPE-1, Libertad 1, AeroCube 2, CSTB 1, and MAST) were put into orbit using P-PODs on a Dnepr launch vehicle in April. The CP series CubeSats were built by the California Polytechnic State University, San Luis Obispo, California, CAPE-1 was built by the University of Louisiana at Lafayette, Libertad 1 was built by the Sergio Arboleda University in Columbia, AeroCube 2 was built by The Aerospace Corporation in El Segundo, California, CSTB 1 was built by the Boeing Corporation in Huntington Beach, and MAST was built by Tethers Unlimited and Stanford University [58, 59]. The 2.5 kg mass, $15 \times 15 \times 15 \, \text{cm}^3$ ZDPS 1 nanosatellite, designed and built by Zhejiang University, China to demonstrate small satellite technologies, was launched as a secondary payload on a Chinese CZ-2D launch vehicle on May 26. Most of these nanosatellites were designed and built in California, but South America and China were now entering the field.

The next decade of the "New Space" era saw spectacular growth in nanosatellite launch rates. Nanosatellites were launched at yearly rates of 8 in 2008, 14 in 2009, 19 in 2010, 9 in 2011, 25 in 2012, and 85 in 2013. Four years later in 2017, the yearly nanosatellite launch rate hit an amazing 293 spacecraft. Table 1.3 shows the launches of microsatellites, non-CubeSat nanosatellites, and CubeSats during 2017. Note that the last launch on November 28, 2017 was a failure. The most popular CubeSat size in 2017 was a 3U, and the next most popular size was 2U. What drove this growth? Increasing global CubeSat launch capacity, increased government funding of CubeSats at national laboratories and universities, and commercial companies trying to provide store-and-forward communications and new Earth observation capabilities using CubeSats fueled this growth.

The large number of 2U CubeSats launched in 2017 was an anomaly driven by the QB50 project. QB50 was approved in January 2012 by the European Commission as a multisatellite network to study lower thermospheric physics. Forty 2U CubeSats were to be deployed to fill a complete orbit, plus 10 2U/3U CubeSats for "science and technology demonstration" [60]. Twenty eight were deployed by the ISS in May 2017, and eight more were deployed by an Indian Polar Small Launch Vehicle in June 2017 [61]. These CubeSats were built by universities in Europe, the USA, Australia, South Africa, etc., and typically carried one of the

Table 1.3 Launches of small satellites during 2017.

Date	Country	Micro	Nano	1U	1.5U	2U	3U	3.5U	5U	6U	12U	Total
9 January	China	1				2						3
15 February	India		2	1		1	99					103
2 March	China	1										1
18 April	USA			1		29	6		1	1		38
20 April	China						1					1
3 June	USA			5								5
14 June	Russia	1	1			2						4
15 June	China	1										1
23 June	India	3	1	3		4	18			1		30
23 June	Russia		1									1
14 July	Russia	4		1			59			7		71
14 August	USA	1					1			2		4
26 August	USA		1		2							3
31 October	USA						4					4
12 November	USA			1	2	1	10	1		1		16
14 November	China	1										1
18 November	USA			3			2					5
21 November	China	3										3
28 November*	Russia	3	1				12			2		18
Total		19	7	15	4	37	214	1	1	14		312

*Entries with an asterisk denote a launch failure.

three space science instruments: an ion-neutral mass spectrometer, a flux-Φ-probe (flux of atomic oxygen), or a multineedle Langmuir probe [62]. QB50 created two distributed space weather measurement systems.

1.5.1 Technology Development

Small satellites in any size class go through a series of system, subsystem, and component technology development stages. Smaller satellites need more miniaturization and become possible at later dates in time as technologies mature. Fortunately, ever-smaller transistors have enabled the doubling of integrated circuit density roughly every 2 years [63]. This trend, called Moore's law, has proven true since 1965 for microprocessors, memory, and other electronic circuits. Consider the Intel 4004 4 bit microprocessor introduced in 1971, with 3200 transistors on a 12 mm^2 die, fabricated with 10 000 nm minimum feature size, that processed instructions at an average rate of 60 kHz [64]. Today's 32-core, AMD EPYC 64 bit processor was introduced in 2017 and fabricated using 14 mm minimum feature size, has 19.2 billion transistors on a 768 mm^2 die and processes instructions at 2.2 GHz, per

Table 1.4 Data downlink rates for CubeSats in 2014.

Downlink data rate	Number
<9600 bps (Morse code, 400 baud, 1200 baud, etc.)	85
9600 bps	36
9600 to <1 Mbps	16
1 Mbps and greater	7

Total number of CubeSat transmitters for this dataset was 144.

core [65]. In terms of processing efficiency adjusted for 32 bit operations, the 4004 operated at ~6000 instructions per second per Watt, while EPYC (a high-end server microprocessor) operates at 390 million instructions per second per Watt. Simpler microcontrollers with a smaller memory addressing space and simpler input/output circuits easily exceed several billion instructions per second per Watt. Note that typical spacecraft command and control operations require about 1 million instructions per second, while image processing, autonomous navigation, etc. may require processing speeds in excess of 100 million instructions per second. Even 1/4U CubeSats can now support these processing functions on their limited power budget.

Semiconductor memory has also followed Moore's law. PACSAT, launched in 1990, had 10.25 megabytes of memory, with 8 megabytes used by the store-and-forward payload [66]. Today, we can use 256 gigabyte flash RAM cards to put 1 terabyte or more into a 1/4U or larger spacecraft. Generating 1 TB of data is easy; a high-definition HD video stream at 1920 × 1080 pixel resolution, 24 bit pixel depth, at 30 frames per second would generate 1.6 TB of data per day using 10× data compression. Getting this data rate down to the ground from LEO is hard. If one assumes four ten-minute passes per day for a single ground station, the spacecraft would need a 540 million bit per second download rate. Table 1.4 shows data download rates for CubeSats in 2014 [67]. Less than 5% of the transmitters had a data download rate greater than one million bits per second (bps), and the majority used data rates less than 9600 bits per second. Out of 172 total CubeSat transmitters, 112 used the amateur radio 435 MHz UHF band, but only four used X-band where data rates of 100+ million bps (Mbps) are possible.

Use of X-band downlinks has significantly increased since 2014, driven primarily by commercial Earth imaging CubeSats from Planet Labs. Their latest 2 W output, 8150 MHz transmitter has demonstrated 220 Mbps at X-band [68]. More bandwidth is available at higher frequencies, such as Ku, K, and Ka bands, and ultimately, optical. Future nanosatellites will need these increased bandwidths. The 3U Integrated Solar Array and Reflectarray Antenna (ISARA) CubeSat recently demonstrated a Ka band antenna, with greater than 33 dB gain, patterned on the backside of deployed three-panel solar array that would enable 100+ Mbps communications for LEO CubeSats [69]. Figure 1.11 shows a larger, X-band version of the reflectarray used on the two Mars Cube One (MarCO) 6U CubeSats currently heading for a flyby of Mars to support the InSight Mars lander of NASA's Jet Propulsion Laboratory [70]. These are the first interplanetary CubeSats.

Figure 1.11 Photograph of one of the 12U MarCo CubeSats heading to Mars. Source: Photo courtesy of NASA.

Due to much higher frequencies, optical communication offers downlink rates in the 1–10 Gbps range, even on nanosatellites. Figure 1.12 shows a photograph of AeroCube-OCSD-B CubeSat put into orbit in late 2017. This effort, like ISARA, was funded by NASA's Small Satellite Technology Program. Two almost identical 1.5U CubeSats were launched to test proximity operations using a steam thruster. Each CubeSat also carries a 2 W output, 1064 nm fiber laser capable of producing a 500 Mbps data rate using on/off modulation. The laser is hard-mounted to the CubeSat body, and the on-board attitude control system provides tracking of an optical ground station during a pass [71]. One of the CubeSats recently demonstrated an optical downlink at 200 Mbps to a 40 cm diameter receive telescope, with 400 Mbps expected by mid-2019. Gigabit per second data rates can be achieved on future CubeSats by using larger-diameter ground stations, and/or narrower angular beam widths.

Obtaining accurate geo-registration of Earth images at several meter ground resolution or pointing a 0.05° full-width-half maximum (FWHM) laser beam at a specific ground station requires a three-axis attitude control system with one or more star trackers and at least three reaction wheels. Star trackers were readily available in the early 2000s as 10 cm size and larger modules for normal size spacecraft and microsatellites. A prototype using "new" active pixel sensor technology, ubiquitous today in cellphones and most cameras, was a 6 cm^3 with a 10 cm diameter and 14 cm long sunshade [72]. It had a total mass of about 1 kg, consumed 5 W of power, and used a 1024 × 1024 array of 15 μm square pixels, a 24.5° square field-of-view, and saw stars down to 5th magnitude. The AeroCube-OCSD CubeSats use a 742 × 480 image sensor with 6 μm pixels for each of two star trackers [73]. They are about 2.5 cm on a side, excluding an appropriately scaled sun shade, and would easily fit in a 1/2U CubeSat. Many similar-size designs are available today from a number of vendors, some of which could fit in a $^1/_4$U CubeSat.

Figure 1.12 Photograph of AeroCube-OCSD-B with deployed wings.

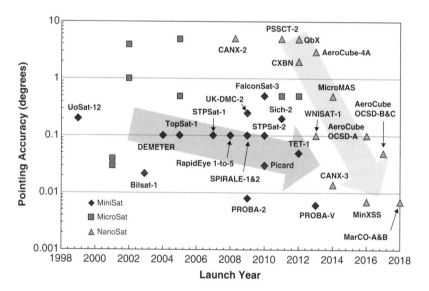

Figure 1.13 Evolution of small satellite pointing accuracy over the last 20 years. (*See color plate section for color representation of this figure*).

The development of cm-scale star trackers and cm-scale reaction wheels enabled nanosatellites with three-axis pointing accuracies typically associated with light minisatellites in the 100–200 kg mass range. Figure 1.13 shows the evolution of pointing accuracy for 100–200 kg mass minisatellites, microsatellites, and nanosatellites as a function of launch year. These data are primarily from reference [74] with extra satellite data added using the European Space Agency's (ESA) eoPortal website. Note the gradual increase in minisatellite pointing accuracy (pink arrow trendline) since 2002 and the rapid increase in pointing accuracy for nanosatellites (green arrow trendline) during the last 6 years.

This was driven by the perceived potential market and the ability to design, build, test, fly, and repeat on a 1-year timescale, as opposed to the traditional 7-year timescale for large satellites, and the 2 year to 3-year timescale for minisatellites.

1.5.2 Commercial Nanosatellites and Constellations

Table 1.2 has eight AprizeSat satellites plus four LatinSat satellites that formed the 12-satellite AprizeSat constellation for low data rate store-and-forward communications using 13 kg mass "almost" nanosatellites. WNISAT-1 was a pioneer commercial nanosatellite with a 10 kg mass and 27 cm^3 shape launched in November 2013 with a visible red, green, blue color camera and a near infrared camera. It was designed for commercial monitoring of northern sea routes [75]. It had a spatial resolution of 500 m per pixel, a three-axis pointing accuracy of 0.1°, and was funded by Weathernews, Inc., of Tokyo, Japan.

Explosive commercial nanosatellite development started with the launch of two 3U imaging CubeSats in April 2013 called Dove 1 and Dove 2 by Planet Labs of San Francisco, California. They carried Maksutov–Cassegrain telescopes of 9 cm diameter with an 11 megapixel RGB + near infrared imager at the focus to image the Earth at 3–5 m resolution [76]. The next generation used a five-element optical system coupled to a 29 megapixel imager. Planet's goal is to image the Earth daily at 3–5 m resolution. By the end of August 2013, Planet Labs had raised more than US$13 million in venture capital funding [77]. By April 2015, they had raised an additional US$118 million [78]. By June 2018, 324 Dove imaging satellites, all 3U CubeSats, had been launched.

Table 1.5 lists commercial CubeSat constellations with at least two functional spacecraft on orbit, as of February 2019 [92]. Spire Global currently has the second largest nanosatellite constellation, also using 3U CubeSats. The first experimental CubeSats, ArduSat-1 and ArduSat-X, were launched in 2013 [93]. The experimental ArduSat-2 followed in February 2014, and the Lemur 1 prototype was launched in June 2014. The Lemur 2 series was launched starting in September 2015. These 3U CubeSats have GPS radio occultation receivers to measure atmospheric temperature, pressure, and humidity profiles, AIS receivers to monitor ship locations, and automatic dependent surveillance broadcast receivers for tracking aircrafts.

Astro Digital has launched five of ten "Landmapper-BC" 6U CubeSats, starting in July 2017. Each CubeSat has separate red, green, and near-infrared cameras that match the Landsat 8 Earth observation satellite of US National Oceanic and Atmospheric Administration (NOAA) [94]. Ground resolution is 22 m, and 1.2 TB per day, per satellite, of data are downloaded at Ka band.

The first two companies listed in Table 1.5 are headquartered in San Francisco, California, while the third is headquartered in nearby Santa Clara, California. The fourth is headquartered in London, The United Kingdom. Sky and Space Global (SAS) is planning a 200-nanosatellite constellation of 3U CubeSats to provide voice, data, and global messaging for near-equatorial regions [95]. Three pathfinder CubeSats, the Red, Green, and Blue Diamonds, were launched in June 2017. Note that the last five entries in Table 1.5 are for space connectivity to the Internet of Things (IoT). This is a recent development which could add over 300 new nanosatellites to LEO.

Table 1.5 Commercial CubeSat constellations with at least two functional spacecraft on orbit as of February 2019.

Organization	Number launched	Target number	Form factor	Funding (million US$)	Purpose	Ref.
Planet Labs	355	>150	3U	183	Daily Earth imaging at 5 m resolution	[79]
Spire	103	>150	3U	150	Maritime tracking, GPS radio occultation data, aircraft tracking	[80]
Astro Digital	6	25	6U/16U	> 16.7	Earth imaging at 22 m resolution, four bands	[81]
Sky and Space Global	3	200	3U	11.5	Mobile communications in the tropics	[82]
GeoOptics	7	TBD	6U	5.15	GPS radio occultation data	[83]
Helios Wire	2	28	16U	4	Store-and-forward communications, IoT	[84]
Swarm Technologies	7	8	1U/4U	28	Store-and-forward communications, IoT	[85, 86]
Kepler Communications	2	140	3U, 6U	21	Store-and-forward communications, Ka band	[87]
Hiber Global	2	48	6U	14.5	Store-and-forward communications, IoT	[88, 89]
Fleet Space	4	100	1.5U, 3U, 12U	5	Store-and-forward communications, IoT	[90, 91]

1.6 Summary

The first nanosatellites were launched near the start of the Space Age between 1958 and 1968. These were experimental spacecraft designed to test new technologies and to provide environmental data for low Earth orbit and beyond. Small satellites were launched from 1968 through 1996, but none of these were active satellites within the 1–10 kg mass range. Miniaturization technologies continued to evolve during this period, and eventually enabled active nanosatellites starting in 1997. CubeSats were invented during the next few years and started flying in 2003. The ability to fly a satellite for US$50 000–300 000 enabled universities and small companies to fly their own spacecraft. The availability of 10 or more launch opportunities per year created a design, build, test, and fly cycle of about 1 year for CubeSats, which spurred rapid development of key spacecraft technologies, such as imaging systems and accurate three-axis attitude control. These developments stimulated venture capital and other funding sources to create "New Space" businesses based on CubeSats. Nanosatellite launch rates have grown exponentially since 1997, similar to Moore's law, with a doubling time of 2.44 years. Based on current plans for new nanosatellite constellations, this exponential growth could continue for at least another 5 years.

References

1 Sweeting, M.N. (1992). UoSAT microsatellite missions. *IEEE Electronics & Communication Engineering Journal* 4 (3): 141–150.

2 Thompson, T.D. (ed.) (1997). *Space Log 1996*. Redondo Beach, CA: TRW Space and Electronics Group.

3 Krebs, Gunter Dirk, 2019. "Chronology of Space Launches" web page, Gunter's Space Page, viewed 12 February 2019, https://space.skyrocket.de/directories/chronology.htm.

4 Encyclopedia Astronautica, 2019. Chronology web pages, 2003, 2004, 2005, 2006, and 2007, viewed 12 February 2019, http://astronautix.com/2/2003chronology.html, http://astronautix.com/2/2004chronology.html, http://astronautix.com/2/2005chronology.html, http://astronautix.com/2/2006chronology.html, and http://astronautix.com/2/2007chronology.html.

5 Buhl, M. (2008). TUBSATe: The Technical University of Berlin Satellite Program. In: *Small Satellites: Past, Present, and Future*, 349–384, Chapter 12. Reston, VA, USA: American Institute of Aeronautics and Astronautics, Inc.

6 Krebs, Gunter Dirk, 2019. "Gunter's Space Page," viewed February 12, 2019, http://space.skyrocket.de/index.html.

7 TRW Systems Group (1972). *TRW Space Log Volume 10 (1970–71)*. Redondo Beach, CA, USA: TRW Incorporated.

8 Heymean, Jos, 2019. "Directory of U.S. Military Rockets and Missiles, Appendix 3: Space Vehicles, ERS," viewed 12 February 2019, http://www.designation-systems.net/dusrm/app3/ers.html.

9 Moore, R.G. (2008). The first small satellites: sputnik, explorer, and vanguard. In: *Small Satellites: Past, Present, and Future*, 1–46, Chapter 1. Reston, VA, USA: American Institute of Aeronautics and Astronautics, Inc.

10 Janson, S.W. (2008). The history of small satellites. In: *Small Satellites: Past, Present, and Future*, 58–59, Chapter 2. Reston, VA, USA: American Institute of Aeronautics and Astronautics, Inc.Table 2.3 in

11 NASA, 2019. NASA Space Science Data Coordinated Archive web page for Pioneer 3, viewed 12 February 2019, https://nssdc.gsfc.nasa.gov/nmc/spacecraftDisplay.do?id=1958-008A.

12 NASA, 2019. NASA Space Science Data Coordinated Archive web page for Pioneer 4, viewed 12 February 2019, https://nssdc.gsfc.nasa.gov/nmc/spacecraftDisplay.do?id=1959-013A.

13 NASA, 2019. NASA Space Science Data Coordinated Archive web page for Vanguard 2, viewed 12 February 2019, https://nssdc.gsfc.nasa.gov/nmc/spacecraftDisplay.do?id=1959-001A.

14 Smith, G.G. (2008). The role of AMSAT in the evolution of small satellites. In: *Small Satellites: Past, Present, and Future*, 119–149, Chapter 4. Reston, VA, USA: American Institute of Aeronautics and Astronautics, Inc.

15 Bilsing, Andreas, 2012. "OSCAR 1 Launched 50 Years Ago," Funkamateur, pp. 12–13, January 2012, English translation, viewed 12 February 2019, http://www.qsl.net/dl2lux/sat/bilsing.pdf.

16 NASA, 2019. NASA Space Science Data Coordinated Archive web page for OSCAR 1, viewed 12 February 2019, https://nssdc.gsfc.nasa.gov/nmc/spacecraftDisplay.do?id=1961-034B.

17 Hillger, Don, and Toth, Garry, "Sub-satellites: Part 1, Early manned-spacecraft-deployed," viewed 12 February 2019, http://rammb.cira.colostate.edu/dev/hillger/pdf/Sub-satellites_part-1_Early_manned-spacecraft-deployed.pdf.

18 Vartabedian, Ralph, 2019. "How a Satellite Called Syncom Changed the World," The Los Angeles Times, July 26, 2013, viewed 12 February 2019, http://www.latimes.com/nation/la-na-syncom-satellite-20130726-dto-htmlstory.html.

19 Congress of the United States (1962). *Public Law 87–624*. U.S. Government Publishing Office viewed 12 February 2019, https://www.gpo.gov/fdsys/pkg/STATUTE-76/pdf/STATUTE-76-Pg419.pdf.

20 NASA, 2019. NASA Space Science Data Coordinated Archive web page for Pacsat, viewed 12 February 2019, https://nssdc.gsfc.nasa.gov/nmc/spacecraftDisplay.do?id=1990-005D.

21 NASA, 2019. NASA Space Science Data Coordinated Archive web page for OSCAR-17, viewed 12 February 2019, https://nssdc.gsfc.nasa.gov/nmc/spacecraftDisplay.do?id=1990-005E.

22 NASA, 2019. NASA Space Science Data Coordinated Archive web page for Lusat, URL: viewed 12 February 2019, https://nssdc.gsfc.nasa.gov/nmc/spacecraftDisplay.do?id=1990-005G.

23 NASA, 2019. NASA Space Science Data Coordinated Archive web page for OSCAR-18, viewed 12 February 2019, https://nssdc.gsfc.nasa.gov/nmc/spacecraftDisplay.do?id=1990-005F.

24 NASA, 2019. NASA Space Science Data Coordinated Archive web page for OSCAR-14, viewed 12 February 2019, https://nssdc.gsfc.nasa.gov/nmc/spacecraftDisplay.do?id=1990-005B.

25 NASA, 2019. NASA Space Science Data Coordinated Archive web page for OSCAR-15, viewed 12 February 2019, https://nssdc.gsfc.nasa.gov/nmc/spacecraftDisplay.do?id=1990-005C.

26 http://Astronautix.com, 2019. "SPOT-1-2-3" web page, viewed 12 February 2019, http://www.astronautix.com/s/spot-1-2-3.html.

27 Horais, B.J. (1995). The Ariane ASAP ring: workhorse launcher for small satellites. In: *Platforms and Systems*, vol. 2317, 124–130. SPIE.

28 King, J.A. et al, 1990. "The in-orbit performance of four Microsat spacecraft", *Proc. 4th Annual AIAA/USU Conference on Small Satellites*, Vol. 1, Logan, Utah, August 27–30.

29 Sinclair, Doug, and Damarin, Chris, 2019. "Flight Results from a Novel Magnetic Actuator on the LatinSat Spacecraft," Paper SSC03-IV-8, 17th Annual AIAA/USU Conference on Small Satellites, Logan, Utah, August, 2003, viewed 12 February 2019, https://digitalcommons.usu.edu/cgi/viewcontent.cgi?article=1773&context=smallsat.

30 (2015). *AprizeSat Home Web Page, Aprize Satellite*. Virginia: Fairfax viewed 12 February 2019 , http://www.aprizesat.com.

31 Hawes, D. and Choi, S.W. (2008). A revolutionary approach to spacecraft design, production, and operations: Orbital's MicroStar satellite. In: *Small Satellites: Past, Present, and*

Future, 407–448, Chapter 14 in . Reston, VA, USA: American Institute of Aeronautics and Astronautics, Inc.

32 eoPortal web page, 2019. "AprizeSat-3 and -4," European Space Agency, viewed 12 February 2019, https://directory.eoportal.org/web/eoportal/satellite-missions/a/aprizesat-3-4#footback7%29.

33 41 99 (RS17 18 19), 2019. web page. viewed 12 February 2019, http://space.skyrocket.de/doc_sdat/sputnik-40.htm.

34 Weebau Space Encyclopedia, 2019. "Sputnik 40" web page, viewed 12 February 2019, http://weebau.com/satellite/S/sputnik%2040.htm.

35 eoPortal, 2019. TUBSAT Web page, European Space Agency, viewed 12 February 2019, https://directory.eoportal.org/web/eoportal/satellite-missions/t/tubsat.

36 Friedman, Assi, Underhill, Brian, Ferring, Shea, Lenz, Christian, Rademacher, Joel, and Reed, Helen, 2000. "ASUSat-1: Low-Cost Student-Designed nanosatellite," paper SSC00-V-2, 14th Annual AIAA/USU Conference on Small Satellites, Logan, Utah, August, 2000.

37 Cutler, James, and Hutchins, Greg, 2000. "OPAL: Smaller, Simpler, and Just Plain Luckier," Proc. Of the 14th annual AIAA/USU Conference on Small Satellites, Ogden, Utah, August 2000.

38 Janson, S.W. (2008). The PicoSat. In: *Small Satellites: Past, Present, and Future*, 74–75, Section 2.3.4. Reston, VA, USA: American Institute of Aeronautics and Astronautics, Inc.

39 Twiggs, B. (2008). Origin of CubeSat. In: *Small Satellites: Past, Present, and Future*, 151–173, Chapter 5 . Reston, VA, USA: American Institute of Aeronautics and Astronautics, Inc.

40 Underwood, C.L. (2008). The UK's first nanosatellite: SNAP-1. In: *Small Satellites: Past, Present, and Future*, 297–325, Chapter 10 . Reston, VA, USA: American Institute of Aeronautics and Astronautics, Inc.

41 Gibbon, D., and Underwood, C., 2001. "Low Cost Butane Propulsion System for Small Spacecraft," paper SSC01-XI-1, 15th Annual AIAA/USU Conference on Small Satellites, Logan, Utah, August 2001.

42 eoPortal, 2019. Munin Web page, European Space Agency, viewed 12 February 2019, https://directory.eoportal.org/web/eoportal/satellite-missions/m/munin.

43 Bruninga, B. (2008). PCSat-1. In: *Small Satellites: Past, Present, and Future*, 231–236, Section 8.3. Reston, VA, USA: American Institute of Aeronautics and Astronautics, Inc.

44 Hinkley, D. (2008). Picosatellites at the aerospace corporation. In: *Small Satellites: Past, Present, and Future*, 635–674, Chapter 20 . Reston, VA, USA: American Institute of Aeronautics and Astronautics, Inc.

45 Greenberg, Andy, 2019. "Nanosatellites Take Off," Forbes Magazine, November 4, 2010, viewed 12 February 2019, https://www.forbes.com/forbes/2010/1122/technology-pumpkin-inc-andrew-kalman-toasters-in-space.html#219ffae51e69.

46 Wells, G. James, Stras, Luke, and Jeans, Tiger, 2002. "Canada's Smallest Satellite: The Canadian Advanced Nanospace eXperiment (CANX-1)," Student Scholarship Competition, 16th Annual AIAA/USU Conference on Small Satellites, Logan, Utah, August 2002, viewed 12 February 2019, https://digitalcommons.usu.edu/cgi/viewcontent.cgi?article=1925&context=smallsat.

47 Alminade, L., Bisgaard, M., Vinther, D. et al. (2004). *The AAU-CubeSat Student Satellite Project: Architectural Overview and Lessons Learned.* Saint Petersburg, Russia: IFAC Automatic Control in Aerospace, viewed 12 February 2019, https://www.sciencedirect.com/science/article/pii/S1474667017323017.

48 eoPortal, 2019. "CubeSat-Launch 1" web page, European Space Agency, viewed 12 February 2019, https://earth.esa.int/web/eoportal/satellite-missions/c-missions/cubesat-launch-1.

49 Funase, R., Nakamura, Y., Nagai, M. et al. (2004). *University of Tokyo's Student Nano-Satellite Project CubeSat-XI and Its On-Orbit Experiment Results,* 901–906. Saint Petersburg, Russia: IFAC Automatic Control in Aerospace viewed 12 February 2019, https://www.sciencedirect.com/science/article/pii/S1474667017322930.

50 eoPortal, 2019. QuakeSat web page, European Space Agency, viewed 12 February 2019, https://directory.eoportal.org/web/eoportal/satellite-missions/q/quakesat.

51 DK3WN Satellite Blog, 2019. Viewed 12 February 2019, http://www.dk3wn.info/p/?page_id=29535.

52 Urlichich, Yu M., Selivanov, A.S., and Stepanov, A.A., 2019. "Two Nanosatellites for Space Experiments," viewed 12 February 2019, http://www.dlr.de/iaa.symp/Portaldata/49/Resources/dokumente/archiv5/1403_Urlichich.pdf.

53 eoPortal, 2019. "TNS" website, European Space Agency, viewed 12 February 2019, https://directory.eoportal.org/web/eoportal/satellite-missions/t/tns.

54 eoPortal, 2019. "UWE-1" website, European Space Agency, viewed 12 February 2019, https://directory.eoportal.org/web/eoportal/satellite-missions/u/uwe-1.

55 eoPortal, 2019. "CubeSat – Launch 2" website, European Space Agency, viewed 12 February 2019, https://earth.esa.int/web/eoportal/satellite-missions/c-missions/cubesat-launch-2.

56 ISIS, 2019. "History" web page, Innovative Solutions in Space, Delft, The Netherlands, viewed 12 February 2019, https://www.isispace.nl/about-us/history.

57 GOMspace, 2019. "The Story" webpage, GOMSpace, Aalborg East, Denmark, viewed 12 February 2019, https://gomspace.com/the-story.aspx.

58 eoPortal, 2019. CSTB-1 (CubeSat Testbed-1) web page, European Space Agency, viewed 12 February 2019, https://directory.eoportal.org/web/eoportal/satellite-missions/c-missions/cstb1.

59 Hoyt, Robert, Voronka, Nestor, Newton, Tyrell, Barnes, Ian, Shepherd, Jack, Frank, S. Scott, Slostad, Jeff, Jaroux, Belgacem, and Twiggs, Robert, 2007. "Early Results of The Multi-Application Survivable Tether (MAST) Space Tether Experiment," paper SSC07-VII-8/048, 21st Annual AIAA/USU Conference on Small Satellites, Logan, Utah, August 2007, viewed 12 February 2019, https://digitalcommons.usu.edu/cgi/viewcontent.cgi?article=1481&context=smallsat.

60 The Von Karman Institute, 2019. "QB50 Project web page", The Von Karman Institute for Fluid Dynamics, Brussels, viewed 12 February 2019, www.vki.ac.be/index.php/news-topmenu-238/318-qb50-project.

61 QB50 Project, 2019. "QB50 Project: Launch Scenario" web page, The Von Karman Institute for Fluid Dynamics, Brussels, viewed 12 February 2019, https://www.qb50.eu/index.php/contact.html.

62 eoPortal, 2019. "ISS: Nanoracks-QB50 web page", European Space Agency, viewed 12 February 2019, https://directory.eoportal.org/web/eoportal/satellite-missions/i/iss-nanoracks-qb50.

63 Moore, G.E. (1965). Cramming more components onto integrated circuits. *Electronics* 38 (8): 114–117.

64 The History of Computing Project, 2019. "Intel 4004 CPU" web page, viewed 12 February 2019, https://www.thocp.net/hardware/intel_4004.htm.

65 WikiChip, 2019. "EPYC 7601-AMD web page", WiKiChip, LLC, viewed 12 February 2019, https://en.wikichip.org/wiki/amd/epyc/7601.

66 Johnson, Lyle V., and Green, Charles L., 2019. "MICROSAT PROJECT–Flight CPU Hardware," pp. 104–106, ARRL Amateur Radio 7th Computer Networking Conference, American Radio Relay League, Columbia, Maryland, October 1, 1988, viewed 12 February 2019, https://ia800805.us.archive.org/3/items/AmateurRadioComputerNetworkingConference7/07DCC1988_Columbia.pdf.

67 Klofas, Bryan, 2019. "CubeSat Radios: From Kilobits to Megabits," Ground Systems Architecture Workshop, Los Angeles, CA, February 2014, viewed 12 February 2019, http://gsaw.org/wp-content/uploads/2014/03/2014s09klofas.pdf.

68 Devaraj, Kiruthika, Kingsbury, Ryan, Ligon, Matt, Breu, Joseph, Vittaldev, Vivek, Klofas, Bryan, Yeon, Patrick, and Kolton, Kyle, 2017. "Dove High Speed Downlink System," paper SSC17-VII-02, 31st Annual AIAA/USU Conference on Small Satellites, Logan, Utah, August 2017.

69 eoPortal, 2019. "ISARA" web page, European Space Agency, viewed 12 February 2019, https://directory.eoportal.org/web/eoportal/satellite-missions/i/isara.

70 NASA-Jet Propulsion Laboratory/California Institute of Technology, 2019. "Mars Cube One (MarCO) web page, viewed 12 February 2019, https://www.jpl.nasa.gov/cubesat/missions/marco.php .

71 Janson, Siegfried, Welle, Richard, Rose, Todd, Rowen, Darren, Hardy, Brian, Dolphus, Richard, Doyle, Patrick, Faler, Addison, Chien, David, Chin, Andrew, Maul, Geoffrey, Coffman, Chris, La Lumondiere, Stephen D., Werner, Nicolette I., and Hinkley, David, 2016. "The NASA Optical Communications and Sensors Demonstration Program: Initial Flight Results," paper SSC16-III-03, 29th Annual AIAA/USU Conference on Small Satellites, Logan, Utah, August 2016.

72 Dong, Ying, You, Zheng, and Zing, Fei, 2004. "Design of an APS-Based Star Tracker for Microsatellite Attitude Determination," paper IAC-04-U.2.05, 55th International Astronautical Congress, Vancouver, Canada.

73 Janson, Siegfried W., Welle, Richard P., Rose, Todd S., Rowen, Darren W., Hinkley, David A., Hardy, Brian S., La Lumondiere, Stephen D., Maul, Geoffrey A., and Werner, Nicolette I., 2015. "The NASA Optical Communications and Sensors Demonstration Program: Preflight Update," paper SSC15-III-1, 29th Annual AIAA/USU Conference on Small Satellites, Logan, Utah, August 2015.

74 Frost, Chad, and Agasid, Elwood, 2014. "Small Spacecraft Technology: State of the Art," p. 61, NASA Technical Report TP-2014-216648/REV-1, NASA-Ames Research Center, viewed 12 February 2019, http://cmapspublic3.ihmc.us/rid=1NG0S479X-29HLYMF-18L7/Small_Spacecraft_Technology_State_of_the_Art_2014.pdf.

75 eoPortal, 2019. WNISAT-1 web page, European Space Agency, viewed 12 February 2019, https://directory.eoportal.org/web/eoportal/satellite-missions/v-w-x-y-z/wnisat-1.

76 eoPortal, 2019. "Flock-1" web page, European Space Agency, viewed 12 February 2019, https://directory.eoportal.org/web/eoportal/satellite-missions/f/flock-1.

77 Wall, Mike, 2013. "Planet Labs Unveils Tiny Earth-Observation Satellite Family," http://space.com, August 31, 2013, viewed 12 February 2019, https://www.space.com/22622-planet-labs-dove-satellite-photos.html.

78 Buhr, Sarah, 2019. "Planet Labs Rockets to $118 Million in Series C Funding to Cover the Earth in Tiny Satellites," http://techcrunch.com, viewed 12 February 2019, https://techcrunch.com/2015/04/13/planet-labs-rockets-to-118-million-in-series-c-funding-to-cover-the-earth-in-tiny-satellites.

79 Planet Labs, 2019. Planet web page, Planet Labs, San Francisco, CA, USA, viewed 12 February 2019, https://www.planet.com.

80 Spire Global, 2019. "Space to Ground Data Analytics web page", San Francisco, CA, USA, viewed 12 February 2019, https://spire.com/.

81 Astro Digital, 2019. Astro Digital, Santa Clara, CA, USA, viewed 12 February 2019, https://astrodigital.com.

82 Sky and Space Global corporate, 2019. Sky and Space Global corporate website, West Perth, Australia, viewed 12 February 2019, https://www.skyandspace.global/corporate/.

83 Geo Optics, 2019. Geo Optics web page, Geo Optics, Inc., Pasadena, CA, USA, viewed 12 February 2019, http://www.geooptics.com.

84 Space Q Media Inc., 2019. Space Q Media, Inc. website, London, Ontario, Canada viewed February 13, 2019, http://spaceq.ca/helios-wire-and-exactearth-small-satellites-set-for-launch-on-smallsat-express.

85 Swarm Technologies, 2019. Swarm Technologies Home Page, Los Altos, California, USA, viewed February 13, 2019, https://www.swarm.space/.

86 Crunchbase Swarm Technologies, 2019. Crunchbase Swarm Technologies web page, San Francisco, California, USA, viewed February 13, 2019, https://www.crunchbase.com/organization/swarm-technologies#section-locked-charts.

87 Kepler Communications, 2019.Kepler Communications home page, Kepler Communications, Toronto, Canada, viewed 12 February 2019, http://www.keplercommunications.com/.

88 Hiber Global, 2019. Hiber Global web page, Amsterdam, The Netherlands, viewed February 13, 2019, https://support.hiber.global/hc/en-us.

89 Crunchbase Hiber, 2019. Crunchbase Hiber web page, San Francisco, California, USA, viewed February 13, 2019, https://www.crunchbase.com/organization/magnitude-space#section-overview.

90 ITNews, 2019. ITNews web page, Australia, viewed February 13, 2019, www.itnews.com.au/news/fleet-space-tech-to-launch-two-commercial-cubesats-514736.

91 Crunchbase Fleet Space Technologies 2019. Crunchbase Fleet Space Technologies web page, San Francisco, California, USA, viewed February 13, 2019, https://www.crunchbase.com/organization/fleet-space-technologies#section-overview.

92 Kulu, Erik, 2019. "CubeSat Tables of the Nanosatellite Database web page", viewed 12 February 2019, http://www.nanosats.eu/tables.html.

93 eoPortal, 2019. "Lemur web page", European Space Agency, viewed 12 February 2019, https://directory.eoportal.org/web/eoportal/satellite-missions/l/lemur.

94 Astro Digital, 2019. "Landmapper Constellation data sheet", Astro Digital, Inc. Santa Clara, California, USA, viewed 12 February 2019, https://astrodigital.com/downloads/brochure-astro.pdf.

95 Sky and Space Global, 2019. "Operations: Technology" web page, Sky and Space Global, West Perth, Western Australia, viewed 12 February 2019, https://www.skyandspace.global/operations-overview/technology.

2 I-2a

On-board Computer and Data Handling

Jaime Estela[1] and Sergio Montenegro[2]

[1]*Spectrum Aerospace Group, Germering, Germany*
[2]*Chair of Computer Science VIII, University Würzburg, Germany*

2.1 Introduction

In the early years of the Space Era, rudimentary data processing units supported the control of all electronic and electromechanical units in satellites. Starting with the first sounding rockets and satellite launchers, analog computers were present in the evolution of space systems [1]. The V-2 German rocket was used for guidance control of an analog computer based on an electronic integrator and differentiator. The early Russian rocket R-7 Semyorka (8K71) was using the board computer "Kvarts" for trajectory calculation since 1958 [2]. At that time, due to the reduced calculation capability available, part of the calculation tasks was done on the ground. With the development of modern electronics and the increasing demand of new space systems, more autonomy was needed in such systems. With the improvement of the nanosatellite technology, such systems can achieve more and more complex space missions. Nowadays, supercomputers are going to be tested on the International Space Station (ISS). The evolution of the board computers is going toward high-performance and high-capacity autonomous systems. In short, the use and proof of nanosatellites, which are small and cheap to build and launch, lead to new technologies and electronics, and these innovations make nanosatellites powerful platforms.

2.2 History

The first satellite was the Russian Sputnik 1. This satellite initiated the Space Era and the spacecraft engineering field. The electronics of the Sputnik 1 had basically a radio transmitter and a battery system [3]. This construction allowed the transmission of radio signals for a period of approximately 1 month. The simplicity of the system was in accordance with the mission objectives. No processing unit was necessary (Figure 2.1).

Nanosatellites: Space and Ground Technologies, Operations and Economics, First Edition.
Edited by Rogerio Atem de Carvalho, Jaime Estela, and Martin Langer.
© 2020 John Wiley & Sons Ltd. Published 2020 by John Wiley & Sons Ltd.

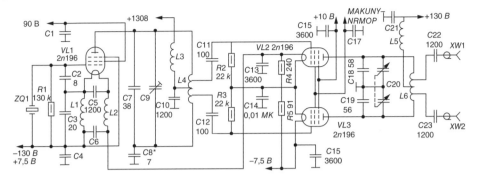

Figure 2.1 Sputnik 1 electronics. Source: Reproduced from First Transmitter IS3, Radio Magazine, N°4, 2013, Russia.

Figure 2.2 Apollo guidance computer (AGC). Source: Photo courtesy of NASA.

The Sputnik 2 satellite, the second satellite in space, carried the dog Laika into space. For the mission objectives, it was necessary to use a more complex system. In this case, a board computer was available together with radio transmitters, telemetry system, and temperature control system [4]. The first transistor was also used in the Sputnik 2 and in the next years transistors won more and more importance in space projects until the integrated circuits appeared [5]. The Gemini Digital Computer (GDC) for the Gemini Mission was one of the first complex computers in space. It was still based on single transistors and had no redundancy. The computer was built by IBM, consumed around 100 W and weighing 27 kg. The Apollo Guidance Computer (AGC) (Figure 2.2) was installed on the Apollo Landers and in the Command Modules. The AGC was designed by the MIT Instrumentation Laboratory and was one of the first space computers based on integrated circuits. The Space Shuttle computer was designed and manufactured by IBM and was called AP-101. The shuttle data processing system (DPS) was a redundant system and had four computers installed in parallel. A failure was detected due to a voting system [6] (Figure 2.3).

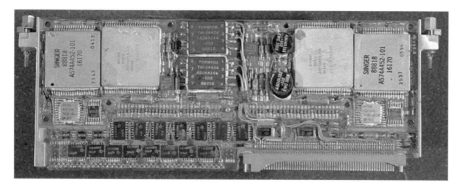

Figure 2.3 IBM AP-101 Space Shuttle computer. Source: Photo courtesy of NASA.

Figure 2.4 Argon-16 (70 kg). Source: Research Institute Argon.

The Soyuz spacecraft used the Argon-16 computer with triple redundancy, 2 kB RAM, and 16 kB ROM since 1974. In 2016, the Argon-16 was replaced by the high-performance computer TsVM-101. The TsVM-101 is smaller, faster, and can execute six million instructions per second (IPS). It has 2 MB RAM and 2 MB ROM as internal memory [7, 8] (Figures 2.4 and 2.5).

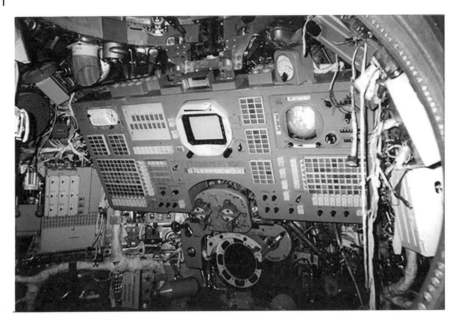

Figure 2.5 Soyuz instrument panel. Source: Reproduced with permission from Roscosmos, Russia.

2.3 Special Requirements for Space Applications

Space systems have different requirements than the ordinary electronics. The environment is extremely hard, and no direct contact exists. Often all activities with the spacecraft have to be done remotely. The following requirements are relevant for space missions:

Reliability: Spacecraft cannot be repaired in space and space missions are expensive. For this reason, the spacecraft has to be robust and work reliably at least for the mission duration.

Self-healing: In case of failures, the spacecraft has to correct errors and recover the operational modus. In extreme cases, human support can solve critical problems. Redundancy is very common in such system and increment the robustness and reliability of space systems.

Fault tolerance: Bit flips succeed in space due to radiation effects. These errors have to be detected and corrected.

Limited resources: Parameters such as space, energy, weight, and temperature are very critical, and in the design all these parameters must be considered and kept as economical as possible.

Temperature: In space, a spacecraft experiences extreme temperature changes, which could range from −170 to +120°C. Inside the satellite, the temperature range goes from +10 to +40°C. For the good function of the electronics, the operational temperature range must be kept. Components can be damaged if the environmental temperature stays beyond the limits for a critical period of time.

Vacuum: There is no air in space, and in vacuum the thermal management is different than to the ground. Metal conductors transport the heat to radiators.

Microgravity: This does not affect the function of the hardware in space.

Vibrations: The space hardware must tolerate extreme vibrations occurred during the launch. If the hardware is not robust enough, it will be destroyed during the launch.

Software complexity: The software in space systems has to comply with strong requirements. It has to optimally exploit hardware resources and must recognize errors and correct them. The spacecraft must operate autonomously, and if the system has a malfunction, the system software must try to recover it.

Software-uploads: The software on-board can be improved or adapted after the launch. During the operations phase, small errors or improvements may be identified and the satellite software should be enhanced. In case of a big part of the satellite, software has to be updated. Many contacts are needed because of the nature of the software-uploads, and the uploaded software has to be checked in the spacecraft before the old software is replaced with the new one. This activity is very risky.

Remote diagnosis: With telemetry data, it is possible to analyze the status of all subsystems of the spacecraft. Not all problems are solved automatically in the spacecraft, and from time to time it is necessary to solve the problem from the ground.

High computing performance: Satellite systems have less computing capacity than their counterparts on the ground. First of all, for most of the missions a low computing capacity is enough. The high reliability needed in space systems requests the use of robust technologies. These technologies guarantee stable functionality in space but do not allow high computing performance. Limited resources such as power consumption favor the use of low power electronics with less processing speed. [9]

2.4 Hardware

2.4.1 Components

The space market is still dominated with High Reliable (Hi-Rel) parts designed and manufactured for military applications, and this matches with space project requirements. Hi-Rel space parts are also often subject to International Traffic in Arms Regulations (ITAR). These special robust components can operate in harsh environments. The demand of this market is small in comparison with the electronic mass market, and for this reason a smaller number of such components are manufactured. The qualification of such parts is also a costly process where the components have to comply with environmental tests. For example, the qualification of one component in the market takes around 2 years and costs around €2 million. All these factors make the single price of such components very expensive. Most of the ITAR components are manufactured in the USA. To export these to the rest of the world, special permission is necessary. It takes around 6 months with no guarantee of success. The prices of Hi-Rel components vary from few thousand Euros to hundred-thousands. One memory component can cost, for example, €10 000, and a high integrated processor can reach a price of €100 000. Hi-Rel parts are extremely robust and do not match with nanosatellite mission requirements. For instance, a nanosatellite mission has an average mission duration from 6 months up to 5 years in low Earth orbit (LEO). High-Rel parts can operate in LEO orbit for more than 100 years. Thus, those components are mostly for highly reliable satellites having a guaranteed lifetime from 5 up to 15 years.

Nanosatellites mainly use commercial components without appropriate qualification and are flown without guarantee for the operation in space. For this reason, such missions have a short duration (e.g. 6 months), and a high default risk is acceptable. Even constellations instead to qualify the components prefer to include a failure rate up to 20%. For the professional use of nanosatellites, qualification of the electronic components has to be done in order to increase the reliability of nanosatellites. Some initiatives in such direction have been taken. One of these initiatives are the so-called Space-COTS [http://space-cots.com], which are selected commercial components qualified for space. Not only must components be qualified, but the qualification processes have to also be adapted to the new markets and new system requirements. Standards for the qualification of Hi-Rel components from the NASA and ESA exist, but they do not match nanosatellite needs. The existing standards are a base for new standards that can support nanosatellites. The International Organization for Standardization (ISO) also started the publication of new qualification standards for space systems where nanosatellite needs are considered.

2.4.2 Brief History of On-board Computers

The Space Era began after the invention of the transistor by Shockley, Bardeen, and Brattain by the end of the 1940s, and this new kind of component also played a very important role in space systems. In the early years (1950s and 1960s) of the space exploration in the USA, no qualified parts for space were available. The first components installed in American space systems were alloy and grown junction transistors, such as the 2N45 and 2N335, alloy diodes, paper capacitors and ceramic capacitors, and carbon composition resistors and relays [10]. The production of semiconductors in the USSR began in 1947. The first transistor was installed in the Sputnik 2. The Explorer 1 was the first system completely built with transistors. The Sputnik 2 set a very ambitious aim of keeping the dog Laika alive during the mission, and few transistors were used in the complex system. The Soviet launchers had a high capacity to carry payloads into space, and the use of valves (tubes) was possible. The situation was opposite in the USA. For the low-carrying capacity of the American launcher, it was necessary to design a lightweight Explorer satellite. This was made possible using transistors. The Explorer 1 was operational for almost 2 months until the batteries ran out. For this reason, it was not possible to know for how long these transistors were capable of operating without failures in space. At least in the Explorer 1 mission, no failures were reported. It is also important to mention the high performance and manufacturing quality of the valves at that time. The price of the valves was low (US$1 each), and the first produced transistors cost around US$20. Also, the operational temperature yielded a difference, whereas the valves ensured much more stability. [5]

The first components used in satellites were mostly commercial and few of them were for military purposes. The main criteria for selection of parts were:

- Good manufacturing
- Stable parts
- Good results of environmental tests
- X-ray inspection of internal structure

The Marshall Space Flight Center published its first Hi-Rel parts specifications in 1962 and after that a Preferred Parts List was also published [10].

2.4.3 Processors

In the end of the 1970s, the first processors were introduced in satellites. A compilation of these is shown in the following text [6]:

- *Viking 1 and 2*: Honeywell HDC 402 (two units)
- *Voyager 1 and 2*: Customized 4 bit Complementary Metal-Oxide-Semiconductor (CMOS) processor (three computers with double redundancy)
- *Space Shuttle*: IBM APA-104S 32 bit processor (five redundancy)
- *Galileo*: RCA 1802 COSMAC processor. 8 bit logic. First radiation hard microcontroller
- *Hubble Space Telescope*: Intel 80386 and math coprocessor 80387 (double redundancy)
- *Sojourner*: Intel 80C85
- *ISS*: Intel 80386SX-20 and 80387 (command computers)
- *Spirit and Opportunity rovers*: IBM RISC Single Chip

In 1980, the standard MIL-STD-1750A was introduced [11]. This standard defined a 16 bit microprocessor architecture. Several semiconductor manufacturers designed their own processors based on this new standard. The main characteristics of the standard based processor are:

- 16 bit logic
- 16 registers with 16 bit word length
- Registers can be grouped in 32 or 48 bits registers length
- Directly addressable memory of 128 kB
- Memory expansible up to 2 MB by using a memory management unit (MMU)

After a decade, it was necessary to reduce the system complexity and the new computer architecture called reduced instruction set computer (RISC) was introduced. Processor designs based on the RISC architecture have very compact low-level instruction set where each instruction can be executed in one clock cycle.

Nowadays, the standard is used for the design of on-board computers. The main characteristic of this standard is the use of serial data bus for data transfer.

Many companies of Hi-Rel components offer Rad-Hard processors. The nanosatellites use commercial microcontrollers. Such COTS microcontrollers already have space flight heritage and have performed short missions in many cases successfully. The following microcontrollers have been flown in nanosatellites:

- *Texas Instruments*: MSP430 (several versions)
- *Microchip*: PIC24 5 PIC32 (several versions)
- *Atmel*: AVR & ARM based microcontrollers (several versions)
- *STMicroelectronics*: STM32 (several versions)

Some of the mentioned microcontrollers were partially tested and a degradation was found. The first degradation observed is always the increase of the power (current) consumption. The increment of the current consumption will be often several times higher than the nominal value [12] (Figure 2.6).

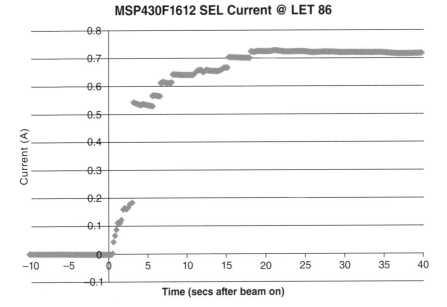

Figure 2.6 MSP430 single event latch-up (SEL) test results.

2.4.3.1 Field Programmable Gate Array (FPGA)

Field programmable gate arrays (FPGAs) are highly integrated components with millions of unconnected logical (AND, OR, XOR, register) units. These logical units can be programed and interconnected per software. Also, complex functions can be implemented with this method.

FPGAs are very interesting for space missions because they can interconnect other components avoiding the use of external low integrated components. Each pin can be also defined as required in system. Operations executed repeatedly can be implemented in the FPGA reducing the system complexity.

The size of the FPGA allows the implementation of complete systems inside one chip. Several CPUs can be synthesized in FPGAs (soft processor). Other designs integrate CPU cores (up to four) inside one FPGA. For space missions, the aim is to integrate as much electronics as possible in few integrated circuits, reducing the system complexity, power consumption, size, and weight.

One limitation of the FPGAs is that the internal configuration of the FPGA logic has to be stored in external memories. Some manufacturers integrated Flash and SRAM cells, trying to make the FPGA more autonomous. Further elements were also implemented in the FPGAs, such as interfaces, drivers, decoders, etc. [9]

Standard CPUs have no protection mechanism in case of bit flipping. The ESA has developed the processor LEON to solve this lack.

The LEON processor was developed specially for safety critical systems. In the LEON, every internal function is implemented triply, and the results are compared immediately after each basic operation. Deviations are corrected immediately where they began.

Memory and FPGAs internal data can be protected by redundancy. Redundant codes recognize corrupted data and correct it. The simplest method to check and correct corrupted data is to replicate the logic two or three times and compare all results.

Inside of LEON every function and every register are replicated three times (TMR, triple modules redundancy). After every operation all three results are compared, if one is different, it is immediately corrected. LEON is robust but slow. Another alternative is to use COTS processors and to build the redundancy externally or by replication in software. By software one can execute every function two or three times and compare results. It is more economical to check only results against plausibility rules. A very simple and economical check is to use watchdogs. For each function, a maximum duration is defined. Not later than this defined time the watchdog is triggered and the complete computer is reset. After the reset, a recovery program is executed [Book DLR].

Application-specific integrated circuit (ASIC) is a customized FPGA. A FPGA has many logical units for lot of combinations. An ASIC is a type of FPGA composed only of the elements needed for a specific application, and, for this reason, an ASIC consumes much less power than the FPGA and is also smaller. The following characteristics summarize the advantages of the ASICs:

- Fully customized
- Lower costs per unit
- Smaller form factor
- Less power consumption

System on chip (SOC) integrates a microcontroller core with memory and peripheral circuits, and builds an embedded system. The most important advantages are the small size of high complex systems and the low power consumption. For space applications, the SOC technology satisfies the increasing demand of processing capacity, reducing automatically the system mass and power consumption.

2.4.4   Mass Memory

The GDC had a magnetic core memory with a capacity of 4096 words. This storage unit was made of ferrite rings woven into a matrix of wires. Each ring was crossed by three wires, two for the address (row/column) and a read inhibit wire.

Magnetic core storage system was used before the semiconductor technology appeared in the 1970s. This type of memory system is robust against electromagnetic effects, high temperature, and radiation. In the Apollo program, magnetic core memories were used. The Space Shuttle missions used magnetic core storage systems in the first phase, but these were upgraded to semiconductor memories.

The Pioneer and Voyager spacecrafts used magnetic tape drives for mass storage. Science data were recorded and transmitted to the ground during the contacts with the Earth. The tape inside the Voyager was 328 m long and could store 536 Mbit of data. Write speed was 115.2 kB/s and the playback speed was 57.6 kB/s.

A tape band is very reliable, because it is very robust against radiation. Its disadvantages are the use of moving parts and the serial access to the data [13, 14].

Semiconductor memories are sensible against radiation. Charged particles coming from space have different energy levels and can interact and corrupt the digital data in the processing system. In the spacecraft design, recovery mechanism is implemented to recover the lost data. Last technology developments are also available in DRAM memories with internal recovery systems. It means extra hardware was implemented in the memory to protect the data and recover it very fast. This hardware solution is faster and more efficient than the software protection. Such components demonstrate already a higher robustness against radiation.

Ferroelectric RAM (FRAM) is the new technology available in the electronic market. Microcontrollers from Texas Instruments assembled with FRAM memories will be flown in nanosatellites to test their reliability. This technology uses capacitors with ferroelectric dielectric instead of traditional capacitors. The advantages of this technology are the higher reliability and faster speed than EEPROM technology. It can store more than 10 years of recorded data and has a higher temperature operation range. The read and write cycles are more than 10^{10}–10^{15} (EEPROM 10^6 cycles). The FRAM requires power only by reading or writing a cell and uses 99% less power than a DRAM memory. The FRAM cells are more tolerant to radiation than FLASH cells and offer more reliability for satellites [15]. Nanosatellite missions are in plan to test FRAM technology.

2.4.5 Bus

In order to reduce costs and expenses, nanosatellites use commercial interfaces and buses. The quality of such buses and interfaces is proved in the electronic market worldwide and represents a good solution for nanosatellites.

USB (Universal Serial Bus): This bus is available in commercial nanosatellite hardware and is used mostly for communication with the PC or external units, programing, debugging, and emulation.

I2C (Interconnected integrated circuit): This two wires bus is available in many integrated circuits (ICs) and requests a minimal wiring to interconnect single ICs or units. This bus was developed by Philips and offers many advantages:
- Requires only two bus lines. The serial data line (SAD) and a serial clock line (SCL).
- Each device connected to the bus is software addressable.
- True multi-master bus.
- Collision detection.
- Transfers data up to 3.4 Mbit s^{-1} (high-speed mode).
- Extremely low current consumption.
- High noise immunity.
- Wide supply voltage range.
- Wide operating temperature range.

RS-232 and RS-485: These traditional serial bus protocols are always present in nanosatellites. The data is transferred with voltage signals. The simplicity of its use and the availability of such interface in lot of IC components make both of these indispensable together with the serial bus. The communication between components or units with a good performance and data rate guarantees the future use of the RS-232 and RS-485 protocols and the serial interface in nanosatellite systems.

CAN (Controller Area Network): This bus was designed by Bosch originally for the auto industry with the aim to reduce the cabling inside a car. The CAN bus offers following advantages:

- Data rate up to 1 Mbps.
- Reduction of wiring in system.
- Reduction of costs.
- Supports protocols from 8 to 64 bytes.
- Retransmission of lost messages.
- Works in several electrical environments.
- Different error detection capabilities.

The "ECSS-E-ST-50-15C CAN bus extension protocol" is an ECSS (European Cooperation for Space Standardization) standard, which covers almost all needs of satellite buses for several mission types.

PC/104: It is an industrial standard for embedded systems and is specialized for rugged environments and small size systems. In most of the nanosatellites, the PC/104 PCI connector is used to stack boards and offers a very stable structure. Although the PCI connector is used, the pin assignment is implemented on its own. In this case, no standard is defined, and each pin has completely different functions for each nanosatellite.

SpaceWire: This standard is used for internal spacecraft communication. Several space agencies such as the NASA, ESA, Japan Aerospace Exploration Agency (JAXA), and Roscosmos adopt this standard in its own designs. SpaceWire is a communication bus based on the IEEE 1355 standard. The SpaceWire allows low-cost nodes interconnection, low latency, and full duplex communication. For hardware, the SpaceWire uses the low-voltage differential signaling (LVDS), which was designed for high-speed video for industrial cameras. The LVDS bus consists of two wires and the data is transferred with current variations. This makes this system more robust than the famous RS-232, driven with voltage signals. The use of SpaceWire succeeds also in coordination with the use of the Consultative Committee of Space Data Systems (CCSDS) standards for space systems communications and services. Nanosatellite platforms started to use SpaceWire for internal communication providing interface ports.

2.5 Design

2.5.1 System Architecture

The mission requirements and the operational needs define the framework of the spacecraft system. The system architecture shows the interaction of the system units and is illustrated in a block diagram. Computer systems like a satellite bus have three main architecture block diagrams:

Data architecture: Describes structure of the data network, the protocol, and the logical interaction between the nodes.

Hardware architecture: Defines the hardware resources available in the system, the connectivity between units, and the instructions that can be executed in the system.

Software architecture: Shows how the processing instructions are executed. The detailed execution of the instructions, routines, interruptions, and services is included in this block diagram.

Several system architectures are available and can be implemented depending on the system requirements [16].

Centralized architecture (star): A central node is connected directly with the remaining nodes of the network. The central node manages the communication and activities in the whole system (Figure 2.7).

Advantages:

- Best solution for small systems.
- Errors will not affect other nodes.

Ring architecture: No managing node is available. All nodes form a circular pattern and each node is connected with only two other nodes. The information gets forwarded from one node to the other until the message arrives to the appropriate node (Figure 2.8).

Advantages:

- Less harness and the data bus can be kept simple.
- New nodes can be added easily.

Bus architecture: All nodes share a common data bus. A protocol manages the communication in a good manner (Figure 2.9).

Advantages:

- High reliability due to easy execution of tests and trouble shootings.
- Loss of one or more nodes does not affect the communication between the remaining nodes.

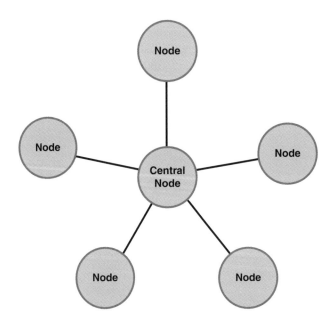

Figure 2.7 Centralized architecture.

Figure 2.8 Ring architecture.

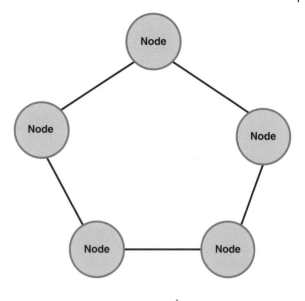

Figure 2.9 Bus architecture.

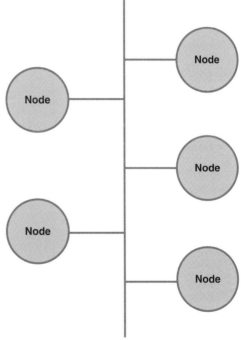

2.5.2 Central Versus Distributed Processing

Similar to the auto industry, a spacecraft has many computers from many different manufacturers. In order to avoid a loss of reliability, the hardware and the software should be compatible. The aim is that the control of any device shall be able to run and be executed on any computer, and not only on a dedicated computer for every device. A resource-sharing software for resource management is necessary. The same hardware platform should be used for the control of many devices and for the execution of applications on the spacecraft bus as well as on the payload side. For example:

- *Platform*: Telemetry, attitude control, navigation, power-control, etc.
- *Payload*: We could have pattern recognition, image processing, classifications, etc.

With distributed processing, a failure does not affect the complete satellite. Other strategies are:

- Separation of vital functions from extra functions.
- The ideal case is that very critical functions should run in different processors to avoid interferences.

With a centralized processing, the board computer can be a processor pool which provides dependable storage and computing performance for all devices and applications. Software tasks can be distributed in this processor pool. Payload tasks and other applications are likewise executed. The processor pool shall consist of several redundant independent computers. Even if some fail, the remaining ones should be able to fulfill the mission [9].

2.5.3 Design Criteria

Following factors have to be considered for the design of a board computer system:

Mission Requirements

- Mission type
- Number of satellites
- Communication link characteristics
- Schedule
- Cost

System Requirements

- Processing capacity
- Memory capacity
- Interfaces
- Power consumption
- Operating system
- Operation temperature range
- Size
- Weight
- Security
- Radiation tolerance
- Development workbench

Additional Requirements

- Testability
- Feasibility
- Usability
- Reliability
- Flexibility
- Maintainability
- Replaceability

2.5.4 Definition of Requirements

For a successful mission, it is necessary to define the correct mission requirements. Following parameters act as guidelines for this purpose:

- Define functional requirements.
- Evaluate the validity of the defined requirements.
- Assess considered system architectures.
- Define the functions of each part in the whole system.
- Check if the needed technology is available.
- Test if defined system complies mission objectives.

2.5.5 Resource Estimation and Data Budget

In order to comply with the mission requirements, it is necessary to dimension the on-board computer with the mission processing and storage needs. It happened already in some nanosatellite missions where the CPU had a lower capacity than the required from the experiment and it was not possible to operate it. In other cases, a powerful computer was selected where the experiment used only 10% of the computing capacity. The remaining calculation capacity was not used. The professional solution is the selection of a CPU that complies all mission requirements, no more and no less. It is difficult to find literature explaining the calculation of the data budget. The ECSS-M-ST-10C (space project management) standard of the ECSS defines the life cycle of space projects, which is divided in seven phases [17]:

- *Phase 0*: Mission analysis/needs identification
- *Phase A*: Feasibility
- *Phase B*: Preliminary definition
- *Phase C*: Detailed definition
- *Phase D*: Qualification and production
- *Phase E*: Utilization
- *Phase F*: Disposal

The satellite designer has to make an accurate calculation of all important parameters such as the mass, power, etc., in order to get the best configuration of the system-saving resources and money. For a satellite mission, following types of budgets have to be calculated in Phase B (preliminary definition):

Mass budget: Calculation of the weight of all structural elements.

Power budget: Most of the nanosatellites generate little power, and this is one of the most critical parameters. An accurate and reliable calculation of the power consumption is crucial. The energy generation and energy storage are the main elements of the power budget, and these two parameters can be summarized in the solar panel budgeting and the battery budgeting.

Link budget: The generated data in the satellite (housekeeping and science data) have to be transferred to the ground. The link budget guarantees the success of this task. The characteristics of the ground station and of the communication subsystem of the satellite define this budget.

Data budget: The generated data of the satellite have to be processed and stored internally. This will be explained in detail in the following text.

Thermal budget: Materials react with the environment and the variation of the temperature changes the characteristics of the materials. Electronic components have an operating temperature and are different for each component. Inside the satellite, the operating temperature of the whole system has to be guaranteed. The thermal budget summarizes the operational temperature of all components in order to keep the right temperature range.

For the design of the on-board computer, the data budget [18] gives relevant information for the calculation of the On-Board Computer (OBC) resources. Each subsystem has a data budget demand and the OBC manages the exchange of information in the satellite. The CPU must have enough resources to achieve its job. Furthermore, the telemetry data is generated with sensors distributed in the spacecraft. In the data budget, all the generated information with telemetry sensors, subsystem services, and payload data is calculated. The following sub-budgets are to be considered:

Telemetry packet budget: Each sensor generates different housekeeping data. It depends on the sensor's nature, measurement accuracy, and sampling rate.

Payload data budget: This data amount estimation depends on the experiment characteristics. For example, a camera is a common payload used in nanosatellites. A camera has following critical parameters:

- Image sensor type: panchromatic, multispectral, hyperspectral
- Resolution
- Frame rate
- Bit per pixel
- Compression rate

And after the integration in the system:

- Telemetry sensors data
- System control signals

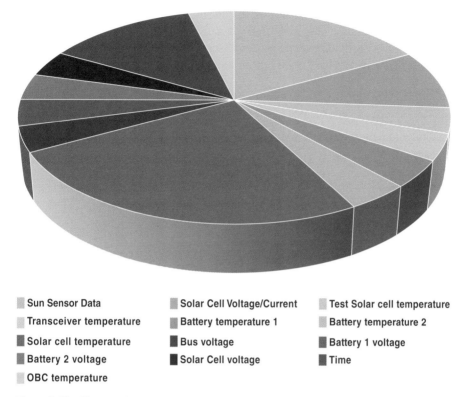

Sun Sensor Data	Solar Cell Voltage/Current	Test Solar cell temperature
Transceiver temperature	Battery temperature 1	Battery temperature 2
Solar cell temperature	Bus voltage	Battery 1 voltage
Battery 2 voltage	Solar Cell voltage	Time
OBC temperature		

Figure 2.10 Nanosatellite data budget example. Source: Photo courtesy of NASA. (*See color plate section for color representation of this figure*).

2.5.5.1 Data Budget Analysis

Considering the mission objectives and the available resources, the data budget is adjusted. The following criteria help to define the needed resources and the future operation of the example payload (camera) (Figure 2.10):

- Telemetry housekeeping data rate
- Science data rate (data takes)
- Data takes (pictures) per day
- Memory size for data takes storage
- Compression rate (image compression)
- Data transmission capacity

2.5.6 Commanding

A satellite is a computer and a computer must be programed. Satellites operate in space and only a radio transmission allows the exchange of data between the spacecraft and the ground. Satellites are not programed at hardware level and inside the satellite computer instruction blocks are defined and fixed in the satellite memory. These blocks of instructions can control the satellite completely and can be activated with (short) commands. With this strategy, many control activities can be done with less information (commands). This has the advantage that less data have to be sent to the satellite and it reduces the expense to uplink information to the satellite.

It is also possible to execute software-uploads from time to time to improve the operating software. Commands can be executed immediately or at a specific time:

- Immediate commands
- Time-tagged commands

2.5.7 Telemetry

The telemetry informs about the status of the satellite. Strategic points are selected to fix sensors (voltage, current, temperature, magnetic field, etc.). Depending on the sensor and the accuracy, a specific telemetry data amount is generated. Also, single digital bits can be monitored with the telemetry. The generated raw data are packetized with the communication protocol and sent to the OBC. The data that have to be sent to the ground are forwarded from the OBC to the communication subsystem.

2.5.8 Time Generation

The on-board computer generates the system clock that serves as synchronization reference. The clock signal is generated with the support of one register counter. The bit-resolution varies from satellite to satellite. The Earth time (GMT) is also synchronized for the generation of telemetry and the execution of time-tagged commands or to trigger events.

2.5.9 Handling of Errors

COTS microcontrollers have a high level of integration, but they do not support internal error detection and handling. Protection mechanism has to be implemented in new microcontrollers or with external hardware.

If the processor crashes, a watchdog timer can detect this status and reset the system. If a more reliable architecture is needed, the hardware of the system can be duplicated or triplicated. With a double redundancy, an error can be detected but not corrected. A triple redundancy allows the detection and correction of an error.

For error detection in memories, it is possible to use several mechanisms, such as:

- Parity
- EDAC (error detection and correction) code
- CRC (cyclic redundant codes) at block level
- Multiple copy of data

The most used and most simple method for external error handling is to replicate the hardware and compare externally their results using a supervisor. As mentioned earlier, a triple redundancy is an effective robust system that allows the detection of errors and the immediate correction of this [9].

2.5.10   Radiation Effects

The cosmic radiation consists of protons, alpha particles, heavy ions, electrons, and neutrons. Three sorts of damages can be caused to all these particles in miniaturized circuits: totally ionizing dose (TID), single event upset (SEU), and single event latch-up (SEL).

The TID causes a quick aging of the silicon components. The power consumption and the internal noise thereby rise slowly, however, continuously, until the signal/noise ratio is so bad that the signal cannot be distinguished any more clearly from the noise. Then the component gets failed permanently. The TID is measured (even today) in Krad (kilo of radiation absorbed dose, or the energy dose). The radiation intensity depends on orbit. For example, for LEOs (below 800 km) 3–10 Krad per year after 2 mm aluminum shield can be counted. Normal electronic components survive approximately 30 Krad and Hi-Rel components are laid out for a dose by 300 Krad. As comparison: 1 Krad is already deadly for people.

SEU and SEL as mentioned are both single event effects (SEE). They are caused by the impact of single particles rich in energy on semiconductor. The adjacent material catches the energy and causes an ion track. If this happens within a flip flop or a memory cell, a stat change (from 1 to 0 or from 0 to 1) may take place. This originates a data corruption (SEU). SEUs cause only data damages. The components do not get damaged on their own. One must count on from 1 to 10 SEUs per day. To prevent these data damages, one has to implement some kind of redundant codes or redundant data.

The SELs are current spikes which can damage the component permanently. A charged particle can cause an ion track in the substrate and thereby so much current can flow that the chip surface is burnt. SELs are extremely rare; several years can pass before one appears. To avoid damages, the circuit has to be interrupted immediately. In less than 1 ms, one has to turn the device (power) off.

TID helps only in screening and the robustness of the hardware components. SEU data redundancy helps to recognize corrupted data and to correct it. In case of SEL, the complete circuit must be switched off within 1 ms. It can be switched on again one or two seconds later.

SEL causes a sudden power spike which lies under circumstances hardly higher than the normal current. The latch-up protection (current monitoring) must recognize such spikes and switch off everything immediately. To be able to recognize the spikes, it is advisable to monitor all components separately [9].

References

1 Hoelzer, H. (1946). *Fully Electronic Analog Computer Used in the German V2 (A4) Rockets*. TH Darmstadt.

2 Russian Space Web. R-7 ballistic missile, http://www.russianspaceweb.com/r7.html, Accessed 2019.

3 Afanasyev I.B., Lavrenov A.N. The first artificial Earth satellite, http://osiktakan.ru/1-isz2_2.html, Accessed 2019.

4 Sputnik 2 – Wikipedia, https://en.wikipedia.org/wiki/Sputnik_2. Accessed 2019.

5 First transistor in space: little-known aspects of space race, https://geektimes.ru/post/ 282258, Max Gorbunov, November 2016.

6 Eickhoff, J. (2011). *Onboard Computers, Onboard Software and Satellite Operations.* Springer Publisher/Institute of Space Systems, University of Stuttgart.

7 A Digital Soyuz, https://spectrum.ieee.org/aerospace/space-flight/a-digital-soyuz, IEEE Spectrum, James Oberg, September 2010.

8 Hall, R.D. and Shayler, D.J. (2003). *Soyuz – A Universal Spacecraft.* Springer Publisher.

9 Ley, W., Wittmann, K., and Hallmann, W. (2009). *Handbook of Space Technology.* Wiley publisher.

10 Hamiter, L. (1991). *The History of Space Quality EEE Parts in the USA.* Hunstville, USA: Components Technology Institute Inc.

11 MIL-STD-1750A, United States Military Standard, 1982.

12 Candidate CubeSat Processors, Steven M. Guertin, Jet Propulsion Laboratory/CALTECH, Pasadena, USA, September 2014.

13 From the Satellite to the Ground, http://teacherlink.ed.usu.edu/tlnasa/reference/ imaginedvd/files/imagine/docs/sats_n_data/sat_to_grnd.html, NASA Goddard Space Flight Center, 2012.

14 Wertz, J.R. and Larson, W.J. (2007). *Space Mission Analysis and Design.* Springer Publisher.

15 Normann, A. (2015). *Hardware Review of an On-Board Controller for a CubeSat.* Norwegian University of Science and Technology, Trondheim.

16 Berlin, P. (2005). *Satellite Platform Design.* Kiruna, Sweden: Department of Space Science, Luleå University of Technology.

17 ECSS-M-ST-10C Rev. 1, European Cooperation for Space Standardization, ECSS Secretariat Requirements & Standards Division, Noordwijk, The Netherlands, March 2009.

18 Krishnamurthy, N. (2008). *Dynamic Modelling of CubeSat Project MOVE.* Kiruna, Sweden: Department of Space Science, Luleå University of Technology.

3 I-2b

Operational Systems

Lucas Ramos Hissa and Rogerio Atem de Carvalho

Reference Centre for Embedded and Aerospace Systems (CRSEA), Innovation Hub, Instituto Federal Fluminense (IFF), Campos dos Goytacazes, Brazil

3.1 Introduction

Adopting a technology into projects in an area with lots of other consolidated technologies and methods is always a challenge. So the real-time operational systems (RTOSs) in the aerospace area can be introduced among all the limitations imposed by the environment where spacecraft operate, such as the exposition to radiation and the communication issues with the Earth. The process of changing the way of how software is developed for spacecraft can bring a lot of benefits at the end, but also creates a series of troubles until achieving stability. This challenge gets even bigger when the focus comes specifically to the small satellites, where the limitations are also bigger, such as the power consumption issue due to the lack of space available to allocate batteries. There is need to adopt smaller and more efficient parts and the different ways to interact with the ecosystem; for example, small space debris that is harmless to huge satellites could compromise a small satellite. The aim of this chapter is to briefly introduce the main RTOS concepts, the current state of their use in nanosatellite missions—pointing out their successes and failures, and presenting at the end a summary of directions that can be taken by teams wishing to adopt this type of technology in their projects.

3.2 RTOS Overview

RTOS is an operational system with its focus on embedded systems that need a precise processing time management. This kind of system usually requires a restricted time deadline for the tasks to execute, allowing the software to succeed in its purpose. That is why scheduling algorithms is one of the most important parts of the RTOS.

The operational system for small satellites is an example of system that demands this kind of precision because of its limitations that go beyond the time scope, reaching even power consumption limitations due to the difficulty of charging the batteries in space, making it necessary to have a system capable of dealing with scarce resources, in general.

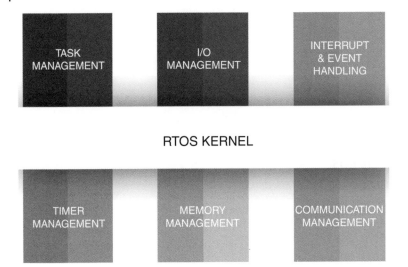

Figure 3.1 Common RTOS kernel services.

Currently, many distinct RTOSs are available in the market, some of them are open source, such as FreeRTOS, others are commercial licensed, such as Micrium's µC/OS-II and µC/OS-III. Also, there are some RTOSs focused on specific purposes, such as Contiki, dedicated to networking. Others are focused on a group of chips, such as TI-RTOS, dedicated to ARMs and microcontrollers of Texas Instruments.

All kinds of RTOSs have one component in common, the kernel, which acts as an abstraction layer between the hardware and the applications, dealing with the services. The most common services are featured in Figure 3.1.

The kernel is complemented by a series of modules that increase the functionalities that the RTOS handles. Generic RTOSs favor to have a lot of distinct modules. On the other hand, purpose-dedicated systems tend to have a small number of modules, bringing only those that are relevant for the desired purpose. An illustrative architecture of a generic RTOS can be viewed in Figure 3.2.

This scenario results in some operational systems being ideal for the aerospace area, and others being almost impracticable for using in orbit. Some small satellites have already been designed using RTOS, proving the feasibility of their use for this kind of spacecraft.

3.3 RTOS on On-board Computers (OBCs): Requirements for a Small Satellite

As is shown in Chapter I-2a, the evolution of OBC processors leads to the system on chip (SoC) concept adoption, which integrates a microcontroller core with memory and peripheral circuits, having their most important advantages as the small size and the low power consumption. Hence, the system engineers have two options, either they develop the applications directly on the microcontroller, but they also have to develop the services, such as

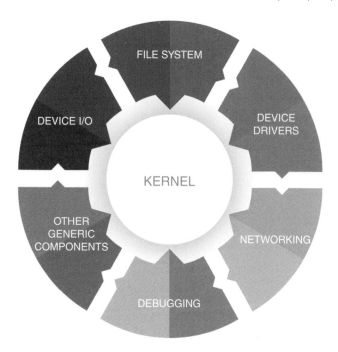

Figure 3.2 Generic RTOS architecture.

scheduling, or they use an operational system, which on the other hand will take part of the memory and computing cycles to run these same services. Thus, in an environment of scarce resources, such as nanosatellites, there will always be the decision of directly dealing with the complexity of developing the services "by hand" versus the cost of employing an operating system. At this point, RTOSs present a solution that balances both factors by providing the necessary basic services while consuming the least possible resources of the host hardware.

In Chapter I-2a, some design criteria are pointed while considering an OBC for small satellites. Among them are some mission requirements, system requirements, and some additional requirements. In most scenarios, RTOS-equipped designs match all those requirements, providing the designers a smaller risk environment. To fit the requirements of OBCs, the RTOS must have its own requirements, ensuring that it will be able to deal with the complexity of a small satellite.

A satellite's OBC must deal with lots of peripherals with many limitations, especially on smaller satellites, such as 1U, where space and power sources are very limited. That represents the main goal of adopting an RTOS, extracting the maximum of the components, doing more tasks, consuming less power, and taking up less memory. Figure 3.3 illustrates a generic OBC architecture using an RTOS.

The versatility of RTOSs allows engineers to adopt ARM processors with multiple cores making it possible to provide great processing power at a reasonable energy cost. This kind of processor also increases the available slots of peripherals interfaces, allowing developers to embed more components without using much space inside the satellite. Most of the

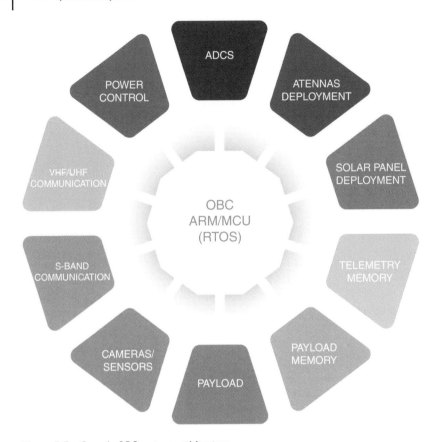

Figure 3.3 Generic OBC system architecture.

RTOSs available also provide a large number of application program interfaces (APIs), making it easy to implement tasks such as networking and memory management. In this way, these systems reduce the time and risk of development, making the OBC + operational system combination more reliable overall.

3.3.1 Requirements

Considering that RTOSs are then a good option in general for small satellites, it is necessary to define what specific requirements they must meet for this type of application. As is known, the limited space brings a series of consequences, one of the main being the reduced space available for batteries. Therefore, one of the major requirements for the RTOS is low power consumption, translated in practice into optimized number of computing cycles and memory accesses. In other words, the RTOS must extract real-time performance from the hardware without pushing it to its limits.

Another important requirement is the reliability of the architecture, in order to reduce the risks of failure, when it is very difficult and costly to correct errors during the mission. Therefore, the RTOS must be capable of ensuring that the basic hardware functions of OBCs, such as communication, Attitude Determination and Control System (ADCS), sensors, and memory management, are working properly.

Other features of RTOSs make them a good choice in terms of the ability of meeting time and resource constraints, in general. Among these features, some distinguished ones are the deterministic system calls, fast switch between processes and threads, fast interrupt responses, support for concurrency with multitasking, real-time and synchronization, and the possibility of the developer taking control of policies, such as scheduling, priority levels, and memory usage.

When trying to find what drives the teams to adopt an RTOS in aerospace area, it is possible to identify three extra key features usually cited by the system engineers:

- Safety
- Reliability
- Multitasking and speed

These three features make sense in terms of high costs and when safety and reliability are crucial to the mission success. Specifically, multitasking and speed are strong requirements given the fact that small satellites are currently used for low Earth orbit (LEO) flight; therefore, they experience small windows for communicating with the ground, making every single period of time valuable during this window.

3.4 Example Projects

A search for recent work in the application of RTOS to small satellites was conducted. The majority of the related works analyzed presented RTOSs as a good solution for OBCs based on ARM processors as well as in microcontrollers. A wide range of possibilities was detected, but the most common architectures used are from Atmel, NXP, and Texas Instruments. The current state of RTOS usage on small satellites is summarized in the following text.

Rajulu et al. [1] present an RTOS implementation for the OBC of STUDSAT-2, the first picosatellite developed in India by the undergraduate students from seven engineering colleges across South India. The decision to choose the implementation of FreeRTOS on STM32 Cortex-M4 controller was taken because utilization of FreeRTOS simplifies software development, enables code modularity, and results in maintainable and expandable high performance that reduces the effort required to develop software for STUDSAT-2A/2B mission. The only drawback found by the team was that tasks are usually designed to enter in infinite loops, being resumed by specific interruptions, while satellite function tasks should function in the specified conditions without going into infinite loops.

Putra et al. [2] show the CooCox CoOS over an LPC1769 microcontroller from NXP. Two types of RTOS schedulers were developed by the team, which are non-preemptive and preemptive. In the non-preemptive scheduler, once the task is started, the task must be executed until completed or until the task is blocked for some reason. In preemptive scheduler, a task can be interrupted by other tasks which are more important (higher priority) and can be resumed. In general, if possible, a critical task that should be allowed to interrupt other tasks is less critical in order to meet the deadline. All functions of the RTOS, including obtaining temperature, current, voltage, and housekeeping data have been successfully demonstrated.

Parkinson [3] focused not on an OBC, but on the RTOS on multicore processors for aerospace in general, addressing interesting points, such as the fact that many aerospace applications are not safety critical, as their failure may not directly impact the safety of the spacecraft, but their failure could impact the success of the mission. Another point is that avionics applications can have strict startup time requirements, meaning that after a power failure, the processor must be reinitialized, run a boot loader, load the operating system, then start the RTOS, and run the applications with meaningful information on the display, all within 1 s. Parkinson (op. cit.) concludes that currently some uncertainty appears about the best choice of processor for individual aerospace application use cases. It is likely that positive experiences gained by early adopters on multicore programmers will result in a virtuous cycle of support, leading to further adoption and success.

Craveiro et al. [4] present a partitioned architecture between a RTOS kernel and a general-purpose GNU/Linux kernel using AIR [5] interpartition communication, calling the project Bullet Linux. Bullet is based on the ARINC 653 specification [6]. The integration of Bullet Linux made available to AIR applications a wide range of utilities, tools, language interpreters (Python, Perl, Tcl, etc.), and device drivers. These specific facilities no longer need to be ported to the RTOS to construct AIR applications.

3.5 Conclusions

A first conclusion is that, currently, there is a tendency of adopting FreeRTOS or Pumpkin's Salvo™ for small satellites, when it is desired to use an out-of-the-box operational system. On the other hand, when it is decided to follow the path of creating a specialized distribution of the RTOS for the mission, the most common line is to base it on the Linux kernel, which is always described as a solid and stable starting point. This can be explained by the reliability of the Linux kernel and by the fact that it is well known by software engineers, who provide help across a huge, open, and global community, hence, increasing the development speed and reducing the risk of getting stuck to a problem with the kernel. The same logic of community support can be used to explain the FreeRTOS and Pumpkin's Salvo adoption tendency. Table 3.1 summarizes the findings of recent small satellite missions that used RTOS.

The examples here presented show how RTOSs can improve small satellite missions, providing a very consistent knowledge base for designing a new satellite implementing RTOS, proving that these types of operational systems are not only viable, but, in some cases, are essential to achieve the expected results for the mission. The RTOS adoption on a small satellite project can be very positive, providing a lot of real-time features for the mission, such as multitasking, threading, and scheduling, which opens a new range of possibilities on what can be done with small satellites.

Table 3.1 RTOS on small satellite projects.

Satellite	RTOS	Architecture	Usage	Advantage	Key features	Distinguishing point	Year
DANDE [7]	Linux kernel based	Atmel NGW100 mkII	Preemptive kernel structure	Quick and predictable response to a large number of tasks	Speed, program flow, and event response	RTOS developed by the team based on Linux Kernel	2012
skCUBE [8]	Team Developed	MSP430	OBC	Focus in reliability and fault tolerance	Safety critical	RTOS developed by the team using MATLAB SIMULINK	2015
HAUSAT-2 [9]	RTOS VxWorks	NXP MPC860 32 bits	Flight software	Multitasking	C&DH, ADCS, Electric Power System (EPS), Therma Control System (TCS), and Communication System (CS)	Acts as the main OS for the system	2006
AAUSAT3 [10]	FreeRTOS	AT90CAN128/ARM7	OBC	Multitasking	Mission-critical tasks		2008
CAT-1 [11]	Xenomai	AT91SAM9G20	OBC	High performance	Linux extended	Linux-based distribution extended with Xenomai for mission-critical tasks	2014
iTÜ-pSAT I [12]	Salvo RTOS	MSP430F1611	OBC	Low power and compact system	Uses tasks as the main sub-block in the operating system	Very low hardware requirements	2007

(Continued)

Table 3.1 (Continued)

Satellite	RTOS	Architecture	Usage	Advantage	Key features	Distinguishing point	Year
Aalto-2 [13]	FreeRTOS	Texas Instruments RM48L952PGE	Task managing	Flexibility, task control options with priorities and real-time capabilities	Time-based scheduling	Meschach library usage for various algorithms for linear matrix algebra	2014
NanoSat [14]	FreeRTOS	LPC1764 ARM Cortex-M3	OBC	Professional grade, license free, robust, open source real-time kernel	Takes less than 4 kb flash memory	It is configured for both pre-emptive and cooperative schedulers	2015
CubeSat [15]	Salvo RTOS	MSP430	OBC	Timeouts, event flags, messages, queues	Uses very little memory and no stack	Tasks may also include function calls	2009
STUDSAT-2 [1]	FreeRTOS	STM32 ARM Cortex-M4	OBC	Code complexity, time critical nature of the system, testing, and code reuse	Simplifies software development, enables code modularity	Nonfunctional attributes, such as quality control and IP infringement protection	2014
UGMSat-1 [2]	CooCox CoOS	NXP LPC1769	OBC	Multitasking	Ability to work within a specified time interval	The highest priority among the tasks on the system is communication Task	2016

References

1 Rajulu, Bheema, Sankar Dasiga, and Naveen R. Iyer. 2014. "Open source RTOS implementation for on-board computer (OBC) in STUDSAT-2." Aerospace Conference, 2014 IEEE. IEEE.

2 Putra, Agfianto Eko, Tri Kuntoro Priyambodo, and Noris Mestika. 2016. Real-time operating system implementation on OBC/OBDH for UGMSat-1 sequence *AIP Conference Proceedings* (eds. Tri Rini Nuringtyas, et al.) 1755. 1. AIP Publishing.

3 Parkinson, P.J. (2016). *The Challenges of Developing Embedded Real-Time Aerospace Applications on Next Generation Multi-Core Processors*. Munich: Aviation Electronics Europe.

4 Craveiro, Joao, et al. 2009. "Embedded Linux in a partitioned architecture for aerospace applications." Computer Systems and Applications. AICCSA 2009. IEEE/ACS International Conference on IEEE.

5 J. Rufino, S. Filipe, M. Coutinho, S. Santos, and J. Windsor, 2007. "ARINC 653 interface in RTEMS," in Proceedings of Data Systems in Aerospace (DASIA'07), Naples, Italy, June.

6 Airlines Electronic Engineering Committee (AEEC) (2006). *Avionics Application Software Standard Interface (ARINC Specification 653–2)*. ARINC, Inc.

7 Cooke, Caitlyn M. 2012. "Implementation of a Real-Time Operating System on a Small Satellite Platform." Space Grant Undergraduate Research Symposium.

8 Slačka, Juraj, and Miroslav Halás. 2015. "Safety critical RTOS for space satellites." Process Control (PC), 2015 20th International Conference. IEEE.

9 Chang, Young-Keun, et al. 2006. "Low-Cost Responsive Exploitation of Space by HAUSAT-2 Nano Satellite." Proceedings, 4th Responsive Space Conference, AIAA, Los Angeles, CA, USA.

10 Bønding, Jesper, et al. (2008). "Software Framework for Reconfigurable Distributed System on AAUSAT3.".

11 Araguz López, Carles. (2014). "Towards a modular Nano-Satellite Software Platform: Prolog Constraint-based Scheduling and System Architecture."

12 Kurtuluş, Can, et al.. 2007. iTÜ-pSAT I: Istanbul Technical University Student. Pico-Satellite Program 3rd Recent Advances in Space Technologies Conference, Istanbul.

13 Jovanovic, Nemanja. (2014). "Aalto-2 satellite attitude control system."

14 Regi, Janani Sivaraman. 2015. "A Remote Sensing Nano-Satellite running Hard Real Time Control and Communication System Sending Camera Images to Ground Control Station and Over-the-Air Upgrade."

15 Peters, D., Raskovic, D., and Thorsen, D. (2009). An energy efficient parallel embedded system for small satellite applications. *ISAST Transactions on Computers and Intelligent Systems* 1 (2): 8–16.

4 I-2c

Attitude Control and Determination

Willem H. Steyn[1] *and Vaios J. Lappas*[2]

[1] *Department of Electrical & Electronic Engineering, University of Stellenbosch, South Africa*
[2] *Applied Mechanics Lab, Department of Mechanical Engineering and Aeronautics, University of Patras, Greece*

4.1 Introduction

The use of active attitude control and determination on nanosatellites is growing, but still less than 25% of all orbiting nanosatellites are three-axis stabilized. In a CubeSat Attitude Determination and Control System (ADCS) survey in 2006 by École Polytechnique Fédérale de Lausanne (EPFL) Space Center, out of 18 CubeSats only nine (50%) used active magnetic control, seven used passive magnetic control, and two had no control. Only four Cube-Sats could be stabilized to a pointing attitude: one using a momentum wheel, one was spin-stabilized, and two used a gravity-gradient boom. In a more recent 2014 review [1] of 42 small satellites (excluding CubeSats) from 6.5 to 94 kg launch mass over the past 25 years, 5% had no control, 19% had only passive control, and 76% had some form of active control. The list of ADCS sensors comprised of: magnetometers (90%), sun sensors (80%), earth sensors (10%), Global Positioning System (GPS) (33%), rate sensors (40%), and star trackers (35%). The list of ADCS actuators comprised of: permanent magnets (20%), magnetic torques (80%), momentum wheels (8%), reaction wheels (40%), propulsion systems (18%), gravity-gradient booms (15%), and control moment gyros (3%). It is therefore clear that most small satellites today make use of active magnetic control, and about 50% will compliment this with active reaction/momentum wheel control for three-axis stabilization.

4.2 ADCS Fundamentals

The ADCS stabilizes, controls, and positions a satellite in a desired orientation despite any external or internal disturbances acting on it. The satellite's payload requires a specific pointing direction whether the payload is a camera, science instrument, or an antenna. Satellites also require orientation for thermal control or to acquire the sun for their solar panels. The ADCS system uses sensors in order to determine a satellite's attitude and actuators to control the vehicle to a required direction. The ADCS systems need to achieve the various mission and payload objectives, such as pointing accuracy, stability, rotational rate

Nanosatellites: Space and Ground Technologies, Operations and Economics, First Edition.
Edited by Rogerio Atem de Carvalho, Jaime Estela, and Martin Langer.
© 2020 John Wiley & Sons Ltd. Published 2020 by John Wiley & Sons Ltd.

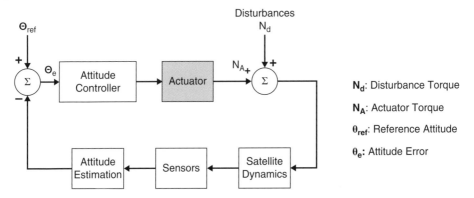

Figure 4.1 ADCS block diagram.

(slew), and sensing with many physical constraints, such as mass, power, volume, computer power/storage, space environment, robustness/lifetime, and cost. The ADCS is a synthesis of two subsystems, the attitude determination system (ADS) and the attitude control system (ACS), which controls the attitude/motion of a satellite as depicted in Figure 4.1.

4.3 ADCS Requirements and Stabilization Methods

The need for active attitude control is determined by the nanosatellite's mission and its attitude requirements. As mentioned in the introduction, most nanosatellite missions have some specific pointing functionality and stabilization requirement, and thus ADCS systems with some capability are needed. In other cases, passive attitude control is employed by mounting permanent magnets to the structure. The magnets create a torque that strives to align them with the local magnetic field. This does not result in a stabilized attitude, but provides a predictable attitude pattern as the satellite moves in its orbit. Another form of passive attitude stabilization is achieved using gravity-gradient-induced torque, usually from a deployable boom with a tip mass.

Active attitude control also comes in different flavors. Simple tumbling control modes can be implemented with the minimum hardware and power requirements. Stabilized attitude (roll, pitch, and yaw angles controlled to zero) can be achieved using a momentum wheel and full three-axis control with the ability to perform commanded slew maneuvers placing the most demanding requirements on volume, mass, and power resources. A list of requirements that are usually considered for satellite ADCS is summarized in Table 4.1.

The methods for attitude stabilization and control are briefly elaborated in the following paragraphs:

Gravity-gradient

This technique exploits Newton's law of general gravitation, and through the use of gravitational forces it can always keep a spacecraft pointing nadir. This is achieved by using a boom extending a small distinct mass (usually a magnetometer in order to minimize magnetic interference) from the spacecraft (which becomes the second distinct mass) by some distance. These two masses which are connected by a thin and light boom can then

Table 4.1 ADCS requirements.

Requirement	Definition
Determination/sensing	
Accuracy/attitude knowledge	How well a satellite's orientation is with respect to an absolute reference
Range	Range of angular motion over which accuracy must be met
Control	
Accuracy	How well the satellite attitude can be controlled with respect to a commanded direction
Range	Range of angular motion over which control performance must be met
Operating conditions	Parts of the orbit and/or mission that attitude control is needed, such as eclipse/daylight
Stability/jitter	A specified angle bound or angular rate limit on short-term, high-frequency motion
Slew rate/agility	Slew or angular rate required to perform a rapid maneuver
Drift	A limit on slow, low-frequency vehicle motion
Settling time	Allowed time to recover from maneuvers or upsets

be used to exploit the difference in gravitational pull on the main satellite platform and the additional mass (magnetometer) due to the difference in their distance from Earth. This small difference can be sufficient to enable the satellite/additional mass system to be aligned with the radius vector at all times as an orbiting pendulum. The gravity-gradient stabilization scheme can be beneficial for coarse pointing (~5°) around the nadir axis, while still the other two axes are needed to be stabilized.

Magnetic

Approximating the Earth's magnetic field as a dipole, it is possible to have a satellite use a magnetometer to track the Earth's magnetic field lines in what is called a "compass mode," which allows a spacecraft to be passively stabilized but with coarse attitude (5°–10°) due to the various irregularities and harmonics of the Earth's magnetic field.

Spinners

Spinning a satellite generates an angular momentum vector, which remains nearly fixed in inertial space. The angular momentum generated provides gyroscopic stiffness to a spinning satellite, making it less prone to external disturbances and more stable for thruster/apogee motor firings.

Bias momentum

As a three-axis system, momentum bias systems use one actuator-momentum wheel aligned about the pitch axis normal to the orbit place. Gyroscopic stiffness is used in order to control the vehicle by keeping the momentum wheel spinning at a constant rate. Small variations in wheel speed allow the control of the pitch axis. Yaw-roll coupling for nadir-pointing bias momentum systems can be used to control the other two axes.

Zero momentum/three axis control

In these systems, reaction wheels are used for each axis in order to compensate for external disturbances and in order to complete various commanded maneuvers. A pointing error is used to make the reaction wheels accelerate from an initial zero value, and then the wheels move to a small spin rate which keeps increasing due to the maneuvers required and due to secular disturbances, making them reach their saturation limits. The increase in angular momentum to saturation levels requires a desaturation strategy, which is called "momentum dumping" or unloading. This is achieved by using magnetorquers and to a much lesser extent thrusters (for larger satellites), thus enabling the wheels to go to zero values.

Figure 4.2 and Table 4.2 show some examples of CubeSats making use of the various attitude-stabilization methods discussed earlier.

Delfi-C3 (TU Delft)

Dove-2 (Planet Labs Inc.)

nSight-1 (SCS Aerospace)

Figure 4.2   Examples of CubeSats.

Table 4.2   Examples of CubeSat missions for various attitude control scenarios.

Attitude control	Missions	Status
Passive attitude control		
Permanent magnet	NanoSail-D2	Launched in 2010
Permanent magnet and magnetic hysteresis damping	Delfi-C3	Launched in 2008
Gravity-gradient boom	IceCube 1,2	Failed launch (2006)
Active attitude control		
Tumbling mode only	CUTE-1.7 + APD II	Launched in 2008
Momentum stabilized	CanX-2	Launched in 2008
	nSight-1	Launched in 2017
Full three-axis arbitrary pointing	Dove-1 to -4	Launched in 2013
	Delfi-n3Xt	Launched in 2013
	MarCO A and B	Launched in 2018

4.4 ADCS Background Theory

4.4.1 Coordinate Frame Definitions

A satellite's attitude is normally controlled with respect to the orbit coordinates (ORCs), where the Z_O axis points toward nadir, the X_O axis points toward the velocity vector for a near circular orbit, and the Y_O axis along the orbit plane antinormal direction. The aerodynamic N_{Aero} and gravity-gradient N_{GG} disturbance torque vectors are also conveniently modeled ORCs. The satellite body coordinates (SBCs) are nominally aligned with the ORC frame at zero pitch, roll, and yaw attitude. Since the sun and satellite orbits are propagated in the J2000 earth-centered inertial (ECI) coordinate frame, we require a transformation matrix from ECI to ORC coordinates. This can easily be calculated from the satellite position $\bar{u}_I$ and velocity $\bar{v}_I$ unit vectors (obtained using the position and velocity outputs of the satellite orbit propagator):

$$A_{I/O} = \begin{bmatrix} (\bar{u}_I \times (\bar{v}_I \times \bar{u}_I))^T \\ (\bar{v}_I \times \bar{u}_I)^T \\ -\bar{u}_I^T \end{bmatrix} \tag{4.1}$$

4.4.2 Attitude Kinematics

The attitude of an earth-orbiting satellite can be expressed as a quaternion vector $\boldsymbol{q}$ to avoid any singularities to determine the orientation with respect to the ORC frame. The ORC reference body rates, $\omega_B^O = [\omega_{xo} \ \omega_{yo} \ \omega_{zo}]^T$, must be used to propagate the quaternion kinematics as:

$$\begin{bmatrix} \dot{q}_1 \\ \dot{q}_2 \\ \dot{q}_3 \\ \dot{q}_4 \end{bmatrix} = 0.5 \begin{bmatrix} 0 & \omega_{zo} & -\omega_{yo} & \omega_{xo} \\ -\omega_{zo} & 0 & \omega_{xo} & \omega_{yo} \\ \omega_{yo} & -\omega_{xo} & 0 & \omega_{zo} \\ -\omega_{xo} & -\omega_{yo} & -\omega_{zo} & 0 \end{bmatrix} \begin{bmatrix} q_1 \\ q_2 \\ q_3 \\ q_4 \end{bmatrix} \tag{4.2}$$

The attitude matrix to describe the transformation from ORC to SBC can be expressed in terms of quaternions as:

$$A_{O/B} = \begin{bmatrix} q_1^2 - q_2^2 - q_3^2 + q_4^2 & 2(q_1 q_2 + q_3 q_4) & 2(q_1 q_3 - q_2 q_4) \\ 2(q_1 q_2 - q_3 q_4) & -q_1^2 + q_2^2 - q_3^2 + q_4^2 & 2(q_2 q_3 + q_1 q_4) \\ 2(q_1 q_3 + q_2 q_4) & 2(q_2 q_3 - q_1 q_4) & -q_1^2 - q_2^2 + q_3^2 + q_4^2 \end{bmatrix} \tag{4.3}$$

The attitude is normally presented as pitch θ, roll φ, and yaw ψ angles, defined as successive rotations, starting with the first rotation from the ORC axes and ending after the final rotation in the SBC axis. If a Euler 213 sequence (first θ around Y_O, then φ around X', and finally ψ around Z_B) is used, then the attitude matrix and Euler angles can be computed as:

$$A_{O/B} = \begin{bmatrix} C\psi\,C\theta + S\psi\,S\varphi\,S\theta & S\psi\,C\varphi & -C\psi\,S\theta + S\psi\,S\varphi\,C\theta \\ -S\psi\,C\theta + C\psi\,S\varphi\,S\theta & C\psi\,C\varphi & S\psi\,S\theta + C\psi\,S\varphi\,C\theta \\ C\varphi\,S\theta & -S\varphi & C\varphi\,C\theta \end{bmatrix} \tag{4.4}$$

with,

$$C = \text{cosine function}, S = \text{sine function}$$

and

$$\theta = \arctan 4(A_{31}, A_{33})$$

$$\varphi = -\arcsin(A_{32}) \tag{4.5}$$

$$\psi = \arctan 4(A_{12}, A_{22})$$

This Euler angle representation allows unlimited rotations in pitch and yaw, but only maximum $\pm 90°$ rotations in roll.

4.4.3 Attitude Dynamics

The attitude dynamics of an earth-orbiting satellite can be derived using the Euler equation:

$$I\dot{\omega}_B^I = N_{GG} + N_D + N_W + N_{MT} - \omega_B^I \times (I\omega_B^I + h_W) \tag{4.6}$$

with $\omega_B^I = \omega_B^O + A_{O/B}\begin{bmatrix} 0 & -\omega_o & 0 \end{bmatrix}^T$ as the inertially referenced body rate vector, $N_{GG} = 3\omega_o^2(z_o^B \times I\,z_o^B)$ as the gravity-gradient disturbance torque vector, with $z_o^B = A_{O/B}\begin{bmatrix} 0 & 0 & 1 \end{bmatrix}^T$ as the orbit nadir unit vector in body coordinates, N_D is the external disturbance torques (e.g. from aerodynamic and solar pressure), $N_W = -\dot{h}_W$ is the reaction or momentum wheel torque vector, with h_W the wheel angular momentum vector, N_{MT} is the magnetic control torque, ω_o is the orbit angular rate, and I is the inertia matrix of the satellite.

4.5 Attitude and Angular Rate Determination

To enable the attitude and angular rate controllers of the next section to calculate the control torques, measurements or estimates of the orbit referenced angular rate vector and attitude quaternion must be known at sampling instances. A quaternion error can be calculated if the reference attitude quaternion and an estimated quaternion representing the current satellite attitude are available. The current satellite quaternion can be calculated every sampling period using a TRIAD algorithm [2] from measured $\bar{v}_B$ (in SBC) and modeled $\bar{v}_O$ (in ORC) unit direction vectors from two different types of attitude sensors, for example, magnetometer/sun or sun/earth (nadir) sensors.

As an example, CubeSense [3] is a sensor which can be used to measure the sun unit (direction) vector in SBC $\bar{s}_B$ and the nadir unit vector in SBC $\bar{n}_B$. If the sun and satellite orbit is modeled the sun unit vector in ORC, $\bar{s}_O$ can also easily be calculated on-board (4.7), and the nadir unit vector in ORC $\bar{n}_O$ will simply be $\begin{bmatrix} 0 & 0 & 1 \end{bmatrix}^T$, the direction of the ORC Z_O axis.

In the CubeSense sensor, these measured vectors are accurately obtained from two small visible spectrum CMOS cameras, by calculating the centroids of the illuminated circular images (Sun and Earth):

1) For the sun sensor, the optics can be a filtered fisheye lens or pinhole lens deposited on a 1% neutral density filter and mounted with boresight aligned with the body Y_B axis (+ or − depending on sun incidence to the orbital plane) or in the nominally zenith direction, the body $-Z_B$ axis.
2) For the nadir sensor, the optics can also utilize a 180° fisheye lens and mounted with boresight along the body $+Z_B$ axis. We can ignore the small errors caused by the Earth's oblateness, but compensate for the lens distortion in the earth image.

The modeled (ORC) sun-to-satellite unit vector can be calculated from simple analytical sun and satellite (e.g. SGP4) orbit models in ECI coordinates. The ECI referenced unit vector can then be transformed to ORC coordinates using the known current satellite Keplerian angles:

$$\bar{s}_O = A_{I/O} \, \bar{s}_I \tag{4.7}$$

with

$\bar{s}_I$ = ECI sun-to-satellite unit vector from sun and satellite orbit models.

4.5.1 TRIAD Quaternion Determination

Two orthonormal TRIADs are formed from the measured (observed) and modeled (referenced) vector pairs as presented earlier:

$$\begin{aligned}
\bar{o}_1 &= \bar{n}_B, \bar{o}_2 = \bar{n}_B \times \bar{s}_B, \bar{o}_3 = \bar{o}_1 \times \bar{o}_2 \\
\bar{r}_1 &= \bar{n}_O, \bar{r}_2 = \bar{n}_O \times \bar{s}_O, \bar{r}_3 = \bar{r}_1 \times \bar{r}_2
\end{aligned} \tag{4.8}$$

The estimated ORC to SBC transformation matrix can then be calculated as:

$$A_{O/B}(\hat{q}) = \begin{bmatrix} \bar{o}_1 & \bar{o}_2 & \bar{o}_3 \end{bmatrix} \begin{bmatrix} \bar{r}_1 & \bar{r}_2 & \bar{r}_3 \end{bmatrix}^T \tag{4.9}$$

and

$$\begin{aligned}
\hat{q}_4 &= \sqrt{1 + A_{11} + A_{22} + A_{33}}/2 \\
\hat{q}_1 &= [A_{23} - A_{32}]/(4\hat{q}_4) \\
\hat{q}_2 &= [A_{31} - A_{13}]/(4\hat{q}_4) \\
\hat{q}_3 &= [A_{12} - A_{21}]/(4\hat{q}_4)
\end{aligned} \tag{4.10}$$

4.5.2 Kalman Rate Estimator

To accurately measure low angular rates as experienced during three-axis stabilization, a high performance IMU will be required; this will neither fit onto a nanosatellite nor be cost effective. Low cost microelectromechanical systems (MEMS) rate sensors currently are still too noisy and also experience high bias drift. A modified implementation of a Kalman rate estimator can be used for the gyroless estimation of the nanosatellite body rates. This estimator was successfully used in many small satellite missions, such as the SNAP-1 nanosatellite mission [4]. It used magnetic field vector measurements that are continuously available and the body-measured rate of change of the geomagnetic field vector direction can be used as a measurement input for this rate estimator. However, this vector is not inertially fixed, as it rotates twice per polar orbit. The estimated inertially referenced body rates will therefore have errors contributed by the magnetic field vector rotation rate. A more accurate estimated rate vector can be determined by measuring the sun vector, which only rotates inertially once per year. As the sun vector measurements are only available during the sunlit part of each orbit, the Kalman estimator must propagate the angular rates during eclipse. A nadir-pointing nanosatellite nominally rotating once per orbit within the ORC (around the body $-Y_B$ axis) and full observability using the sun vector measurement is typically ensured. The only exception is when the satellite body rate vector is always aligned with the sun vector direction; else the angular rate vector with respect to an almost inertially fixed sun direction can be estimated as $\omega_B^I = [\hat{\omega}_{xi} \ \hat{\omega}_{yi} \ \hat{\omega}_{zi}]^T$. The expected measurement error will therefore include the sun sensor measurement noise and a negligibly small satellite-to-sun inertial rotation.

4.5.2.1 System Model

The discrete Kalman filter state vector $x(k)$ is defined as the inertially referenced body rate vector $\omega_B^I(k)$. From the Euler dynamic model of Eq. (4.6) without wheel actuators, the continuous time model becomes

$$\dot{\omega}_B^I(t) = I^{-1}(N_{MT}(t) + N_{GG}(t) - \omega_B^I(t) \times I\omega_B^I(t))$$
$$\dot{x}(t) = F x(t) + G u(t) + s(t)$$

(4.11)

with

$$F = [0], G = I^{-1}, u(t) = N_{MT}(t) = \text{Control input vector}$$
$$s(t) = I^{-1}(N_{GG}(t) - \omega_B^I(t) \times I\omega_B^I(t)) = \text{System noise vector}$$

The discrete system model will then be

$$x(k+1) = \Phi x(k) + \Gamma u(k) + s(k)$$

(4.12)

with

$$\Phi = [1_{3x3}], \quad \Gamma = I^{-1}T_s$$
$$T_s = \text{Kalman filter sampling period}$$
$$s(k) = N\{0, Q(k)\} = \text{Zero mean system noise vector with covariance matrix } Q$$

4.5.2.2 Measurement Model

If we assume the satellite-to-sun vector as "inertially fixed" due to the large distance from the Earth to Sun compared to the Earth to satellite and the slow rotation of the Earth around the Sun, the rate of change of the sun sensor measured unit vector can be used to accurately estimate the inertially referenced body angular rates. The magnetometer unit vector, as an orbit rotating vector, can also be used for continuous measurement updates, but with expected rate estimation errors of approximately twice the orbit rate ω_o. For the rest of this discussion and the derivation of the measurement model, we assume the sun vector measurements will be used during the sunlit part of each orbit for updating the Kalman rate estimator. Successive sun vector measurements will result in a small angle discrete approximation of the vector rotation matrix:

$$\bar{s}(k) = \Delta A(k)\, s(k-1) \tag{4.13}$$

with

$$\Delta A(k) \approx \begin{bmatrix} 1 & \omega_{zi}(k)T_s & -\omega_{yi}(k)T_s \\ -\omega_{zi}(k)T_s & 1 & \omega_{xi}(k)T_s \\ \omega_{yi}(k)T_s & -\omega_{xi}(k)T_s & 1 \end{bmatrix}$$

$$\approx [1_{3x3}] + \Lambda\{\omega_B^I(k)\} \tag{4.14}$$

The Kalman filter measurement model then becomes

$$\Delta s(k) = \bar{s}(k) - \bar{s}(k-1) = \Lambda\{\omega_B^I(k)\}\,\bar{s}(k-1)$$
$$y(k) = \Delta s(k) = H(k)\,x(k) + m(k) \tag{4.15}$$

with

$$H(k) = \begin{bmatrix} 0 & -s_z(k-1)T_s & s_y(k-1)T_s \\ s_z(k-1)T_s & 0 & -s_x(k-1)T_s \\ -s_y(k-1)T_s & s_x(k-1)T_s & 0 \end{bmatrix} \tag{4.16}$$

and $m(k) = N\{0, R(k)\}$ as zero mean measurement noise, with a covariance matrix R.

Kalman Filter Algorithm Define $P_k \equiv E\{x_k . x_k^T\}$ as the state covariance matrix, then the following steps are executed every sampling period T_s between measurements (at time step k):

1. Numerically integrate the nonlinear dynamic model of Eq. (4.11);

$$\hat{x}_{k+1/k} = \hat{x}_{k/k} + 0.5T_s(3\Delta x_k - \Delta x_{k-1}) \; \{\text{modified Euler integration}\} \tag{4.17}$$

   with

$$\Delta x_k = I^{-1}(N_{MT}(k) - \hat{\omega}_B^I(k) \times I\hat{\omega}_B^I(k)) \tag{4.18}$$

2. Propagate the state covariance matrix:

$$P_{k+1/k} = \Phi\, P_{k/k}\, \Phi^T + Q = P_{k/k} + Q \tag{4.19}$$

   across measurements (at time step $k+1$ and only in sunlit part of orbit).

3. Gain update, compute H_{k+1} from Eq. (4.16) using previous vector measurements $\bar{s}(k)$:

$$K_{k+1} = P_{k+1/k}H_{k+1}^T[H_{k+1}P_{k+1/k}H_{k+1}^T + R]^T \tag{4.20}$$

4. Update the system state:

$$\hat{x}_{k+1/k+1} = \hat{x}_{k+1/k} + K_{k+1}(y_{k+1} - H_{k+1}\hat{x}_{k+1/k}) \tag{4.21}$$

   with

$$y_{k+1} = \bar{s}(k+1) - \bar{s}(k).$$

5. Update the state covariance matrix

$$P_{k+1/k+1} = [1_{3x3} + K_{k+1}H_{k+1}] P_{k+1/k} \tag{4.22}$$

Finally, the estimated ORC angular rate vector can be calculated from the Kalman filtered estimated ECI rate vector, using the TRIAD result of Eq. (4.9) or the full state estimator of Section 4.5.3:

$$\hat{\omega}_B^O(k) = \hat{\omega}_B^I(k) - A_{O/B}(\hat{q}(k)) \begin{bmatrix} 0 & -\omega_o & 0 \end{bmatrix}^T \tag{4.23}$$

Figure 4.3 shows typical Kalman rate estimation results obtained as real-time telemetry data points during commissioning of a small satellite.

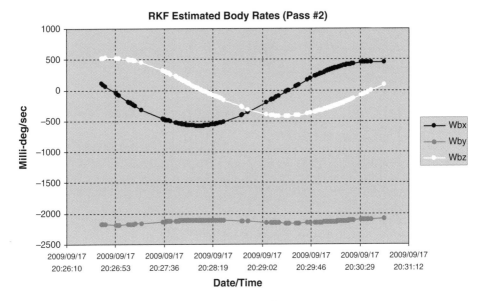

Figure 4.3 On-board Kalman rate filter telemetry results.

4.5.3 Full-State Extended Kalman Filter Estimator

Before any of the wheel control modes can be applied to a nanosatellite, more accurate and continuous angular rate and attitude knowledge is required. An extended Kalman filter (EKF) is implemented to estimate the full attitude state of the satellite from all attitude sensor SBC measurements (e.g. from magnetometer, sun, nadir, and star sensors) and the corresponding ORC modeled vectors, see reference [5] for a detailed derivation. The seven-element discrete state vector to be estimated is defined as:

$$\hat{x}(k) = \begin{bmatrix} \hat{\omega}_B^I(k) \\ \hat{q}(k) \end{bmatrix} \tag{4.24}$$

The innovation used in the EKF is the vector cross-product of a measured body reference unit vector and a modeled orbit reference unit vector, transformed to the body coordinates by the estimated attitude transformation matrix$A[\hat{q}(k)]$:

$$e(k) = \bar{v}_B(k) \times A[\hat{q}(k)]\, \bar{v}_O(k) \tag{4.25}$$

with

$$\bar{v}_B(k) = B_{\mathrm{magm}}(k)/\|B_{\mathrm{magm}}(k)\| \text{ or } S_{\mathrm{sun}}(k)/\|S_{\mathrm{sun}}(k)\|$$

, for magnetic and sun vector pairs.

$$\bar{v}_O(k) = B_{\mathrm{igrf}}(k)/\|B_{\mathrm{igrf}}(k)\| \text{ or } S_{\mathrm{orbit}}(k)/\|S_{\mathrm{orbit}}(k)\|$$

The on-board magnetometer measurements must first be calibrated offline by comparing the measured B-field magnitude to the international geomagnetic reference field (IGRF) model's magnitude. This is can be done by sampling at least a full orbit's raw or prelaunch calibrated magnetometer vector measurements and the corresponding IGRF modeled magnetic vectors. These data samples can then be further ground-processed by using an attitude-independent three-axis magnetometer calibration method [6] to estimate the gain (scaling and orthogonality) matrix G_{cal} and offset (bias) vector O_{cal}.

Thereafter, calibrated magnetometer measurements can easily be calculated on-board to be used in the EKF:

$$B_{\mathrm{magm}}(k) = G_{\mathrm{cal}}B_{\mathrm{raw}}(k) - O_{\mathrm{cal}} \tag{4.26}$$

Figure 4.4 shows the typical magnetometer measurement accuracy after ground calibration when the on-board calibrated magnetic magnitude is compared to an IGRF model output for a small satellite in a 500 km polar orbit.

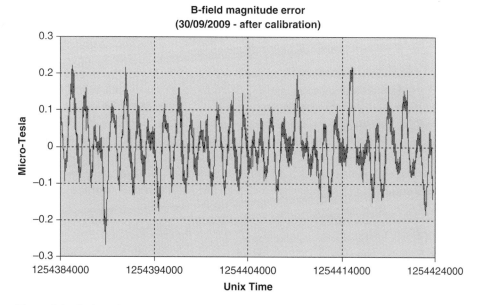

Figure 4.4 On-board magnetometer calibration result.

4.6 Attitude and Angular Rate Controllers

4.6.1 Detumbling Magnetic Controllers

After release from the launcher stage or a CubeSat PicoSatellite Orbital Deployer (POD), the nanosatellite is first detumbled using minimum ADCS resources and power to bring it to a controlled spin rate and/or spin attitude, typically a Y-Thomson spin [7]. A Y-Thomson spin ensures that the satellite will align its body Y_B axis normal to the orbit plane, that is, with the satellite spinning within the orbit plane. This not only results in a controlled spin rate, but also in a known spin attitude, without the need to estimate on-board the satellite's attitude. The only requirement for a stable Y-Thomson spin is that the body Y_B axis must have the largest moment of inertia (I_{yy} MOI parameter) and small (<3% MOI) products of inertia parameters. A simple B-dot [8] magnetic controller quickly dumps any X_B and Z_B axes angular rates and aligns the Y_B axis normal to the orbit plane. Using measurements from a single MEMS rate sensor, the Y_B spin rate can then be magnetically controlled to an inertially referenced spin rate of typically $-2°$ s^{-1} (the reference rate depends on the magnitude of the external disturbance torques and must be high enough to ensure a sufficient gyroscopic stiffness). The magnetic detumbling controllers require only the measured magnetic field vector components (from a three-axis magnetometer) and the inertially referenced Y_B body rate (from the Kalman rate estimator in Section 4.5.2 or a MEMS rate sensor), and can be utilized continuously. The magnetic-only controllers used during detumbling can be:

$$M_y = K_d d\beta dt \quad \text{for} \quad \beta = \arccos{(B_{my} \| B_{meas} \|)} \qquad \{\text{B-dot controller}\}$$

$$M_x = K_s(\omega_{yi} - \omega_{yref})\, \text{sgn}(B_{mz}) \quad \text{for } |B_{mz}| > |B_{mx}| \quad \{\text{Y-spin controller}\} \qquad (4.27)$$

$$M_z = -K_s(\omega_{yi} - \omega_{yref})\, \text{sgn}(B_{mx}) \quad \text{for } |B_{mx}| > |B_{mz}| \quad \{\text{Y-spin controller}\}$$

with β the angle between the body Y_B axis and the local B-field vector, K_d and K_s are the detumbling and spin controller gains, and ω_{yref} the reference Y_B body spin rate. $M_x, M_y,$ and M_z are the magnetorquer moments in Am^2 units that can be scaled to pulse width modulated (PWM) outputs $M_{PMW\_x,y,z}$, as most magnetorquers on satellites are current-controlled via discrete switching amplifiers. As the magnetorquer magnetic moments can disturb the local magnetic field measurements, we typically limit the magnetorquer on-times to 80% of the discrete magnetic controller period T_s to leave a window for magnetometer sampling.

The pulse output of the magnetorquers is therefore saturated to 80% of the controller period T_s,

$$\text{sat}\{M_{PWM_i}\} = \text{sgn}(M_{PWM_i}) \min\{|M_{PWM_i}|, 0.8T_s\} \quad \text{for } i = x, y, z \qquad (4.28)$$

The average magnetic moment and torque vector during a controller period can then be calculated as:

$$M_{avg} = \frac{M_{max}}{T_s} \text{sat}\{M_{PWM}\}\ Am^2 \qquad (4.29)$$

$$N_{MT} = M_{avg} \times B_B \qquad (4.30)$$

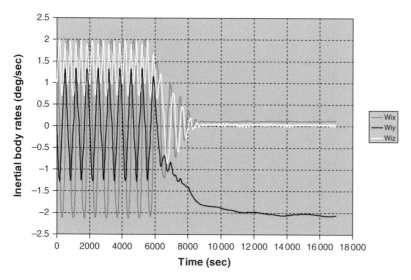

Figure 4.5 Magnetic detumbling to a Y-Thomson spin.

with M_{max} the maximum "on" magnetic moment of the magnetorquer, and B_B the true magnetic field vector in body coordinates.

Figure 4.5 shows a typical detumbling performance from an initial angular rate of $\omega_B^I(0) = \begin{bmatrix} 2 & 0 & 1 \end{bmatrix}^{T\circ}$ s^{-1}. During the first orbit, no control was done and from 6000 s the detumbling and Y-spin controller of Eq. (4.27) was enabled. Within less than an orbit, the satellite was controlled to a $-2°$ s^{-1} Y-Thomson spin using only the magnetorquers. The body angular rate was estimated by the Kalman rate filter of Section 4.5.2 utilizing only the raw magnetometer vector measurements.

4.6.2 Y-Momentum Wheel Controller

From the Y-Thomson body spin of the previous section, a momentum wheel aligned to the Y_B body axis can be used to absorb the Y-body momentum and control the pitch angle with zeroed roll and yaw angles, for example, to maintain a nadir-pointing attitude for earth imaging payloads and directional antennae for ground station communications. The Y-momentum wheel controller can be implemented with attitude and rate estimations from the EKF of Section 4.5.3, as:

$$N_{wy}(k) = K_{py} \arcsin\left(\hat{q}_2(k)\, \text{sgn}(\hat{q}_4(k))\right) + K_{dy}\, \hat{\omega}_{yo}(k) \tag{4.31}$$

with K_{py} and K_{dy} the proportional and derivative gains.

To maintain the Y-wheel momentum at a certain reference level (corresponding to the initial Y_B body momentum during the Y-spin mode) and to damp any body nutation rates in the X_B and Z_B axes, a magnetic cross-product control law can be utilized [4]:

$$M(k) = \frac{e(k) \times B(k)}{\|B(k)\|} \tag{4.32}$$

with

$$
e(k) = \begin{bmatrix} K_n \, \widehat{\omega}_{xo}(k) \\ K_h \, (h_{wy}(k) - h_{wy-ref}) \\ K_n \, \widehat{\omega}_{zo}(k) \end{bmatrix}
\tag{4.33}
$$

where K_n is the nutation damping gain, K_h is the Y-wheel momentum control gain, and h_{wy-ref} is the Y-wheel reference angular momentum.

The cross-product controller of Eq. (4.32) is applied continuously. During initial commissioning, the Y-momentum control mode is normally used to calibrate and determine the alignment of all the accurate attitude sensors, that is, sun and earth horizon (nadir) sensors and star tracker. After the in-orbit calibration and alignment parameters have been determined, the measurements from these sensors can then be included in the EKF of Section 4.5.3 to improve the attitude and rate estimation accuracy. Next, the nanosatellite is ready for a three-axis reaction wheel control mode, if required for full three-axis rotation-pointing capability.

4.6.3 Three-axis Reaction Wheel Controller

From the Y-momentum wheel mode, the X_B and Z_B reaction wheels can be activated and a three-axis reaction wheel controller implemented using the estimated attitude and angular rates from the EKF of Section 4.5.3. The globally stable quaternion feedback controller of Wie [9] can be modified to become an orbit-referenced pointing control law. The quaternion and rate reference vectors can be generated from a sun orbit model for a sun-pointing attitude (to maximize solar energy generation), or it can be zero vectors for a nadir-pointing attitude or any specified constant attitude reference for a specific roll, pitch, or yaw requirement (see Figure 4.6). The three-axis reaction wheel control law (wheel torque vector) to be used for all these cases is:

$$
N_w(k) = K_{P1} I q_{err}(k) + K_{D1} I \widehat{\omega}_B^O(k) - \widehat{\omega}_B^I(k) \times (I \widehat{\omega}_B^I(k) + h_w(k))
\tag{4.34}
$$

with $K_{P1} = 2\omega_n^2$, $K_{D1} = 2\zeta\omega_n$ are the pointing gains for a required controller bandwidth and damping factor. I is the satellite moment of inertia matrix, $h_w(k)$ is the measured angular momentum of the reaction wheels, $\widehat{\omega}_B^O(k) = \begin{bmatrix} \widehat{\omega}_{xo}(k) & \widehat{\omega}_{yo}(k) & \widehat{\omega}_{zo}(k) \end{bmatrix}^T$ is the body orbit reference angular rate estimate, $q_{err}(k) = \begin{bmatrix} q_{1e}(k) & q_{2e}(k) & q_{3e}(k) \end{bmatrix}^T$ is the vector part of error quaternion q_{err}, where

$$
q_{err}(k) = q_{com}(k) \oplus \widehat{q}(k)
$$

$$
\begin{bmatrix} q_{1e} \\ q_{2e} \\ q_{3e} \\ q_{4e} \end{bmatrix} = \begin{bmatrix} q_{4c} & q_{3c} & -q_{2c} & -q_{1c} \\ -q_{3c} & q_{4c} & q_{1c} & -q_{2c} \\ q_{2c} & -q_{1c} & q_{4c} & -q_{3c} \\ q_{1c} & q_{2c} & q_{3c} & q_{4c} \end{bmatrix} \begin{bmatrix} \widehat{q}_1 \\ \widehat{q}_2 \\ \widehat{q}_3 \\ \widehat{q}_4 \end{bmatrix}
\tag{4.35}
$$

with $q_{com}(k)$ the commanded reference quaternion, for example, a sun-direction quaternion, and $\oplus$ for q quaternion division.

A nominal reaction wheel control mode can, for example, do sun-pointing in the sunlit part of the orbit and nadir-pointing, that is, $q_{com}(k) = \begin{bmatrix} 0 & 0 & 0 & 1 \end{bmatrix}^T$, in eclipse. The nadir-pointing attitude ensures optimal antenna coverage for ground communication during eclipse and thermal stability for the imager telescope. Continuous momentum

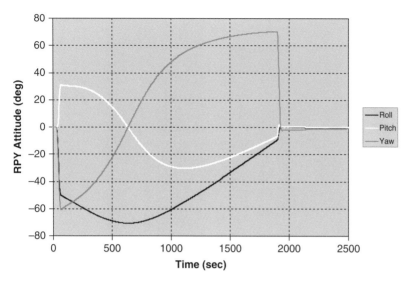

Figure 4.6 Three-axis pointing control, first sun then nadir.

management of the reaction wheels can be done using a simple cross-product magnetic controller [5]:

$$M(k) = K_m \frac{h_w(k) \times B(k)}{\|B(k)\|} \qquad (4.36)$$

with K_m the momentum dumping gain.

4.7 ADCS Sensor and Actuator Hardware

The ADCS sensors typically used on nanosatellite are limited by the mass, volume, and power constraints. Over the last couple of years, the available technology has improved and become more compact and lower power consuming due to an increase in the density of semiconductor integrated circuits, and advances in MEMS technology and nanomechanics. Therefore, most types of ADCS sensors and actuators flying on larger satellites can now be found in miniaturized form for use in nanosatellites. Although their accuracy performance in some aspects is still not the same as their larger and more power hungry bigger brothers, the gap is slowly closing, that is, where the laws of physics do allow it. The next section will present some examples of these nanosatellite ADCS components that are available commercially and have been successfully commissioned in various nanosatellite missions.

4.7.1 Three-Axis Magnetometers

Although accurate three-axis fluxgate magnetometers have not been produced for very small nanosatellites yet, various magnetoresistive [10] and magnetoinductive [11] MEMS type sensors with built-in bias and temperature correction circuitry have been used successfully on nanosatellites. Their noise level, linearity, and temperature sensitivity are still slightly worse compared to fluxgate sensors. However, for most magnetic control ADCS systems, where accuracy is not a hard requirement, these magnetometers are ideally suited with their inherent small size and low power specifications. See Figure 4.7 for a deployable

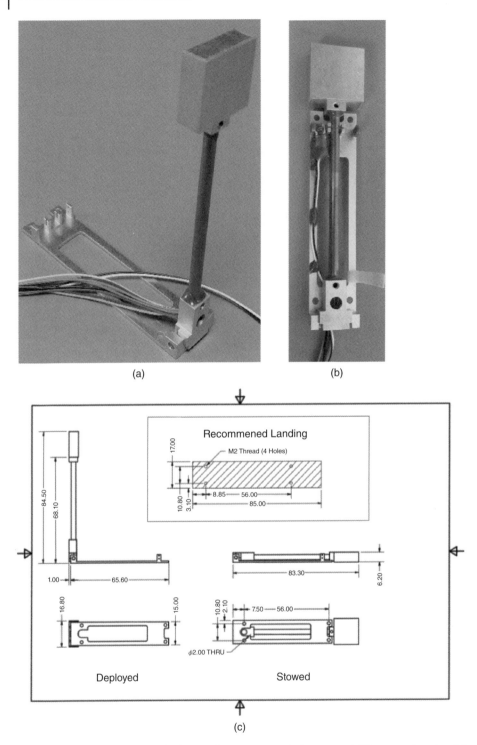

Figure 4.7 Burn-wire deployable three-axis magnetorestrictive nanosatellite magnetometer: (a) deployed, (b) stowed, (c) dimensions.

nanosatellite three-axis magnetometer [12] to ensure smaller magnetic disturbances from the main satellite bus with a total mass of only 13 g and power consumption of 33 mW.

4.7.2 Sun Sensors

The sun as a bright inertial object in the celestial sky is perfectly suited for accurate attitude vector measurements using a relative low cost, mass- and power-sensing device. Sun sensors vary from planar photodiodes or solar cells where the short-circuit current is measured to get a value proportional to the cosine of the sun angle to the surface normal. Six of these sensors mounted with unobstructed hemispherical view each on the facets of a box-type satellite always give the components of the sun-direction vector from up to three sensors facing the sun. Due to earth albedo absorption and a nonideal cosine response (due to reflections at low angles from the sensor surface), the sun vector accuracy from these coarse sensors is at best about ±5°, but their mass and power are negligible. Higher-accuracy sensors often make use of MEMS position sensitive detector (PSD) and optical windowing to give a sun-direction vector measurement from the sun azimuth and elevation measurement angles. The typical root mean square (RMS) accuracy is <0.1° for a ±60°–±70° field-of-view (FOV), mass of 25–35 g, and average power consumption of about 35 mW. See Figure 4.8 for typical nanosatellite coarse and fine sun sensors [12–14].

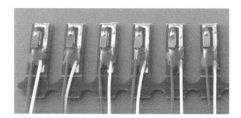  

Figure 4.8 Analog photodiode-type coarse sun sensors and two MEMS two-axis digital fine sun sensors.

4.7.3 Star Trackers

The most accurate attitude sensors used on larger satellites are star trackers. They are sensors with very sensitive light detectors, typically charge-coupled devices (CCDs); in some cases, these sensors are also cooled to reduce the thermal noise for an increased signal-to-noise ratio. The FOV of these sensors depends on the visual magnitude (M_v) stars that can be detected, for example, for a CCD detector sensitive enough to see a 6.5 M_v star, a FOV of 15° ensures at least three stars to be visible for more than 99% of the celestial sphere. The stars detected in the FOV are slightly defocused to enable the star centroid to be accurately determined using a center-of-gravity method. The separation distances (angles) between all the measured stars are then matched to reference stars in an on-board star catalog. For example, a single match is detected when a matching triangle can be found of three measured and reference stars. All other visible star separation angles are then matched to generate the maximum number of measured stars in the SBC frame and reference stars in the ECI frame for star tracking.

After the initial intensive "lost-in-space" matching process, successive measurements are only used to track their matched reference stars by searching in a small region around their previous position, assuming a slow rotating satellite. The tracking process is less intensive than star matching and this enables most star trackers to generate vector pair solutions typically between 1 and 10 Hz for further attitude and rate determination use in an EKF estimator. Star trackers on larger satellites have typical RMS accuracies of less than 5 arcsec in the bore-sight direction and 15 arcsec in bore-sight rotation. This performance is made possible by high-quality low-distortion optics and very sensitive star detection. For nanosatellites, the high CCD power requirement and large optics with sun- and earth-blocking baffles are inhibiting factors. However, a few nanosized star trackers have already been developed and some also qualified the flight successfully [15]. The typical performance specifications claimed by these nanosized star trackers are RMS accuracies of 12–36 arcsec in bore-sight direction, 65–108 arcsec for bore-sight rotation, rotation rates of $1°$–$3°$ s^{-1} at 1–2 Hz update rates, 220–500 mW average power, and 50–85 g mass (without radiation shielding and baffle). See Figure 4.9 for some typical commercially available nanosatellite star trackers.

Figure 4.9 Nanosized star trackers, left to right: ST-16 [16], ST-200 [17], and CubeStar [18].

4.7.4 MEMS Rate Sensors

Accurate measurement of the inertially referenced low angular rates of nanosatellites during three-axis stabilization is still not possible, without using low measurement noise and low bias drift fiber optic gyroscopes (FOGs) due to their high power, volume, and mass requirement. Although the development of MEMS rate sensors is continuously improving, it is still not matching the performance of FOGs. However, for measuring the initial relatively high tumbling rates of nanosatellites, MEMS rate sensors can be utilized effectively. The typical performance of MEMS rate sensors shows low-pass-filtered RMS measurement noise levels of $0.008°$–$0.015°$ s^{-1} and bias drift of $5°$–$28°$ h^{-1}. The improvement in performance is at the penalty of a mass increase of 1–55 g and power increase of 45–1500 mW for a three-axis rate measurement. See Figure 4.10 for these MEMS rate sensors.

Figure 4.10 MEMS rate sensors: (left) STIM300 a three-axis IMU [19], and (right) the CRM100 and CRM200 [20] single-axis units.

4.7.5 Magnetorquers

Actuators to generate magnetic moments for interaction with the geomagnetic field can easily be scaled for nanosatellite use. Magnetorquer rods are preferred due to their smaller volume, power, and mass compared to torquer coils, but sometimes due to layout problems, air-core coils are used. Torquer rods make use of a low remanence ferromagnetic core, for example, MuMetal or Supra-50 alloys are suitable. Torquer rods give a magnetic moment amplification of 80–100 compared to an air-core coil, therefore, they use less current and power and a smaller enclosed area for a similar magnetic moment. The physical placing of the torquer rods is critical, as they can influence each other, and the direction of the generated magnetic moment can rotate, especially if they are separated by distances less than a rod length (except for a symmetric T-configuration, see Figure 4.11). By pulse-width modulation (see Section 4.6.1) of the X_B, Y_B, and Z_B magnetorquer currents, a magnetic moment vector in any desired direction and size can be produced.

The maximum magnetic moment required for nanosatellites varies typically between 0.1 and 0.5 Am². Figure 4.11 shows a typical torquer rod and air-core coil [21]. The produced torquer rod has a maximum magnetic moment of 0.24 Am² at 200 mW, mass of 22 g, length of 60 mm, and diameter of 10 mm. The torquer coil has a maximum moment of 0.24 Am² at 625 mW, and mass of 50 g in a PC104 form factor for a CubeSat.

  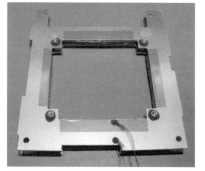

Figure 4.11 Left to right: single 0.24 Am² torquer rod, T-configuration of two torquer rods, and a PC104-sized torquer coil.

4.7.6 Reaction/Momentum Wheels

Reaction/momentum wheels are actuators that operate using the principle of preservation of angular momentum, to exchange the controlled momentum change in the wheel disc rotation speed to the body of the satellite. A reaction wheel assembly normally consists of a brushless direct current motor (BDCM) with a shaft-mounted disc acting as a flywheel. The flywheel's speed is accurately measured with a shaft encoder to enable a feedback speed control system for accurate angular momentum control. The flywheel's size is chosen according to the momentum storage requirements of a satellite in a specific orbit. The BDCM torque is selected to meet the agility requirements during rotation maneuvers, that is, how fast the satellite body must rotate during these maneuvers. Precise speed control with optimized low power requirements and small volume and mass are the driving factors for the wheel choice on nanosatellites.

The difference between reaction wheels and momentum wheels lies only in the application of the wheel's angular momentum to control the satellite attitude. A reaction wheel operates in a near-zero momentum bias configuration, that is, to limit the gyroscopic torques caused by an angular momentum vector during three-axis attitude rotations. A momentum wheel operates around an offset speed to give an angular momentum bias to the satellite's body for gyroscopic stiffness. This means a momentum wheel controls the satellite's attitude actively in the wheel spin axis direction through momentum exchange (by varying the wheel speed) and passively through gyroscopic stiffness by keeping the attitude in the other two axes.

For full three-axis rotation capability, a minimum of three reaction wheels are required and the wheel speeds are controlled around zero average speed. For redundancy reasons and to enable offset wheel speeds (to avoid the wheel torque disturbances at zero speed crossings), more than three reaction wheels can be used while still ensuring a zero average momentum vector applied to the satellite body. For example, four reaction wheels can be used in a fourth skewed wheel, tetrahedral or pyramid configuration.

Nanosatellite wheels that are commercially available are high-performance micromechanical devices; for example, they are balanced to low static and dynamic specifications to limit wheel vibrations affecting the payload performance, they use special vacuum-rated bearings to ensure a long life in space, and they have to survive the launcher forces and vibrations. The typical performance specifications of these nanosatellite wheels vary according to their momentum storage capability, for example, 0.58–30 mNm (higher momentum means larger wheels) and maximum wheel torque, for example, 0.023–2 mNm (higher torque means higher power). The mass range of these nanowheels is 20–185 g each, the flywheel diameter range 21–55 mm, and the average power at a constant speed of 2000 rpm varies from 100 to 620 mW per wheel with maximum power at full torque from 510 to 1800 mW. An important performance parameter for accurate attitude control, but not always given by the manufacturers is the wheel speed (momentum) control accuracy especially near zero speed for a three reaction wheel configuration. Figure 4.12 shows some available nanosatellite wheels, from left to right the RW-0.007-4 with built-in drive electronics [22], the CubeWheel with momentum control RMS error of <1 μNm [23], and the RW-1 in a tetrahedral four-wheel configuration [24].

Figure 4.12 Nanosatellite-sized reaction/momentum wheel examples.

4.7.7 Orbit Control Sensors and Actuators

Although not many nanosatellites are currently doing orbit maneuvers, future formations and constellations of nanosatellites will need to phase and maintain their relative orbit positions. Space GPS receivers will be required to accurately measure the nanosatellite's orbit position and supply accurate timing during these maneuvers. The GPS receivers currently used in nanosatellite missions must satisfy a low power requirement and they are mostly software-modified standard terrestrial GPS receivers. The software modifications imply the removal of the velocity and altitude limits and an increased bandwidth allowance for the larger Doppler shift experienced on these low Earth orbit (LEO) nanosatellites.

Nanosatellite propulsion systems are mostly developed using MEMS technology using cold gas or electric propulsion, for example, pulse plasma thruster (PPT) or field-emission electric propulsion (FEEP) [25]. Most of these systems are still experimental and only available at high cost. See Figure 4.13 for some nanosatellite propulsion systems developed for CubeSat missions; right to left: NANOPS cold gas thruster of CanX-2 [26], $T^3\mu PS$ cold gas thruster for DelFFi QB50 mission [27], and a FEEP thruster proposed for a CubeSat [25].

Figure 4.13 Nanosatellite-sized propulsion system examples.

4.7.8 Integrated ADCS Modules

A few complete ADCS solutions are available for nanosatellite missions:

A) The iADCS100 [28] was initially developed to be the most compact high performance ADCS for 1–3U CubeSats. The launching customer was the Aalto-1 satellite, launched in June 2017. The unit's layout is shown in Figure 4.14. It has a total mass of 250–380 g, depending on the wheel momentum storage. The size of $95 \times 90 \times 32\,mm^3$

Figure 4.14 iADCS100 Integrated ADCS unit for 1U to 3U CubeSats [28].

fits into a 0.3U CubeSat volume. The nominal power consumption is 1400 mW. It is a three-reaction wheel unit with three-axis target, nadir- and sun-pointing capability. It uses three magnetorquers and a built-in magnetometer for attitude detumbling and safe-mode control. For ADCS sensors, a ST200 star tracker, three-axis MEMS rate sensors, and plug-in sun sensors are used.

B) The Y-Momentum ADCS bundle [12] was originally developed for the QB50 mission. It is currently flying successfully on several 2U, 3U, and 6U CubeSats. A three-axis reaction wheel integrated ADCS bundle [29] was also developed for high accuracy pointing capability. The Y-Momentum unit's layout and a three-axis flight unit are shown in Figure 4.15. It has a total mass of 360 g and size of $95 \times 90 \times 58$ mm^3 to fit into a 0.6U CubeSat volume. The average power consumption is 336 mW, with a peak power of 644 mW. It is a Y-momentum wheel unit with nadir-pointing and pitch rotation capability. It uses Y_B and Z_B magnetic torquer rods plus a X_B torquer coil, a MEMS rate sensor, and deployable magnetometer for attitude detumbling and safe-mode control. For attitude measurement, a CubeSense [3] accurate sun and nadir sensor and six coarse sun sensors are used. A space GPS receiver and CubeStar star tracker can also be integrated to these units.

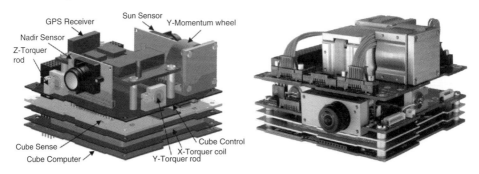

Figure 4.15 The Y-Momentum ADCS bundle's layout and a three-axis CubeADCS flight unit [29].

References

1 G.H. Janse van Vuuren, 2015. The Design and Simulation Analysis of an Attitude Determination and Control System for Small Earth Observation Satellites, MEng Thesis, University of Stellenbosch, March 2015.

2 Shuster, M.D. and Oh, S.D. (1981). Three-axis attitude determination from vector observations. *AIAA Journal of Guidance Control and Dynamics* 4 (1): 70–77.

3 Cubesense Online document: https://cubespace.co.za/cubesense

4 W.H. Steyn & Y. Hashida, 2001. In-Orbit Attitude Performance of the 3-Axis Stabilised SNAP-1 Nano-satellite, 15th Annual AIAA/USU Conference on Small Satellites, Utah State University, Logan, Utah, USA, August 2001, Proceedings SSC01-V-1.

5 W.H. Steyn, 1995. A Multi-Mode Attitude Determination and Control System for Small Satellites, PhD Dissertation, University of Stellenbosch, December 1995.

6 Crassidis, J.L., Lai, K.-L., and Harman, R.R. (2005). Real-time attitude-independent three-axis magnetometer calibration. *AIAA Journal of Guidance Control and Dynamics* 28 (1): 115–120.

7 Thomson, W.T. (1962). Spin stabilisation of attitude against gravity torque. *Journal of Astronautical Science* 9: 31–33.

8 Stickler A.C. & K.T. Alfriend, 1974. An Elementary Magnetic Attitude Control System, AIAA Mechanics and Control of Flights Conference, Anaheim California, August 1974.

9 Wie, B., Weiss, H., and Arapostathis, A. (1989). Quaternion feedback regulator for spacecraft eigenaxis rotations. *AIAA Journal of Guidance, Control, and Dynamics* 12 (3): 375–380.

10 Online web page: https://aerospace.honeywell.com/en/products/sensors/non-inertial-sensors/magnetic-field-sensing-and-sensor-solutions/magnetic-sensor-components

11 Online web page: http://www.pnicorp.com/products/geomagnetic-sensors

12 Online web page: https://cubespace.co.za/y-momentum

13 Online document: http://www.newspacesystems.com/wp-content/themes/formation-child/datasheets/NSS_Fine_Sun_Sensor_Datasheet_2b-.pdf

14 Online document: http://www.solar-mems.com/smt_pdf/commercial_brochure_SSOC-A.pdf

15 C.C. Grant, K. Sarda, M. Chaumont & R.E. Zee, 2014. On-Orbit Performance of the BRITE Nanosatellite Astronomy Constellation, 65th International Astronautical Congress, Toronto, Canada, September/October 2014, Proceedings IAC-14-B4.2.3.

16 Online web page: http://www.sinclairinterplanetary.com/startrackers

17 Online document: http://www.berlin-space-tech.com/fileadmin/media/BST_ST-200_Flyer.pdf

18 Erlank, A. and Steyn, W.H. (2014). Arcminute attitude estimation for CubeSats with a novel nano star tracker. *IFAC Proceedings Volume* 47: 9679–9684.

19 Online document: http://www.sensonor.com/media/91313/ts1524.r8%20datasheet%20stim300.pdf

20 Online document: http://www.siliconsensing.com/media/253549/pinpoint_gyro-eval-boards-rev3-draft-3-.pdf

21 Online document: https://cubespace.co.za/cubetorquer

22 Online web page: http://www.sinclairinterplanetary.com/reactionwheels

23 Online web page: https://cubespace.co.za/cubewheel

24 Online document: http://www.astrofein.com/2728/dwnld/admin/Datenblatt_RW1.pdf

25 Scharlemann, C., Tajmar, M., Vasiljevich, I. et al. (2011). Propulsion for nanosatellites. In: *32nd International Electric Propulsion Conference*, Wiesbaden, Germany, Proceedings Paper IEPC-2011, Vol. 171, 1–11.

26 S. Mauthe, F. Pranajaya & R.E. Zee, 2005. The Design and Test of a Compact Propulsion System for CANX Nanosatellite Formation Flying, 19th Annual AIAA/USU Conference on Small Satellites, Utah State University, Logan Utah USA, August 2005, Proceedings SSC05-VI-5.

27 A. Cervone, B. Zandbergen, J. Guo, E. Gill, W. Wieling, F.T. Nardini & C. Schuurbiers, 2012. Application of an Advanced Micro-Propulsion System to the DELFFI Formation-flying Demonstration within the QB50 Mission, 63rd International Astronautical Congress, Naples Italy, September/October 2012, Proceedings IAC-12 C4.6.2.

28 Online document: http://hyperiontechnologies.nl/products/iadcs100

29 Online web page: https://cubespace.co.za/3-axis

5 I-2d

Propulsion Systems

Flavia Tata Nardini[1], Michele Coletti[2], Alexander Reissner[3], and David Krejci[3,4]

[1] Fleet Space Technologies, Beverley, Australia
[2] Mars Space Ltd, Unit 61, Basepoint Enterprise Centre, Southampton, UK
[3] ENPULSION, Wiener Neustadt, Austria
[4] Space Propulsion Lab, Massachusetts Institute of Technology, Cambridge, USA

5.1   Introduction

One of the major shortcomings in the capabilities of nanosatellites in the past has been the lack of efficient propulsion systems. Capable propulsion systems can allow the satellites to control their orbit better or perform orbit changes, achieving ambitious missions in Earth orbit or elsewhere in the solar system. This can extend nanosatellite missions to scientific space explorations, long-life commercial applications, or constellations of small satellites flying in formation. Efficient propulsion systems can therefore revolutionize the use of nanosatellites, enabling universities and industries to transition from concept demonstrator missions to real space missions that were considered prohibitive in terms of costs until now.

Integrating a propulsion system enables wider applications for missions requiring precise attitude control, drag compensation, controlled deorbiting, rendezvous maneuvers for orbital debris removal and very low perturbation compensations, orbital transfers, such as maneuvers from low Earth orbit (LEO) to medium Earth orbit (MEO), and transfers from geostationary transfer orbit (GTO) to the Moon as well as interplanetary missions, such as the recent MarCO mission to Mars [1]. Such orbit control capability can allow single instrument-based space exploration missions or constellation in precise formation with synthetic aperture strategy. However, miniaturized propulsion systems, able to fit into and efficiently propel nanosatellites, are highly advanced technologies that require a long lead time for development and must overcome several challenges. Some of the challenges include reducing the current technologies into less bulky packaging, downscaling the current technology to the power levels and physical dimensions available on-board nanosatellites while maintaining a level of performance that is high enough to meet the mission requirements.

This chapter focuses on the implementation of propulsion technologies on on-board nanosatellites, a subject that requires a meticulous systems engineering approach which is often neglected or underestimated. A discussion of propulsion technologies must not only consider the thruster component, but the entire propulsion system. The complete system

Nanosatellites: Space and Ground Technologies, Operations and Economics, First Edition.
Edited by Rogerio Atem de Carvalho, Jaime Estela, and Martin Langer.
© 2020 John Wiley & Sons Ltd. Published 2020 by John Wiley & Sons Ltd.

includes feed system components and electrical subsystems that require a high level of integration and should demonstrate a high degree of modularity and miniaturization in order to fulfill the strict requirements of a nanosatellite mission. The same holds for performance parameters, with total system input power as a key parameter in miniaturized spacecraft, as auxiliary propulsion systems, such as neutralizers, valves, and heaters, can form a substantial part of the total power consumption.

5.2 Propulsion Elements

This section introduces some fundamental propulsion equations. These notions are essential to understand and compare the performance of different propulsion systems.

The specific impulse defines how efficiently a system converts the propellant into thrust, as shown in Eq. (5.1):

$$I_{sp} = \frac{F}{mg_0} = \frac{v_e}{g_0} \tag{5.1}$$

where I_{sp} is the specific impulse (s), F is the thrust magnitude (N), m the propellant mass flow rate (kg s^{-1}), g_0 is the Earth's gravitational acceleration constant (9.807 m s^{-2}) and ve the effective exhaust velocity (m s^{-1}).

The total impulse is defined as the integral of the thrust over the total thruster firing time, as shown in Eq. (5.2)

$$F = m_p I_{sp} g_0$$

$$I_{tot} = \int_0^{t_b} F dt \tag{5.2}$$

where t_b is the total time of firing (s) and m_p the propellant mass (kg). Equation (5.3) shows that for any space maneuver (or journey involving a number of maneuvers) the Tsiolkovsky rocket equation is valid, from which it is possible to derive the delta-v requirements:

$$\Delta V = v_e \ln \frac{m_0}{m_0 - m_p} \tag{5.3}$$

where m_0 is the initial total spacecraft mass including propellant (kg), mp is the mass of the propellant consumed in the maneuver (kg), and ΔV (pronounced delta-v) is the change or difference in velocity (m s^{-1}) required by the maneuver. From the previous equation, it is also possible to derive the propellant mass required for a desired delta-v, as shown in Eq. (5.4):

$$m_p = m_0 \left(1 - e^{\frac{-\Delta V}{I_{sp} g_0}}\right) \tag{5.4}$$

When analyzing the performance of an electric propulsion system, power consumption parameter should also be considered, and is calculated as shown in Eq. (5.5):

$$P = \frac{F I_{sp} g_0}{2\eta} \tag{5.5}$$

where P is the power input to the thruster (W) and η is the thruster efficiency (%). The thruster efficiency is defined in this case as the beam power divided by the input power.

Another important parameter in an electric propulsion system is the power-to-thrust ratio. It measures how much power is required to produce a certain amount of thrust, given a certain propellant exhaust velocity, as shown in Eq. (5.6):

$$\frac{P}{F} = \frac{v_e}{2\eta} \tag{5.6}$$

5.3 Key Elements in the Development of Micropropulsion Systems

Rigorous systems engineering is often neglected in the development of nanosatellites due to the fact that the systems are often relatively simple and there is rarely a client wanting specific objectives to be met. Propulsion systems are however complex systems requiring multiple objectives to be met. They have numerous interfaces and links between the various subsystems and require considerable verification and validation to ensure they will operate in the harsh remote environment of space. Applying a systems engineering approach from the outset of a micropropulsion development program ensures traceability among the operational space, the requirements, the system functionality, and the verification and validation.

Some key considerations and challenges in a typical micropropulsion development process are listed in the following text. These challenges could be overcome with the use of a systems engineering approach in the development.

- *Understanding the potential application:* The development of a propulsion system can be a long process, potentially taking many years. It is crucial to understand the potential users, applications, and the missions of interest; this key element is often not given due consideration. The process should be refined in conjunction with the client developing the mission. Many desirable missions are dismissed due to the lack of a propulsion system fit for purpose, and conversely, many propulsions system are developed without fully understanding the client's mission and were not well received by the client or the market. A clear understanding of the mission's operational orbits, attitude control, and primary propulsion requirements, as well as the characteristics and constraints associated with the type of spacecraft, costs involved, and end user requirements, helps develop a propulsion system that is suitable to enter the market.
- *Type of technology:* The choice between electrical or chemical propulsion on-board a nanosatellite must be considered on a case-by-case basis. Both classes include additional mass overhead and present different advantages and complexity. Chemical propulsion systems include storage tanks, feed systems, and valves, while electric propulsion systems include a medium-to-high voltage power supply unit. The use of chemical propulsion systems for such small devices leads the design to the use of microelectromechanical systems (MEMS) in most cases. However, the performance, especially regarding nozzles, in such small devices is still lacking and fairly complicated models are yet to be developed. In addition, some systems such as the cold gas thruster are relatively simple

Table 5.1 Mission assumptions.

Parameter	Value
Total spacecraft mass	5 kg
Delta-v	300 m s^{-1}
Propellant required	30.6 g if I_{sp} = 5000 s (electric propulsion)
Propellant required	926 g if I_{sp} = 180 s (chemical propulsion)
Overhead for chemical system	300 g (assumed constant)
Overhead for electric system[a]	$m_{ppu} = \dfrac{T\alpha}{\beta}$

a) m_{ppu} is the mass of the power supply, α is a scale factor for the PPU that it is on average 30 g W^{-1}, and β is the thrust-to-power ratio, average value 10 μN W^{-1} [3].

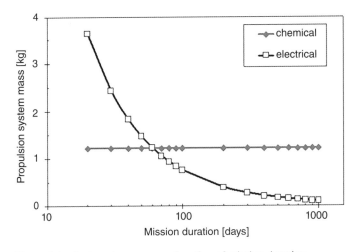

Figure 5.1 System dry mass as a function of mission duration.

in terms of integration and functioning, however, the relative low exhaust velocity leads to a necessity to carry significant amount of propellant to achieve large delta-v [2].

It is possible to perform a comparative analysis between the uses of chemical or electrical propulsion systems depending on the mission requirements. As an example, we use the mission assumptions shown in Table 5.1 to calculate the system dry mass as a function of mission duration for electrical and chemical systems and plot them in. For shorter missions, the chemical propulsion system is more favorable. The crossover point in this case is at 60 days. This value relates to an average thrust of 300 μN [3] (Figure 5.1).

- *Traceability:* During the development process, it is important to trace the requirements from system level right through to subsystem level. Micropropulsion systems present some system requirements that are very demanding and difficult to achieve and include miniaturization, modularity, flexibility, and a high level of integration. These requirements have to translate from system level to component level to allow novel design methodologies for subsystems such as propellant storage and distribution, power unit, and thrusters while still achieving the level of integration required on such small satellites.

- *Modularity:* The current driver for nanosatellites is modularity: Commercial off-the-shelf (COTS) components are designed to be easily integrated on-board the spacecraft using printed circuit boards (PCBs), fulfilling dimensional and electrical standards that allow satellites to reach a high degree of modularity. Adapting this concept to micropropulsion systems can be a significant challenge due to the number of components to be integrated and the difference in design from one propulsion concept to the other. The objective of a modular design for propulsion is to simplify mission planning, system selection, and satellite integration, to the point that any level of nanosatellite's user can consider a propulsive mission. This would make it possible to consider more challenging missions where the number of modules can be varied in order to meet as many applications and requirements as possible. The modular propulsion system would need to be compact, not complex, and relatively lightweight, and should attempt to utilize as many existing propellants as possible.

- *Flexible mission design:* The modularity and flexibility of a propulsion system determines how adaptable it is to different missions and spacecraft configurations, and is a critical factor in the design of micropropulsion systems. However, this concept is difficult to adapt to a nanosatellite mission, since the requirements, in terms of available mass, volume, power, as well as required performance, could be very restrictive. In the past, this has driven the development of micropropulsion systems to "single-use" systems that are difficult to adapt to other missions. However, in recent years, a modularity approach is noticeable in selected electric propulsion systems, proving the feasibility of using the same propulsion system as a building block to adapt to different mission sizes. It is still highly recommended to consider any modularity that can give the propulsion system more flexibility and adaptability to different mission scenarios.

- *Subsystems development:* The development of subsystems should follow the modularity and flexibility approach, designing the circuits at PCB level. It is recommended to develop all the subsystems with a systems engineering approach, since the thruster itself is only one part of the entire micropropulsion system. The entire micropropulsion system may include other system "overheads," such as power unit, tanks for propellant storage, and valves to regulate the propellant flow. There are currently several technologies that benefit from the use of MEMS, which significantly reduces the mass of the thruster and/or valves. Nevertheless, high-voltage power supply and tanks have considerable masses that might have a significant impact on the overall design and weight of the micropropulsion system. Compact and self-contained units are recommended for the power units, considering that high-voltage cables should be kept short within such a small satellite. It has been shown in Table 5.2 that the mass of the power unit is a function of the average power handled by the system, defining a specific mass which as a rule of thumb scales approximately $30\,\mathrm{g\,W^{-1}}$ (α) [3]. The impact of this mass in an electric propulsion system is therefore critical.

- *Technology breakthrough:* Real challenges in the development of a micropropulsion system are faced during the process of miniaturization, which pushes the boundaries of the propellant flow process and fundamental physical concepts to a point that often a total new understanding of these mechanisms is required. This often leads to investigating advanced approaches in micromanufacturing, micromachined component assembly, and packaging technologies, obtaining some amazing technology breakthroughs.

- *Speed and low cost:* The development of a micropropulsion system could be a long process in which extensive funds are required to reach the final goal. It is important to plan the development so that the propulsion technology is advanced quickly to the key position of having tested a breadboard model, in which all the subsystems, including the electronics, are included.

5.4 Propulsion System Technologies

The aim of this section is to provide a quick overview of the existing propulsion technologies (both chemical and electrical) that can be of interest to nanosatellites. A quick description of each of these technologies is presented together with their main advantages and disadvantages.

A list of the propulsion systems currently being developed or under development specifically for nanosatellites with masses below 10 kg is reported in Tables 5.4 and 5.5.

It must be noted that an in-depth discussion of the functioning and issues of each of these technologies is outside of the scope of this book. Readers wanting to explore the technologies reported in this chapter in further detail can refer to the reference list [3, 7–9].

5.4.1 Chemical Propulsion Technologies

5.4.1.1 Cold Gas Thruster

Cold gas thrusters have been used in space since the 1960s and are a simple and proven technology. In principle, a cold gas thruster consists of a pressurized gaseous propellant vented through a valve and a nozzle to produce thrust. Any gas can be used as propellant, but in practice mostly nitrogen (N2), helium (He), butane, and sometimes xenon are used. The main advantage of cold gas thrusters is that they are the simplest propulsion technology currently available, consisting only of a tank, a valve, and the necessary electronics to open and close this valve. Thanks to their simplicity, cold gas thrusters are very easily scalable to the dimensions required by nanosatellites. The main drawback of this technology is the low specific impulse that they produce, hence, the relatively high propellant mass needed to perform a given maneuver. Nevertheless, since they rely only on the propellant pressure to produce thrust, they can produce forces of the order of several mN even when scaled down to the sizes needed by nanosatellites (Figure 5.2).

5.4.1.2 Monopropellant Engines

Small high-performing monopropellant thruster systems would enable greater capability for nanosatellites with comparable high thrust levels. However, the introduction of a liquid propulsion system on-board a nanosatellite adds a level of complexity affecting launch processing, range safety, mission assurance, and handling procedures for some classes of propellants.

Traditional spacecraft primarily use hydrazine propulsion systems for attitude control in primary and secondary propulsion systems. However, hydrazine and its derivatives are

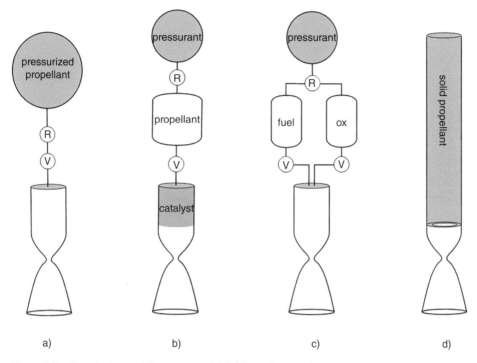

Figure 5.2 Chemical propulsion systems: (a) Cold gas thruster, (b) monopropellant engine, (c) bipropellant engine, and (d) solid propellant engine.

considered reactive and flammable, and pose a significant health hazard. The propellant therefore requires special handling procedures, increasing the total cost of the system. In a hydrazine thruster, the propellant is passed through a preheated catalyst bed and decomposed. The products of this chemical reaction are nitrogen, hydrogen, and ammonia, obtained in two different stages: an exothermic reaction and an endothermic reaction. The size, weights, and thrust levels allow easy integration of a monopropellant thruster in nanosatellites. The favorable application is for small/intermediate delta-v maneuvers (less than 1000 m s^{-1}). However, if higher delta-v is required, they become heavy due to the amount of propellant required [2].

The real challenge in the development of such a system comes from the microvalves that are usually the bulky item in the configuration. On the other hand, hydrazine thrusters have appealing advantages, since they do not suffer from propellant leakage, have good propellant densities, and feature good specific impulse. Moreover, they require relatively low input power.

In recent years, the "green propellants" for low-thrust engines have become attractive possible substitutes for hydrazine. These propellants are less toxic and therefore they do not require elaborate handling procedures and mostly do not pose health issues. High-energy green propellants considered good are ammonium dinitramide (ADN), hydroxylamine nitrate (HAN), hydrazinium nitroformate (HNF), and hydrogen peroxide (H_2O_2). Hydrogen peroxide and ADN are currently considered promising green monopropellant and bipropellant options for nanosatellites. ADN-based monopropellant is usually selected due to its benign properties and possibility to be bought commercially.

Hydrogen peroxide is nontoxic, can be diluted with water, and typically does not require complicated handling procedures. Although hydrogen peroxide is lower in performance than hydrazine, its relatively benign nature makes it ideal for many nanosatellite applications. This propellant, when subjected to a suitable catalyst, decomposes into water and oxygen in an exothermic reaction. It could be used for multiple propulsive roles, such as monopropellant attitude control or bipropellant or hybrid primary propulsion. The elimination of a separate oxidizer/attitude-control propellant tanks leads to saving mass, volume, and cost. However, catalyst lifetime in miniaturized systems is an issue. In addition, another disadvantage is that the choice of materials in the tanks, valves, and feeding systems has to be carefully considered, since this green propellant shows several compatibility issues.

The development of MEMS-based monopropellant thrusters could be an interesting alternative for nanosatellite propulsion systems. The challenges involved in the design include machining the very small nozzle areas and maintaining the divergence angle of the nozzle. A drawback of MEMS systems is that they typically having to be made of silicon, which is not compatible with propellants such as hydrazine, therefore, other propellant choices are more appealing.

5.4.1.3 Bipropellant Engines

Bipropellant engines have complex separated feeding lines for fuel and oxidizer and possibly pressurant, therefore, larger dry masses for the complete system. On the other hand, they show higher specific impulse that leads to propellant mass savings. They are usually normally used for missions requiring delta-v higher of hundreds of m s^{-1} or km s^{-1}.

Challenges of designing small bipropellant engines (i.e. less than 22N class engine) include the potential of combustion efficiency losses into the engine structure, thermal control issues of the chambers, nozzle throats, and injectors, mixture ratio control, and contamination issues. In general, better vaporization is achieved in longer chambers while small-diameter engines show lower chamber flow Reynolds numbers that lead to less turbulent flow, reducing the propellant-mixing performance. Chamber wall cooling would be a challenge in such a small engine as well. The fuel used for film cooling could go up to 30–40% for engines in the 22N class causing performance losses [2].

5.4.1.4 Solid Propellant Engines

Solid propellant engines are normally used for kick-stage or orbit insertion. Fuel, oxidizer, and organic binder are combined into a composite to form the solid propellant. The concept is based on the high rate combustion of one single propellant or explosive stored in a combustion chamber. The gas generated by the combustion of the propellant is accelerated in a nozzle, thus, delivering a thrust. This technology usually presents compact size, relatively good specific impulse, no leakage because of the absence of liquid fuel, no moving parts, and the ignition requires low power consumption. However, these engines are not restartable, therefore, they are used in missions that need one single burn and do not require very high accuracy for delta-v [2]. The lack of restartability could be offset by the fabrication of arrays of microrockets.

5.4.2 Electric Propulsion Technologies

5.4.2.1 Resistojet

Work on resistojets began in the 1960s, and the concept belongs to the simplest electric thruster. Resistojets are electrothermal thrusters wherein heat is transferred to the propellant from electrically heated surfaces, such as the chamber wall or a heater coil, and then use a conventional nozzle to generate thrust. A sketch of a resistojet is shown in Figure 5.3.

Figure 5.3 Sketch of a resistojet.

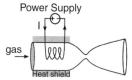

In principle, any liquid or gaseous propellant can be used; the most common choices are water, ammonia, nitrogen, and nitrous oxide. It is important to consider the molecular weight on the propellant choice. Low molecular weight yields higher specific impulse, but results in higher heating requirements due to the high heat of vaporization. Liquid propellant storage significantly reduces system mass and size by using lighter-weight and smaller propellant tanks compared to high-pressure gaseous storage. Additionally, leakages are less severe for liquid propellants than for high-pressure gaseous propellant. Chamber temperature is limited by the materials of the wall and heater coils (to an absolute theoretical maximum of some 3000 K or less), and, therefore, the exhaust velocities, even with hydrogen, cannot exceed $10\,000\,\mathrm{m\,s^{-1}}$ [10].

Resistojets are a good alternative to cold gas, being not much more complex from the system point-of-view and being able to deliver higher specific impulse if enough power is available. Moreover, the use of MEMS machining allows a significant miniaturization of this type of technology, making it compatible with the size of nanosatellites [11]. The main issue with the use of resistojets on on-board nanosatellites is the potential heat dissipation problems in such compact spacecraft. Nevertheless, they promise to deliver specific impulse in the order of 100 s with power consumption of the order of 1 W.

5.4.2.2 Gridded Ion Engine (GIE)

A GIE is an electrostatic type of electric thruster that produces thrust by accelerating a beam of ions by means of electric fields. GIEs consist of three basic components: the plasma generator, the ion accelerator, and the neutralizer cathode, as shown in Figure 5.4.

Figure 5.4 Schematic of gridded ion engine showing the grids, plasma generator, and neutralizer cathode [7].

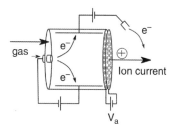

There are three types of plasma generators: these utilize either direct current (DC) electron discharges, radio frequency (RF) discharges, or microwave discharges to produce the plasma. In all three cases, electrons are energized either using electric (DC generators) or oscillating electromagnetic fields (RF and microwave discharges). These electrons collide with the neutral propellant (normally a noble gas, such as xenon) causing ionization. The ion accelerator normally consists of two electrically biased multi-aperture grids placed at one end of the plasma generator. The first grid is biased at a high positive potential (>1 kV), whereas the second grid is biased at a negative potential (~-200 V). Such potentials generate an electric field that extracts ions from the plasma generator and accelerates them producing an ion beam.

A neutralizer cathode is positioned outside the thruster body to provide the electrons necessary to neutralize the space charge of the emerging ion beam, and to avoid the satellite becoming electrically charged by the expulsion of only positively charged particles.

The main advantage of GIEs is that, among all propulsion technologies, they deliver one of the highest specific impulses (normally in the range 3000–5000 s), hence, require very little propellant in comparison to other technologies. Their main drawbacks are the generally high-power requirements (due to the high specific impulse), low thrust-to-power ratio, and the need for complex flow control and power processing units that also tend to increase the overall propulsion system cost. For this reason, the use of GIEs on nanosatellites is probably limited to those with mass larger than about 5 kg that require very high delta-v to perform complex missions. Nevertheless, various international projects are currently underway, developing nanosatellite-sized gridded ion engines. The most promising efforts are reported in Tables 5.4 and 5.5.

5.4.2.3 Hall Effect Thruster

Hall discharge plasma accelerators, also called Hall thrusters, were originally developed in Russia in the 1960s and since then have been extensively studied and used in real space applications (Figure 5.5).

The propellant (usually xenon) is injected through the anode into an annular space and is ionized by counterflowing electrons which are part of the current produced by the external hollow cathode. Permanent magnets or coils are embedded within the thruster structure and generate a magnetic field that forces electrons to move in a spiraling motion (consisting of a fast azimuthal drift and a slow axial motion toward the anode) to increase the ionization efficiency.

The main advantage of HETs is the higher thrust-to-power ratio that they can deliver at relatively attractive specific impulse (normally in the range 1000–3000 s) compared to GIEs. Their main drawbacks are the complexity of their power supply (that is slightly less complex

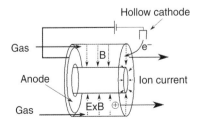

Figure 5.5 Schematic of HET showing the radial magnetic field and the accelerating electric field [7].

than a GIE power supply) and the complexity associated with their downscaling to sizes of interest to nanosatellites.

The challenges in Hall Effect Thruster (HET) miniaturization are related to the plasma generation process. Scaling to low power requires a decrease in the thruster channel size and an increase in the required magnetic field strength. This leads to problems related to the saturation of the miniaturized inner components of the magnetic circuit and to nonoptimal fields leading to a reduction in the efficiency and an increase in the heating and erosion of thruster components. Nevertheless, several developments involving miniaturization of Hall effect thrusters are currently ongoing worldwide.

5.4.2.4 Pulsed Plasma Thruster (PPT) and Vacuum Arc Thruster (VAT)

PPTs are electromagnetic devices that accelerate plasma particles to velocities of the order of $10 \, \mathrm{km \, s^{-1}}$ to produce thrust. A pulsed, high-current arc discharge lasting a few microseconds is produced between two electrodes; it ionizes, heats, and finally accelerates the propellant particles through the Lorentz force generated by the interaction of the discharge itself and the self-induced magnetic field.

A typical design of PPT is shown in Figure 5.6. It consists of a pair of electrodes charged by one or more capacitors to generate an electrical potential difference, which can range from hundreds of volts to a few kilovolts. The stored energy level can vary from a few Joules for the smallest thrusters to many hundreds of Joules, with the related power consumption ranging from fraction to hundreds of Watts, depending on the pulse repetition rate.

The discharge is usually initiated by a spark plug, which is connected to its dedicated energy storage device. The charged particles injected by the spark plug reduce the dielectric strength of the interelectrode space, producing a spontaneous discharge between the charged electrodes that is initially localized at the thruster breech, minimizing the inductance of the circuit. In solid propellant PPTs, the discharge is originated across the surface of a solid insulator, usually Teflon, and is sustained by the particles it ablates and ionizes from this surface, thereby providing the propellant to be accelerated.

An appealing feature of PPT technology is the use of a solid nontoxic propellant feed system, which can eliminate safety, handling, and leakage issues common to on-board fluids and their systems. Moreover, a tank is not needed, hence, reducing the propulsive system dry mass. Another advantage of the PPT is its high scalability, relative simplicity, and its ability to work over a wide range of input power and produced thrust, by changing the thruster pulse rate. These features offer very fine pointing, station-keeping, and attitude-control capabilities.

Over the past few years, PPTs became a technology of interest for microsatellites and nanosatellites. Several developments are being directed toward miniaturization of PPT technology, operation at lower pulse energies, mass reduction, and lighter capacitor designs. These developments showed that PPT for nanosatellites normally delivers specific impulse

Figure 5.6 Schematic of a solid propellant pulsed plasma thruster [12].

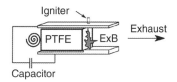

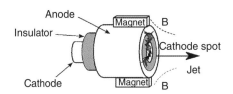

Figure 5.7 Schematic of a vacuum arc thruster [13].

in the range 500–1000 s with thrust-to-power ratios of about 20 μN s^{-1}. The most promising PPT developments are reported in Tables 5.4 and 5.5.

VATs (see Figure 5.7) are pulsed electromagnetic devices. During each pulse, an arc is generated between a cathode and an anode. The arc ablates part of the material of which the cathode is manufactured and ionizes it. The combination of electromagnetic forces generated by the arc causes the acceleration of the ionized cathode material. The performances of this type of device have been found to be significantly enhanced by the application of an external magnetic field [13].

VATs are relatively similar to PPT; the main difference between the two is that VATs normally are "trigger-less," hence, no trigger electrode (or spark plug) is needed to ignite the arc. Moreover, this "trigger-less" operation can be achieved at low voltages in the order of 100 V, hence, allowing the use of inductive storage and removing the requirements for high voltage (HV) electronics. Due to the use of inductive storage, VATs normally tend to operate with shot energies much lower than 1 J, and are therefore fired at much higher frequency than PPTs. The use of cathode material as propellant is advantageous given the high density of the material normally used (copper, tungsten), but at the same time might pose concerns regarding the effect that impingement of back-streaming particles coming from the plume might have on the solar panels and/or instruments of the spacecraft.

VATs are among the technologies that lend themselves best to miniaturization to levels applicable to nanosatellites and have to date demonstrated specific impulse in the order of 2000 s with thrust-to-power ratio of about 10 μN W^{-1}. Some of the most relevant VAT developments for nanosatellites applications are reported in Table 5.5.

5.4.2.5 Colloid/Electrospray and Field-emission Thruster

Field-emission and colloid thrusters are electrostatic accelerators which do not rely on ionization in the gas phase and can therefore be accomplished in small volumes. Colloids consist of a thin emitter which supplies a conducting fluid to its tip from an upstream reservoir. The emitter can be a sharp needle, a capillary, or a narrow slit. Field-emission electric propulsion (FEEP) thrusters use a liquefied metal propellant suspended over sharp needles or can be executed as porous structures with the propellant entrenched in the pores and fed by capillary forces.

An electric field is established between the emitter and an accelerating electrode placed close to the emitter orifice, which is often in a needle-like shape to increase local field strength, as the field is intensified near the emitter tip. The high local-field strength causes the fluid surface to become unstable and deform into a so-called Taylor cone. Ion, and droplet, emission occurs from a jet located at the apex of this Taylor cone. FEEP thrusters and colloid emitters differentiate by the composition of emitted species, with FEEP thrusters emitting single charged ions together with quasi-neutral droplets, whereas electrosprays emit a combination of solvated and nonsolvated ions together with droplets,

Figure 5.8 Schematic of a FEEP or colloid thruster.

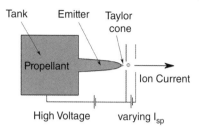

and colloid thrusters emit charged droplets. These ions and charged droplets are electrostatically accelerated to high speeds and are neutralized by an external cathode producing thrust by reaction. To increase produced thrust, emitters are often multiplexed and operated in parallel in linear, circular, or array configuration. FEEP and electrospray thrusters can be designed as completely passive systems, in which propellant transport occurs through capillary forces only. A schematic of a FEEP or colloid thruster is shown in Figure 5.8.

Colloid thrusters were actively studied during the 1960s and 1970s in the USA, but work ceased due to problems with low charge-to-mass ratios and the resulting need for high voltages (~10 kV) [4]. Since about 1990, interest in colloid, or electrospray, thrusters for microsatellite and nanosatellite applications has returned for a couple of reasons. The first reason is the development of high-conductivity liquids giving higher charge-to-mass ratios and hence lower voltages to achieve reasonably high specific impulses; and the second reason is the potential use of MEMS technology which reduces both the dry mass and also offers the possibility of further reductions in voltage due to the decrease in electrode separation [4]. Since the thrust is principally a function of the number of individual elements in the array, with the use of MEMS this thruster technology does not suffer the reduction in efficiency as thrust level and thruster size are reduced, typical of many other thruster technologies. Usage of ionic liquids as propellant allowed for both positive and negative ion extraction, and pairs of emitters operated in opposite polarity have been used to balance spacecraft charge avoiding the need for a dedicated neutralizer with generally lower efficiency of a propulsion system [14].

Such electrospray thrusters exhibit small dimensions and can operate generally at very high efficiencies and very low operating power levels. Other advantages of the technology are that they can be actively controlled to deliver precise impulse bits and operate on nonreactive and noncorrosive propellants. The colloid system does not use a reactive metal as propellant, so colloid thrusters are more benign in terms of contamination of spacecraft compared to similar technologies.

In the FEEP concept, thrust is generated by accelerating ions by means of electrostatic forces which are in general higher compared to electrosprays due to the liquid metal propellant. Ions are produced by field emissions from a liquid metal surface with the metal acting as the propellant, fed into the thruster by capillary forces. There exist different field emitter configurations, such as needle, capillary, and slit emitter types. The principle of operation is the same in all cases.

The advantages of the FEEP are mainly the high-performance and the high specific impulse that is achievable and the wide range of controllability of the specific impulse, which is primarily determined by the applied emitter potentials. The impulse bit can be accurately controlled by adjusting the electrical field, which allows flexible throttling in the entire thrust and specific impulse range. The system design is compact and simpler compared with other technologies. As with most electrospray thrusters, the feed system of a FEEP can be executed entirely passive by utilization of capillary forces to feed the propellant from the tank to the emission sites, making the propellant system less complex without need for valves, piping, and gas tanks. On the other hand, the high voltage required by this technology (up to 10 kV) leads to complex power processing units. In addition, due to the use of metals as propellant, these thrusters require stay-out zones to avoid spacecraft contamination. An indium-based FEEP thruster was one of the first high delta-v electric propulsion devices that was successfully demonstrated on a nanosatellite of only 3U reaching specific impulse levels beyond 3500 s [14].

5.5 Mission Elements

Spacecraft require propulsion to either change their motion (e.g. perform an orbit change, deorbiting, maintaining formation flying, changing the satellite attitude), to maintain their motion counteracting external forces (e.g. drag compensation), or to provide means to help control spacecraft attitude in the absence of planetary magnetic fields. In this section, some typical values for the delta-v required by various maneuvers for nanosatellites are presented.

5.5.1 Orbit Change

Spacecraft are required to change their orbit for various reasons; for example, a nanosatellite might need to move to its operational orbit from the orbit on which the launcher left it (varying either the orbit radius or inclination), or a nanosatellite might need to travel from an earth orbit to a lunar orbit or to another planet to perform a scientific mission.

The typical delta-v required to perform such maneuvers depends on the level of thrust available. If the thrust available is so high (relative to the spacecraft mass) that the spacecraft orbital velocity can be changed in a length of time that is much shorter than the orbital period, a so-called impulsive transfer is performed. If instead the amount of thrust available is relatively low, then a spiraling transfer will have to be carried out. In general terms, impulsive transfers require less delta-v than spiraling transfers. However, the requirement for high thrust imposed by the impulsive transfer prevents the use of electric propulsion, hence, limiting the maximum specific impulse to few hundreds of seconds. On the contrary, a spiraling transfer can be carried out using electric propulsion (hence, high specific impulse). The conclusion is that the cost of an orbit change can not only be quantified in terms of its delta-v, but the technologies that can be used to perform such tasks must also be taken into account.

The delta-v to reach other planets on its own is currently evaluated unfeasible for the micropropulsion systems of on-board nanosatellites, and therefore, is not included in this section.

Typical delta-v values for small orbit changes and for a trip to the moon or outer planetary bodies are reported in Table 5.2 [15].

Table 5.2 Orbit change [4].

Maneuvers	ΔV (km s^{-1})
Impulsive shot	
LEO to GEO[a]	3.95 (no plane change)
GTO to GEO	1.5 (no plane change)
LEO to earth escape	3.2 (for jet exhaust to initial circular velocity ratio = 10) [5]
LEO to lunar orbit	3.9
GTO to lunar orbit	1.7
Low thrust	
LEO (200 km altitude) to GEO (no plane change)	4.71 (~55 days of transfer time)
LEO to MEO (19 150 km; no plane change)	3.83 (~44 days of transfer time)
LEO to earth escape (for initial values of acceleration to local gravitational acceleration, 10^{-2} and 10^{-4})	5.82/7.08
LEO to lunar orbit	~8 (months–year) [6]
GTO to lunar orbit	3.6–4.5 (250–450 d)

a) Calculated using Edelbaum's equation.

5.5.2 Drag Compensation

Drag compensation (sometimes referred to as station keeping) is a maneuver where the thrust produced by the propulsion system is used to counteract the effect of the atmospheric drag that naturally tends to lower the orbit of a satellite.

Most nanosatellites are placed in LEO with orbital altitudes lower than 600 km, with a significant number of the nanosatellites used for commercial purposes launched into lower LEO orbits with altitude of about 300 km. At these altitudes, the lifetime of a nanosatellite is normally in the order of a few weeks to a few months, before it disintegrates re-entering the Earth's atmosphere. It is therefore of interest to commercial nanosatellite operators to be able to extend their in-orbit operation.

The natural lifetime of a nanosatellite on a LEO orbit, the drag on its initial orbit, and the delta-v required to keep it on orbit for an extra week and for a 50% lifetime increase are reported in Table 5.3. The data were derived using the NRLMSISE-00 atmospheric model, assuming a drag coefficient $c_D = 2.2$ and no deployable panels. In order to calculate the deorbiting time of a satellite, its mass, shape, and attitude need to be known. We assumed that satellites with masses of 1 and 4 kg adhered to the CubeSat standard, having a cross-section of 100×100 mm^2 and heights of 100 (1U) and 300 mm (3U), respectively. For satellites of 8 and 10 kg, we assumed they were cubes with side lengths of 180 and 200 mm, respectively. In all cases, we assumed the spacecraft flew with the smallest surface facing the flight path direction. Using the assumptions stated earlier, it should be noted that an 8 kg satellite in

Table 5.3 Drag compensation.

Orbit altitude (km)	Mass (kg)	Lifetime (y m d)	ΔV for 1 extra week (m s^{-1})	ΔV for 50% lifetime increase (m s^{-1})
200	1	1.3 d	99.89	9.28
	4	4.4 d	25.20	7.92
	8	2.8 d	44	8.80
	10	3 d	40	8.57
300	1	21.8 d	7.68	11.96
	4	2 m 26 d	1.90	11.67
	8	1 m 22 d	3.17	11.77
	10	1 m 26 d	3.94	15.76
400	1	6 m 13 d	1.03	14.20
	4	2 y 1 m 11 d	0.25	13.77
	8	1 y 3 m 12 d	0.42	14.01
	10	1 y 4 m 18 d	0.39	14.01
500	1	3 y 5 m 17 d	0.17	15.32
	4	13 y 10 m 3 d	0.04	14.42
	8	8 y 4 m 18 d	0.07	15.29
	10	9 y 0 m 9 d	0.06	14.12
600	1	19 y 2 m 10 d	0.03	17.86
	4	76 y 9 m 8 d	0.01	16.81
	8	46 y 6 m 2 d	0.01	17.82
	10	50 y 1 m 2 d	0.01	16.45

a 6U CubeSat configuration ($100 \times 200 \times 300$ mm^3) has a similar lifetime as a 4 kg satellite (as can be seen in the table), since the two spacecraft have the same ballistic coefficient.

Table 5.3 demonstrates that the lifetime increases very quickly with increasing orbit altitudes. This is due to the strong nonlinear trend of the atmospheric density at these altitudes. It is worth noting that nanosatellites orbiting 300 km in altitude or less typically deorbit in less than 2 months and approximately 1 year at an altitude of 400 km. Another interesting point is that the amount of delta-v needed to extend the lifetime of a nanosatellite by 50% does not vary significantly between satellites of any mass orbiting at altitudes in the range 200–400 km. As can be seen from Table 5.3, a 50% lifetime extension "costs" a delta-v of about 10–15 m s^{-1}. These delta-vs are well within the range of capabilities offered by propulsion systems currently being developed and/or under development for nanosatellites.

5.5.3 Deorbiting

Deorbiting can be treated in the same way as a change in orbit. Deorbiting has become particularly important due to the latest international and national regulations requiring spacecraft to deorbit and burn in the Earth atmosphere within 25 years from the end of their operational life. As is evident from Table 5.3, most nanosatellites placed in orbit up to 500 km will comply naturally with this requirement. Nevertheless, if a nanosatellite is placed on a higher orbit, it will need to demonstrate its compliance with the 25-year deorbiting rule.

Let us take the example of a 10 kg spacecraft. This spacecraft will require a thrust of about 3 N to perform an impulsive transfer, and more importantly, this transfer will require a delta-v of about 160 m s^{-1}.

If we perform a spiraling deorbit instead, the thrust requirement will drop to about 5 mN, however, the delta-v will rise to 330 m s^{-1}. It must also be noted that an impulsive deorbiting maneuver will take about 1 h, whereas a spiraling deorbiting maneuver will take several weeks. Moreover, the longer deorbiting duration increases the chance of the nanosatellite colliding with some other spacecraft during the deorbiting phase.

Moreover, in the last few years, several research efforts have gone into the design, manufacture, and test of propellant-less deorbiting devices for nanosatellites. In general, these devices mange to increase the natural drag on the nanosatellite by either increasing its surface area by deploying a sail-like device, or by inducing an electromagnetic drag, normally using an electrostatic tether.

5.5.4 Attitude Control

Attitude control is that family of maneuvers that involves rotating the spacecraft along one of its axis to obtain the desired pointing of the payload on the spacecraft. The exact requirement for such maneuvers is highly dependent on the mission, the pointing error that can be tolerated by the payload and the number of attitude changes that need to be performed during the nanosatellite mission. These requirements are generally not significant, since nanosatellites can be stabilized using differential drag techniques and/or magnetorquers if the payload is tolerant to pointing errors, or for higher mission requirements, miniaturized reaction wheels can be used.

In general, a delta-v in the order of 1–10 m s^{-1} per year is required by nanosatellites for attitude control according to reference [16]. Nevertheless, it must be noted that till date, most of the nanosatellites with mass lower than 5 kg do not employ propulsion to perform attitude control maneuvers.

5.6 Survey of All Existing Systems

A considerable set of reviews on micropropulsion technologies is available [5, 6, 15]. With a small number of exceptions, the majority of systems are in development and therefore still at relatively low technology readiness level (TRL), and performance figures published for specific technologies are therefore subjected to frequent changes.

A review of systems currently under development worldwide has been summarized and collated in Tables 5.4 and 5.5. Development is also taking place at Surrey SSTL (quad confinement), at the Laboratoire de Physique and Plasmas (ion engine), for the CAPE project, at the University of Bologna (monopropellant and cold gas), on plasma thrusters in Australia,

Table 5.4 Existing systems information table.

#	Name of technology	Developers	Type of organization	Nation	Price	TRL	Remarks
			Resistojet				
1	Micro-resistojet	Busek	Private company	USA	—	5	Integrated primary and attitude control systems
2	Water-fed resistojet	TU Delft	University	NL	—		MEMS thruster
3	Free Molecule Micro-Resistojet (FMMR)	University of South Carolina and US Air Force Research Lab	Defense and industry	USA	—	8	MEMS thruster
4	Low power resistojet	SSTL Surrey	Private company	UK		8–9	Good heritage in space
5	CHIPS	CU Aerospace/VACCO	Private companies	USA		7	Warm and cold operational modes
			Cold gas thruster				
6	T³μPS	TNO	Research center	NL		9	Demonstrated on-board Delfi-Next.
7	Advanced μPS	TNO	Research center	NL		6–7	More capable version of the T³μPS
8	Modular cold gas	University of Texas	University	USA	—	Low	3D plastic printing manufacturing techniques INSPIRE mission
9	INPS	Microspace	Private company	Singapore	€81 000	9	Operation demonstrated on-board POPSAT
10	MEPSI	Aerospace Corporation	Private company	USA	—	7	
			Water electrolysis				
11	HYDROS	Tethers Unlimited	Private company	USA	—	6	Modular and scalable tank, designed to be augmented with a cold gas system
12	Electrolysis MPS	Cornell University	University	USA	—	Low	Ni electrodes and 0.5 M KOH as electrolyte

No.	Name	Institution	Type	Country	Cost	TRL	Notes
13	NANOPS	University of Toronto	University	Canada	C$20 000	9	Successfully demonstrated on CanX-2
14	SNAP-1 MPS	SSTL Surrey	Private company	UK	—	9	Demonstrated on-board SNAP-1
15	MEMS module	NanoSpace	Private company	Sweden	—	5	Four thrusters
16	MiPS (MEPSI)	VACCO	Private company	USA	—	8–9	Five thrusters Using patented chemically etched micro system (ChEMS) technology
17	PALOMAR	VACCO	Private company	USA	—		Reaction control system with use of 6 DoF
		Monopropellant					
18	MPS110/120/130/120XW/120XL/160	Aerojet	Private company	USA	—	8–9	3D-printed engines demonstrated in 2014
19	SE	Stellar Exploration	Private company	USA	—		
20	NKS	TNO-ISIS-APP	Private companies	NL	—	4–5	Solid motor to deorbit a 3U from a 1000 km orbit
21	M005HP to M100	Micro Aerospace Solutions	Private company	USA	—		Innovative metal gauze catalyst design
22	ADN MiPS	VACCO	Private company	USA	—		Four high specific impulse thrusters
23	Green propellant thruster	Busek	Private company	USA	—	5	
24	CAPS-3	Digital Solid State Propulsion, Inc.	Private company	USA	US$50 000	8	Possible to fire up to 12 microthruster elements Ignition power delivered via capacitor discharge
25	ADN-based mPS	NanoAvionics	Private company	Lithuania	—	7	Ready to fly on one of the QB50 satellites

(continued)

Table 5.4 (Continued)

#	Name of technology	Developers	Type of organization	Nation	Price	TRL	Remarks
			Pulsed plasma				
26	Nano-PPT	Mars Space	Private company	UK	—	4/5	
27	PPTCUP	Mars Space	Private company	UK	£12 000	8	Compliant with CubeSat standard
28	μPPT	Busek	Private company	USA	—	5	Based on technology developed at AFRL Direct flight heritage (FalconSAT-3)
29	L-μPPT	JMP Ingenieros, Najera Aerospace, IPPLM, NanoSpace, KCI	Companies and universities	Europe	—	4–5	The operation of the syringe pump has been successfully tested in vacuum
30	Micro PPT module	Fotec	Private company	Austria	—	8	Flown on one of the QB50 satellites
31	Propulsion module (VAT)	Comat Aerospace	Private company	France			
			Colloid/electrospray				
32	PUC electrospray	Busek	Private company	USA	—	5	ST7-DRS, cathode (flight heritage), thruster, and valve (design heritage)
33	HARPS	Busek	Private company	USA	—	3–4	Passive electrospray thruster, no moving parts, low power
34	MicroThrust	NanoSpace, TNO, EPFL, Systematic, and QMUL	Companies and universities	EU	—	5	Based on colloid technologies developed at QMUL and EPFL
35	MAX-1 PS	Accion Systems	Private company	USA	—	7	Based on MIT technology iEPS

		FEEP				
36	IFM nano thruster	ENPULSION	Private company	Austria	—	8–9	Several thrusters in-orbit from 3U to 100 kg spacecraft. Specific impulse and thrust-controllable inflight. Modular approach
37	IL-FEEP	Alta/Sitael	Private company	Italy	—	5	
			Ion engine				
38	BIT-1	Busek	Private company	USA	—	5	Fast start-up and rapid thrust response to input frequency (FR) power modulation
39	MIXI	JPL	Research center	USA	—	5	
40	RIT μX	ArianeGroup	Private company	USA	—	5	Significant thrust range available
41	NPT30	ThrustMe	Private company	France	—	4	Potential for operation without cathode in later versions
			Hall thruster				
42	ExoMG-nano	Exotrail	Private company	France	—	4	
			Arcjet				
43	mVAT	Alameda Applied Sciences Corporation	Private company	USA	—	5	Four thrusters
			Microgravity discharge				
44	PUC	CU Aerospace/VACCO	Private companies	USA	—	7	

Table 5.5 Existing systems technical data.

#	Name of technology	Propellant	Specific impulse (s)	Total impulse (Ns)	Thrust (mN)	System mass (g)	Propellant mass (g)	Size (mm)	Storage pressure (bar)	Power usage (W)	Remarks
					Resistojet						
1	Micro-resistojet	Methanol	150 primary 80 each ACS thrusters	40 primary 23 ACS	2–10 primary 0.5 each ACS thruster	<1250	280 ml reservoir	90×90×100 <1U	—	3–15	4 kg satellite 60 m s^{-1} primary 6 m s^{-1} ACS 8 ACS thrusters
2	Water-fed resistojet	Water	—	—	1.4–0.8	<0.2 (thruster)	—	—	4.5–2.5	7–3.6	Heating T_{max} >900 °C
3	FMMR	Nitrogen (water vapor)	65 @ 575 K (85)	1	12	—	—	—	—	—	—
4	Low power resistojet	Xenon/ nitrogen or butane	48 xenon 99 N$_2$ 95 butane	—	Up to 100	90 without valves	—	—	10	10	Operating temperature up to 500 °C
5	CHIPS	R-134a	82 primary 47 ACS thrusters	258/0.5U 563/1U 805/1.5U	30	—	—	95×95×141.8	—	—	Three-axis ACS 3 kg satellite→ 91 m s^{-1}/0.5U 214 m s^{-1}/1U 326 m s^{-1}/1.5U
					Cold gas thruster						
6	T³μPS	Nitrogen solid storage	69	0.8	1–100	119	1.2	94×94×18	N/A	10.4 ignition 0.37 valve	Cold gas technology and no-pyrotechnic ignition patented
7	Advanced μPS	Nitrogen solid storage	69	36	—	438	61	90×90×80	N/A	—	—

8	Modular cold gas	Dupont Suva 236	89.5 @ 85°C	—	150 @ 85°C	380	90	100×90×40	—	1.5 (max)	—
9	INPS	Nitrogen (argon available)	50	—	0–1	>300	—	40×100×100 85×40×100 PCB	10–20 max	2 peak 0.5 standby	8 nozzles, three-axis control, redundant system for formation flying
10	MEPSI	Xenon	50	26.6	—	250	500	—	—	—	—
					Water electrolysis						
11	HYDROS	Water electrolyzes into H_2 and O_2	300	—	Up to 800	—	5 kg satellite 100 ml water → 100 m s^{-1}	0.5U to 1U	N/A	0.5–10	100–300 m s^{-1} for 3U 50–150 m s^{-1} for 6U scalable to >2 km s^{-1}
12	Electrolysis MPS	Water electrolyzes into H_2 and O_2	350	—	—	—	1.5 ml of water	—	—	—	—
					Liquefied gas						
13	NANOPS	Sulfur hexaflouride (100 new version)	46.7	8 (175)	50 (50–100)	500	0.02 (0.18)	100×100×50	34.5	—	—
14	SNAP-1 MPS	Butane	—	22.3	45–120	455	33	25×100×100	4 @ 40°C	3.5–6	—
15	MEMS module	Butane	90–110	40	—	250	33	100×100×20	5 @ 60°C	<2	—

(continued)

Table 5.5 (Continued)

#	Name of technology	Propellant	Specific impulse (s)	Total impulse (Ns)	Thrust (mN)	System mass (g)	Propellant mass (g)	Size (mm)	Storage pressure (bar)	Power usage (W)	Remarks
16	MiPS	Liquid isobutane	65	23	53 @ 20°C	456 (dry)	53	29 cm^3 expandable	10.3 (max)	—	Operating temperature 0–60°C
17	PALOMAR	Isobutane (other propellant available)	—	85	35	1.063	173	—	10.3 (max)	<5 peak	Operating temperature 0–50°C
					Monopropellant						
18	CHAMPS		3000 1.5U tank 1500 1U tank					1U to 2U for XL			3U per 4 kg 6U per 10 kg
	MPS-110	Inert gas	—	—		Variable			—	—	10 m s^{-1} cold gas
	MPS-120	Hydrazine	—	—	0.26–2.8 N (×4)	<1.3 kg dry <1.6 kg wet	300		—	—	209 m s^{-1} 81 m s^{-1}
	MPS-130	AF-M315E	—	—	Up to 1 N (×4)	<1.3 kg dry <1.6 kg wet	300		—	—	30 m s^{-1} 130 m s^{-1}
	MPS-120XW	Hydrazine	—	—	0.26–2.8 N (×4)	<2.4 kg dry <3.2 kg wet	800		—	—	440 m s^{-1} 166 m s^{-1}
	MPS-120XL	Hydrazine	—	—	0.26–2.8 N (×4)	<2.4 kg dry <3.2 kg wet	800		—	—	539 m s^{-1} 200 m s^{-1}
	MP-160	Xenon	—	—					—	—	80 W solar electric power >2000 m s^{-1}
19	SE	Hydrazine	220	1260	1500 (×4)	955	522	1U	11	—	—

20	NKS	Ammonium perchlorate based	—	590	180 N	—	—	1U	—	—	Class 1.3 propellant with no-pyrotechnic ignition system patented by TNO
21	Series from M005HP to M100	ADN	—	1808	400	1800	720	—	16	—	Operating temperature 0–50°C
22	ADN MiPS	ADN	—	1828	400	1800	720	From 0.5 to >1U	16.8	—	3 kg satellite→697 m s^{-1}
23	Green propellant thruster	AF-M315E no toxic propellant	220	—	500	<1500 (3U) <2700 (6U)	—	0.5U to 1U (3U sat) <2U (6U sat)	—	<15	3U, 4 kg satellite→130 m s^{-1} 6U, 10 kg satellite→>251 m s^{-1}
24	CAPS-3	Non-metallized HIPEP	Up to 900	—	300 average	530	—	PCB stack, 5.71 cm in height	—	—	High-power, short duration ignition impulse
25	ADN-based mPS	ADN	—	—	Scalable, 3 N 900 dry for a 3U		270 (3U)	CubeSat compatible	—	8–10 Cat-bed preheat	200 m s^{-1} 0.035 Ns impulse bit. Burning time 1800 s
					Pulsed plasma						
26	Nano-PPT	PTFE	640	188	0.09 @ 5 W	440	30	150 × 100 × 70	—	5	—
27	PPTCUP	PTFE	600	44 (63 version available)	0.04 @ 2 W	280	7	100 × 100 × 3	N/A	2	—

(continued)

Table 5.5 (Continued)

#	Name of technology	Propellant	Specific impulse (s)	Total impulse (Ns)	Thrust (mN)	System mass (g)	Propellant mass (g)	Size (mm)	Storage pressure (bar)	Power usage (W)	Remarks
28	μPPT	Solid Teflon	443	—	0.5 primary 0.13 ACS	<550	—	110×101×110 <0.5U	—	2	4 kg satellite→ 63 m s^{-1} primary 65 m s^{-1} ACS
29	L-μPPT	Low vapor pressure liquid propellant	—	500	—	276 (dry)	50 cm^3	Volume tank 100×100×24	—	—	—
30	Micro PPT thruster module	PTFE	904	7	8 μN @ 1 Hz 2 μN @ 1/4 Hz	300	—	920×920×350	—	<2.5	5.1 m s^{-1} 1U 2.6 m s^{-1} 2U
31	Propulsion module (VAT)	—	2000	4000	0.3 mean thrust 30 @ 100 Hz	300 (dry)	200	100×100×100	—	—	200 m s^{-1}, 20 kg satellite
					Colloid/electrospray						
32	PUC electrospray	Ionic liquid	800	675 (50 ml tank)	0.7	<1150	—	0.5U 85×85×60	—	<9	Propellant-less field-emission cathode. 151 m s^{-1} with 50 ml and 4 kg satellite
33	HARPS	Safe, not toxic, not volatile	750	304	0.1	<400	—	<0.4U	—	0.75	76 m s^{-1} for 4 kg satellite

34	MicroThrust	Ionic liquid EMI-BF4 or (EMI-Tf2N)	2000–3000 depending of operation mode	—	—	820	520	1U	—	—	—
35	MAX-1 PS	Ionic liquid	2000	—	0.12	110 (wet mass)	30	0.2U 100 × 100 × 20	—	5	Operating temperature −10 to +500°C
						FEEP					
36	IFM nano thruster	Indium	2000–6000	Up to 10 000	0.01–1 0.35 nominal	<900	250	100 × 100 × 80	0	25	No chemicals, no pressurized components, no liquids
37	IL-FEEP	Ionic liquid EMI-BF4	2000	2000	0.1	500	—	1U of volume	—	5	Variant of the Alta Cesium FEEP 2U→1 $km\ s^{-1}$ 3U→0.5 $km\ s^{-1}$
						Ion engine					
38	BIT-1	AF-315 or xenon (compatible also with Ar, Kr, H_2, He, I)	2250	—	0.15	<1000	—	<1.5U	—	10	100 $m\ s^{-1}$, 4 kg satellite
39	MIXI	Xenon	3200	—	0.1–1.5	Thruster 250 2100 for 7000 $m\ s^{-1}$ Feeding system and tank 300	—	—	—	40	7000 $m\ s^{-1}$ in LuMi configuration
40	RIT μX	Xenon	300–3000	>10 000	0.05–0.5	440 (thruster head)	—	<1U	—	<50	
41	NPT30	Xenon	—	<5000	0.3–0.9	2	—	2U	—	65	

(continued)

Table 5.5 (Continued)

#	Name of technology	Propellant	Specific impulse (s)	Total impulse (Ns)	Thrust (mN)	System mass (g)	Propellant mass (g)	Size (mm)	Storage pressure (bar)	Power usage (W)	Remarks
						Hall thruster					
42	ExoMG-nano	Xenon	—	3000	1	—	—	2U		40	
						Arcjet					
43	mVAT	Solid (Al/tungsten)	1110 (Ta) 3000 (Al)	—	10 nN–300 μN	500	—	—	—	10	Ti as cathode material
						Microgravity discharge					
44	PUC	SO_2	70 warm fire 47 cold fire	184 warm fire 124 cold fire	4.5 warm fire 5.5 cold fire	718	268	0.25U 89 × 89 × 67	3.3 nominal	15 warm 8 cold	Fully integrated system

at CNRS (Hall effect thruster), in the Miles team for the NASA CubeQuest Challenge (ConstantQ™ plasma thruster), the NASA Microfluidic Electrospray Propulsion (MEP) project (electrospray and FEEP thruster), at the University of Michigan (Plasmadynamics & Electric Propulsion Laboratory [PEPL] ambipolar thruster), and at the George Washington University (VATs). However, for lack of publicly available technical information, these propulsion systems and probably several others have not been included in the tables.

5.7 Future Prospect

While in the past the majority of nanosatellites have not used propulsion, we currently see an increased demand and first completed in-orbit tests of high-performing propulsion systems specifically designed for nanosatellites, including complex electric propulsion systems. In line with the nanosatellite market evolving into a real commercial market, the need for propulsion is expected to further increase and will most likely be used on-board of the majority of commercial platforms to perform orbit changes, lifetime extension, and comply with deorbiting requirements. This increase in demand has triggered an increase in the economic interest in nanosatellite propulsion capabilities, establishing new propulsion companies and forcing more of the companies working in "conventional" spacecraft propulsion to move toward the nanosatellite market. At the same time, more ambitious and challenging nanosatellite missions are being considered and implemented worldwide that require propulsion to achieve formation flying and/or to travel away from LEO orbits toward the moon and other celestial bodies.

Moreover, the successful production of nanosatellite propulsion systems capable of delivering attractive performance at a price that aligns with the tight nanosatellite cost requirements has already shown an impact on the propulsion market. In fact, nanosatellite-sized relatively inexpensive technologies have already gained significant flight heritage and can form the basis for scaled-up versions of these technologies, built with the same cost-effective approach. This can, in turn, attract the interest of heritage satellite operators, hence, having an impact on the "conventional" propulsion players.

References

1 Andrew Klesh, Brian Clement, Cody Colley, John Essmiller, Danielle Forgette, Joel Krajewski, Anne Marinan, Thomas Martin-Mur, Joel Steinkraus, David Sternberg, Thomas Werne, Brian Young, 2018. "MarCO: Early Operations of the First CubeSats to Mars," SSC18-WKIX-04, Small Satellite Conference.

2 Alexandra Bulit, Matthias Gollor, Pierre Lionnet, Jean-Charles Treuet, Ines Alonso Gomez, (2015). "D2.1 Database on EP (and EP-related) technologies and TRL", Electric Propulsion Innovation & Competitiveness (EPIC) project.

3 Micci, M.M. and Ketsedever, A.D. (2000). *Micropropulsion for Small Spacecraft*, AIAA Progress in Astronautics and Aeronautics (AIAA), vol. 187 (ed. P. Zarchan) Editor-in-Chief. Reston, VA.

4 Barry Zandbergen, lectures notes belonging to the TUDelft course" Aerospace Design and Systems Engineering Elements I, part Spacecraft design and sizing", https://www .tudelft.nl/onderwijs/opleidingen/masters/ae/msc-aerospace-engineering/master-tracks/ space-flight/programme/space-engineering-profile/

5 Levchenko, I., Bazaka, K., Ding, Y. et al. (2018). Space micropropulsion systems for Cubesats and small satellites: From proximate targets to furthermost frontiers. *Applied Physics Review* 5: 011104.

6 Mueller J., 1997. "Thruster Options for Microspacecraft: A Review and Evaluation of Existing Hardware and Emerging Technologies", AIAA-97-3058, 33rd AIAA/ASME/SAE/ASEE Joint Propulsion Conference, Washington, USA, July 6–9, 1997.

7 Osiander, R., Ann Garrison, M., and Darrin, J.L.C. (2006). *MEMS and Microstructures in Aerospace Applications*. CRC Press, Taylor & Francis Group.

8 R. Intini Marques, (2009). "A Mechanism to Accelerate the Late Time Ablation for the Pulsed Plasma Thruster", PhD Thesis, Astronautics Research Group, University of Southampton.

9 "Small Spacecraft Technology State of the Art", NASA/TP–2014–216648/REV1, Mission Design Division Staff Ames Research Center, Moffett Field, California (2015).

10 J. Mueller, R. Hofer, J. Ziemer, (2010). "Survey of Propulsion Technologies Applicable to NanoSats", JANNAF, Colorado Springs, Colorado, May 3–7, 2010.

11 Jahn, R.G. and Choueiri, E.Y. (2001). Electric propulsion. In: *Encyclopedia of Physical Science and Technology*, 3e, vol. 5. San Diego: The Academic Press.

12 Keidar, M. et al. Magnetically enhanced vacuum arc thruster. *Plasma Sources Sci. Technol.* 14: 661–669, 205 doi:https://doi.org/10.1088/0963-0252/14/4/004.

13 Goebel, D.M. and Katz, I. (2008). *Fundamentals of Electric Propulsion: Ion and Hall Thrusters*", ISBN: 978-0-470-42927-3. Wiley.

14 David Krejci, Alexander Reissner, Bernhard Seifert, David Jelem, Thomas Horbe, Florin Plesescu, Pete Friedhoff, Steve Lai, 2018. "Demonstration of the IFM Nano FEEP Thruster in Low Earth Orbit", The 4S Symposium, Sorrento.

15 Lemmer, K. (2017). Propulsion for CubeSats. *Acta Astronautica* 134: 231–243.

16 Krejci, D. and Lozano, P. (2018). Space Propulsion Technology for Small Spacecraft. *Proceedings of the IEEE* 106 (3): 362–378.

6 I-2e

Communications

Nicolas Appel[1], Sebastian Rückerl[1], Martin Langer[1,2], and Rolf-Dieter Klein[3]

[1] *Institute of Astronautics, Technical University of Munich, Garching, Germany*
[2] *Orbital Oracle Technologies GmbH, Gilching, Germany*
[3] *Multimedia Studio Rolf-Dieter Klein, München, Germany*

6.1 Introduction

The communication subsystems are among the most critical parts of satellites, as they provide the only access to the spacecraft. During the satellite's operations, the communication links are used for commanding, retrieving telemetry, tracking and ranging, and tasks such as applying software updates. Additionally, communication links relay and broadcast signals and downlinking payload data. The architecture of each communication link depends on its destined application. Figure 6.1 shows the two common types of nanosatellite communication links: space-to-ground links and intersatellite links. Space-to-ground communication is the default for virtually all communication services provided by satellites. Intersatellite links are an emerging technology among nanosatellites, as a means to coordinate and relay data within satellite swarms and constellations.

The space segment involved in both scenarios consists of a radio module and one or more antennas. The radio module is the active part of the space segment and provides the translation from digital baseband signals to high-frequency signals and vice versa. Most nanosatellite radio modules provide both receiver and transmitter functionalities and are therefore called transceivers. Transceivers provide bidirectional communication for telemetry and telecommand (TM/TC) links, ranging, tracking, or intersatellite links. Transmitters provide unidirectional communication and thus commonly find their use in pure data downlinks, which commonly are associated with very high data rates.

The purpose of the respective radio link directly affects many of its properties. TM/TC usually require a high level of robustness, as they are the lifeline to the satellite. In order to reduce the probability of a loss of the system, TM/TC link designs often do not rely on active attitude control systems. A common way to achieve attitude independence is to use omnidirectional antennas. One drawback of these antennas is a lower efficiency compared to directional antennas. This drawback becomes dominant in links that need to cover large distances or have a wide bandwidth. This is because of the free-space path loss, which is an attenuation effect of wireless transmissions that rises with the square of frequency and distance. The advantages of high frequencies are higher available bandwidth and smaller size of directional antennas.

Nanosatellites: Space and Ground Technologies, Operations and Economics, First Edition.
Edited by Rogerio Atem de Carvalho, Jaime Estela, and Martin Langer.
© 2020 John Wiley & Sons Ltd. Published 2020 by John Wiley & Sons Ltd.

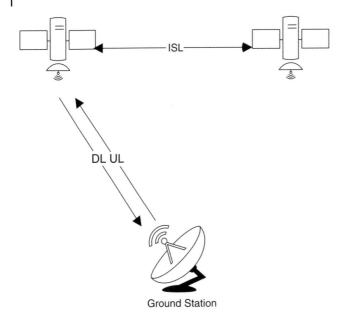

Figure 6.1 Illustration of nanosatellite links.

6.2 Regulatory Considerations

The used radio frequencies fall under the authority of the International Telecommunication Union (ITU). The ITU enacts regulations for the harmonization in frequency allocation and coordination of frequency use. Moreover, the ITU provides recommendations (ITU-R) and standards for telecommunication (ITU-T). The ITU divides the frequency spectrum into different bands. In this scope, three bands are relevant: Very high frequency (VHF) (30–300 MHz), Ultra high frequency (UHF) (300 MHz–3 GHz), and Super high frequency (SHF) (3–30 GHz). The band designators of the Institute of Electrical and Electronics Engineers (IEEE) are used, as they provide a better resolution and are more common. Since nanosatellites are up until now almost exclusively deployed to low Earth orbit (LEO), the regulations of the Fixed Satellite Service (FSS) are not discussed in here. Table 6.1 lists all different channels relevant for nanosatellites. The frequency bands are divided into channels, depending on the designated service. For nanosatellites, the typical services are:

- Space Research Service (SRS): Links for spacecraft with scientific or technological research purposes.
- Space Operation Service (SOS): Links exclusively used for spacecraft TM/TC.
- Earth Exploration Service (EES): Links which carry information from earth observation between ground and spacecraft and between spacecraft.
- Amateur Satellite Service (ASS): Links which are used for amateur purposes.

For satellite communications, the national authorities act as proxy for coordination with the ITU. It is mandatory to register all transmitting space segments with the ITU and adhere to its regulations. Typically, the ITU coordinates new allocations with national agencies to

Table 6.1 Frequency bands.

Name	Frequency range (MHz)	Service	Direction (UL/DL/Any)
VHF	137–138	SOS, SRS	DL
	144–146	ASS	Any
UHF	400.15–401	SOS, SRS	DL
	401–402	SOS, EES	DL
	402–403	EES	UL
	410–420	SRS	ISL
	430–440	ASS	Any
	449.75–440	SOS	UL
L-Band	1215–1340	SRS	DL
	1427–1429	SOS	UL
	1525–1535	SOS	DL
S-Band	2025–2110	SOS, SRS, EES	UL, ISL
	2200–2290	SOS, SRS, EES	DL, ISL
C-Band	5840–5850	ASS	Any
X-Band	7190–7250	EES	UL
	8025–8400	EES	DL
	10 450–1050	ASS	Any

avoid interference. Frequency allocations in SRS, SOS, and EES satellite bands usually come with a standard fee, which might inhibit nonprofit organizations from using these bands. Additionally, the radio spectrum remains a shared medium and interference may occur. The ITU regulates bandwidth expansion, transmitter power, and out-of-band emissions [1].

The International Amateur Radio Union (IARU) coordinates transmitters in the amateur radio bands and also represents the concerns of amateur radio users to the ITU. The amateur radio service is self-regulated and the amateur radio community regulates the band allocations and their use, which results in inconsistent channel allocations and power restrictions within the amateur bands. The main advantage of the amateur radio bands is the low cost, as frequency allocations are free to all holders of amateur radio licenses. Generally, due to the nature of the amateur radio service, the rules are more relaxed compared to nonamateur frequencies. Therefore, this grants simpler access but also higher risk of interference. In some countries, the amateur radio service does not have exclusive access to a certain frequency band, which may cause higher levels of noise in the spectrum.

6.3 Satellite Link Characteristics

A channel is the medium through which communication takes place and it is defined by its physical makeup, capacity, and error characteristics. As one of the first determining the characteristics of communication channels, Harry Nyquist [2] linked the rate of the information to the bandwidth expansion of the signal:

$$f_s \leq B \cdot 2 \text{ Hz} \tag{6.1}$$

where f_s is the number of information symbols per unit time and B is the required bandwidth of the signal. Later, Ralph Hartley showed in reference [3] that in an error-less channel, the maximum information rate R is limited to:

$$R \leq B \cdot 2 \cdot \log_2 M \text{ bit s}^{-1} \tag{6.2}$$

where M is the number of bits per symbol or modulation order. In practice, this upper bound is not very useful, as virtually all communication channels exhibit some sort of error characteristic. The most common model is that of additive white Gaussian noise (AWGN). Mathematically, this is described by:

$$y_i = x_i + n_i \tag{6.3}$$

where y_i is the received symbol, x_i is the transmitted symbol, and n_i is a normal distributed random variable. Since in AWGN $n_i \sim \mathcal{N}(0, \sigma^2)$, the received symbol y_i is also normal distributed: $y_i \sim \mathcal{N}(x_i, \sigma^2)$. The influence of the noise depends on the ratio between signal power P and noise power density N_0, where the noise is expressed by the variance $\sigma^2 = N_0/2$. For digital transmissions, the signal power is commonly normalized by the number of bits per symbol, denoted as E_b, to simplify calculations with different modulations and symbol rates. The normalized signal-to-noise ratio (SNR) is therefore referred to as E_b/N_0:

$$\frac{E_b}{N_0} = \frac{P}{R \cdot N_0} \tag{6.4}$$

Claude Shannon developed the theoretical framework to calculate the maximum possible information rate in noisy channels, at which error-free communication is possible [4, 5]. This rate is often referred to as channel capacity or Shannon limit. In the AWGN channel, the channel capacity C is given by:

$$C = B \cdot \log_2 \left(1 + \frac{P}{N_0 \cdot B} \right) \text{ bit s}^{-1} \tag{6.5}$$

This implies that by increasing the bandwidth indefinitely, the channel capacity approaches a limit:

$$\lim_{B \to \infty} C = \frac{P}{N_0} \log_2 e \text{ bit s}^{-1} \tag{6.6}$$

Figure 6.2 shows the channel capacity over a bandwidth range, with a steep increase in capacity for small bandwidth values. This region is therefore called the bandwidth-limited regime. Beyond the bandwidth-limited regime, the graph flattens out and the impact of increasing bandwidth diminishes, as the capacity approaches the limit. This region is called the power-limited regime, as significant increases in capacity can only be achieved by improving the SNR. This is a very important observation, as it shows the connection between these properties and gives the designer the tools to decide on the trade-off between power and bandwidth to maximize capacity. For example, in a real communication channel with limited usable bandwidth, it is reasonable to maximize bandwidth utilization by improving SNR and increasing the modulation order.

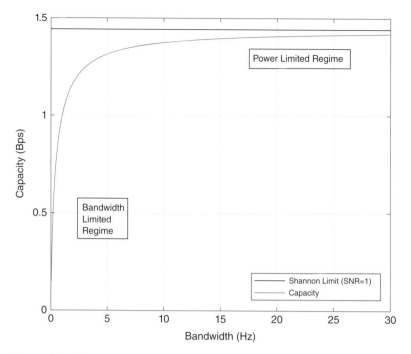

Figure 6.2 Illustration of the channel capacity over bandwidth for given SNR = 1.

To calculate the quality and capacity of a radio link, a link budget is used. The link budget is the sum over all logarithmic gains and losses of a communication system, which results in the carrier-to-noise (C/N) ratio. The E_b/N_0 can be calculated from the C/N by the connection:

$$\frac{E_b}{N_0} = \frac{C}{N} \cdot \frac{B}{R} \tag{6.7}$$

The link budget is a good tool to determine the technical design parameters of the radio frequency (RF) equipment. In its simplest form, the link budget is given by the equation:

$$\frac{C}{N} = P_{TX} + G_{TX} - L_{FSPL} - L_{ATM} + G_{RX} - T_{RX} - k_{dB} - B \tag{6.8}$$

where

P_{TX} The transmitter output power.

G_{TX} The transmitter antenna gain in direction of the receiver, including feed and waveguide losses.

L_{FSPL} The free-space path loss.

L_{ATM} The atmospheric attenuation, including losses due to water vapor.

G_{RX} The receiver gain in direction of the transmitter, including feed and waveguide losses.

T_{RX} The receiver noise temperature.

k_{dB} The logarithmic Boltzmann constant: $10 \log(k) = -228.6$ dB W/ K Hz.

B The signal bandwidth.

In satellite communication, the major component of the link budget is the free-space path loss. This component accounts for the decrease of the received signal power flux density (PFD) with increasing distance r:

$$\mathrm{PFD}(P_{\mathrm{TX}}, r) = \frac{P_{\mathrm{TX}}}{4\pi r^2} \frac{W}{m^2} \tag{6.9}$$

In the link budget, all antenna gain values are relative to an isotropic radiator. Therefore, the free-space path loss is calculated from the PFD by multiplying it with the effective area of an isotropic antenna. With the transmitter power given separately in the link budget, this results in:

$$L_{\mathrm{FSPL}} = \frac{1}{4\pi r^2} \cdot \frac{\lambda^2}{4\pi} = \left(\frac{\lambda}{4\pi r}\right)^2 \tag{6.10}$$

where λ is the radiated carrier wavelength. In a typical LEO configuration, the free-space path loss can vary over a range of 12 dB or more. This dynamic is often used to select higher data rates when the satellite is near to a ground station.

The second source of signal attenuation in the link budget stems from the atmosphere. Frequencies below 3 GHz are affected by the highest atmospheric layer, the ionosphere. The ionosphere is filled with ionized gas and plasma, which affects the radio wave as it travels through it. The total level of ionization is quantified by the total electron content (TEC) or electrons per cubic meter. This property fluctuates with, among other things, sun activity, daytime, and time of year. Small, rapid fluctuations in the TEC result in scintillation or fading effects, perturbations that can affect the received signal phase and amplitude and the refractive index, leading to multipath signal propagation. In weak cases, scintillations may reduce the signal strength at the receiver by multiple decibels. Another issue is the rotation of the plane of polarization of the radio wave, which is usually referred to as Faraday rotation. This effect especially affects VHF communication. The rotation depends on the TEC and carrier frequency, and reaches its maximum during the day.

Tropospheric effects mostly affect frequencies above 3 GHz. This includes attenuation due to atmospheric gases, rain, clouds, and fog, and also tropospheric scintillation. However, the impact of these effects rarely exceeds more than 2 dB for frequencies below 10 GHz. More information on this topic is available in ITU-P.676 [6].

It is apparent that the link budget analysis is rather static and cannot account for other detrimental contributions, such as modulation quality, intermodulation products, electromagnetic interference (EMI), third party interference, adjacent channel emissions, and receiver saturation. Depending on the quality of the available equipment, it is sensible to include wide margins in the link budget, which additionally account for signal fading, performance fluctuations, and atmospheric attenuation. For professional equipment, a margin of 6 dB is reasonable. Users of low-cost and amateur systems should plan with margins of up to 10 dB.

6.3.1 Digital Modulation

Modulation is the process of modifying a carrier wave to carry information. A general electromagnetic wave can be described as:

$$y(t) = A * \sin(\omega * t + \phi) \tag{6.11}$$

where

A Amplitude of the transmitted signal
ω Carrier frequency (angular frequency); can be expressed as $\omega = \frac{1}{2*\pi*f}$
ϕ Phase offset

All of these are candidates for modulation. Three different basic types of modulation are available: amplitude modulation (AM) modifies the amplitude A, frequency modulation (FM) modifies the carrier frequency ω, and phase modulation (PM) modifies the phase ϕ of the signal. All of these basic modulation types are suitable for analog and digital signals, where only the digital modulations play a significant role in satellite communication.

Amplitude-shift keying (ASK) represents the digital information as different amplitude levels of the signal waveform, and thus is the common name for digital AM. Even though ASK has a high bandwidth efficiency, it is not commonly used, as it is sensitive to disturbances such as fading of the signal. One exception is the use of continuous wave (CW) Morse call signs, a special kind of on–off keying (OOK), which are especially important if amateur radio frequencies are used.

Frequency-shift keying (FSK), the digital FM, encodes the digital information as a shift of the carrier frequency. The spectral efficiency of this modulation highly depends on the specific used encoding scheme, especially on the used carrier frequency shift and subsequent signal filter. Minimum-shift keying (MSK) and MSK with an additional Gaussian filter (GMSK) reduce the required bandwidth of the signal at the cost of a reduced noise immunity and a higher (de)modulator complexity. FSK modulations are quite common for small satellite communication, especially if combined with amateur radio frequencies, as the amateur radio community has been using these modulations for a long time. A special and more exotic case of frequency modulation is audio frequency-shift keying (AFSK). For this modulation, an audio signal is frequency modulated similar to a regular FSK modulation. The resulting signal is again frequency-modulated to the desired carrier frequency. This modulation is very simple to implement but has a poor spectral efficiency.

Phase shift keying (PSK) represents the digital information as a change in the phase of the carrier wave. The simplest form of PSK is the binary phase-shift keying (BPSK) where only two different phases are used, and each symbol represents a single bit. To achieve a higher data rate, it is also possible to use four or even eight different phases (Quaternary Phase-Shift

Keying [QPSK]/8-PSK), and thus, transmit two or three bits of information per symbol. An additional special form of QPSK is offset-QPSK (OQPSK), where the phase changes representing the two encoded bits per symbol are shifted to avoid a sudden change of the phase by 180°. Due to this property, the receiver of the signal has a lower probability to lose the lock on the signal. To improve the bits per symbol, it is also possible to combine modulation of the amplitude and phase of the signal. The resulting modulation, called quadrature amplitude modulation (QAM), allows the use of more than 3 bits per symbol. The disadvantage of QAM is its high sensitivity to noise and low signal quality, as well as the increased receiver complexity.

Figure 6.3 shows the different modulation waveforms for ASK (OOK), FSK, and PSK (BPSK) without any additional filtering. Table 6.2 gives an overview over the most common digital modulations used for satellite communication.

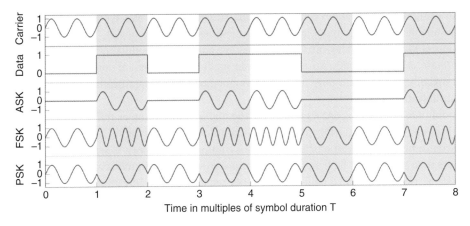

Figure 6.3 Modulation waveforms.

Table 6.2 Overview over digital modulations.

Name	Modified parameters	Bit/symbol	Bandwidth efficiency	Sensitivity to nonlinear distortions
BPSK	ϕ	1	High	Low
QPSK/OQPSK	ϕ	2	High	Medium
8-PSK	ϕ	3	High	High
FSK	ω	1	Low	Medium
MSK	ω	1	Medium	Medium
GMSK	ω	1	High	Medium
QAM	A, ϕ	3+	High	High
AFSK	ω	1	Low	Low
OOK	A	1	High	Low

An important feature of a modulation is the expected bit error rate (BER) at a given E_b/N_0. Modulations with a smaller BER should be preferred. Figure 6.4 compares the most common modulation schemes accordingly.

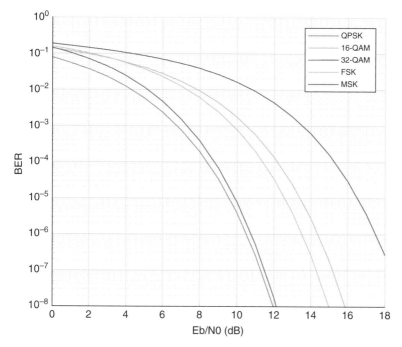

Figure 6.4 Expected bit error rate (BER) for common modulation schemes in an AWGN channel without additional channel coding. (*See color plate section for color representation of this figure*).

6.4 Channel Coding

Shannon provided the theoretical work, which allows us to calculate the maximum channel capacity. However, his work does not cover the problem of achieving channel capacity. All previously discussed modulation types do not come close to achieving channel capacity when used without channel coding. Channel coding is a technique that adds redundancy to the transmitted information. The receiver can use this redundant information to restore erroneous symbols of the message. Channel coding can be seen as a way to add additional energy to the message symbols and to distribute the energy of individual symbols among various symbols. During decoding, this additional energy from correct symbols can be used to increase the SNR of erroneous symbols and extract the correct information. An important property of channel codes is the increase in channel capacity because their use outweighs the added redundancy. A tuple (n, k) generally describes the size of a channel code, where k is the message length and $n-k$ is the length of the additional redundancy. The ratio $r = k/n$ is referred to as code rate. Code rates with $r > 1/2$ are considered high, else low.

There are two classes of channel codes:

- Block codes operate fixed-size blocks of data.
- Convolutional codes work on a potentially infinite stream of data.

Convolutional codes and block codes offer similar performance. The main difference between block and convolutional codes lies in the decoding stage. Block codes require the entire encoded message before decoding can take place. This introduces a delay, which may not be desired in some cases. Furthermore, this raises the need for large buffers and constrains the required hardware.

The Consultative Committee of Space Data Systems (CCSDS) has standardized a number of different codes that cover virtually all channel coding needs of current satellites, probes, and rovers [7, 8]. CCSDS is an organization formed by major space agencies of the world. CCSDS is concerned with developing solutions to common problems in space data systems, which are published in the form of recommended standards. Table 6.3 offers an overview of the typical codes used in satellite communications.

Table 6.3 Overview over typical CCSDS channel codes, their performance and complexity.

Code	Gain (BER = 10^{-6})	Code rate	Complexity
Convolutional	4–6 dB	1/2	Medium
Reed–Solomon	3–4 dB	7/8	Low
SCCC (convolutional and RS)	6.5–7.5 dB	7/16	Medium
LDPC	6–9 dB	7/8–1/2	High
Turbo	9–11 dB	1/2–1/6	High

Figure 6.5 shows the BER of the respective codes over the SNR. It can be seen that low-density parity-check codes (LDPC) and turbo codes offer similar performance in limited power situations, while Reed–Solomon (RS) codes and regular convolutional codes are outperformed.

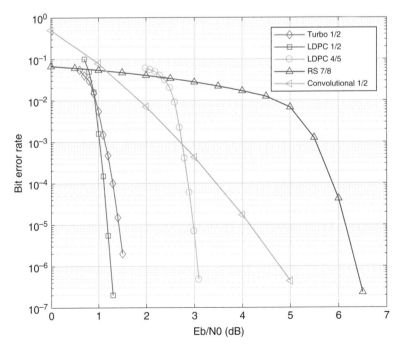

Figure 6.5 Bit error rate for CCSDS turbo with rate 1/2 and block length $k = 1784$, LDPC with block length $k = 4096$, the (255, 223) Reed–Solomon code, and the (7, 1/2) convolutional code. (*See color plate section for color representation of this figure*).

6.4.1 Convolutional Codes

Convolutional codes are a family of codes that operate on a continuous sequence of input data, with virtually no limitation on message length. Peter Elias developed convolutional codes in 1955 in order to find a class of error-correcting code for the AWGN channel [9]. The name stems from the mathematical operation used for encoding, which closely resembles a convolution. The encoding operation performs a binary convolution of k message bits of a sliding window, to form a coded bit. For every message bit, the window advances by one. The value k is called constraint length and strongly influences the correcting capability of the code. A higher constraint length means better time-diversity of the code and better correcting capability, at the expense of increased decoding complexity. Another important factor regarding the correcting capability is the generator polynomial. The generator polynomial determines which window positions are used for the convolution. CCSDS defines a standard convolutional code with parameters ($k = 7$, $r = 1/2$). A hardware schematic of the corresponding encoder is shown in Figure 6.6. Note the feed-forward architecture of the encoder. Pure convolutional codes have code rates of $1/n$, where n is an integer greater than 1. Higher code rates are possible using puncturing, at the cost of performance loss. Convolutional codes are typically decoded using the Viterbi algorithm, a maximum likelihood decoder with good parallelization properties. If the error-correcting capabilities of a convolutional code are exceeded, the Viterbi decoder yields a wrong solution for $k + 1$ bit. Reed–Solomon codes excel at correcting this sort of burst error and are often used to complement the single-bit correcting capabilities of convolutional codes. This principle is called serial concatenated convolutional code (SCCC) [10].

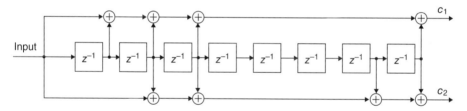

Figure 6.6 CCSDS convolutional encoder.

Turbo codes are high-performance channel codes developed at the beginning of the 1990s by Berrou and Glavieux [11]. Turbo codes can closely approach the channel capacity and find their use in many mobile- and satellite-communication standards. Turbo codes are parallel concatenated convolutional codes (PCCC). A typical turbo encoder consists of two identical, parallel, recursive convolutional encoders. Typically, the constraint length of these codes is short, for example, 4 or 5. An interleaver before the second convolutional encoder introduces additional time diversity of the information. Like convolutional codes, turbo codes can be punctured to achieve higher code rates. Turbo codes exhibit an error floor below a certain BER. This means that the SNR–BER curve begins to flatten out and the code loses power efficiency. An important property of turbo codes is their good performance even with smaller block lengths.

6.4.2 Block Codes

Typical candidates of block codes are RS codes and the more modern LDPC codes. Reed and Solomon invented RS codes in 1960 [12]. RS codes have experienced great popularity in the past, as they have the best possible error-correction ability of their size. An RS code can tolerate $t = \left\lfloor \frac{n-k-1}{2} \right\rfloor$ erroneous symbols. RS codes are nonbinary codes, meaning that their error-correction operates on symbols, which typically consist of 8 bits. Symbols are erroneous if one or more bits of a symbol are incorrect. This means, as a corollary, that RS codes are very tolerant to burst errors but also very vulnerable to random bit errors. The encoding and decoding of RS codes is of very moderate complexity, allowing their use on even the smallest devices. Due to their widespread use, many implementations of encoders and decoders exist in the public domain. Reed–Solomon codes are cyclic, meaning that cyclically left-shifted code words are also valid code words. This renders RS codes vulnerable to transmission errors, where bits are inserted or deleted. A good solution to this issue is to scramble the code blocks with a pseudo-random sequence before transmission. In satellite communications, RS codes have found their use as secondary code in SCCC. Today, RS codes are of diminishing relevance, as they perform worse than the more modern LDPC codes.

Robert Gallager developed the LDPC codes in 1960 [13]. Although Gallager showed the good performance of LDPC in many channel models, they became forgotten because of their high complexity. In 1996, MacKay rediscovered LDPC [14, 15], which subsequently experienced ever-increasing popularity due to their many advantages. LDPC codes can outperform the standard (7, 1/2) convolutional code, despite their higher code rate. LDPC codes are used in the current DVB-S2 standard and in standards by CCSDS. LDPC codes have been shown to approach channel capacity or even achieve channel capacity in some cases, and they generally do not suffer from an error-floor. LDPC codes can be constructed to have almost arbitrary code rates and are efficient choices in both the power- and bandwidth-limited regime. The performance of LDPC strongly depends on their size. The longer the code, the better the performance, despite unchanged code rate. The encoding and decoding complexity of LDPC codes is linear to their block length. Despite this fact, the complexity of LDPC is often very high, since most codes have block lengths of 1024 bits or more. For links with medium to high data rate, encoding and decoding of LDPC is impractical on standard microcontrollers or CPU.

6.5 Data Link Layer

The previous layers have defined the link on a purely physical level. In the Open Systems Interconnection (OSI) model, the data link layer is the first logical layer. A data link layer protocol defines the communication between two or more nodes using the underlying physical connection. The data link layer provides an abstraction of the physical link to upper layer protocols of the network and transport layers. Data link layer protocols provide a definition of the bitstream, using data structures called frames. Frames contain additional information bits, which allow the receiver to retrieve the original payload using synchronization techniques and provide additional metadata to ensure error-free delivery. Further tasks of a data link layer protocol are medium access control (MAC) and logical link control (LLC).

MAC mechanisms provide defined access to a shared medium among multiple communication partners of the same network. This is the case, for example, in half-duplex links, where collisions happen when more than one partner is active. In satellite communication, this aspect is of lower priority, as most links use physical methods, such as frequency multiplexing, to prevent collisions. If a half-duplex link is used, several multiple access schemes exist. The decision depends on the available hardware, resources, and desired complexity. The earliest and simplest type of multiple access protocol was ALOHA, developed at the University of Hawaii at the beginning of the Internet era [16]. ALOHA nodes transmit messages at any point in time, without consideration of the channel status. A node retransmits its message when it detects a collision. This method requires no coordination between the nodes but suffers from severe throughput degradation with increasing number of participants. Slotted ALOHA alleviates this problem to a certain extent. Slotted ALOHA divides the channel into discrete time slots, at the beginning of which each node may begin transmission. This reduces the probability of collision significantly, but requires coordination of the slot timing. Another solution to shared medium access is carrier-sense multiple access (CSMA). CSMA nodes probe the channel before transmission. If another transmission is in progress, the node waits until the medium is idle. Several different additions to CSMA strive to avoid collisions once the medium falls idle. The most notable of these being CSMA with collision avoidance (CSMA/CA). CSMA/CA nodes transmit once the medium is idle and wait for an acknowledgment of receipt. Should the acknowledgment not arrive within a certain timeframe, the node waits a certain amount of time before trying to retransmit. A disadvantage of CSMA is the hidden node problem. It requires all nodes to share the same broadcast domain, meaning that nodes, which are out of range of each other, may cause interference at a third node.

LLC is concerned with the services made available to upper layers. The LLC may allow multiplexing of different network layer protocols or message flows, providing flow control and error-control mechanisms. Depending on the protocol stack, these functionalities may be shared with higher-level protocols or not exist at all. In networks with Internet technology, flow control is typically provided by transport layer protocols. In wireless communication links, such as satellite links, error control often complements channel coding. The simplest form of error control is error detection using parity bits or checksums. Systems with bidirectional communication have the option to use automatic repeat request (ARQ), a technique that triggers retransmissions in case of transmission errors. ARQ requires frame error rates of few percent or less to provide benefit to the system. In practice, it is therefore often used in combination with channel coding. This technique of error control is called hybrid ARQ (HARQ). There are two types of HARQ. The simpler type-I HARQ (re)transmits both the original message and error control information. The more complex type-II HARQ only transmits the error control information if a transmission error has been detected. Type-II HARQ requires communication between channel coding and ARQ mechanisms, which may be undesirable from a design point of view, as it introduces coupling between the components. Generally, ARQ is a good error-control mechanism in channels with inconsistent error conditions, such as spontaneous burst errors. A drawback is that ARQ requires a storage for the message frames, which may be undesirable in links with a high bandwidth-delay product.

The requirements of satellite communication differ from those used in mobile communication or other terrestrial communication. CCSDS has developed a number of data link layer protocols for space systems. Virtually all larger satellites, probes, and ground stations are capable of CCSDS transmission standards and protocols. The protocols by CCSDS are complex and provide more services than are normally needed by nanosatellites and may strain already limited communication resources. An example of a protocol developed specifically for the requirements of nanosatellite links is Nanolink [17]. Nanolink defines the communication on a full duplex point-to-point radio link between satellite and ground control. A core feature of Nanolink is a hybrid error-control scheme based on channel coding and cumulative/selective ARQ. This renders the communication more reliable and efficient by reacting to varying channel conditions more dynamically than the individual techniques. The expenditure for ARQ is reduced using extension headers. This flexible header concept allows a dynamic frame header length and thus reduces the overhead of static headers. Moreover, Nanolink supports virtual channels. These virtual channels can be used for prioritizing traffic, thus, avoiding blockage of the link by low-priority data.

6.6 Hardware

The communication hardware has two major components, the transmitter and a receiver. For the transmitter, a power amplifier (PA) is needed, which is then connected to the antenna. The receiver can share the antenna with the transmitter, or if using a different band, make use of an entirely different antenna. Both receiver and transmitter need a very stable frequency, which is usually derived from a stable clock source. The stability requirement of the clock thereby depends on the RF used. Usually, an oscillator working on a much lower frequency than the RF used produces the clock source. Therefore, the signal needs to be multiplied by a phase-locked loop (PLL) device to the desired carrier frequency, resulting also in a multiplication of the precision of the oscillator. Besides the oscillator, usually a mixer is used for up- and down-conversion of the signal to or from a suitable frequency, which then can be processed either directly by a field programmable gate array (FPGA) or Central processing unit (CPU) after digitizing or in a special decoder- or encoder-circuit.

6.6.1 Antennas

The antenna is a critical part of each communication system. A basic understanding of the functionality of antennas is necessary to choose a suitable candidate for a satellite mission. This includes gain, directional characteristics, polarization, and physical dimensions of an antenna. The radiated power of an ideal isotropic antenna is the reference value for the gain of any given antenna. Typical antennas for nanosatellite missions (ground and space) and their characteristics are depicted in Table 6.4.

The power radiated by an isotropic antenna in the direction of the azimuth and elevation angles θ and φ is given as $P(\theta,\varphi) = P_0/4\pi$, where P_0 is the total radiated power. In relation to the isotropic antenna, the gain of an arbitrary antenna is given as $G(\theta,\varphi) = P(\theta,\varphi)/P_0/4\pi$. In most antenna datasheets, the gain of the main lobe of the resulting pattern is given. The unit for the gain is dBi, which is used to calculate the link budget as given in Eq. (6.8). In some datasheets, the gain of an antenna is given in relation to a half-wave dipole antenna. In this case, the unit dBd is used, where the value in dBd is 2.15 dB smaller than the corresponding value in dBi.

Table 6.4 Typical antennas for nanosatellite missions (ground and space) and their characteristics.

Type	Gain typ.	HPBW typ.	Dimensions typ.	Polarizations
Half-Wave dipole	2.15 dBi	Almost omnidirectional	$\sim \frac{\lambda}{2}$ long wire	Linear
Patch antenna	5–9 dBi	$\sim 30°–60°$	$\sim \frac{\lambda}{2} \times \frac{\lambda}{2}$, flat	Linear, circular
Helix	10–15 dBi	$\sim 15°–60°$	$\sim \frac{\lambda}{4} \times \frac{\lambda}{4} \times \frac{N\lambda}{4}$	Circular
Yagi–Uda	$\sim$5–18 dBi	$\sim 15°– 60°$	$\sim \frac{\lambda}{2} \times \frac{\lambda}{2} \times [0.5 \dots 5] \times \lambda$	Linear, circular
Parabolic	$\sim 10 \log 0.5 \times \left(\frac{\pi \times D}{\lambda}\right)^2$ dBi	$\sim \frac{70*\lambda}{D}$	Dish with diameter D, feed needed in front of dish	Linear, circular
Horn	$\sim$10–30 dBi	$\sim 5°–60°$	Typically, multiples of λ in all directions	Linear, circular

The antenna gain of all antennas usually changes with the angle of incidence. The beamwidth of an antenna is the angular separation of two identical points on opposite directions of the antenna pattern. The most common beamwidth is the half-power beamwidth (HPBW), where the radiated power is -3 dB from the maximum. This angle also often directly relates to the maximum pointing error allowed within the limits of the calculated link budget. For example, if an antenna system with a given pointing accuracy α is used in combination with an antenna with a HPBW of $2 \times \alpha$, the maximum loss due to the pointing error is 3 dB. This should be considered while calculating the link budget. Note that the half-power beamwidth can be different for elevation and azimuth.

The polarization of an antenna describes orientation of the electromagnetic wave, as it moves through space. Antennas are either linear or circular polarized. Linear antennas are simpler to build, but it is necessary to align the sending and receiving antenna accordingly. Circular polarized antennas on the other hand are more complex to build, but do not require a specific rotation. Right-hand circular polarized (RHCP) and left-hand circular polarized (LHCP) antennas are not able to receive each other's signal (this usually results in an attenuation of more than 30 dB). It is possible to receive a circular polarized signal with a linear polarized antenna and vice versa. In this case, a loss of 3 dB must be expected. The advantage of this combination is the use of a relatively simple linear polarized antenna on the spacecraft and no additional constraint regarding the rotation of the spacecraft along the axis of the communication link (whereas the orientation of the antennas toward each other is still necessary).

On the spacecraft, it is important to use a proper matching circuit in order to connect the antenna to a commonly used asymmetric 50 Ω port. This includes matching the impedance and converting from the asymmetric connector to a symmetric antenna feed. Impedance matching can be done with a simple transformer or similar to a π-filter setup. A balun is normally used to convert an asymmetric signal to a symmetric signal and vice versa. In all cases, the specific setup required highly depends on the used antenna (Table 6.4).

6.6.2 Oscillators

As described, each transmitter or receiver system needs a stable clock source. The clock is used either for the CPU or for generating a stable high frequency carrier. The clock is usually generated by a crystal stabilized oscillator. The complexity of this signal generation strongly depends on the carrier frequency used by the transmission or reception. The stability of oscillators is usually defined in parts per million (ppm). For a typical 50 MHz oscillator with a standard value of 10 ppm, this means a frequency value of ±500 Hz. Since, usually the carrier frequency is up-converted by a PLL to get the carrier frequency. This, for example, results in a factor of 44, that is, the carrier is within ±22 000 Hz, when using 2200 MHz. This precision has to be summed up for each oscillator. In addition, oscillators normally have a temperature drift, which also has to be considered in all design decisions.

Multiple types of crystal oscillators, also available in integrated circuit modules, come with different properties of precision, phase noise, and temperature stability. In this work, we will present selected examples of typical circuits that could be found inside oscillator modules, but also could be built as discrete versions. The Pierce type of oscillator (see Figure 6.7) is a classic example, where instead of integrated circuits, like the Schmitt trigger depicted, transistors could also be used if a very compact design is required. The output of oscillators is a square wave and their frequency is around up to 20 MHz for a fundamental crystal. A typical series resonant crystal oscillator is depicted in Figure 6.8. Matthys [18] describes the different types of crystal oscillators in full depth. In addition, different types of oscillators used to get more precision and less temperature drift exist. Temperature compensated oscillators (TCXO) and oven compensated oscillators (OXCO) are very resilient against temperature variations. Table 6.5 summarizes the different types of oscillators and their precision. More information on the precision of oscillators can be found in reference [19].

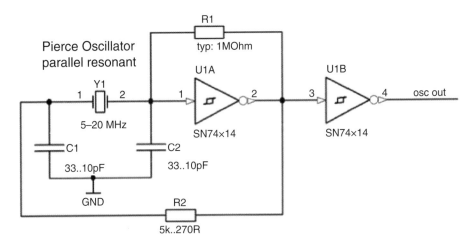

Figure 6.7 Example of a Pierce crystal oscillator (parallel resonant).

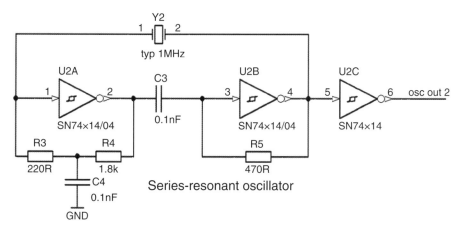

Figure 6.8 Typical series resonant crystal oscillator.

TCXOs use a temperature sensor signal that is fed into a compensation network. This will correct the temperature drift of the crystal and improve the temperature stability of the TCXO in a certain range. The circuit is very generic (see Figure 6.9, left) and usually uses two sensor devices, for example, Negative Temperature Coefficient (NTC) resistors, which change their resistance depending on the temperature. The resistor network adjusts this for a voltage generated at a varactor diode, which in turn changes the capacitance depending on the voltage. This is used to shift the center frequency of the crystal to compensate for its temperature drift. Therefore, the sensors must be close to the crystal to measure the proper temperature. Such TCXOs are available as integrated modules, which can have a very small footprint (see Figure 6.9, right).

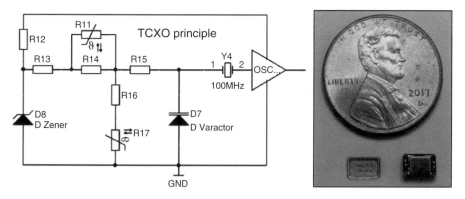

Figure 6.9 General diagram of a temperature compensated oscillator (left). Different TCXOs with small footprint. Source: Right: Photo courtesy of Rolf-Dieter Klein.

OXCOs have the highest temperature stability of these types of circuits. This comes with the drawback that a heater is required, which can consume quite a considerable amount of current to reach a defined temperature. The Abracon AOCJYR series, for example, needs

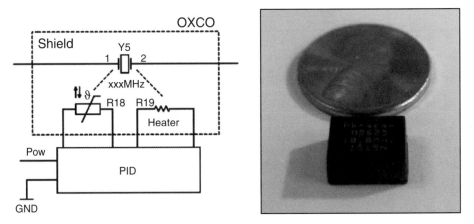

Figure 6.10 Oven-stabilized crystal oscillator (left). OXCO with small footprint as fully integrated Surface-mounted device (SMD). Source: Right: Photo courtesy of Rolf-Dieter Klein.

1 W for warming up and 400 mW for steady state operation (handbook values—need to be determined for space application). Usually a resistor is used as heater. In principle, cooling is also possible using a Peltier element. However, in practice, this is not done and only oven crystal oscillators with resistors are available in a small integrated module case (see Figure 6.10, right). Again, a temperature sensor measures the temperature directly of the crystal. This is fed into a proportional-integral-derivative (PID) controller, which directly uses this signal to heat up a resistor if the temperature is too low (see Figure 6.10, left), and thus, stabilizes the temperature of the crystal. For the heater to work, of course the overall temperature of the printed circuit board (PCB) must be lower than the selected crystal temperature. Otherwise a Peltier system must be chosen which can also cool the crystal. In vacuum, thermal isolation is much simpler than in the atmosphere. IR radiation must be kept away to avoid accidental heating and the integrated modules usually contain proper isolation.

Table 6.5 Overview of the different types of oscillators.

	Clock oscillator (ppm)	TCXO (ppm)	OCXO (ppm)
0–70°C	±10	±0.5	±0.003
−20–70°C	±25	±0.5	±0.003
−40–85°C	±30	±1	±0.02
−55–125°C	±50	N/A	N/A

6.6.3 PLLs and Synthesizers

PLL synthesizers are needed in transmitters or receivers to synthesize a stable frequency from a constant oscillator source. There are different types of synthesizers: integer synthesizers (depicted in Figure 6.11) or fractal synthesizers with different properties in step size and phase noise. Main elements of a PLL synthesizer are a reference clock (F_x), a voltage-controlled oscillator (VCO), which oscillates at the target frequency, and a divider, to divide the high frequency, so it can be compared to the low-frequency reference oscillator.

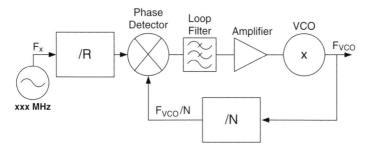

Figure 6.11 Integer synthesizer PLL.

A phase detector determines how much the difference in frequency and phase is and via a so-called loop filter generates an analog signal that is fed into the VCO. An additional divider (R) allows smaller steps than the reference frequency. By using different values for N and R, different output frequencies are possible with the resulting frequency $f = F_x \times N/R$. With $R = 1$ integer, multiples of the reference frequency are possible. Typically, the reference oscillator is around 10–100 MHz. When $R = 1$, these are rather large steps. Therefore, a value higher than 1 is used for R. But the drawback of this is a slow loop filter. For example, for 10 MHz with $R = 10\,000\,000$, a step of 1 Hz is possible. But as the phase detector only has 1 Hz as comparison frequency, the setting to a stable frequency is very slow, as the decision if the VCO is too high or too low can be done only every second.

A major obstacle of the PLL is the proper loop filter. The phase detector sends out short pulses, where positive or negative voltages indicate an increase or decrease for the VCO. This must be filtered to provide a stable analog voltage control of the VCO. Any jitter will result in a phase noise of the synthesized frequency. A simple low-pass circuit of a RC combination, as shown in Figure 6.12 (left), can be used for this purpose, but has very slow response, which might not result in a lock of the PLL.

Also, a single-stage filter can result in oscillations. This can be solved by a stage with lossy low-pass behavior (see Figure 6.12 (right)). If there are up-pulses at the input (IN), both C9 and C10 are charged. But R22 limits the current to C10. Therefore, the voltage increases slower at C10 than at C9. Thus, C9 has a higher voltage after the pulse ends as C10. This voltage can be higher than the wanted tuning voltage for the VCO. C9 now discharges through R22 to reduce this overshoot until the voltage of both are the same. This is the same for following up pulses if the VCO is still too low or in the case of down-pulses. Therefore, R22 and C10 form a second low-pass filter.

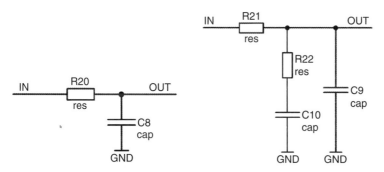

Figure 6.12 Simple RC low-pass as loop filter (left). Loop filter with lossy low-pass behavior (right).

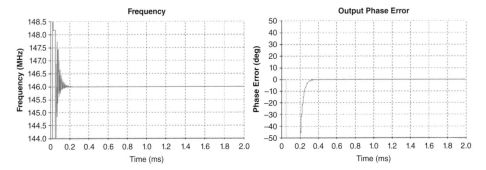

Figure 6.13 Simulation results for the important PLL parameters created by the ADIsimPLL tool.

To calculate the exact design values of the PLL, it is recommended to use manufacturer tools, as for example the ADIsim tool from Analog Devices, as shown in Figure 6.13. Such tools not only calculate the values for different types of selectable loop filters but also generate diagrams that show the response of tuning steps and settling time. Figure 6.13 depicts an example for a 146 MHz signal. After the device is switched on at around 0.4 ms, a stable signal is available.

As previously discussed, the disadvantage of an integer synthesizer is the trade-off between the frequency step resolution and loop settling time. To avoid this, a fractal synthesizer can be utilized (see Figure 6.14). The critical F_{vco}/N, which is used in an integer synthesizer for comparison, is switched between N and $(N+1)$. Therefore, a fraction of N can be used, for example, $N + a$ for $0 \le a < 1$, where a is the fractal component. There are different ways to switch between N and $N+1$, and it is necessary to avoid a regular pattern, as this will result in spurious frequencies.

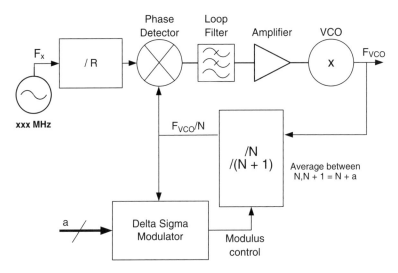

Figure 6.14 Typical fractal synthesizer PLL.

One possible way to avoid regular patterns is to use a delta–sigma modulator to generate a rather random sequence to switch between N and $N + 1$. This helps reducing the spurious signals and keeping phase noise low. With this modulator, R can be lowered, so a high frequency is possible at the phase detector though having very small steps (for example, 1 Hz) to train the synthesizer. Modern PLL chips have both integer and fractal components with digital settings to allow different N and $N + 1$ and use digital values. Furthermore, they can often be switched between integer and fractal modes [20, 21].

6.6.4 Mixers

Mixers are needed for frequency up- and down-conversion. An ideal mixer is a multiplier—it multiplies (mathematically) one signal with a different signal. This is typically a frequency source multiplied with either a low frequency source for up-conversion or a high frequency source for down-conversions. Mixers are a crucial part of all types of transmitters or receivers. With ω_0 being the RF frequency, for example, and ω_1 the local oscillator (LO) frequency:

$$y(t) = A \times \sin(\omega_0 \times t) \times B \times \sin(\omega_1 \times t) \tag{6.12}$$

$$y(t) = A \times B \times \frac{1}{2} \times \cos(\omega_0 \times t - \omega_1 \times t) - A \times B \times \cos(\omega_0 \times t + \omega_1 \times t) \tag{6.13}$$

With the multiplication of the two signals, ω_0 *and* ω_1, additional frequencies appear in the result. This is the sum and the difference (see Figure 6.15, left). A mixer can also be a diode, used in microwave applications. In this case, the transfer function can be described as Taylor sum, derived for example from an exponential function. This results in terms of ×3, ×4, etc., which are multiple of the original signals and must be filtered out.

For up-conversion, usually one of the resulting frequencies must be filtered out using a high- or low-pass filter. The higher the intermediate frequency (IF) is, the easier it is to filter out any unwanted signal. When the IF is the same as the baseband frequency, filtering can be very hard. In this case, the lowest component of the base signal is close to LO. Sometimes both sidebands are wanted, which makes this process simpler. Figure 6.15 (right) shows the

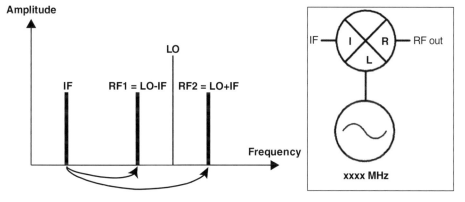

Figure 6.15 Up-conversion with IF and LO (left). Schematics of mixer with oscillator for up-conversion (right).

typical diagram used for a mixer. I is the intermediate frequency and L the LO frequency for mixing. It is derived from a synthesizer or oscillator to tune the desired RF-out frequency.

For down-conversion, the RF signal is mixed with a LO frequency close to the RF signal (see Figure 6.16 (left)). This results in intermediate frequency with the difference but also with the sum. In this case, the difference signal is wanted. In this process, it is relatively easy to filter out the unwanted sum of the frequency by using a low-pass filter to pass the low IF component for further processing.

For down-conversion, the same schematic as in Figure 6.15 (right) can be used, switching only IF and RF (see Figure 6.16 (right)). The LO signal also is derived from a synthesizer or oscillator to mix the signal down to the intermediate frequency or even directly to the base band.

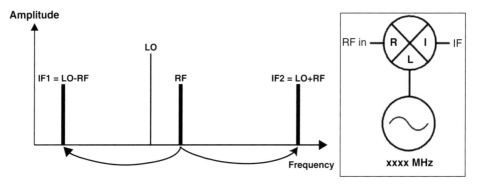

Figure 6.16 Down-conversion with given RF and LO (left). Schematics of mixer used for down-conversion (right).

Finally, also the quadrature mixing of signals is an important aspect to be explained in the following. In quadrature mixing, the RF signal is mixed with two different LO signals for down-conversion (see Figure 6.17 (left)). Both signals thereby have a phase shift of 90°. This allows the generation of so-called In-Phase-&-Quadrature (IQ) signals. The IQ signals carry phase information of the original RF signal, which might be important to decode the signal, and helps to eliminate the sidebands.

For up-conversion of the signals, a quadrature mixer allows feeding the IQ signals, which are generated for example by a D/A converter. By mixing the IQ signals with LO signals of 90° phase shift, the RF out signal carries both IQ signals in one RF signal and thus incorporates phase information in the resulting signal.

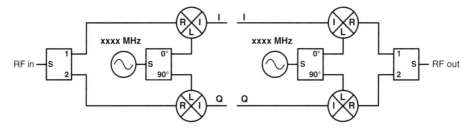

Figure 6.17 Generating IQ signals (left). IQ signals used to generate the HF signal (right).

6.6.5 Receiver

With the mixer and the synthesizer defined, only a few more components are needed to construct a simple RF receiver (see Figure 6.18 (left)). As seen before, an antenna has to be used and the RF signal of the antenna is usually feed onto a band-pass. This reduces unwanted signals, which might saturate the following amplifier stages. A variable amplifier allows different signal levels to be adjusted. It depends on the following stages, if a fixed value can be used. A fixed value is often possible if the dynamic range of the processing system, especially the A/D converter, is large enough. The sample rate of the converter determines the maximum frequency that can be sampled on the IF. Sometimes, also an additional mixer stage is necessary for down-conversion. Finally, antenna inputs can carry high voltages for a short time, which will lead to destruction of the semiconductor devices coming after the antenna. A simple circuit, shown in Figure 6.18 (right), which uses two antiparallel diodes, can protect the input from high voltages. The diodes become conductive for voltages above around 0.3 V, depending on the types of diodes used. To safely protect the input, the diodes must be very fast and be able to prevent high current flow. Thus, special protection circuits using diodes capable of accepting a large current (up to several 100 A) for a very short time should be used for that purpose. Also, all leads to the diodes must be short, so they do not act as inductors and the current can flow through for short pulses. A multistage protection, shown Figure 6.18, can be used to prevent flow into the receiver stages.

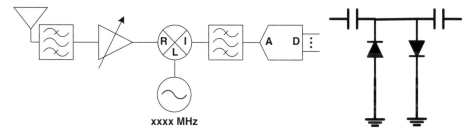

Figure 6.18 Simple receiver with down mixer and A/D converter (left). Simple input protection (right).

6.6.6 Transmitter

A simple transmitter, shown in Figure 6.19, can be designed by a D/A converter followed by a filter. This filter can be a band-pass or a low-pass filter. Depending on the signal used as

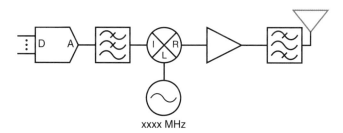

Figure 6.19 Simple transmitter with up mixer and D/A converter as source.

baseband, the filters can be rather complex. In our simple case, a mixer follows to up-convert the intermediate frequency or baseband to the RF. After this, a PA is used to generate the output power for transmission. After the PA, an additional filter stage, used to limit the spectrum to the allowed transmission spectrum, follows.

6.6.7 Transceivers

Receiver and transmitter can be combined into one device, which is then called a transceiver. There are several integrated solutions in the market. As an example, we have chosen the AD9364 from Analog Devices [22] for the nanosatellite mission MOVE-II [23]. The chip is used for the SBAND transceiver for the mission, as it allows for directly handling the SBAND frequency and has a high bandwidth for the chosen modulation. Those integrated solutions often contain all the critical parts of a receiver and a transmitter, featuring a broad frequency range and often including separate fractal synthesizers for the receiver and for the transmitter, so both parts can be used on different frequencies at the same time. Furthermore, the mixers for down- and up-conversion of the signal are nowadays included in those integrated circuits.

6.7 Testing

After a satellite system is built, it is necessary to test all the parameters to fulfill all boundary conditions. It must be guaranteed that all legal aspects for RF transmissions are satisfied and the function of all components must be checked. Power of the transmitter and the sensitivity of the receiver as well as the modulation must be verified in addition to the complete system either using simulated or real ground stations.

6.7.1 Modulation Quality

The quality of the transmitter plays an important role in the design of a communication system: The signal quality at the receiver can never be better than directly at the modulator. Measuring and knowing the quality of the modulated signal is crucial for systems intended to operate with low link margins or higher modulation orders. Modern signal analyzers often provide the possibility to measure imperfections and effects that deteriorate modulation quality, such as IQ mismatch, phase noise, and magnitude errors. However, these properties are more important to RF engineers than to satellite system engineers. A common measure that combines all these effects into a single value is the error vector magnitude (EVM) [24]. Figure 6.20 shows an illustration of the EVM in the IQ constellation diagram. The error vector measures the difference between the ideal constellation point of a modulated symbol and a measured symbol.

Thereby, the EVM is defined as the average ratio of a series of error vectors amplitudes and ideal symbols amplitudes, relative to the peak signal amplitude. The EVM is closely related to the SNR and it can be shown that in linear units:

$$\text{SNR} \approx \frac{1}{\text{EVM}^2} \tag{6.14}$$

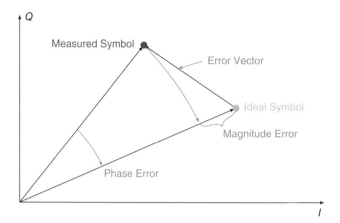

Figure 6.20 Illustration of the error vector magnitude and its components.

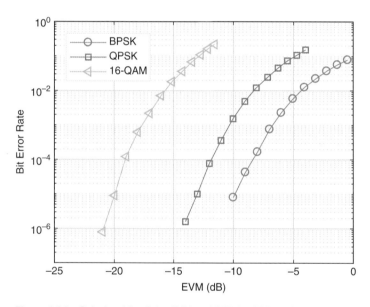

Figure 6.21 Relationship of the EVM and BER for BPSK, QPSK, and 16-QAM.

Thus, the EVM sets the lower limit for the BER. Figure 6.21 shows this relationship. Generally, a professional transmitter system should exhibit a root mean square (RMS) EVM of −30 dB or lower. Higher values indicate the presence of implementation imperfections.

6.7.2 Power Measurement

RF power measurement is an important issue for testing the output power of the transmitter. True RMS measurement can be done by two methods: thermocouples and diodes. On the instrument side, many different options in the market can be used for simple measurements. The results of simple power measurements can be further expanded by using a spectrum analysis. In such case, not only a mean value can be measured but also complex tasks, like using masks for time windows during the transmission can be utilized. For all power measurements, it is important to use a power attenuator, so the devices are not overloaded. This attenuator also serves as a dummy load to absorb most of the power during transmissions. Thus, in principle, the dummy load serves as a resistor. For higher frequencies, the dummy load can be a rather complex device to have a precise flat transmission curve over all measured frequencies.

6.7.3 Spectrum Analysis

For spectrum analysis, a known or unknown signal is usually measured in terms of frequency, power, distortion, harmonics, or bandwidth, often with the magnitude versus the frequency. Early instruments made use of a receiver, similar as the receiver depicted in Figure 6.18 (left), but with the ability to sweep through a range of frequencies, for example, by tuning the LO frequency of the receiver. Then a power-meter determines the power of each spectral component and both are transferred to a display. Another way is to calculate the fast Fourier transform (FFT) of the signal. Most modern digital oscilloscopes allow this, but the problem then is the available bandwidth. Thus, it is often easier to measure a high frequency range with the simple receiver and then down-convert the signal than directly converting the signal to be analyzed. On the other hand, modern high-end systems can do direct digitizing up to 100 GHz, often with a high bandwidth, which also helps in traditional sweep systems and allows for analysis of the transmitted baseband signal. To conclude, a variety of tools are available in modern spectrum analysis systems for rigorously testing the satellite communication hardware.

References

1 International Telecommunication Union, 2015. "SM.1541: Unwanted emissions in the out-of-band domain."

2 Nyquist, H. (1928). Certain topics in telegraph transmission theory. *Transactions of the American Institute of Electrical Engineers* 47 (2): 617–644.

3 Hartley, R.V.L. (1928). Transmission of information. *Bell System Technical Journal*.

4 Shannon, C.E. (1949). *A Mathematical Theory of Communication*. University of Illinois Press.

5 Shannon, C.E. (1949). Communication in the presence of noise. *Proceedings of the Institute of Radio Engineers* 37 (1): 10–21.

6 International Telecommunication Union, 2016. "ITU-P.676–11 Attenuation by atmospheric gases." 2016.

7 Consultative Committee for Space Data Systems (2017). *TM Synchronization and Channel Coding*, 3e. Washington, DC, USA: CCSDS Secretariat, National Aeronautics and Space Administration.

8 Consultative Committee for Space Data Systems (2012). *TM Synchronization and Channel Coding – Summary of Concept and Rationale*, 2e. Washington, DC, USA: CCSDS Secretariat, National Aeronautics and Space Administration.

9 Elias, P. (1954). Error free coding. *IRE Professional Group on Information Theory*: 29–37.

10 Forney, G.D. (1966). *Concatenated Codes*. MIT Press.

11 C. Berrou and A. Glavieux, 1993. "Near Shannon limit error-coding and decoding: Turbocodes," in Proceedings of ICC '93 – IEEE International Conference on Communications.

12 Reed, I. and Solomon, G. (1960). Polynomial codes over certain finite fields. *Journal of the Society for Industrial and Applied Mathematics* 8 (2): 300–304.

13 Gallager, R. (1963). *Low-Density Parity-Check Codes*. Cambridge, MA: MIT Press.

14 MacKay, D. and Neal, R. (1997). Near Shannon limit performance of low density parity check codes. *Electronics Letters* 32 (18): 457–458.

15 MacKay, D. (1999). Error-correcting codes based on very sparse matrices. *IEEE Transactions on Information Theory* 45 (2): 399–431.

16 N. Abramson, 1970. "The ALOHA System: Another Alternative for Computer Communications," Fall Joint Computer Conference, November 1970.

17 N. Appel, S. Rückerl and M. Langer, 2016. Nanolink: A robust and efficient protocol for small satellite radio links, The 4S Symposium, Malta.

18 Matthys, R.J. (1983). *Crystal Oscillator Circuits*. United States: Wiley.

19 https://www.4timing.com/techoscillator.htm

20 Behzad Razavi, Electrical Engineering Department, University of California, Los Angeles, Integer-N and Fractional-N Synthesizer, http://www.seas.ucla.edu/brweb/teaching/215C_W2013/Synthesizers13.pdf

21 Fractional/Integer-N PLL Basic, Technical brief SWRA029, http://www.ti.com/lit/an/swra029/swra029.pdf

22 AD 9364: http://www.analog.com/en/products/ad9364.html Datasheet Analog Devices.

23 M. Langer et al., 2017. "MOVE-II – The Munich Orbital Verification Experiment II," IAA-AAS-CU-17-06-05, Proceedings of the 4th IAA Conference on University Satellite Missions & CubeSat Workshop, Rome, Italy, December 2017.

24 R. A. Shafik, M. S. Rahman, A. R. Islam and N. S. Ashraf, 2006. "On the error vector magnitude as a performance metric and comparative analysis," in International Conference on Emerging Technologies.

7 I-2f

Structural Subsystem

Kenan Y. Şanlıtürk[1], Murat Süer[2], and A. Rüstem Aslan[3]

[1]Mechanical Engineering Department, Istanbul Technical University, Turkey
[2]GUMUSH AeroSpace & Defense, Istanbul, Turkey
[3]Astronautical Engineering Department, Istanbul Technical University, Turkey

7.1 Definition and Tasks

The structure of a satellite is one of the satellite subsystems, and it is designed to provide a volume or a "shelter" where all subsystems are assembled and housed. The primary missions of this "shelter" are to protect all the satellite subsystems against external disturbances and to provide other subsystems a platform for healthy operation in space. Considering the overall mission of a typical satellite, the structural subsystem must be capable of meeting certain requirements as listed in the following text.

1. The structural subsystem must have sufficient internal volume to accommodate the satellite subsystems. If a canisterized deployment system, also called PicoSatellite Orbital Deployer (POD), is to be used, the interface geometry of the structure should be compatible with the deployment system. Otherwise, the structure must have the necessary mechanical interfaces compatible with that of the launch vehicle.
2. It is highly desirable that the structural subsystem is a modular design, allowing safe and easy access to subsystems, easy integration, disassembly, and servicing of the satellite components.
3. The structural subsystem should provide mechanical interfaces for solar panels, antennas, and sensors.
4. Considering the mass properties and thermal management of subsystems, the structural subsystem should allow flexible and configurable internal design.
5. The structural subsystem must withstand and protect the satellite against actual loads during ground transportation, launch, deployment, and mission operations.
6. The material used for the structure must be lightweight and appropriate for ground, launch, and space environments.
7. The structure must be thermally stable to endure orbital temperature variations in orbit configuration.
8. The structure must provide a common electrical grounding point.

Nanosatellites: Space and Ground Technologies, Operations and Economics, First Edition.
Edited by Rogerio Atem de Carvalho, Jaime Estela, and Martin Langer.
© 2020 John Wiley & Sons Ltd. Published 2020 by John Wiley & Sons Ltd.

Once the mission is defined for a satellite and the launch vehicle is selected, then the subsystems are to be identified. At this stage of the design process, the constraints for structural subsystem and the structural requirements are determined. The structure subsystem parts are usually grouped as primary, secondary, and tertiary structures [1]. This classification and the structural functions of individual groups are summarized in Figure 7.1 [2].

The primary structure is the backbone of the satellite, which acts like a structural interface between other satellite subsystems and the launch vehicle or POD, in the sense that the mechanical loads are transferred from launch vehicle to the satellite subsystems through the primary structure. The primary structure must have sufficient strength to hold all the subsystems together safely in ground, launch, and space environments. The primary structure usually consists of beam-like elements, frames, and plates. For CubeSats, the primary structure is usually composed of one or more cubical structures with square cross-sections [1]. The primary structure must be designed to allow easy access to install

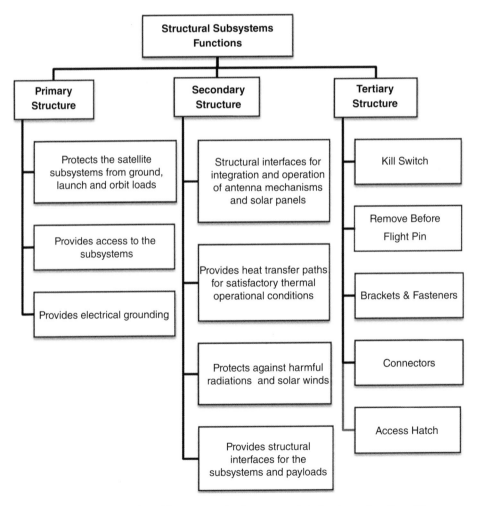

Figure 7.1 Classification of satellite structural subsystem and their primary functions [2].

and service satellite components during the satellite development phases up to the launch. The secondary structure of a typical satellite is made up of side panels, solar panels, booms, antenna, and structural interfaces, which are used to assemble other subsystems with the primary structure. Some coatings used for protecting the satellite against adverse effects of space environment—such as radiations, solar winds, and other thermal effects—are also considered as secondary structures. The tertiary structure, on the other hand, comprises various brackets, fasteners, connectors, pins, and switches. The so-called "kill switch" and the "remove before flight pin" are used to switch off the satellite (to cut off the electric power) during its development, testing, and launch prior to its deployment from the launch vehicle. These switches minimize the risk of accidental damage through particularly electromagnetic interference to other payloads on the launch vehicle.

7.2 Existing State-of-the-Art Structures for CubeSats

Starting in 2003 till 2008, about 30 nanosatellites employing different structures were built and launched into low Earth orbit (LEO) toward gaining space heritage. Since then, some of them have become commercially available for other satellite developers. Pumpkin [3] and Innovative Solutions in Space (ISIS) [4] were the first two educational CubeSat team-based spin-off companies to provide commercial off-the-shelf (COTS) nanosatellite structures. Today, over 1200 nanosatellites are being launched (see www.nanosat.eu); a few more companies such as GUMUSH AeroSpace & Defense Ltd. [5] of Turkey provide COTS structures. Some others offer a COTS platform rather than just the structure, such as Tyvak of USA [6], UTIAS SFL of Canada [7], and GomSpace of Denmark [8]. Many other structures from CubeSat developers, such as XI-IV of University of Tokyo (see http://makesat .com), have not been commercialized yet. Most of these designs comply with the Cube-Sat Design Specification (CDS), the so-called CubeSat standard, developed and updated by Cal Poly of USA [9]. Commercial structure subsystem developers are steadily improving their products to meet increased nanosatellite mission requirements while aiming mass and cost efficiency. Satellite developers can choose one of them considering the volume requirements, ease of use, and space heritage. Some of the most significant criteria about the structure subsystem set by the CubeSat standard are summarized in Table 7.1.

The CDS provides details for 1U, 1.5U, 2U, 3U, 3U+, and multiples of 3U structures (e.g. 6U [10]) that can be used to develop nanosatellites. Custom design nanosatellites not following the CDS may be developed based on regular standards [11, 12]. In most applications, the shape of the structure subsystem for nanosatellites is chosen as a rectangular box with square cross-sections [1]. This shape is in compliance with the POD used to deploy the CubeSat. Moreover, the shape being symmetrical is a desirable feature for nanosatellites so as to increase the stability of the satellite and to minimize the adverse effects of external disturbances caused by air drag and solar activities [13]. Furthermore, this type of geometry makes the subsystem integration easier compared to other alternatives. As the size is increased, hexagonal or octagonal shapes are adopted to minimize the mass and to maximize the volume efficiency, since a POD is not used. Currently, the mass of the structure subsystem including secondary and tertiary elements can be as little as 10–15% of the total mass of a satellite.

Table 7.1 Brief summary about CubeSat structural subsystem standards [2, 9].

Outer dimensions of one unit (1U)	Maximum 100 mm × 100 mm × 113.5 mm
Mass	The mass of each unit should not exceed 1.33 kg
Center of mass	The distance between the geometric center of the cube satellite and the center of mass shall be less than 2 cm in the 10 cm × 10 cm plane. It may be higher than 2 cm in the longer direction (up to 3 cm for 1.5U, up to 4.5 cm for 2U, and up to 7 cm for 3U)
Material compatibility	Materials used in CubeSat should be compatible with launch adaptor (POD) material
Dynamic behavior	Structural stiffness of the structure has to ensure that the natural frequency of the structure is greater than 45 and 90 Hz along horizontal and vertical axes, respectively

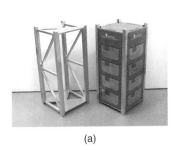   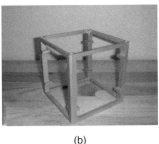   

(a)   (b)   (c)

Figure 7.2 Examples of monoblock structures: (a) CubeSTAR [16], (b) SwissCube [2], (c) ESTCube-1 [15].

The nanosatellite structures can be designed and built as monoblock or multiblock. These two alternative design options naturally result in different assembly and integration procedures. The main philosophy behind the monoblock design is to reduce overall mass of the satellite, increase the stiffness of the structure, and minimize the time required for assembling the structure subsystem [14]. However, the monoblock design results in significantly higher complexity of satellite's assembly procedure. Some examples of monoblock structures are presented in Figure 7.2.

The multiblock design, on the other hand, aims to reduce the design and manufacturing costs, increase accessibility/ergonomics, and simplify the assembly and integration procedure of a satellite. However, in this case, individual blocks need to be assembled using many fasteners, such as screws and rivets, leading to increased mass and longer time for the building process of the structural subsystem. In spite of the fact that the use of many fasteners is usually considered as a drawback, it is worth mentioning here that the fasteners and the associated contact surfaces can be very valuable sources of dry friction damping for controlling resonance vibration amplitudes. Although both designs are being used in practice, multiblock designs appear to be preferred over the monoblock ones. Figure 7.3 shows two examples of multiblock structures, one of which is Nano-JASMINE from "National

(a)   (b)

Figure 7.3 Examples of multiblock structures: (a) Nano-JASMINE [17], (b) ITU 3U [18].

Astronomical Observatory of Japan (NAOJ) and University of Tokyo" and the other is ITU 3U from Istanbul Technical University.

Pumpkin's [3] designs range from 0.5U to 3U and are based on precision sheet-metal fabrication. They are made of 5052-H32 aluminum sheet-metal that is hard-anodized and alodyned in order to comply with the CDSs. ISIS [4] structures range from 0.5U to 8U, and they are also in compliance with the CubeSat standard. Avionics and payload modules are mounted onto the primary load-carrying components. Typical components of an ISIS structure listed in the following text provide examples of various parts in a CubeSat structure:

- Primary structure:
  - 2× side frames, black hard-anodized
  - Ribs, blank alodyned
  - 2× kill switch mechanisms
  - Supplied with inserted phosphor bronze HeliCoils
  - Fasteners
- Secondary structure:
  - 6x aluminum shear panels, blank alodyned
  - M3 threaded rods, M3 hex nuts, M3 bus spacers
  - Boards are supported using M3 washers

Another most recently available CubeSat structure is the n-ART ST of GUMUSH AeroSpace & Defense [5], which is a generic and modular nanocube satellite structure, also based on the CubeSat standard. Each cube cell (1U) can be connected with others by the so-called "link plates" to build up 1U–6U platforms, providing easy assembly and integration operations. Tyvak Inc. [6] provides a platform rather than just a structure, as mentioned before. Figure 7.4 shows some examples of existing prefabricated 1U structures commercially available in the market.

Whether a satellite structure is to be designed in-house or procured from the market, a number of choices, such as primary and secondary structure types and the external shape of the satellite structure, outlined in Figure 7.5, should be made according to the mission and payload requirements. Moreover, system engineer should perform a trade-off analysis by considering many factors including the spacecraft size, satellite stabilization, power

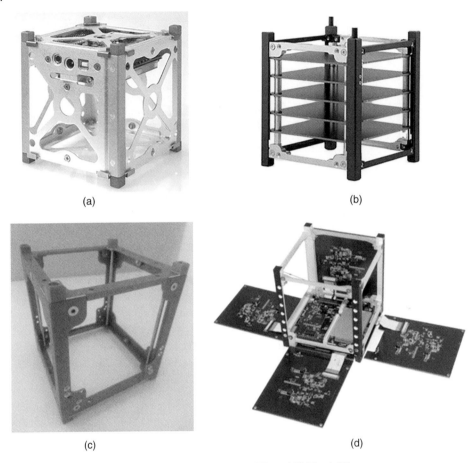

Figure 7.4 (a) Pumpkin [3], (b) ISIS [4], (c) GUMUSH [5], and (d) Tyvak [6] structures.

Figure 7.5 Options for structure selection or design architecture.

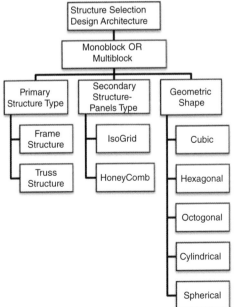

requirement, ease of assembly and integration, and access to the subsystems. In what follows, the options listed in Figure 7.5 are described briefly.

As depicted in Figure 7.5, there are two types of primary structures, namely, frame and truss types. Frame type of primary structure is probably the most suitable for nanosatellites. It may have metal sheets or panels, which can be attached to a metal frame using bolts or rivets. This type of structure is known to resist bending, torsion, and axial loads very well. Another advantage of this type is that it allows using intermediate frames that can be used for mounting subsystems as well as contributing to the overall strength. Truss type of primary structure, on the other hand, has extruded tube polygons or sheet frames machined as polygons. As the name implies, individual elements of this type of primary structure are designed to carry axial load only. Again, individual parts are joined together using bolts or rivets, welding or bonding, whichever is appropriate depending on the material type [19].

The so-called isogrid panel is a plate with triangular integral stiffening ribs [20]. Similar to triangular trusses, the isogrid pattern of triangular shape is known to be very efficient in terms of carrying loads with minimum amount of structural mass. The term "isogrid" is used since the structure acts like an isotropic material. The honeycomb panels, on the other hand, allow carrying axial loads, bending moments and in-plane shears while the core carries the normal flexural shears [20, 21].

As mentioned before, the shape of the structure subsystem for nanosatellites is mostly preferred to be a symmetrical rectangular box so as to increase the stability of the satellite and to suppress the disturbance torque caused by air drag and solar activities. Furthermore, such a configuration also brings ease of subsystem integration. Table 7.2 lists the advantages and disadvantages of various types of geometrical shapes for nanosatellites, justifying why rectangular shape is the most widely used in practice.

Table 7.2 Summary of various shapes for satellite structures.

Shape	Advantages	Disadvantages
Rectangular	Simple design Simple manufacturing process Simple solar panel attachment Easy installations of subsystems Less joints and link elements Fully compatible with the POD	Less surface area toward the Sun
Hexagonal	Increased surface area per unit volume exposed to the Sun	Complex design, more joints, and link elements. May be convenient for larger nanosatellites
Octagonal	Greatly increased surface area per unit volume exposed to the Sun	Many additional joints and link elements, complex design, difficult to use the volume efficiently. May be convenient for larger nanosatellites
Cylindrical	Maximized volume per unit surface area and solar panel mounting area	Complex design and manufacturing process. Difficult to attach solar panels or cells
Spherical	Maximized volume per unit surface area and solar panel mounting area Increased surface area per unit volume exposed to the Sun	Most complex design, most complex manufacturing process, high cost, difficult to use the volume efficiently, difficult to attach solar panels

7.3 Materials and Thermal Considerations for Structural Design

The properties of various materials used in nanosatellite applications must be compatible with ground, launch, and space environments spanning the life cycle of the satellite. The ground environment is where the satellites are fabricated, assembled, integrated, tested, and transported to the launch area. The materials used for structural subsystem and other subsystems must be able to withstand the operational loads during ground transportation. The so-called launch environment is the harshest environment where the structure subsystem (and other subsystems) usually experiences the highest structural loads due to acoustic and transient excitations. The third environment is the vacuum space environment where the satellites spend the rest of their lives and experience thermal as well as operational loads. The space environment has a variety of extreme conditions that can cause failure of structures and contamination of sensitive subsystems. Vacuum space surrounding the Earth ranges from a pressure of 1.3×10^{-7} kPa at 200 km to less than 1.3×10^{-12} kPa beyond 6500 km, and temperature ranging from -160 to $+180°$C [22]. There are also emerging new materials such as Haynes 230 withstanding very high temperatures (over 1000°C, see www .hightempmetals.com). In summary, it is of paramount importance to select the appropriate materials for the structural and other subsystems.

The materials for individual components of the structural subsystem should be chosen by considering availability, cost, manufacturability, effect on mass budget, strength/fatigue/ fracture properties, thermal properties for heating and cooling, resistance against vacuum conditions, outgassing in vacuum, corrosion resistance, and assembling/attachment issues [23, 24]. Table 7.3 lists the operational factors in three environments and the corresponding material properties to be taken into account during the material selection phase of the structural design.

Considering the operational requirements, typical expectations from materials used for satellite applications are summarized in Table 7.4. The suggested values for the expected properties can be found in the related standards, such as Goddard Space Flight Center (GSFC) [11] or European Cooperation for Space Standardization (ECSS) [12].

Materials for structural subsystem are evaluated according to the requirements given in Table 7.4, and appropriate materials satisfying the requirements are selected. Metal alloys

Table 7.3 Operational environments, operational factors, and important material properties for material selection.

Environment	Operational factors	Engineering properties
• Ground/ transportation • Launch • Space	• Temperature limits • Mechanical and thermal loads • Contamination • Life expectancy • Exposure to moisture or other fluid media	• Strength • Corrosion • Conductivity, thermal expansion • Thermal and mechanical fatigue • Moisture resistance • Wearing, fretting • Vacuum outgassing • Toxic off-gassing

Table 7.4 Requirements from satellite materials.

Considerations	Expected, desirable properties
Structural	High strength-to-density ratio
Thermal	High thermal conductivity
	Low thermal expansion
	Appropriate specific heat capacity for the purpose
Electrical	Good electrical conductivity for grounding
Dispenser interaction	Material compatibility for anodizing, which prevents cold welding as well as corrosion and moisture effects
Toxicities	Lowest toxic gas emission
Vacuum	Lowest outgassing

Table 7.5 Advantages and disadvantages of typical satellite materials [1].

Alloys	Desirable properties	Drawbacks	Usage areas
Aluminum	• High strength-to-mass ratio • Easy manufacturing • Cost effective • High ductility • High thermal conductivity • Relatively low density	• Low hardness • High coefficient of thermal expansion • Low strength-to-volume ratio	Primary and secondary structures, heat sink. General usage
Steel	• High stiffness • Low volume strength • High resistance against degradation • Easy manufacturing	• Low thermal conductivity • High density	Fasteners, joints, and links. Elements such as rods. Coating against corrosion
Titanium	• High strength-to-mass ratio • Low thermal expansion • High corrosion resistance	• Difficult to machine • Highest cost • Low thermal conductivity	Primary and secondary structures
Beryllium	• High thermal conductivity • High radiation • Low thermal expansion	• Toxic off-gassing • Difficult to machine • Low ductile material	Antennas

and advanced composite materials are known to be the most widely used ones for the structural subsystems [4]. Table 7.5 lists the most commonly used metal alloys, their desirable properties, drawbacks, and the usage areas in satellite applications. As can be seen from this table, metal alloys have highly desirable properties from the viewpoint of structural subsystem requirements. Not surprisingly, aluminum alloys are the most widely used materials for structural subsystems (Table 7.6). The properties of materials used commonly for satellite structures are given in the related literature (reference [25]). For the sake of completeness, material properties of the most widely used metal alloys are tabulated in Table 7.7.

Table 7.6 Examples of selected materials for existing prefabricated nanosatellite structures.

Suppliers	Materials
Pumpkin [3]	Structure: AL-5052-H32; Joints and attachments: stainless steel Rails: hard-anodized Other parts: yellow-alodyned to prevent against corrosion effect
ISIS [4]	Structure: AL-6061-T6; Joints and attachments: stainless steel Primary structure: fully black anodized
GUMUSH [5]	Structure: AL-6061-T6; Joints and attachments: stainless steel, fully hard-anodized

Table 7.7 Material properties of the most widely used metal alloys.

	Density (kg m^{-3})	Young's modulus (GN m^{-2})	Yield stress (MN m^{-2})	Thermal expansion μm (m °C)$^{-1}$	Thermal conductivity W (m °C)$^{-1}$	Specific heat J (kg °C)$^{-1}$
Aluminum 6061-T6	2700	68.9	310	23.6	167	896
Aluminum 7075-T6	2810	71.7	572	23.6	130	960
Stainless steel (304L)	7900	200	241	16	15	500
Aluminum 5052-H32	2680	70	193	23.8	138	880
Titanium alloy (Ti-6Al-4V)	4430	113.8	827	8.6	6.7	526
Beryllium copper	8250	125–130	1030–1340	16.7	105–130	420

7.4 Design Parameters and Tools

The structure subsystem design is an integral part of the satellite design process, which starts at a conceptual level and is based on design specifications derived from mission requirements. These specifications usually include accommodation of payload and the other subsystems, deployment and launch requirements, environmental protection, thermal and electrical paths, strength, and cost. Trade-off studies are carried out to choose the best option from various types of materials and structure configurations that are considered as candidates. In the preliminary design phase, the type of structural design, its materials and geometric shape, outer and inner dimensions, method of joining/fastening the structural members/modulus, structure mass, and the power budget of the satellite are all determined. Moreover, preliminary static and dynamic analyses are usually carried out to verify the appropriateness of the selections made at this stage, such as the types of materials, bolts and fasteners, and their numbers and locations. The final stage of the design process is the detailed design phase, which is an iterative development and optimization process to arrive at the final design. "Design verification and validation" standards tailored for nanosatellites such as reference [26] can be followed. Figure 7.6 presents a flowchart indicating the main steps to follow in order to reach the final product.

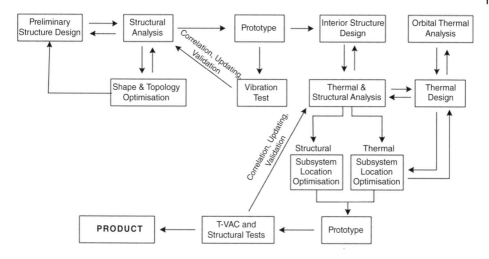

Figure 7.6 Steps from preliminary design of satellite structure to the final product.

In the following two subsections, first, the structural analyses and tools needed for the design are described, and then the thermal design considerations are briefly outlined.

7.4.1 Structural Design Parameters

Having completed the preliminary design, detailed structural design is carried out using available tools in order to optimize stiffness and mass properties as well as the structural interfaces between the structure and other subsystems. Finite element (FE) method is the most widely used analysis tool for this purpose [1], and there are powerful commercial packages such as NASTRAN, ABAQUS, and ANSYS to perform the necessary computations. The information gathered at the end of the preliminary design phase for the structural and other subsystems is enough to establish the necessary three-dimensional FE models which are used to predict static as well as dynamic behavior of the system including the vibration amplitudes under worst load cases. Analyses under worst load cases are used to guarantee that the satellite will function as planned during its lifespan.

Static analyses are usually carried out first in order to check the static strength of the structural subsystem. After a representative FE model is established, a static equilibrium equation given in the following text is solved under appropriate boundary conditions to simulate the effects of various static and quasi-static loads:

$$[K]\{x\} = \{F\} \tag{7.1}$$

where $[K]$ and $\{F\}$ are the stiffness matrix and the load vector, respectively, and $\{x\}$ stands for nodal displacements. The mechanical loads that can be simulated via static analyses are the following quasi-static loads:

(i) Loads due to gravitational and/or steady accelerations during ground transportation.
(ii) Loads due to steady acceleration of the launch vehicle during launch, expressed as longitudinal and axial G-loads.
(iii) Operational loads due to mechanical or thermoelastic loads.

The first two of the static loads in the preceding text are identified from available standards and from the mechanical load information of the launch vehicle [27]. The third type of static load is specific to individual satellites. The results of the static analyses are examined from strength point of view and, if required, necessary changes and modifications are made accordingly.

Another type of analysis—normal mode analysis—is performed in order to satisfy the requirements set by the launch vehicle or applicable standards. The equation of motion for the undamped case is:

$$[M]\{\ddot{x}\} + [K]\{x\} = \{0\} \tag{7.2}$$

where $[M]$ and $\{\ddot{x}\}$ stand for mass matrix and vector of nodal accelerations, respectively. Assuming harmonic motion leads to generalized eigenvalue problem as:

$$[K]\{X\} = \omega^2[M]\{X\} \tag{7.3}$$

The solution of the above equation yields the natural frequencies ω_r and the corresponding mode shapes $\{\emptyset\}_r$ for individual modes r, and these results are used to judge whether the satellite has the required stiffness and/or natural frequencies. Such analyses are needed for nanosatellites to avoid any significant dynamic interaction between the launch vehicle and the satellite during launch. For example, as stated before, it is expected that the lowest natural frequencies of a satellite structure corresponding to mode shapes along horizontal and vertical axes must be greater than 45 and 90 Hz, respectively. Strictly speaking, these values must be taken from the launch vehicle provider for the launch vehicle to be used. Another very important reason for computing the natural frequencies and mode shapes of the satellite system is that modal data obtained with such analyses are used for harmonic and random response calculations as part of the detailed design. Moreover, assessing the stiffness requirement via natural frequencies of a satellite has more implications than the stiffness of the system alone. In Eq. (7.3), inserting $\{\emptyset\}_r$ for $\{X\}$, and then multiplying both sides of this equation by $\{\emptyset\}_r^T$ and finally rearranging the resulting expression yields:

$$\omega_r^2 = \frac{\{\emptyset\}_r^T[K]\{\emptyset\}_r}{\{\emptyset\}_r^T[M]\{\emptyset\}_r} \tag{7.4}$$

The right-hand side of this expression is the ratio of modal stiffness to modal mass for mode r. One possible interpretation of this equation, which has direct impact for the structural subsystem optimization, is that higher natural frequencies for a system can be achieved by optimizing the stiffness and mass distributions within the structure, leading to higher ratios of modal stiffnesses to modal masses. Maximizing the natural frequencies of a satellite can achieve two objectives at the same time, namely, maximizing the stiffness and minimizing the mass; hence, it is a very valuable parameter during detailed design phase. This is very significant for satellite applications as the mass minimization of the system is of great importance.

Before making computationally intensive calculations during preliminary or detailed design phases, it is very practical and convenient to develop relatively simpler FE models for the structural subsystems using beams and shell elements. Such models are used for static as well as dynamic analyses and can provide valuable information for testing some ideas and options affecting the detailed design, including material trade-off, mass distributions, thickness variations, and effects of various fastening options. FE models for structural subsystem using beam and shell elements are also very useful for maximizing the natural frequencies with least amount of effort and CPU time via maximizing the ratio of modal stiffness to the modal mass. This is usually done effectively by computing the modal strains and then adding more materials around locations where there are higher concentrations of modal strains and removing the material from regions where the modal strains are low. This process can yield higher natural frequencies, higher strength, and optimized mass for the structural subsystem.

Although the beam/shell models for the structural subsystem provide valuable information toward design optimization, finalizing the design requires more detailed models and analyses. With the help of existing FE tools, designers nowadays are able to create much more realistic FE models using 3D solid elements for primary load-carrying components, which allow including very fine geometric and assembly details. A typical FE model of a 3U structure built using 3D solid elements and the predicted mode shapes for the first two modes is given in Figure 7.7.

The next step in detailed design phase is to create a detailed model for the whole satellite system including all subsystems and perform the necessary analyses to make sure that the design meets the mission requirements. Appropriate FEs—solid as well as beams, shells, and others—are used to create detailed models for individual subsystems, and substructuring technique can also be employed during this process in order to create an assembly model with reduced degrees of freedom [1]. The assembly model is then used to compute the natural frequencies and mode shapes. Typical results corresponding to the first two modes of an assembly are presented in Figure 7.8.

(a)

(b)

Figure 7.7 First two vibration modes of a 3U structure alone. (*See color plate section for color representation of this figure*).

(a)

(b)

Figure 7.8 First two vibration modes of a 6 kg 3U satellite assembly. (*See color plate section for color representation of this figure*).

Having optimized the mass and stiffness of the structural subsystem and the assembly via modal analyses, it is required to verify that static and the dynamic stresses due to excitations during ground transportation and launch are within allowable limits. This is done by carrying out static, harmonic, and random vibration analyses. Dynamic analyses are performed by solving the equations of motion, given by Eq. (7.5), in terms of relative displacement vector, $\{z\} = \{x\} - \{u\}$, where $\{u\}$ represents the imposed base motion [28]:

$$[M]\{\ddot{z}\} + [C]\{\dot{z}\} + [K]\{z\} = \{F_{eff}\} \tag{7.5}$$

where the additional symbols, $[C]$ and $\{F_{eff}\}$, represent the damping matrix and the effective force vector due to base motion, respectively. The ultimate aim of performing these analyses is to predict the maximum vibration levels and corresponding stresses so as to guarantee that the structural subsystem will not fail during transportation and launch or in space. Maximum vibration levels are expected to occur at resonance situations, that is, when the excitation frequencies get closer to or coincide with the natural frequencies of the system. Whether the excitation is harmonic or random, reliable predictions of the resonance levels require good knowledge about the damping characteristics of the system. However, reliable modeling of the damping characteristics of a typical nanosatellite with existing theoretical (FE) tools is an extremely difficult task. Although it is possible to make some simplifying assumptions, for example, damping level being constant or proportional to the mass and stiffness matrices, making such assumptions is very unlikely to be sufficient for accurate determination of the vibration levels due to dynamic loads. Therefore, the use of modal damping approach is recommended, which allows setting modal damping values based on previous satellite experiences or test results if a satellite prototype is already available. Once the modal properties, including damping information, of a satellite are available, FE packages allow users to compute the response of the system due to

Table 7.8 Example of harmonic vibration load specification
($g = 9.81\,\mathrm{m\,s^{-2}}$) [29]. Source: Obtained from ALCATEL SPACE/CNES.

Start frequency	Amplitude	End frequency	Amplitude
5 Hz	11 mm	21 Hz	11 mm
21 Hz	20 g	30 Hz	20 g
30 Hz	20 g	50 Hz	5 g
50 Hz	5 g	100 Hz	5 g

Table 7.9 Example of random vibration load specification
($g = 9.81\,\mathrm{m\,s^{-2}}$) [29]. Source: Obtained from ALCATEL SPACE/CNES.

Frequency (Hz)	Input level	Global
20–100	+6 dB per octave	
100–1000	$0.05\,\mathrm{g^2\,Hz^{-1}}$	9g root mean square
1000–2000	−3 dB per octave	(RMS)

harmonic and random base excitations. These analyses are aimed to simulate the vibration tests that are to be applied to the satellite prototype during qualification tests. The harmonic response simulations are carried out to assess whether the system is capable of carrying the steady-state harmonic loads applied to the system along three orthogonal directions during ground transportation and launch. A sample description of harmonic vibration load is given in Table 7.8. Similarly, random vibration analyses need to be carried out to simulate the random loads applied to the satellite during ground transportation and launch. Random loads due to ground transportation are available in relevant standards. However, random vibration loads during launch vary depending on the launch vehicle, hence, the random loads specific to a particular launch vehicle should be used in the simulations. A sample random load specification, power spectral density (PSD), is given in Table 7.9. As for the harmonic analysis case, random vibration loads are applied sequentially along three orthogonal directions, and results are assessed in terms of peak stresses and fatigue life. Some sample static and random analyses results are presented in Figure 7.9 for illustration purposes. Interested readers are referred to references [1, 21] for more information about FE analyses of satellites.

7.4.2 Thermal Design Considerations

Another design requirement for the structure subsystem arises from thermal management of the satellite. The structure should provide a controllable volume or a shelter for thermal protection of the satellite subsystems. The objectives of thermal control of a satellite are twofold: (i) to guarantee that the thermoelastic behavior of the satellite subsystems is acceptable, and (ii) to achieve an acceptable temperature distribution for various subsystems for their successful operations. Thermoelastic analysis is carried out to investigate the

(a)

(b)

Figure 7.9 Examples of displacement and stress distributions of a satellite assembly when loaded along vertical direction. (*See color plate section for color representation of this figure*).

Table 7.10 Operational temperature limits for nanosatellite subsystems.

Subsystem	Temperature range
Structure	−40 to +85°C
Altitude determination and control system (ADCS)	−40 to +80°C
Batteries	−10 to +40°C
Solar panels	−100 to +100°C
Electrical power system (EPS)	−40 to +85°C
Comment and data handling system (CDHS)	−40 to +80°C
Antenna	−40 to +85°C
Communication	−10 to +45°C

amount of possible expansion and contraction due to temperature fluctuations of the structural subsystem. The temperature distribution within this volume needs to be controlled in order to keep the temperatures of the subsystems within operational limits [27], which are tabulated for various subsystems in Table 7.10. It should be noted that performance of certain subsystems such as batteries may degrade at about the limit values.

There are two methods for thermal management of satellites: active or passive. Due to limited power, cost, and size of nanosatellites, passive thermal design is more frequently used. Some examples of thermal management system are tabulated in Table 7.11.

To satisfy the operating temperature limits given in Table 7.10, the first requirement is to quantify the amount of heat flux from various sources, as depicted in Figure 7.10. As can be seen, there are three external heat sources: (i) direct solar flux, (ii) Earth infrared radiation, and (iii) albedo (solar reflection). These three heat sources significantly depend on the orbital parameters and the orientation of the satellite. For example, satellite may be shadowed by Earth for some time during its orbital motion, and this can be the major cause of transient temperature distributions and cyclic thermoelastic loads. Detailed thermal analyses must also consider the internal heat generation of the electrical components of a satellite, which is considered as internal heat source. The thermal design process also requires identification of the heat sinks, that is, where the thermal energy is emitted to,

Table 7.11 Examples of components of thermal management with some of its features and functions.

Thermal management system	Type	Function
Smart thermal coating	Active	Emissivity and absorptivity are electrically controlled
Heat exchanger	Active	Heat transfer based on phase changing
Thermal fins	Active	Deployable fins act as radiator
Heater	Active	It converts electrical energy to thermal energy
Surface finishing	Passive	Changing material reflectivity
Painting	Passive	Changing emissivity and absorptivity
Multilayer isolation	Passive	It provides stable temperatures for subsystems

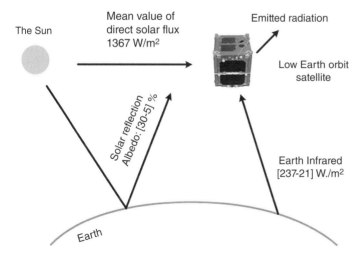

Figure 7.10 Thermal sources and sinks for a satellite [27].

from a satellite. Thermal radiation is the only way of heat rejection in space, hence, all the externally and internally generated thermal energy is eventually radiated to deep cold black space at 3 K. Earth is also considered as a black body emitting at 255 K [30].

As described in detail in reference [27], thermal design is also an iterative and rather complicated process as outlined in Figure 7.11. The complication is due to the fact that the multimode (conduction–radiation) heat transfer problem has time-varying boundary conditions. Therefore, some simplifying assumptions are usually made to be able to solve the problem.

Given the heat sources and the sink, satellite temperatures are controlled by the mass-weighted material-specific heat capacity of the system and the absorption and radiation properties of the satellite surfaces, hence, significant amount of effort is spent for optimizing these properties. At the beginning phase of the satellite design, orbital thermal analysis is carried out in order to select materials, coating types, and to decide the interior design. Orbital thermal analysis is actually based on conservation of energy, and it is used to model multimode heat transfer phenomena, thermal environment of Earth satellite, and orbital dynamics of the satellite, which provides varying thermal boundary conditions depending on orbit and time. The conservation of energy equation for quasi-steady-state is expressed as:

$$Q_{\text{Solar}} + Q_{\text{Albedo}} + Q_{\text{Earth}} + Q_{\text{Internal}} = Q_{\text{Radiated}} \tag{7.6}$$

where Q represents thermal power (Watt) and the subscripts indicate the origin of radiation sources. Thermal radiation terms in Eq. (7.6) can be written more explicitly as:

$$J_S \cdot S_S \cdot \alpha + J_A \cdot S_A \cdot \alpha + J_E \cdot S_E \cdot \alpha + Q_{\text{Internal}} = S_{\text{Total}} \cdot \sigma \cdot \varepsilon \cdot T^4 \tag{7.7}$$

where J is radiation flux (Watt m^{-2}), S is surface area, σ is Stefan–Boltzmann constant (5.67×10^{-8} W $\cdot m^{-2} \cdot K^{-4}$), α is absorptivity coefficient, ε is emissivity coefficient, T is temperature, J_S is solar flux, J_A is albedo flux, and J_E is Earth flux. Following the determination of orbit- and time-dependent thermal boundary conditions, node-based element model is established and its conductive (GL) and radiative (GR) connections are calculated according to the following equations [31]:

$$Q = \frac{kA\Delta T}{L} \tag{7.8a}$$

$$GL = \frac{Q}{\Delta T} = \frac{kA}{L} \tag{7.8b}$$

$$GR = \varepsilon_i \varepsilon_j A_j F_{ji} \tag{7.8c}$$

where A is cross-sectional area, k is thermal conductivity, L is length, ΔT is temperature difference, F is view factor. mC values (mass-weighted specific heat capacity of the individual subsystems) are also needed for the determination of transient subsystems' temperatures ($\Delta Q = mC\Delta T$). A finite difference method is then used to establish a set of equations that can be expressed as [30]:

$$Q_{\text{Radiated}_i} + \sum_{j=1, i=1}^{n} GL_{i,j}(T_i - T_j) + \sigma \sum_{j=1, i=1}^{n} GR_{i,j}(T_i^4 - T_j^4) = (mC)_i \frac{T_i(t + \Delta t) - T_i(t)}{\Delta t}$$

$$\tag{7.9}$$

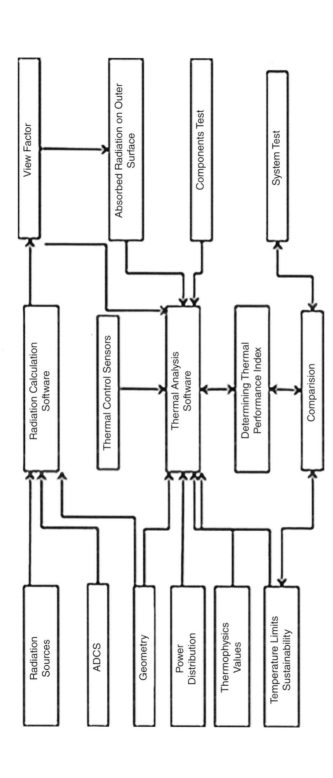

Figure 7.11 Thermal design parameters and tools [27].

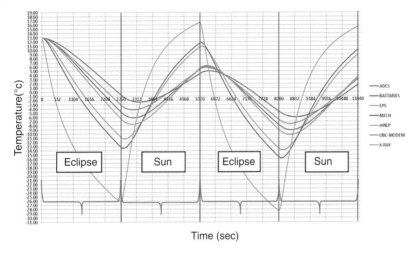

Time (sec)

Figure 7.12 Temperatures of some subsystems as a function orbital position for the BeEagleSat [18]. (*See color plate section for color representation of this figure*).

The lumped parameter in Eq. (7.9) is written under its general form for a transient multiple nodes model. Q_{Radiated_i} is the radiated power for the node i, $GL_{i,j}$ and $GR_{i,j}$ are respectively the conductive and radiative links between nodes i and j. T_i is the temperature of the node i and C_i is its heat capacity. The solutions of the equations resulting from such formulations yield the temperatures of subsystems as a function of time or orbit position. Sample results are presented in Figure 7.12. Based on either FE or finite difference approach, there are various software packages, such as ESATAN-TMS, Thermica-Sinda/G, and COMSOL, available for thermal computations.

7.5 Design Challenges

The structural design of a typical satellite requires many considerations including stiffness and mass optimization, system integration, manufacturability, and cost. Each of the topics mentioned brings its own design challenges.

There are various difficulties and challenges engineers face during the design phase, the most significant ones being related to uncertainties in the theoretical models developed for analyses and optimization. The state-of-the-art tools currently available allow the analysts to perform various simulations in order to predict the static and dynamic stresses caused by the loads experienced by satellite structure and other subsystems. Obviously, one of the essential requirements to get reliable predictions is that the model developed for a satellite should represent the structure—that will eventually be manufactured—as closely as possible. The other requirement is that the loads subjected to a satellite should be known with acceptable accuracy so as to predict the resulting response levels. These two requirements bring their own challenges for structural and mass optimization. As far as developing a representative model is concerned, the main challenge, in fact applicable to most complex structures, is to be able to model so many screws, rivets, and contact surfaces associated with them with acceptable accuracy. Although some very simplifying assumptions can be made to model these contacts, realistic modeling of the contact interfaces between individual elements of the structural system is very challenging indeed.

As far as the contact interface modeling is concerned, analysts are faced with very challenging tasks especially when the dynamic response predictions are to be made. This is not only due to the fact that the flexibility of the contact interfaces can exhibit a very nonlinear behavior, a large proportion of the damping may also be originating from those contact interfaces, especially if there is no special damping treatment applied to the satellite. This makes accurate modeling of the damping properties of the structure very difficult. Damping modeling can become even more complicated when the satellite assembly including all the subsystems is considered. This in turn makes reliable predictions of resonance response levels of nanostructures an extremely difficult task. It is believed that this challenge will remain for quite some time in the future. This is one of the reasons why extensive and expensive tests are performed on satellite structures before they are deployed. Once a satellite prototype is available, the modal damping values can be estimated using experimental data and this can improve the accuracy of the predictions of the theoretical models very significantly [32].

Another challenge in the design of nanosatellites is to control the transmission of the vibrations from the structural subsystem to the other subsystems. Although the vibration isolation is a well-known concept [33] for minimizing the transmission of vibration, there appears to be hardly any application of this concept for nanosatellites, probably due to the very limited space available.

The structural and mass optimizations also introduce their own challenges in the sense that optimizing the strength and optimizing mass appear to conflict with each other. This is usually resolved via a compromise solution, that is, by setting a maximum allowable mass and then optimizing the mass distribution so as to yield maximum strength or rigidity.

As far as the system integration is concerned, probably the most significant challenge is about optimizing the cabling needed for electric/electronic interfacing of the subsystems. The main difficulty here is that final configurations of some cables for some subsystems may have to be set or changed when most of the subsystems are in fact finalized. It is a challenge to minimize, ideally avoid, any need for modifications once the prototype is ready. In fact, the ultimate challenge is to get the product design right so as to achieve the objectives with the first prototype.

7.6 Future Prospects

Increased variety of missions and the use of nanosatellites along with continuous efforts to reduce costs are the main drivers toward advanced nanosatellite structures. Currently, the geometry of the satellite dispenser, so-called canisterized satellite dispenser (CSD), is the determining factor for the rectangular structural geometry, particularly for the CubeSats. The nanosatellites have just been used as constellations, such as the CanX-4 and CanX-5 missions of UTIAS SFL [7] of Canada, Doves of Planet Labs [34], Lemurs of Spire [35], and the QB50 [www.qb50.eu]. Another trend is using mother nanosatellites to dispense picosatellites [36] or femtosatellites [37]. As the size is increased to meet more demanding mission needs, hexagonal or octagonal shapes may be adapted toward volume and mass reduction. The use of various composites, including carbon fibers, is expected to increase the mass and strength efficiency for demanding complicated missions, such as interplanetary ones. Advances in additive manufacturing, particularly emerging use of metals as 3D printing material, is opening the way to produce complex nanosatellite structures faster and cheaper. These products may be directly employed as the structure of a nanosatellite. Currently, 3D printed parts are used for fit-checks or molds for the actual components. Particularly, the use of multifunctional side panels including certain subsystems such as antennas, Altitude Determination and Control System (ADCS) elements, or wire paths will increase, as they will be easily manufactured with the use of 3D printers. 3D printing may also reduce part numbers, particularly removing the need for fasteners, resulting in more rigid and monoblock structures. NASA's Ames Research Center continuously evaluates the state of the art of small satellites and provides future projections on nanosatellite structures [38] as well.

References

1 Abdelal, G.F., Abuelfoutouh, N., and Gad, A.H. (2013). *Finite Element Analysis for Satellite Structure*. Springer.

2 SwissCube, 2006. *"S3-A-STRU-1-4 Structure and Configuration"* EPFL.

3 Pumpkin, Inc., *"Cubesat Kit:* http://www.cubesatkit.com"

4 ISIS: Innovative Solutions in Space B.V., *"Webpage:* http://www.isispace.nl"

5 GUMUSH AeroSpace & Defense, *"Webpage:* www.gumush.com.tr"

6 TYVAK: Tyvak Nano-Satellite Systems INC., *"Webpage:* http://www.tyvak.eu"

7 UTIAS SFL: University of Toronto Institute for Aerospace Studies Space Flight Laboratory, *"Webpage:* http://utias-sfl.net"

8 GomSpace: GomSpace ApS, *"Webpage:* http://gomspace.com"

9 *"CubeSat Design Specification Rev. 13",* Cal Poly SLO, 2014.

10 *"6U CubeSat Design Specification, Rev. PROVISIONAL",* Cal Poly SLO, 20 April 2016.

11 *"General Environmental Verification Standard (GEVS) For GSFC Flight Programs and Projects (GSFC-STD-7000A)",* NASA Goddard Space Flight Center, 2014.

12 European Cooperation for Space Standardization (ECSS) *"Space Engineering Mechanical–Part 2: Structural (ECSS-E-30_Part-2A)"* ESA, 2000.

13 Sako N., Hatsutori Y., Tanaka T., Inamori T., Nakasuka S. 2007. *"Nano-JASMINE: A Small Infrared Astrometry Satellite"* 21st Annual AIAA/USU Conference on Small Satellites.

14 Cote, K., Gabriel, J., Patel, B. et al. (2011). *Mechanical, Power, and Propulsion Subsystem Design for a CubeSat*. Worcester Polytechnic Institute.

15 ESTCube-1: Estonian Student Satellite Program, *"Webpage:* http://www.estcube.eu/en/home"

16 CubeSTAR, *"Webpage:* http://www.cubestar.no"

17 Nano-JASMINE 1, *"Webpage:* http://www.jasmine-galaxy.org/index-en.html"

18 *"Webpage:* http://usttl.itu.edu.tr/en"

19 Rani, B., Babu, T., and Srinivasan, M., 2010. *"Comparative Study of Different Structures of Nanosatellite and its Analysis."* Sathyabama University, India.

20 Loos, A.C. (ed.) (2013). *ASC Series on Advances in Composite Materials, Vol. 6: Manufacturing of Composites.* DEStech Publications Inc.

21 Stevens, C.L. (2002). *Design, Analysis, Fabrication, and Testing of a Nanosatellite Structure.* Virginia: Virginia Polytechnic Institute and State University.

22 Annarella, C. (1995). *Spacecraft Structures.* Reston: Microspacecraft.

23 NASA 2008 *"Standard Materials and Processes Requirements for Spacecraft (NASA-STD-6016)"* NASA.

24 Musgrave, G.E., Larsen, A.(.S.).M., and Sgobba, T. (eds.) (2009). *Safety Design for Space Systems.* Elsevier.

25 MatWeb: Material Properties Database, *"Webpage:* http://www.matweb.com"

26 *"Tailored ECSS Engineering Standards for In-Orbit Demonstration CubeSat Projects,"* Estec, ESA, 24.11.2016.

27 Wertz, J.R., Everett, D.F., and Puschell, J.J. (2011). *Space Mission Engineering: The New SMAD,* 1e, vol. 28. Space Technology Library.

28 Wijker, J. (2004). *Mechanical Vibrations in Spacecraft Design.* Berlin Heidelberg: Springer-Verlag.

29 *"Proteus Chapter 5: Payload Environment Requirements, Issue 06 Rev. 03,"* ALCATEL SPACE/CNES.

30 Jacques L., 2009. *"Thermal Design of the Oufti-1 Nanosatellite",* MSc Thesis, University of Liège.

31 *"ThermXL 4.10 User Manual,"* ITP Engines UK LTD., 2010.

32 Ewins, D.J. (2000). *Modal Testing: Theory, Practice and Application,* 2e. Research Studies Press Ltd.

33 Inman, D.J. (2013). *Engineering Vibration,* 4e. Pearson Publishing.

34 Planet Labs Inc: *"Webpage:* https://www.planet.com/"

35 Spire: *"Webpage:* https://spire.com"

36 UniSat by Group of Astrodynamics for the Use of Space Systems Srl: *"Webpage:* www.gaussteam.com"

37 KickSat 1,2 at Cornell University by Zac Manchester: *"Webpage:* http://space.skyrocket .de/doc_sdat/kicksat-1.htm/"

38 *"Webpage:* https://sst-soa.arc.nasa.gov"

8 I-2g

Power Systems

Marcos Compadre[1], Ausias Garrigós[2], and Andrew Strain[3]

[1] *Power Systems Design Authority, Clyde Space LTD, Glasgow, UK*
[2] *Industrial Electronics Research Group, Miguel Hernández University of Elche, Spain*
[3] *CTO, Clyde Space LTD, Glasgow, UK*

8.1 Introduction

The power system of a satellite performs some major functions within the mission: power generation, power distribution, energy storage, and power control. These functions are vital to the spacecraft, as they provide the electrical needs for the rest of the spacecraft subsystems, including the payload.

A simplified architecture of such a power system is shown in Figure 8.1. The main blocks are:

- The solar arrays (SAs) convert sunlight energy into electrical energy.
- The battery charge regulators (BCRs) are used to charge the battery from the solar arrays.
- The power conditioning modules (PCMs) are used to provide regulated and unregulated voltages to the spacecraft subsystems.
- The power distribution module (PDM) is used to safely distribute power to the different subsystems.
- The battery (BAT) stores the energy required when the power generated by the solar arrays is scarce, especially during eclipse.

The power system of a nanosatellite, shown in Figure 8.2, is also known as the electrical power system (EPS). In a typical nanosatellite, the basic elements of the EPS should be organized in an electrical architecture to obtain high efficiency, low volume, low mass, and low price. These challenging requirements force a power system based on commercial off-the-shelf (COTS) elements and advanced integrated circuits, instead of using discrete electronics, which have been traditionally used in larger spacecraft.

Nanosatellites: Space and Ground Technologies, Operations and Economics, First Edition.
Edited by Rogerio Atem de Carvalho, Jaime Estela, and Martin Langer.
© 2020 John Wiley & Sons Ltd. Published 2020 by John Wiley & Sons Ltd.

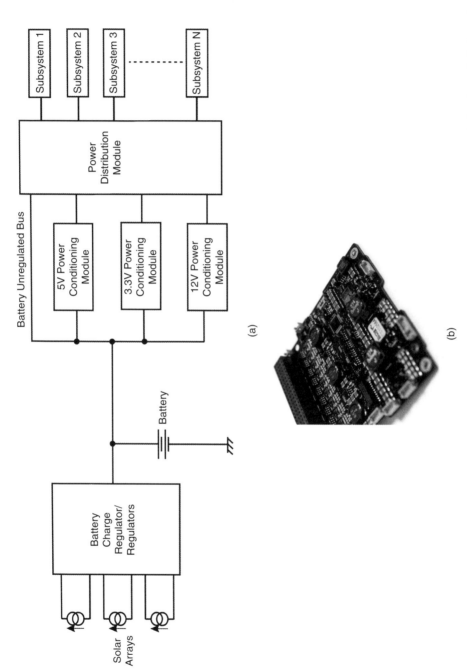

Figure 8.1 (a) Architecture of a nanosatellite power subsystem. (b) CubeSat power system. Source: Reproduced with permission from Clyde Space Ltd.

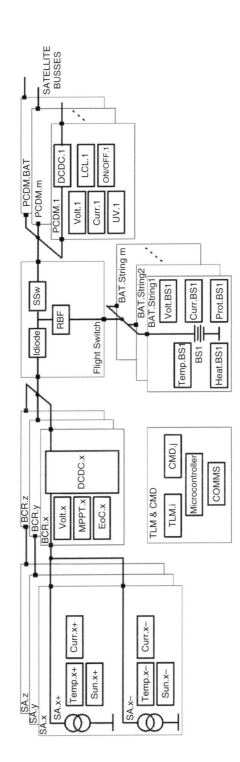

Figure 8.2 Typical nanosatellite electrical power system.

The acronyms shown in Figure 8.2 denote the following functionalities:

SA.x+: Solar array in x-axis and "+" side

   Temp.x+: Temperature telemetry sensor in SA.x+

   Curr.x+: Current telemetry sensor in SA.x+

   Sun.x+: Sun detector in SA.x+

SA.x−: Solar array in x-axis and "−" side (opposite side to SA.x−)

   Temp.x−: Temperature sensor in SA.x−

   Curr.x−: Current sensor in SA.x−

   Sun.x−: Sun detector in SA.x−

BCR.x: Battery charge regulator connected to SA.x

   Volt.x: Input voltage telemetry sensor

   MPPT.x: Maximum power point tracker circuit of BCR.x

   EoC.x: End of charge circuit of BCR.x

   DCDC.x: DC/DC converter of BCR.x

   Idiode: Ideal diode

   RBF: Remove before flight switch

   SSw: Separation switch

PCDM.m: Power conditioning and distribution module "m"

   DCDC.m: DC/DC converter of PCDM.m

   Volt.m: Output voltage telemetry sensor of PCDM.m

   Curr.m: Output current telemetry sensor of PCDM.m

   UV.m: Undervoltage detection circuit of PCDM.m

   ON/OFF.m: ON/OFF telecommand interface for PCDM.m

   LCL.m: This block represents different latching current limiters connected to PCDM.m

BAT.String$m$: Battery string "m"

   BSm: Battery cells ("s" cells in series)

   Volt.BSm: Voltage telemetry sensor of BAT.String "m"

   Curr.BSm: Current telemetry sensor of BAT.String "m"

   Temp.BSm: Temperature telemetry sensor of BAT.String$m$

   Heat.BSm: Heater of BAT.String "m"

   Prot.BSm: Cell level protection circuit of BSm

TLM and CMD: Telemetry and telecommand module of the EPS

   TLM_i: Input telemetry interface

   CMD_j: Telecommand interface

   COMMS: Communications interface

8.2 Power Source: Photovoltaic Solar Cells and Solar Array

Photovoltaic solar cells are the most common power source in nanosatellites, although there have been some developments for using other types of primary power sources, for example, radioisotope power source in interplanetary missions [1, 2]. From the available solar cell technologies, triple junction (GaInP/GaAs/Ge) is widely accepted for miniaturized satellites. Apart from presenting higher efficiency than other technologies, such as silicon, the larger operating voltage provides a more suitable solar array voltage for interfacing to power converters, within limited space (refer to Table 8.1).

A widely accepted electrical model of a solar cell is the single-diode model, depicted in Figure 8.3, where I_{ph} represents the photocurrent, D_c is the cell PN junction and R_p and R_s are the cell shunt and series resistance, respectively. At cell level, the bypass diode D_{bp} is commonly used, because it allows current flowing in case of solar cell reverse biasing, thereby being an effective cell protection against cell shadowing, string current mismatch, and hot-spot heating. The PV string is an arrangement of several solar cells in series to provide higher voltage, while solar array is the connection of several PV strings in parallel to increase the current capability.

Table 8.1 Comparison of electrical parameters of GaInP$_2$ and Si solar cells from Spectrolab–Boeing.

Type	V_{oc} (V)	V_{mpp} (V)	J_{sc} (mA cm^{-2})	J_{mpp} (mA cm^{-2})	Efficiency (%)
GaInP$_2$/GaAs/Ge— Spectrolab XTJ	2.633	2.348	17.76	17.02	29.5
Si—Spectrolab K4702	0.585	0.490	39.20	36.80	13.3

AM0 sunlight, 135.3 mW cm^{-2}, 28°C, Beginning of life (BOL).

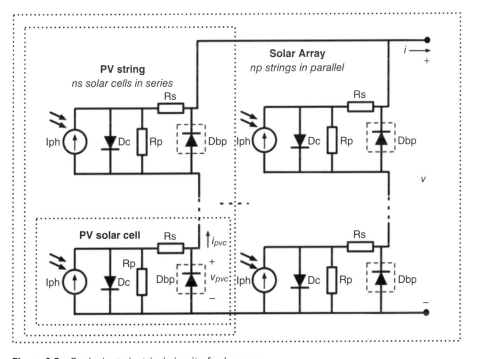

Figure 8.3 Equivalent electrical circuit of solar array.

A mathematical representation of the photovoltaic solar array is given by Eq. (8.1), where n_p is the number of paralleled strings, I_{sat} is the saturation current of the D_c cell, n_s is the number of PV solar cells in series, q is the charge of the electron, k the Boltzmann constant, n is the diode ideality factor, and T is the temperature:

$$i = n_p \cdot \left[I_{ph} - I_{sat} \cdot \left(e^{\frac{\left(\frac{V}{n_s} + \frac{I \cdot R_S}{n_p}\right) q}{nkT}} - 1 \right) \right] - \frac{\frac{V}{n_S} + \frac{I \cdot R_S}{n_p}}{R_p} \tag{8.1}$$

PV solar cell power is mainly affected by three parameters, solar irradiance (G), Sun angle (θ), and temperature (T). Photogenerated current is directly proportional to solar irradiance, while Sun angle is affected by the cosine law ($\cos\theta$) in the range from 0° to 50°. Beyond this limit, the Kelly cosine law represents more accurately the actual response.

Temperature dependence is often characterized by using a coefficient variation from the nominal conditions (AM0, 28°C) given in the product datasheets. The provided nominal performance data such as short-circuit current (I_{sc}), open-circuit voltage (V_{oc}), maximum power point voltage (V_{mpp}), maximum power point current (I_{mpp}), and the maximum power point (P_{mpp}) can then be adjusted accordingly.

Space-qualified solar cells also include radiation related factors for I_{sc}, V_{oc}, and P_{mpp}. At solar array level, Figure 8.4, the operating voltage could be approximated to $v = n_s \cdot v_{pvc}$, and the operating current to $i = n_p \cdot i_{pvc}$.

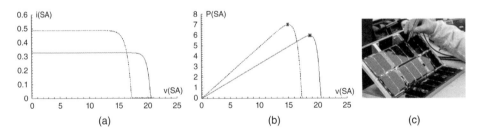

Figure 8.4 7s1p SA−XTJ Spectrolab (dashed line: $G = 1353\,\text{W m}^{-2}$, $\theta = 0°$, $T = 60°\text{C}$; solid line: $G = 1353\,\text{W m}^{-2}$, $\theta = 45°$, $T = -20°\text{C}$). (a) I−V curves, (b) P−V curves, (c) Spectrolab UTJ Solar Array. Source: Reproduced with permission from Clyde Space Ltd.

8.3 Energy Storage: Lithium-ion Batteries

Batteries are used to store the energy required for consistent operation of a satellite, allowing peak power delivery greater than solar array capability and operation during eclipse periods frequently encountered by nanosatellites in low Earth orbit (LEO). Selection of battery technology and battery sizing should consider a number of factors including the number of cycles required, depth of discharge (DoD), temperature ranges, and power delivery requirements, both average and peak.

Lithium-ion (Li-ion) technology is widely used in nanosatellite missions due to the relatively high energy density and high voltage per cell it offers. Li-ion technology has replaced nickel cadmium (NiCd) technology for most nanosatellite missions [3].

The nominal capacity of a battery cell, C, defines the integral of the discharge current over time, and it is expressed in Ampere-hour or Ah. It is defined as the amount of electric charge, at room temperature, that can be delivered before the battery reaches the voltage at which there is no remaining usable energy. It ideally means that a battery is able to provide C amperes for 1 hour or delivers C/n amperes for n hours. With Li-ion batteries, the capacity is degraded with the number of charge/discharge cycles and the amount of charge involved in each cycle. It also depends on the temperature and discharge current rate, delivering less charge at lower temperatures or high discharge currents. The maximum charge and discharge currents of a battery cell are often referred to as the C-rate; being $1C$ equal to the current it would take the battery to charge or discharge in 1 hour. Usually, the maximum charge and discharge current of current Li-ion cells is equal to $1C$, and however, some manufacturers produce cells that can withstand discharge rates of $2C$ during short periods of time. The capacity is often multiplied by the nominal voltage to express the nominal energy, expressed in Watt-hour or Wh.

The typical nominal voltage of a Li-ion cell is 3.7–3.8 V. Higher bus voltages can be achieved by connecting several cells in series. The number of battery cells in series is referred to as the battery string length. In order to increase the capacity of a battery, several strings are connected in parallel.

The typical series–parallel connection of different battery cells is shown in Figure 8.5. The usual nomenclature to define the number of strings and the string length of a battery is X_sY_p, being X the number of cells in series and Y the number of strings in parallel.

As shown in Figure 8.5, the equivalent circuit model of a Li-ion battery is an ideal voltage source and an impedance in series that is commonly known as the internal resistance. The

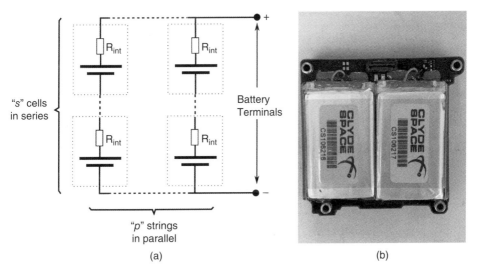

Figure 8.5 Nanosatellite battery. (a) Schematic of a series–parallel connection of battery cells. (b) Typical nanosatellite battery subsystem. Source: Reproduced with permission from Clyde Space Ltd.

ideal voltage source represents the EMF of the cell and can only be measured at zero load or open circuit (OC). The internal resistance depends on different factors, such as the size, the ohmic impedance of the interconnection materials, the conductivity of the electrolyte, temperature, and age.

The typical nominal voltage of a nanosatellite unregulated battery bus is 7.6 V, which corresponds to a battery configuration having two Li-ion battery cells in series. Other nominal voltages could be used, which will require a different string length.

The sizing of the battery depends on the power demand of the spacecraft, and has to be chosen depending on the maximum charge and discharge C-rates as well as the average and peak current demands of the system. When designing the power system of a spacecraft, especially the BCRs, special attention should also be given to these parameters. These values are usually specified by the manufacturer. If the charge rate of the battery is lower than the maximum charge current the BCRs could deliver, based on the solar array input power and the load demands, current controlled feedback loops must be implemented to ensure that the maximum charge current is not exceeded.

When specifying a battery for a mission, it is important to consider the end-of-life performance of the battery, which is influenced by a number of factors. The number of charge and discharge cycles has an impact on the lifetime of the battery, especially in very low altitude LEO missions where the number of cycles would reach up to 16 per day. Another important factor is the DoD, which defines how much charge energy has been used from the battery. It refers to the delivered charge over the nominal capacity, $\text{DoD} = C_{\text{delivered}}/C$. An alternative method is to describe the state of charge of the battery, SoC, which indicates how much charge is left (over the nominal capacity) $\text{SoC} = C_{\text{remaining}}/C$. A short DoD helps to increase the lifetime but increases the size of the battery. DoD values between 20 and 40% are typically selected, depending on the recommendations of the manufacturer. Other important values are the maximum charge voltage, also known as end of charge (EOC) voltage, V_{EOC}, and discharge (cut-off or end of discharge) voltage, V_{EOD}, because exceeding them could produce alterations in the electrochemistry of the battery that severely reduce the lifetime of the battery.

Figure 8.6 shows the typical charge–discharge curve of a Li-ion battery. The flat part of the curve dictates the nominal voltage with rapid changes in voltage around 0 and 100% DoD. Three different voltage definitions are significant for the electrical characterization of

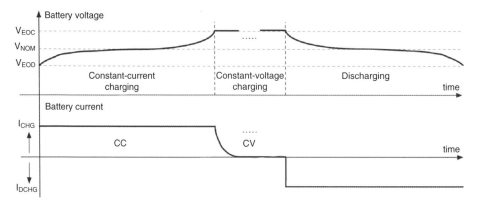

Figure 8.6 Typical Li-ion battery charge/discharge curve with constant-current (CC) and constant-voltage (CV) charging phases.

the Li-ion battery, $V_{EOC} = 4.2$ V per cell, $V_{NOM} = 3.8$ V per cell, and V_{EOD} varies from 2.7 to 3.3 V per cell depending on the load and temperature.

During the charge and discharge processes some energy is lost, due to the electrochemical processes involved, and not all the energy used to charge the battery is available during discharge. On this basis, we can define the energy efficiency, η_{BAT}, which represents the amount of extracted (discharged) electrical energy over the injected (charged) electrical energy. Finally, gravimetric energy density (Wh kg^{-1}), defined as how much energy a battery contains in comparison with its weight, and volumetric energy density (Wh l^{-1}), that compares the energy over volume, are two figures of merit widely accepted to compare different chemistry and technology.

Li-ion batteries require two-step charging process for proper operation. The first part involves constant-current injection with a typical upper limit of C-rate/2. Once the battery reaches the V_{EOC}, the battery should switch to constant-voltage mode and stop the injection of current into the battery to avoid permanent damage to the battery otherwise. In the same way, to protect the battery, discharging current is usually limited to values lower than C-rate [4].

Unlike solar panels, which have a limited current, the batteries can provide large current peaks in the event of a short circuit, therefore overcurrent protections, such as current-limited distribution switches or resettable fuses, commercially known as polyfuse or polyswitch, are required. Battery temperature control is also crucial during charge and discharge; hence, temperature-monitoring circuits are implemented to turn off the distribution switches in case of overtemperature and turn on the heaters in case the temperature is below the operating limits.

8.4 SA-battery Power Conditioning: DET and MPPT

Solar panels provide the energy required for the operation of the spacecraft. They produce energy only when the satellite is in sunlight and this energy must be used to power the rest of the spacecraft and charge the battery so that energy is available during the period of eclipse. There are two methods commonly used to provide charge to the battery: direct energy transfer (DET) and maximum power point tracking (MPPT) [5].

In DET electrical configurations, a direct connection is made between the solar arrays and the battery by using power semiconductors. The two most common implementations of DET are switching shunts, which comprise a shunt MOSFET and a power diode per solar array; and switching series, which uses only a power MOSFET in series. The second is the preferred in nanosatellite applications because of lower number of switches, power loss, and the low open-circuit voltage of the solar array, which is not critical when considering temperature variations or any potential electrostatic discharge issue. In order to be able to operate in this mode, the solar array open circuit voltage must be higher than the battery voltage. This mode of operation is very efficient due to the low power dissipation across the switches. The main disadvantage of this method is that, due to the direct connection between the solar arrays and the battery, the operating point of the solar arrays is dictated by the battery voltage and it cannot be controlled. As a result, solar arrays are often operated at a point that does not maximize the amount of available power.

A widely known DET regulation method, which has extensively been used in large geostationary orbit (GEO) satellites, is the sequential-switching regulator [6]. A diagram of sequential-switching series regulator is presented in Figure 8.7 where the switching

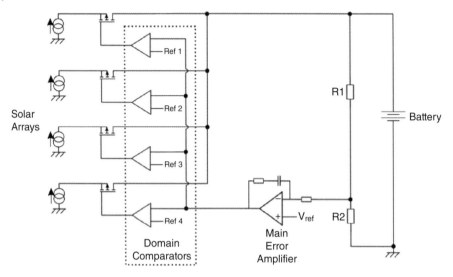

Figure 8.7 Sequential switching series regulator for unregulated battery bus.

elements are connected in series with each solar array. The reference voltage, V_{ref}, monitors the battery voltage and the potential divider R1–R2 is selected to regulate the battery voltage at the EoC voltage. The EoC voltage is the maximum voltage the battery can be charged to. When the battery voltage is below the EoC limit, the main error amplifier (MEA) is saturated, and the series switches are turned fully on thus connecting all the solar arrays to the battery. When the battery voltage reaches the EoC limit, the MEA starts to regulate the battery voltage so that it does not increase beyond the EoC voltage. This charging mode is performed by turning the switches off in a sequential fashion. When the power converter operates in this mode, some switches are turned on, some are turned off, and only one switches between the on and the off state. A common choice for the MEA is an operational amplifier implementing proportional-plus-integral control. The efficiency achieved with this topology is very high, as only one switch element is switching, which only occurs at EoC.

As mentioned previously, DET may result in power delivery that is less than optimal in some conditions. In LEO orbits, a number of these conditions will have a significant impact, that is, several orbits per day, variable illuminated solar array area, and remarkable solar array temperature variation. By decoupling the solar array operational point from the battery voltage, the solar array can be operated at a point that will maximize power delivery in all of these conditions when illuminated.

This mode of operation is called MPPT and corresponds to the P_{MAX} (or P_{MPP}) point shown in the solar array I–V characteristic curve in Figure 8.8. On the left, the current–voltage characteristic (solid line) and the power–voltage one (dotted line) are shown. On the right-hand side, there is a zoomed-in view around the maximum power point (MPP). The MPP corresponds to the point where the solar array operates with maximum efficiency and delivers the maximum power at its terminals.

The energy delivered by the solar arrays has to be used to charge the battery and provide energy to the rest of the spacecraft subsystems. This is done using a BCR. A BCR is usually

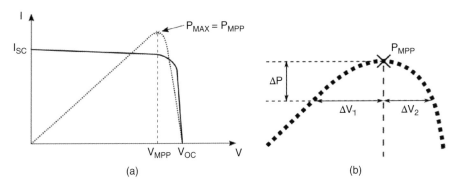

Figure 8.8 (a) Solar array *I–V* characteristic. (b) Zoomed-in view around the MPP.

implemented using a DC/DC converter connected between the solar array, at the input, and the battery of the spacecraft, at the output. DC/DC converters are operated in such a way that one of the converter's variables, usually a voltage or a current, is regulated via a control loop. The control loop of the BCR must regulate the input; commonly the solar array voltage, at the MPP. This can be done using different well-known techniques, such as perturb & observe (P&O), open-circuit measurement (OC), or incremental conductance (IC) [7].

In the P&O method, the controller changes the voltage of the solar array (it "perturbs" its operating point) and it then measures the power delivered. If the power delivered is higher than the power measured during the previous iteration, then the controller continues changing the operating point in the same direction. If, on the other hand, the measured power is lower than before, then the controller changes the direction of the perturbation.

In the open-circuit method, as the name suggests, the open circuit voltage of the solar array is measured, and the controller regulates the solar array voltage at a fixed percentage of that value. The open-circuit measurements must take place every few seconds so that illumination and temperature changes in the solar array can be tracked. This technique might not seem very accurate at first but, as it can be seen in the typical solar panel characteristic curves, the voltage at the MPP, V_{MPP}, varies very little with respect to the open circuit voltage under illumination and temperature changes.

P&O and IC find the MPP of a solar array very accurately, but they require the measurement of voltage and current as well as a processing unit to find the MPP. The OC method is less accurate, but it just requires the measurement of the solar array voltage and a simple voltage or current control loop to regulate the solar array at a fixed percentage of the solar array open-circuit voltage. It exploits the fact that the MPP of a solar array is found at around 80–90% of the open-circuit voltage with relatively small variations in the level of illumination and temperature. This method is usually preferred due to its simplicity and low cost, and an adequate selection of the operating point of the solar array can compensate for the limited accuracy of this MPPT method. As it can be observed from Figure 8.8, the power curve of the solar array presents a rather steep decrease in power to the right of the MPP. When moving to the left of the MPP, the power curve does not present such a steep decrease in power and small variations in the solar array voltage do not produce large variations in the generated power. Based on these observations, the solar array might operate at a voltage slightly smaller than V_{MPP}.

Also, the control loop of this type of MPPT method can be implemented using basic discrete electronics which are robust and cheaper than a solution based on programmable integrated circuits that are required for other common MPPT techniques.

Even though the operation at the solar array MPP optimizes power extraction from the solar arrays, it is important to take into consideration the efficiency of the power converters and the tracking method itself. The efficiency of the power converters is usually lower than that of a DET topology, so the overall energy delivered from the solar arrays to the battery must be computed in both cases before an electrical configuration is selected.

Another simple but effective analog MPPT circuit that has already been used in a larger spacecraft [8] is shown in Figure 8.9. The circuit forces to operate the SA between two points that contain the MPP. When V_{sa}, which follows the triangular waveform given by V_{ref}, reaches P1, the $COMP_i$ forces a change on the R–S flip-flop that inverts the slope of V_{ref} and $S\&H_V$ holds the voltage $K_v \cdot V1$. As V_{ref} decreases, V_{sa} moves toward P2 and once V_{sa} equals $K_v \cdot V1$, $COMP_V$ forces again the change on the R–S flip-flop and $S\&H_i$ holds the current $K_i \cdot I2$, starting a new cycle again. The MPPT frequency and ripple are mainly determined by the comparator limits, but also because of the deviation from ideal operation of different circuits.

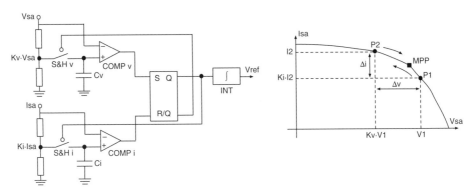

Figure 8.9 Analog S&H maximum power point tracking.

8.5 Battery Charging Control Loops

Batteries have an absolute maximum charge voltage beyond which the battery can suffer permanent damage. It is important to stop the charge process once the maximum battery voltage, V_{EOC}, is reached.

In case of using an MPPT electrical configuration, before the EoC voltage is reached, the BCR operates in MPPT mode and the battery is charged with a current value that depends on the power generated by the solar array and the load power. This mode of operation is known as constant current mode. When the EoC voltage is reached, the battery voltage must be regulated so that it does not increase beyond the EoC voltage. This is accomplished by monitoring the battery voltage and reducing the charge current, as the battery voltage reaches the EoC voltage. During this mode of operation, the battery voltage remains constant due to the action of the regulator and it is known as constant voltage mode. When in MPPT mode, it can be considered that the battery is charged with constant current so long as the solar array delivers constant power. During the EoC phase, the battery is charged with constant voltage across its terminals. Figure 8.6 shows the two charging phases.

During the MPPT phase, the battery voltage increases until it reaches the EoC voltage. The battery is charged with the energy difference between the energy delivered by the BCRs and the energy consumed by the loads. Note that the battery voltage increases during this phase and therefore the current does not remain constant, but it decreases as the battery voltage increases. During the EoC phase, the voltage remains almost constant (at the EoC value) and the current decreases exponentially.

In order to implement the two charge regimes, the BCR must implement two different control loops, one for MPPT and one for EoC. When in MPPT, the EoC control loop is saturated and does not have any influence on the control signal. When the battery reaches the EoC voltage, the EoC control loop takes over and overrides the MPPT control loop. A diagram showing a BCR with the two control circuits (MPPT and EoC) is shown in Figure 8.10.

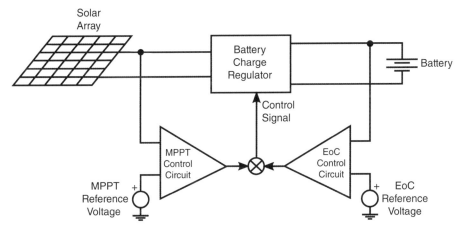

Figure 8.10 BCR, MPPT, and EoC control circuits.

8.6 Bus Power Conditioning and Distribution: Load Converters and Distribution Switches

The power conditioning and distribution modules (PCDMs) perform two main functions: power conditioning to provide regulated and unregulated voltages (from the battery) to the spacecraft subsystems, and controlled power distribution which safely distributes electrical energy to those power buses.

Traditionally, satellite power is distributed on a single bus. Each subsystem then either uses the voltage of this power bus directly or implements point-of-load conversion to the voltages required for operation. However, on a nanosatellite where volume and mass are major considerations, it is often more suitable to have centralized PCMs and distribution of a range of power buses both regulated and unregulated. A regulated voltage only changes in value between a very narrow window, so it is considered as fixed and independent of the load value or load transients generated by the subsystem. Typical regulated voltages, but not limited to, are 3.3, 5, and 12 V. An unregulated voltage corresponds to the raw battery voltage of the spacecraft.

Because of limited battery voltage swing, there is no need for galvanic isolation, high efficiency, and reliability, and low cost, DC/DC topologies employed on the PCDM are usually

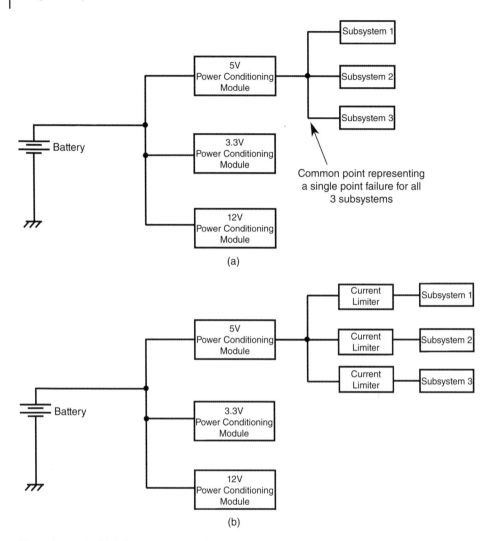

Figure 8.11 (a) 5 V PDM presenting a single point failure. (b) 5 V PDM with overcurrent protection for each output.

limited to buck for step-down voltage conversion ratio, for example, 3.3 and 5 V, and boost for step-up voltage conversion ratio, for example, 12 V.

Since these voltages are shared between all the spacecraft subsystems, they must be protected so that the failure of a subsystem does not propagate across the others. For instance, Figure 8.11 shows a 5 V regulated voltage being distributed to three different subsystems. If there is a short circuit at the input of subsystem 1, then, since the short circuit is happening at a common point to all three subsystems, subsystems 2 and 3 will also be lost. In order to avoid failure propagation, the PDM must protect and isolate the common point of the spacecraft from the failure of any other subsystem.

Distribution switches are controllable solid-state switches with protection. Commercially, they can be found under different names, but also different functionality, electronic fuses, hot-swaps, load switches with protection, circuit breakers, current limiters, and solid-state power controllers are typical terms used. They can provide different protections as overcurrent, current limiting, short circuit, overvoltage and undervoltage, and different fault response actions, such as auto-retry or latching off.

Latching current limiter (LCL) is a widely accepted term in the space sector and it corresponds to an electronic circuit that detects an overcurrent event and limits the output current to a maximum preset value [9]. If the overcurrent event persists for a certain length of time, the output is disconnected from the input, it "latches off," and the overcurrent is cleared.

LCLs are used to distribute power to different loads connected to the same power bus protecting the bus from faulty loads. By doing so, a fault does not propagate to the common power bus allowing the operation of the rest of the loads connected to it. A schematic diagram and the typical response of an LCL are shown in Figure 8.12.

An LCL uses a current sensing element, R_S, a MOSFET to limit the current and a timer to latch off the MOSFET if the overcurrent event is still present after a preset length of time.

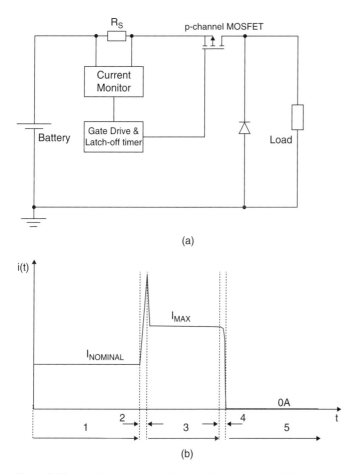

(a)

(b)

Figure 8.12 Latching current limiter. (a) Block diagram. (b) Typical response.

Figure 8.12b shows the typical response of an LCL to an overcurrent event. Initially, during interval 1, the LCL delivers the nominal current, which is lower than the current limit of the LCL I_{MAX}. During this phase, the gate drive circuit turns the MOSFET fully on. This situation corresponds to a normal operation of the load. At some point, an overcurrent event occurs on the load side and an initial inrush current is supplied to the load. This inrush appears because the MOSFET was fully on, and there is a delay before it reacts to the overcurrent event. This phase dictates the reaction time of the LCL (interval 2). Once the MOSFET is driven into current limit mode, the current is limited to the preset value equal to I_{MAX} (interval 3). During the current limit phase, the MOSFET operates as a variable resistor, the value of which depends on the output impedance seen by the LCL, and the power dissipation of the device is relatively high. For this reason, it is important to turn the MOSFET off, otherwise the power dissipated could destroy the device. So, if the overcurrent event persists after a preset value of time, the LCL is commanded to turn off. Interval 4 indicates the time it takes the MOSFET to turn off, after which the output current drops to zero (interval 5).

An example of a commercially available circuit that could perform the function of an LCL, if external timer is added, is the current-limited, high-side integrated P-MOSFET switch with thermal shutdown, MAX890L from Maxim, is represented in Figure 8.13. It operates with input voltage from +2.7 to +5.5 V, making it suitable for +3.3 and +5 V power rails. A current-controlled current source (CCCS) provides a replica current of the main switch, $i_r = i_o/1110$, which in turns creates a proportional voltage drop across R_{set}, $V_{set} = (i_o/1110) \cdot R_{set}$. The current-limit error amplifier compares V_{set} to the internal 1.24 V reference and regulates the switch current to $I_{limit} = 1376/R_{set}$. The maximum current limit is $I_{max} = 1.2$ A, and in case of short-circuit event an internal fast control loop limits to $1.5 \cdot I_{limit}$ before switch disconnection. Additionally, the switch turns off when junction temperature exceeds +135°C and turns back on when device cools by 10°C. An open-drain fault indicator goes low either in current limit or during thermal shutdown.

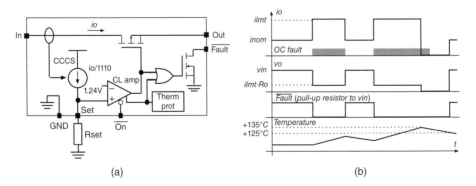

(a) (b)

Figure 8.13 MAX 890L distribution switch. (a) Circuit diagram. (b) Typical waveforms.

8.7 Flight Switch Subsystem

The flight switch subsystem provides a method of isolation of the battery and the rest of the electrical power subsystem during storage, transportation, and launch. It contains two protection devices, the remove before flight (RBF) pin and the separation switch (SS). The RBF pin allows a physical interface to isolate the battery from the power buses. Ideally, the RBF is removed after the nanosatellite is integrated into the pod, but it can be removed just prior to integration. The SS isolates BCRs and battery from the satellite power buses effectively switching the nanosatellite off. When in the launch vehicle the SS is compressed, and the satellite remains off. When the satellite is deployed the SS is depressed, connecting the BCRs and the battery to the power buses. RBF and SS switches typically are designed with solid-state switches in order to avoid large currents through mechanical switches.

A simple solid-state switch is represented in Figure 8.14. The voltage divider formed by R1 and R2 forces a source-to-gate voltage when the control switch (SW) is turned on, which in turn switches on the pass P-channel MOSFET. An additional capacitor C1 could be used to control the voltage slew-rate across the gate and drain, being useful to limit inrush currents in capacitive loads. This type of circuit can be used as a SS or an RBF switch. When used as a SS, the control switch SW represents the mechanical switch that is compressed against the pod during launch and releases after launch. When used as an RBF, the control switch SW represents the pull-pin that is removed when the nanosatellite is integrated in the launcher. In this case, SW should be normally closed and, when the pull-pin is inserted, SW becomes open.

Figure 8.14 P-channel MOSFET solid-state switch with inrush current limiting.

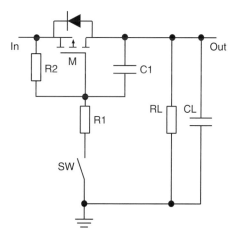

8.8 DC/DC Converters

As described previously, DC/DC converters are basic building blocks for BCRs and PCDMs. A DC/DC converter is an electronic circuit that converts a DC voltage at its input to a different DC voltage at the output by changing the duty-cycle of the switching element. DC/DC converters are used in nanosatellites, which implement MPPT, as BCRs because of their high efficiency and the possibility they offer to control the solar array voltage. Although the battery voltage is not fixed and varies across the orbit (from 6.2 to 8.2 V for the case of a two-series cells Li-ion battery), it can be considered fixed from the control loop standpoint, as the voltage varies very slowly. With this in mind, the controlled variable becomes commonly the input voltage (the solar array voltage). The controller can use any MPPT control method to operate the solar array at its MPP. An example is shown in Figure 8.15a, where a buck (or step-down) converter is used as a BCR. The solar array is connected to the input and the battery to the output. The controller monitors the input voltage and sets a reference voltage that is used to control the duty cycle and regulate solar array voltage.

Two examples of topologies used in BCRs in nanosatellites are the buck (or step-down) and the single-ended primary-inductor converters (SEPICs). The buck converter is used when the solar array voltage is higher than the battery voltage, and the SEPIC converter is used when the solar array voltage is either higher or lower than the battery voltage. These two topologies are described in the following sections.

DC/DC converters are also used to generate the regulated power buses. In this type of applications, the output voltage is the variable that is monitored and controlled to provide the required voltage. The most common topologies used as PCMs are the buck and the boost (or step-up) converters.

Monolithic (integrated) switching regulators (MSRs) are good candidates to be used as DC/DC converters, since they can offer a very compact solution. They generally include integrated power switch and driver, pulse width modulated (PWM) logic, error amplifier, and ancillary functions, such as undervoltage, overcurrent and thermal protections, soft-start, synchronization, or enable input among others. Additionally, MSRs are able to run at high switching frequency, up to few MHz in some cases, which makes reactive parts (capacitors and inductors) smaller.

The following sections provide the input-to-output transfer functions for the three main power converter topologies used in nanosatellites. These transfer functions are obtained assuming continuous current through the inductor, that is, it never reaches zero amperes, and they are obtained using the small ripple approximation [10].

8.8.1 Buck Converter

The buck converter, shown in Figure 8.15, is used to convert an input voltage to a lower voltage at the output. It is also called "step-down" converter.

The steady-state analysis of this converter provides the following transfer functions:

$$\frac{V_{out}}{V_{in}} = D \qquad \frac{I_{out}}{I_{in}} = \frac{1}{D} \tag{8.2}$$

where D is the so-called duty-cycle. The duty-cycle represents the ratio of time between the MOSFET being in the on state to the switching period, that is:

$$D = \frac{t_{on}}{T} \tag{8.3}$$

The buck converter is used for both power generation and power distribution, performing the tasks of a BCR and a PCDM (see Figure 8.15a,b). The difference between the two different applications is the controlled variable. Figure 8.15a does not show the EoC control circuit for simplicity. The buck converter is used as a BCR when the solar array voltage is always higher than the battery voltage and the control circuit regulates the solar array at the MPP. The output diode D_{OUT} is required to avoid reverse conduction from the battery to the solar array through the body diode of the power MOSFET. Conversely, the buck converter is used as a PCM when a regulated bus voltage is required. In this case, the control circuit regulates the output voltage (Figure 8.15b).

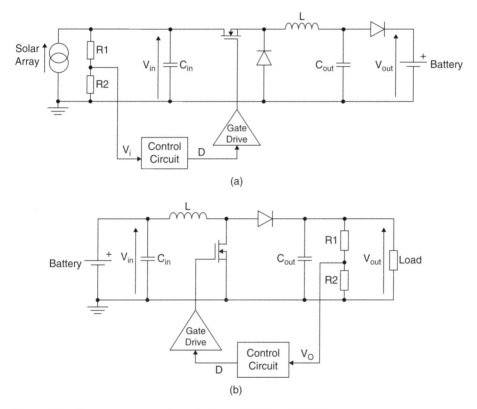

Figure 8.15 Buck converter implemented as a (a) BCR and (b) PCM.

8.8.2 Boost Converter

The boost converter, shown in Figure 8.16, is used to convert an input voltage to a higher one at the output. It is also called "step-up" converter.

The steady-state analysis of this converter provides the following transfer function:

$$\frac{V_{out}}{V_{in}} = \frac{1}{1-D} \quad \frac{I_{out}}{I_{in}} = 1 - D \tag{8.4}$$

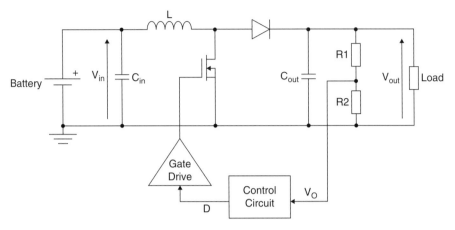

Figure 8.16 Boost converter used as a PCM.

with D being the duty-cycle.

The boost converter is mainly used in power distribution when the required bus voltage is higher than the battery bus.

8.8.3 SEPIC Converter

The SEPIC, shown in Figure 8.17, is used to convert an input voltage to either a higher or a lower one at the output.

The steady-state analysis of this converter provides the following transfer functions for voltage and current:

$$\frac{V_{out}}{V_{in}} = \frac{D}{1-D} \qquad \frac{I_{out}}{I_{in}} = \frac{1-D}{D} \tag{8.5}$$

with D being the duty-cycle. From the previous equations, it can be observed that the output voltage can be either higher or lower than the input voltage. The inductors L1 and L2 can be either coupled inductors, as shown in Figure 8.17, or not coupled.

The SEPIC converter is mainly used in power generation whenever the solar array voltage is either higher or lower than the battery voltage.

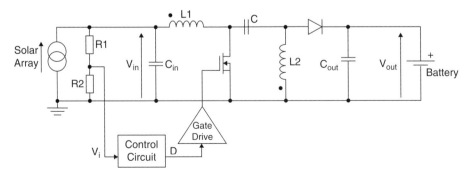

Figure 8.17 SEPIC converter used as a BCR.

8.9 Power System Sizing: Power Budget, Solar Array, and Battery Selection

A nanosatellite has restricted time to generate power, only during sunlight, but has to supply the loads during the entire orbit. The battery subsystem is essential, since it stores the energy during sunlight and delivers it during eclipse, and actually, a positive energy balance (with some margin) in the battery in one orbit remains as the main design criteria for power system sizing.

In terms of general power, the EPS of a nanosatellite must meet the energy balance given by Eq. (8.6), where E_{SA} is the solar array energy, $E_{LOADsun}$ is the consumed energy by the loads during sunlight, E_{CHG} is the energy introduced to the battery during charging to reestablish the initial energy, and $E_{LOSSsun}$ are the losses associated to the power conversion and distribution during sunlight:

$$E_{SA} = E_{LOADsun} + E_{CHG} + E_{LOSSsun} \tag{8.6}$$

Energy balance is typically accounted for a single orbit, so, the energy balance might be expressed in terms of power:

$$E_{SA} = \int_{t_0}^{t_0+t_{sun}} P_{SA}(t) \, dt$$

$$E_{LOADsun} = \int_{t_0}^{t_0+t_{sun}} P_{LOADsun}(t) \, dt$$

$$E_{LOSSsun} = \int_{t_0}^{t_0+t_{sun}} P_{LOSSsun}(t) \, dt$$

$$E_{CHG} = \frac{E_{DCHG}}{\eta_{BAT}} \cong \frac{E_{LOADecl} + E_{LOSSecl}}{\eta_{BAT}}$$

$$E_{LOADecl} = \int_{t_0+t_{sun}}^{T_{orbit}} P_{LOADecl}(t) \, dt$$

$$E_{LOSSecl} = \int_{t_0+t_{sun}}^{T_{orbit}} P_{LOSSecl}(t) \, dt \tag{8.7}$$

where E_{DCHG} is the energy drawn by the battery during the eclipse interval, $E_{LOADecl}$ is the consumed energy by the loads during eclipse, and $E_{LOSSecl}$ are the losses associated to the power conversion and distribution during eclipse.

To perform a preliminary EPS sizing, five main points have to be considered:

1. Identify orbit period (T_{orbit}) and eclipse time ($t_{eclipse}$). These two inputs are very important to size the solar array and the battery. From the power system sizing view, an initial starting point for LEO circular orbits comes from the third Kepler's law of planetary motion, which provides the orbit period in terms of the orbit radius, and the trigonometric identity when the Sun is in the orbit plane for the eclipse time:

$$T_{orbit} = \sqrt{\frac{4\pi^2(R_E + h)^3}{\mu}} \qquad t_{eclipse} = \left[\frac{1}{2} - \frac{\cos^{-1}\left(\frac{R_E}{R_E + h}\right)}{\pi} \right] \cdot T_{orbit} \tag{8.8}$$

where R_E is the Earth's mean radius 6.378×10^6 m, h is orbit altitude, and μ is the standard gravitational parameter. For the Earth, μ is 3.986×10^{14} m^3 s^{-2}.

In circular LEO orbits, typical for nanosatellites, the eclipse duration is dependent on the orbit altitude, inclination, and the sunlight incidence angle on the orbit plane, which varies seasonally. The eclipse duration could vary in factor two in LEO orbits. The ratio between the eclipse time and orbit period also becomes important, since it impacts directly on solar array, battery, and BCR requirements. Fine-tuned orbit and eclipse periods are inputs from the mission analysis and they have to be considered to assure EPS survivability in all conditions.

2. Load power allocation. The first task of power system sizing is obtaining a detailed load power profile of each subsystem to be supplied. Independent load power profiles, for sunlight and eclipse periods, are recommended in order to determine power generation and energy storage requirements, as well as energy losses associated to different power flow paths in the EPS. Average and peak power demands are required for each component. Duty ratio, defined as the ratio of average power in a certain time interval over peak power, Eq. (8.9), is generally employed to characterize the load power profile:

$$D_{\mathrm{LOAD}} = \frac{\int_{t_0}^{t_0+T} P_{\mathrm{LOAD}}(t)dt}{P_{\mathrm{LOADpk}}(t)T} \tag{8.9}$$

3. EPS power flow and losses estimation. Electrical power conversion and distribution bring associated power losses. In order to obtain a preliminary assessment of the EPS energy losses, rough efficiency estimation should be assigned to each power managing subsystem and the possible power flow paths during sunlight and eclipse intervals are evaluated. In an unregulated battery bus EPS, three basic power paths with their related efficiency could be identified: $\eta_{\mathrm{SA/BUS}}$ is the efficiency linked to the power path from solar array electrical output to the main bus, comprising the following lossy elements, solar array blocking diodes (if used), BCRs, and battery diodes; $\eta_{\mathrm{BUS/LOAD}}$ accounts for the efficiency from the main bus to the loads, and it mainly includes PCMs and distribution switches; $\eta_{\mathrm{BUS/BAT}}$ corresponds to the efficiency from the main bus to the battery and vice versa, and it embraces battery protection switches principally.

Attending the simplified EPS block diagram represented in Figure 8.18, three different power flow paths (from 1 to 3) are possible. During sunlight phase, the following combinations are possible simultaneously: power flow paths 1 and 2 when supplying load and charging the battery, power flow path 1 when supplying load and battery is fully charged, and power flow paths 1 and 3 during transient overloads, that is, solar array and battery supplying the load at the same time. During eclipse periods, only power flow path 3 is likely. As it can be observed in Figure 8.18, each power flow path includes two blocks, power flow path 1 contains $\eta_{\mathrm{SA/BUS}}$ and $\eta_{\mathrm{BUS/LOAD}}$, power flow path 2 contains $\eta_{\mathrm{SA/BUS}}$ and $\eta_{\mathrm{BUS/BAT}}$, and power flow path 3 contains $\eta_{\mathrm{BUS/BAT}}$ and $\eta_{\mathrm{BUS/LOAD}}$.

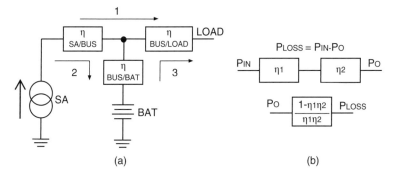

Figure 8.18 EPS power flow paths (left side). Power loss ideal transfer function (right side).

An approximation of the energy losses is given by Eq. (8.10), where the most probable scenario has been considered, that is, power flows through paths 1 and 2 simultaneously during sunlight phase and power flows through path 3 in eclipse:

$$\alpha_1 = \frac{1 - \eta_{SA/BUS} \cdot \eta_{BUS/LOAD}}{\eta_{SA/BUS} \cdot \eta_{BUS/LOAD}}$$

$$\alpha_2 = \frac{1 - \eta_{SA/BUS} \cdot \eta_{BUS/BAT}}{\eta_{SA/BUS} \cdot \eta_{BUS/BAT}}$$

$$\alpha_3 = \frac{1 - \eta_{BUS/BAT} \cdot \eta_{BUS/LOAD}}{\eta_{BUS/BAT} \cdot \eta_{BUS/LOAD}}$$

$$E_{LOSSsun} = \frac{(1 + \alpha_3)\alpha_2}{\eta_{BAT}} E_{LOADecl} + \alpha_1 E_{LOADsun}$$

$$E_{LOSSecl} = \alpha_3 E_{LOADecl} \tag{8.10}$$

Power conversion efficiency is highly dependent on electrical parameters as voltage, current, or temperature, therefore, detailed electronic design is required to obtain refined efficiency models that could be included later on when those parameters are known. Other losses as harnessing, resistive current sensors, or integrated circuit supply current could also be incorporated in the power system sizing tool for better model accuracy.

4. Battery sizing. Once the power load allocation and power losses estimation are clearly identified, the preliminary design of the battery could be made (8.11). First, the battery energy requirements are estimated taking into account the allowed DoD. Next, the number of battery cells per string is selected according to the desired bus voltage, while the required capacity of the battery is defined by the energy of the battery, which is calculated based on the energy to be supplied during the eclipse period, with respect to the battery voltage. Finally, the number of parallel battery strings results from the total capacity over the cell capacity:

$$E_{BAT}(Wh) = \frac{E_{DCHG}(Wh)}{DoD}$$

$$N_S = \frac{V_{BUSnom}}{V_{cellnom}}$$

$$C_{BAT}(Ah) = \frac{E_{BAT}(Wh)}{V_{BUSnom}(V)}$$

$$N_P = \frac{C_{BAT}}{C_{cell}} \tag{8.11}$$

In case of high-power pulsed loads, battery must comply with high discharge currents, requiring more battery strings to increase battery capacity.

5. Solar array sizing. The photovoltaic source must meet, at the end of life, the energy balance given by Eq. (8.12) plus an additional margin. Solar array energy is computed by the integral of the solar array power during the sunlight period, the latter being a function of the average solar radiation in space, also known as the solar constant, S_{SUN} (AM0 conditions), the solar array equivalent illuminated area, A_{SA_EQ}, and the solar array conversion efficiency, η_{SA}:

$$E_{SA} = \int_{t_0}^{t_0+t_{sun}} P_{SA}(t)\, dt \cong S_{SUN}\left(\frac{W}{m^2}\right) \cdot \eta_{SA} \cdot A_{SAEQ}\ (m^2\ s)$$

$$A_{SAeq} = \int_{t_0}^{t_0+t_{sun}} A_{SA}(t)\, dt \tag{8.12}$$

Considering that S_{SUN} and η_{SA} will remain constant during one orbit, the most critical factor to properly estimate the solar array energy is the equivalent solar array area. The A_{SAeq} is the equivalent average area filled with solar cells that point perpendicularly to the Sun during one orbit. This area mainly depends on solar array size and geometry, satellite stabilization method, and orbit characteristics, being difficult to obtain an accurate estimation with simple calculations at the early design stage.

In the case of a cubic-shaped spacecraft, with body-mounted solar cells in all facets, and using the projected areas of a cube, some very simple approximations could be considered to have a preliminary estimation. In that case, the equivalent solar array illuminated area will vary from the area of a single facet, A_{FACET}, when the Sun impinges perpendicularly on one facet, to $\sqrt{3}A_{FACET}$ if one vertex points to the Sun. An intermediate value, but also an interesting case happens when two consecutive facets form the maximum projected area, $\sqrt{2}A_{FACET}$. Simple averaging of those values will give a rude approximation of E_{SA}:

$$E_{SA}(Wh) \cong 1.38 \cdot A_{FACET}(m^2) \cdot S_{SUN}(W/m^2) \cdot \eta_{SA} \cdot t_{SUN}(h) \tag{8.13}$$

On the other hand, S_{SUN} also varies with the day of year, that is, Earth orbit eccentricity and solar activity. A conservative S_{SUN} value, widely used, is 1353 W m^{-2}. Finally, a proper estimation of η_{SA} must include, as a minimum, conversion efficiency at end-of-life (EOL) and temperature drifts (dI/dT and dV/dT):

$$\eta_{SA} = \eta_{BOL}\frac{P_{mppEOL}}{P_{mppBOL}}\left[1 + \left(\frac{\left.\frac{dI_{mpp}}{dT}\right|_{EOL}}{I_{mppref}|_{EOL}} + \frac{\left.\frac{dV_{mpp}}{dT}\right|_{EOL}}{V_{mppref}|_{EOL}}\right)\Delta T\right] \tag{8.14}$$

Once the required solar array area is determined, the solar array configuration must be defined according to the available area and cell physical dimensions to obtain the desired voltage and current per solar array.

8.10 Conclusions

This chapter has given an overview of the fundamentals of all stages of a nanosatellite power system. We have looked at:

- Power generation from the solar panels.
- Power conditioning architectures to transfer energy from the solar panels to the battery.
- Power storage in the battery.
- Power conditioning topologies for DC/DC conversion.
- Power distribution and protection for effective power delivery to the rest of the satellite.

As can be seen from the descriptions in this chapter, there are several options and trade-offs to be considered within the power system design itself. However, it is also vital to understand the requirements of the mission at hand in order to ensure that the best system design is achieved.

For example, solar panels can be optimized for power delivery using an MPPT system, but this system is more complex than a DET system. By considering the orbit, thermal conditions, and power requirements of the mission, one must decide if the benefit of additional power is worth the additional complexity. Similarly, a larger battery will allow larger energy reserves for eclipse operations, peak loads, and battery lifetime. But this comes at the cost of additional mass and volume.

As with many other systems, the power system is vital to the survival of the spacecraft and a well-designed system can be the difference between mission success and failure.

References

1 R. L. Cataldo, 2016. "A concept for a radioisotope powered lunar cubesat," Annual Meeting of the Lunar Exploration Analysis Group (LEAG 2016), USRA, November 1–3, 2016.

2 E. Wertheimer, L. Berthoud, M. Johnson, 2015. "PocketRTG – a cubesat scale radioisotope thermoelectric generator using COTS fuel," iCubeSat Workshop.

3 Robyn, M., Thaller, L., and Scott, D. (1995). Nanosatellite power system considerations. In: *Proceedings of the International Conference on Integrated Micro/Nanotechnology for Space Applications*, Oct 30–Nov 3, 259–265.

4 C. S. Clark, E. Simon, 2007. "Evaluation of lithium polymer technology for small satellite applications," 21st Annual AAIA/USU Conference on Small Satellites, Advanced Technologies 1, SSC07-X-9, https://digitalcommons.usu.edu/smallsat/2007/all2007/66

5 A. K. Hyder, R. L. Wiley, G. Halpert, D. J. Flood, S. Sabripour, 2003. "Spacecraft power technologies," Imperial College Press, ISBN 1-86094-117-6.

6 A. H. Weinberg, D. O'Sullivan, 1980. "Limit cycling regulator apparatus for plural parallel power sources," US patent 4,186,336, January 29, 1980.

7 Salas, V., Olias, E., Barrado, A., and Lázaro, A. (2006). Review of the maximum power point tracking algorithms for stand-alone photovoltaic systems. *Solar Energy and Solar Cells* 90: 1555–1578.

8 W. Denzinger, 1995. "Electrical power subsystem of Globalstar," Proc. 4th European Space Power Conference, ESA SP-369, 4–8 September 1995.

9 C. Delepaut, T. Kuremyr, M. Martín, F. Tonicello, 2014. "LCL current control loop stability," Proc. 10th European Space Power Conference, ESA SP-714, 15–17 April 2014.

10 Erickson, R.W. (1997). *Fundamentals of Power Electronics*. Chapman and Hall.

9 I-2h

Thermal Design, Analysis, and Test

Philipp Reiss[1], Matthias Killian[1], and Philipp Hager[2]

[1] *Institute of Astronautics, Technical University of Munich, Garching, Germany*
[2] *European Space Agency, Keplerlaan 1, Noordwijk, the Netherlands*

9.1 Introduction

The thermal control system (TCS) of a satellite interfaces with nearly all other subsystems. The TCS has to manage external heat loads from the environment, as well as internal heat loads from components inside the satellite. The goal of the TCS is to control all heat loads in order to meet the thermal requirements of all components. Thermal requirements are temperature ranges, temperature gradients, and temperature stability. The TCS has to ensure that the temperatures of all components are kept within the respective operational and nonoperational temperature ranges throughout the lifetime of the satellite. The thermal design of each individual satellite is therefore unique and adapted to the particular requirements and boundary conditions of the system and its mission. It evolves from an increasingly detailed thermal analysis over the entire process of the satellite design, and finds its verification with the thermal test of the integrated satellite. A TCS may imply methods of thermal control, either of passive or active nature, to fulfill its goal.

Inappropriate thermal design of a satellite might equivalently lead to overheating or undercooling of components, assemblies, or instruments. In the best case, this only leads to an altered characteristic of the respective component, such as parameter drift of electronics (e.g. changing capacitance and resistance) or a temperature-dependent frequency shift of radio equipment. In the worst case, it might cause severe damage to a component so that it loses its functionality, such as freezing of the battery, or reduces instrument performance, such as thermally induced stresses that distort optical systems. Both cases lead to at least partial malfunction, decreased performance, or complete loss of the mission. Other examples of thermal failure are related to the differential thermal expansion of structural elements. Thermal cycling supports such problems, resulting in mechanical fatigue and failures, such as breaking solder joints and therefore open electrical circuits. Higher temperatures in particular can accelerate degradation processes that weaken materials and support outgassing from polymers, which can lead to the contamination of sensible thermal control coatings or optical components [1].

Nanosatellites: Space and Ground Technologies, Operations and Economics, First Edition.
Edited by Rogerio Atem de Carvalho, Jaime Estela, and Martin Langer.
© 2020 John Wiley & Sons Ltd. Published 2020 by John Wiley & Sons Ltd.

The risk for the aforementioned failures can be mitigated by supporting the entire satellite design process with thermal analysis in combination with extensive thermal-vacuum testing. However, this requires appropriate facilities, budget, personnel, and time, which are resources that are often not available or limited for nanosatellite missions. The implementation of a thorough and reliable thermal analysis is therefore essential to reduce the risk of thermal failures during the mission.

9.1.1 Thermal Challenges

Nanosatellites often contain highly integrated electronics that dissipate a significant part of their electrical power as heat. Furthermore, the nanosatellite surface in most cases is mainly covered with solar cells. Hence, there is only little space available for installing thermal control components, inside or on the surface of nanosatellites. Additionally, the surface properties can only be modified in a small range and attaching radiators is often not possible. Nanosatellites often use commercial off-the-shelf (COTS) parts that are not designed or qualified specifically for application in the space environment. This leads to a rather small temperature range in which they can reliably operate and subsequently to more stringent requirements on thermal control.

Since nanosatellites are predominantly launched into low Earth orbits, they are subject to a large number of thermal cycles during their lifetime and hence frequently changing thermal stress is to be expected. For nanosatellites without an attitude control system, the estimation of the expected thermal loads is difficult because of the unknown spin rates. Nanosatellites with attitude control are more challenging, as thermal gradients are more pronounced and temperature swings, especially of external components, are wider.

Thermal requirements need to be defined as early as possible in the design process. Thermal requirements are absolute temperature limits, temperature gradients (differences per dimension), and temperature stability (fluctuations over time). All components have an operational and a nonoperational temperature range that need to be adhered to. Typical operational temperatures for nanosatellite components are in the range of −40 to 80°C [2], whereas the more sensitive components are the onboard electronics with common temperature ranges of −40 to 60°C [3–5]. The most delicate components usually are batteries or optical payloads. Batteries have operational temperatures from −20 to 60°C (discharge) and 0–40°C (charge) [6]. In some cases, batteries and other electronics for nanosatellites might allow −40°C to 85°C (maximum ratings) [7].

9.2 Typical Thermal Loads

Thermal loads on a satellite result from heat transfer between the satellite and its environment. Heat transfer is defined as the energy that flows due to a difference in temperature. It cannot be observed or measured directly, but its effects on a system are commonly known. The physical relation between heat transfer and the change of the internal energy of a system through which this transfer takes place is described by classical thermodynamics [8]. The most relevant relations for the application in space are presented in this section.

Nanosatellites that are operated in Earth orbits are mainly subjected to three heat transfer mechanisms: radiation, conduction, and convection. Any object in space exchanges heat via radiation with its environment, that is, the Sun, planetary bodies, and deep space. Conduction plays a major role within the satellite structure and wherever direct physical contact is present. Due to the very low atmospheric pressure in low Earth orbits (about 10^{-4} mbar at 100 km and 10^{-7} mbar at 300 km [9]), convection is usually not relevant for heat transfer, except for fluids in satellite components, such as tanks (e.g. surface tension tanks) or two-phase systems (e.g. heat pipes). Thus, only conduction and radiation are introduced in this chapter as relevant heat transfer mechanisms for nanosatellites.

Material properties are crucial factors for the calculation of heat fluxes. These include thermooptical properties, thermal conductivity, and heat capacity. While thermooptical properties of technical surfaces have a significant influence on the radiative heat exchange, the thermal conductivity between components as well as their contact resistance is a main parameter for conductive heat exchange. The exact determination of such factors is not trivial and, in most cases, requires testing. However, even testing can only reveal the material property for the respective test condition and extrapolation to other conditions can be problematic. Material properties such as thermal conductivity and heat capacity as well as thermooptical properties are usually temperature dependent.

The following sections introduce the fundamentals of conductive and radiative heat exchange between two exemplary objects. The thermal environment in Earth orbit is described alongside the explanation of how the resulting heat fluxes between environment and satellite can be determined in a simple and practical way.

9.2.1 Heat Exchange Calculation

The conductive heat flux $\dot{Q}_{cond}$ between an object temperature T_i and an object with temperature T_j can generally be described with Eq. (9.1), using the thermal conductivity k, the cross-sectional area A, and the distance x between the two objects:

$$\dot{Q}_{cond} = k\frac{A}{x}(T_i - T_j) \tag{9.1}$$

For convenience, all variables except the temperatures are often substituted with the conductive heat exchange factor or linear conductor G_{ij}, which can be interpreted as the inverse of the thermal resistance R:

$$\dot{Q}_{cond} = G_{ij}(T_i - T_j) \tag{9.2}$$

The heat exchange factor is a combination of the material properties and geometrical properties that affect the path of the heat transfer. If the heat flux is between two objects or areas with different material or geometry, it must be distinguished between serial or parallel connection of both, similar to the calculation of resistance in an electrical network. Figure 9.1 shows a model of two thermal nodes i and j that are in ideal thermal contact (no contact resistance). For such a serial connection of two nodes, the inverse of the total heat exchange factor is the sum of the inverse heat exchange factors of both objects, as described by Eqs. (9.3) and (9.4):

$$G_{ij} = \left(\frac{1}{G_i} + \frac{1}{G_j}\right)^{-1} \tag{9.3}$$

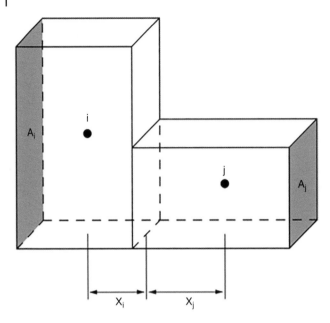

Figure 9.1 Parameters for the calculation of conductive heat exchange factors in a serial connection of two nodes.

$$G_{ij} = \left(\frac{x_i}{k_i A_i} + \frac{x_i}{k_j A_j} \right)^{-1} \tag{9.4}$$

In the case of a parallel connection of two nodes, the conductive heat exchange factor is calculated accordingly, as described by Eqs. (9.5) and (9.6):

$$G_{ij} = G_i + G_j \tag{9.5}$$

$$G_{ij} = \frac{k_i A_i}{x_i} + \frac{k_j A_j}{x_j} \tag{9.6}$$

The radiative heat flux $\dot{Q}_{rad}$ can be described with Eq. (9.7), using the Stefan–Boltzmann constant σ, the emissivity of the emitting body ε_i, the absorptivity of the absorbing body α_j, the emitting surface A_i, and the view factor between both bodies $F_{i \rightarrow j}$:

$$\dot{Q}_{rad} = \sigma \varepsilon_i \alpha_i A_i F_{i \rightarrow j} (T_i^4 - T_j^4) \tag{9.7}$$

The view factor defines the fraction of radiation that leaves the emitting surface A_i and arrives at the absorbing surface A_j. It depends on the orientation between both surfaces and can be determined analytically or numerically for more complex configurations. According to the reciprocity theorem, the term $A_i F_{i \rightarrow j}$ can be replaced with $A_j F_{j \rightarrow i}$. Further view factors can be found in reference [10], the respective online catalog, or the standard ECSS-E-ST-31-HB-1. View factors do not take into account any retro-reflection. If retro-reflections shall be taken into account, for example, for simplified calculations of heat exchange inside a nanosatellite, the so-called Gebhart factors can be used. Note that Eq. (9.7) defines the net radiative heat flux between both objects. Similar to the conductive heat exchange, all variables in Eq. (9.7) except the temperatures can be substituted with the radiative heat exchange factor or radiative conductor R_{ij}:

$$\dot{Q}_{rad} = R_{ij}(T_i^4 - T_j^4) \tag{9.8}$$

For solid materials, the thermal material properties of heat capacity and thermal conductivity are temperature and direction dependent. For technical surfaces, the thermooptical properties are temperature and angle dependent [11]. However, often only mean values for these parameters at room temperature are available. Direction dependence of solid material thermal conductivity can be of importance for printed circuit boards (PCB) in nanosatellites. For nanosatellite applications, room temperature values are often reasonably accurate, but if broader temperature ranges are expected it is inevitable to determine the temperature dependence of the respective material property. Also, in most cases, the directionality of the thermooptical properties is neglected to simplify the analysis.

Another simplification according to Kirchhoff's law is used where applicable, stating that absorptivity and emissivity of a body in thermal equilibrium are equal for a given wavelength [12]:

$$\alpha(\lambda, T) = \varepsilon(\lambda, T) \tag{9.9}$$

Often the absorptivity and emissivity of a surface are given while implicating that this means solar absorptivity and infrared emissivity. Because technical surfaces on a satellite absorb not only solar but also infrared radiation from other objects, the infrared absorptivity is additionally required for heat flux calculations. According to Kirchhoff's law, the infrared absorptivity in this case equals the infrared emissivity because both are at the same wavelength range.

9.2.2 Thermal Environment in Earth Orbit

Due to the absence of atmosphere, any object in space mainly exchanges heat with its environment through radiation. For a satellite in Earth orbit, there are three major radiation sources: direct sunlight, sunlight reflected from Earth (albedo), and infrared (IR) radiation emitted from Earth (Figure 9.2). Deep space acts as a heat sink for any dissipated energy with a constant background temperature of 2.7 K [13]. Particularly low Earth orbits or during reentry, there is also free molecular heating through friction in the atmosphere. However, for most nanosatellites this heat source is negligible during their operational lifetime.

9.2.2.1 Direct Solar Radiation

Solar radiation is the largest heat source in Earth orbit. The mean solar radiation in Earth orbit is defined as the solar constant with a value of $1367\,\mathrm{W\,m^{-2}}$. Although the solar activity varies with an 11-year cycle, the solar radiation can be regarded as constant within 0.1% alteration. The 27-day rotation period of the Sun leads to slightly higher variations of 0.2% [14], which still is negligible for the calculation of heat fluxes. The largest variation of solar radiation in Earth orbit is due to the elliptical trajectory of the Earth around the Sun, amounting to ±3.5% over 1 year, resulting in a radiation intensity between $1322\,\mathrm{W\,m^{-2}}$ at aphelion and $1414\,\mathrm{W\,m^{-2}}$ at perihelion [15].

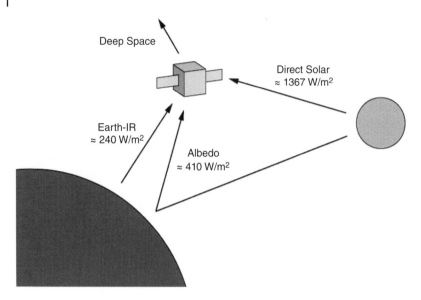

Figure 9.2 Typical radiative heat fluxes for satellites in low Earth orbit (with an albedo factor of 0.3).

With the blackbody temperature of the Sun T_{Sun} (approximately 5770 K), the radius of the Sun R_{Sun}, and the Stefan–Boltzmann constant, the intensity of solar radiation M_{Sun} in dependence of the distance r between Sun and Earth can be calculated with Eq. (9.10):

$$M_{Sun}(r) = \sigma T_{Sun}{}^4 \left(\frac{R_{Sun}}{r}\right)^2 \tag{9.10}$$

On the basis of Eq. (9.7), the resulting heat flux from Sun to satellite is calculated with:

$$\dot{Q}_{Sun} = \alpha_{Sat} A_{Sat} F_{Sat \to Sun} M_{Sun} \tag{9.11}$$

For satellites in Earth orbit, it can be assumed that the solar rays are parallel. This simplifies the computation of the view factor, since for the absorption of solar radiation in this case the projected area of the respective satellite surface can be used (e.g. the circular cross-section for a spherical body). For flat surfaces, the product $A_{Sat}F_{Sat \to Sun}$ can be replaced with the product of A_{Sat} and the cosine of the angle ψ, which is measured between surface normal and Sun vector:

$$\dot{Q}_{Sun} = \alpha_{Sat} A_{Sat} cos(\psi) M_{Sun} \tag{9.12}$$

In order to find the Sun vector for the satellite surface, the so-called beta angle can be used. It defines the angle between the line that connects Earth and Sun and the orbital plane of the satellite. With the orbit inclination i, the right ascension of the ascending node (RAAN) Ω, declination δ_{Sun}, and right ascension Ω_{Sun} of the Sun, the beta angle β can be calculated with Eq. (9.13) [15]:

$$\beta = sin^{-1}(cos(\delta_{Sun})sin(i)sin(\Omega - \Omega_{Sun}) + sin(\delta_{Sun})cos(i)) \tag{9.13}$$

For the thermal balancing of environmental heat fluxes, it is also important to know the duration of the eclipse, during which the satellite is shadowed by the Earth. The eclipse fraction of a circular orbit depends on the beta angle β, the orbit altitude h, and the Earth radius R [15]:

$$f_E = \frac{1}{180} cos^{-1} \left[\frac{\sqrt{h^2 + 2Rh}}{(R+h)cos(\beta)} \right] \tag{9.14}$$

However, Eq. (9.14) only applies if the beta angle is larger than the beta angle at which eclipses actually occur, β^*. It can be calculated with the following simple geometrical relation, assuming that the Earth shadow is cylindrical:

$$\beta^* = sin^{-1} \left[\frac{R}{R+h} \right] \tag{9.15}$$

9.2.2.2 Albedo Radiation

Sunlight reflected from any planetary body is called albedo radiation. Its intensity depends on the albedo of the respective body, which describes the fraction of the incident sunlight that is diffusively reflected from the planet surface. Using the dimensionless albedo factor ρ_{Albedo}, the resulting radiation can be calculated with Eq. (9.16):

$$M_{Albedo} = \rho_{Albedo} M_{Sun} \tag{9.16}$$

For a satellite in Earth orbit, the albedo radiation mainly depends on the surface reflectivity of the satellite's ground track. It is usually higher for cloud-, ice-, or snow-covered regions than for continental and oceanic regions. Therefore, the albedo radiation also tends to be higher for higher latitudes toward the ice-covered poles. Typical values for the albedo factor can be found in literature, for instance, in reference [15]. An albedo factor of 0.3 is often considered as a reasonable mean value for Earth orbits [12]. The resulting albedo heat flux from the Earth to a satellite is highest at the subsolar point and decreases with the cosine when the satellite moves toward the terminator. Over a 5-year period, NASA's Clouds and the Earth's Radiant Energy System (CERES) experiment has measured the albedo between $-85°$ and $+85°$ latitude and determined that the albedo factor in this range varies between 0.22 near the equator and 0.69 at the highest latitudes [16].

Similarly to solar radiation, the heat flux M_{Albedo} resulting from albedo radiation between Earth and satellite is calculated with Eq. (9.17):

$$\dot{Q}_{Albedo} = \alpha_{Sat} A_{Sat} F_{Sat \rightarrow Earth} M_{Albedo} \tag{9.17}$$

9.2.2.3 Earth Infrared Radiation

Earth absorbs solar radiation and re-emits it in the IR wavelength. How much of the Earth's infrared radiation reaches space depends on factors such as local surface temperature and cloud coverage. A simplified approach of calculating the Earth IR radiation is to assume that 30% of the incident solar radiation is reflected as albedo, as described earlier, so that the remaining 70% is absorbed by the Earth. In thermal equilibrium, the Earth emits the same amount of heat as IR radiation to space. If the absorbing surface of the Earth is approximated with its circular cross-section ($A = \pi R^2$) and the emitting surface for IR radiation is approximated by the surface of a sphere ($A = 4\pi R^2$), the average intensity of the Earth IR radiation therefore has to be one fourth of the absorbed solar radiation, or $240\,W\,m^{-2}$.

If the equivalent blackbody temperature of the Earth T_{Earth} below the satellite position is known, the local Earth IR radiation M_{Earth} can be calculated more precisely with Eq. (9.18):

$$M_{Earth} = \sigma T_{Earth}{}^4 \tag{9.18}$$

The average equivalent blackbody temperature of the Earth is approximately 255 K, which results the same radiation intensity of $240\,W\,m^{-2}$ as with the previous simplified calculation approach. The Earth's IR spectrum however cannot be perfectly matched with a blackbody radiation, because of the different wavelength-dependent contributions of both surface and atmosphere. Furthermore, the Earth IR radiation not only varies with local surface and weather features, but also varies between the illuminated and shadowed side of the Earth and between orbit inclinations. Values of the Earth IR radiation lie between 153 and $266\,W\,m^{-2}$ as measured by NASA's CERES experiment over a 5-year period between $-85°$ and $+85°$ latitude [16].

Once the Earth IR radiation intensity is known, the infrared heat flux between Earth and satellite can be calculated with Eq. (9.19):

$$\dot{Q}_{Earth} = \alpha_{Sat} A_{Sat} F_{Sat \rightarrow Earth} M_{Earth} \tag{9.19}$$

9.3 Active and Passive Designs

For the thermal design of a nanosatellite, active and passive elements can be used. Passive elements do not rely on additional power to fulfill their functional purpose, which makes them less prone to failure. Typical passive elements are surface-finished to adjust the thermooptical properties, insulations, such as multilayer insulation (MLI) or low conductivity materials, interface fillers, heat pipes, thermal straps, or heat switches to increase the conductive heat transfer between two parts. In contrast to that, active elements need additional power, such as electrical heaters, active coolers, pumped fluid loops, louvers, and electrochromic paints. In general, a thermal control design based on passive elements is the preferable option, especially for small and low-cost nanosatellites. In addition to passive elements, electrical heaters often find application in nanosatellites because they are compact, light, and reliable. If designed correctly, a thermal design with predominantly passive elements is cheap, robust, and failure-tolerant, as no electronics or control software are involved.

9.3.1 Surface Finishes

The main passive thermal control concept is to control the radiative heat exchange between surfaces by adjusting the IR emittance of all internal and external satellite surfaces, as well as the solar absorptivity of external satellite surfaces. Achieving a particular IR emittance starts with selecting the raw material and its processing (milling, etching, etc.) and ends with surface treatments, such as painting or vapor-deposited coatings. Table 9.1 gives a rough overview of thermooptical properties for commonly used materials. To reduce the IR heat transfer between two surfaces, polished metallic surfaces are usually used, as they

Table 9.1 Thermooptical properties of common surface finishes [15].

Coating	Solar absorptivity	IR emissivity
White paint	0.19	0.89
Black paint	0.95	0.87
Second surface mirror (SSM)	0.10	0.80
Silver polished	0.04	0.02
Gold electroplated	0.23	0.03
Aluminum oxidized	0.13	0.30
Aluminum polished	0.15	0.05

have a low IR emissivity. For high emittance, space-rated white and black paints are most common. The solar absorptance needs to be considered when nanosatellite surfaces are exposed to solar radiation. The ratio between absorptance α and emittance ε of a given surface governs the overall thermal behavior. High α/ε ratios have a heating effect while low α/ε ratios have a cooling effect. For Sun-exposed surfaces, high-emissivity white paint is therefore more suitable than black paint because of the lower α/ε ratio. An even more effective option to achieve high emissivity on Sun-exposed surfaces would be a second surface mirror (SSM) or optical surface reflector (OSR). A SSM is a two-layered surface with a material transparent in the visible wavelength and high emittance at the top and a reflective coating with low absorptance below. All components in space surfaces can degrade over time when they are exposed to the space environment. Depending on the mission lifetime, it can therefore be important to distinguish between so-called beginning-of-life (BOL) values and end-of-life (EOL) values. For example, 2.5 years in orbit can degrade the absorptance of white paint from 0.20 to 0.45 [15]. In Table 9.1, only the diffuse values are shown. Depending on the surface and application, it can be relevant to distinguish between the specular, diffusive, as well as transmissive part of absorptivity and emissivity.

9.3.2 Insulation

An efficient way of insulating against external heat sources as well as to protect from heat losses to space is MLI. MLI consists of multiple layers of low IR emittance surfaces, such as Kapton or Mylar, vapor-deposited with aluminum [15] to minimize the radiative heat exchange between the layers. The conductive heat transfer through MLI is minimized by using meshes with low thermal conductivity, such as Dacron or Nomex. Because the voids between the layers and meshes are evacuated in space, the heat transfer through an MLI is governed by radiation. Figure 9.3 shows a typical layout of a MLI blanket. In the vicinity of seams and fixation points, the blanket is compressed, so that a significant conductive heat flux through the layers can occur at these spots. MLI performance in general depends on the number of layers, the absolute temperature on both sides, the temperature difference between both sides, the surface area of the MLI blanket, and how it is mounted on the

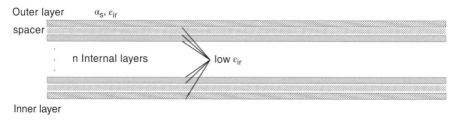

Outer layer α_s, ε_{ir}

spacer

n Internal layers low ε_{ir}

Inner layer

Figure 9.3 Principal layout of MLI.

satellite (number of folds, overlays, cut-outs, stand-offs, etc.). A good first estimation of the MLI performance is given in reference [17]. Prior to testing the MLI in flight configuration, high uncertainties in their thermal performance should be taken into account. Usually, the MLI blanket is designed solely according to the required insulating performance, that is, by selecting the number of layers, the optical coatings, and the spacers. A typical MLI blanket consists of 10–20 single layers to achieve a good insulation. Increasing the number of layers beyond that does not yield a significantly better insulation performance but increases the mass of the blanket. Another important fact to consider is that MLI blankets have to be grounded to the satellite structure in order to avoid static charging [18]. There are many different variations for MLI lay-ups and materials. The outer surface material is usually adapted according to the needs of the α/ε ratio. The lay-up material has to be selected based on temperature requirements. Dacron and Mylar are suitable for continuous temperature above 120°C. In the vicinity of thrusters, more metal foils have to be used. Other means are the embossing of Kapton foils to avoid spacers, for example.

9.3.3 Radiators

Excess heat on nanosatellites can be removed most efficiently by radiating to space, usually through a specific radiator surface for which a high IR emittance (typically >0.8) is required. The simplest radiator design is a metal plate that is in radiative heat exchange with the internal satellite components on one side and deep space on the other side (Figure 9.4a). To enhance the radiative heat exchange, both sides can be treated with high IR emittance coatings, such as black paint. Surface finishes of radiators depend on whether the external side of the radiator is exposed to solar radiation. In this case, a low α/ε ratio is needed, such as it is the case for white paint (compare Section 9.3.1).

A radiator can also be attached directly to internal components or via dedicated interfaces, as displayed in Figure 9.4b,c. The difficulty lies in the interface connection and in spreading the heat over the entire radiator. For radiators used with high power applications, it might be necessary to improve the heat distribution within the radiator surface in order to increase the radiator efficiency. Simply increasing the thickness of the radiator might not be sufficient and increase the mass drastically. Alternatively, doublers or heat pipes can be used. Doublers are alloys with higher thermal conductivity. Heat pipes (see Section 9.3.4) can be mounted either on a radiator plate or within a honeycomb structure. Such embedded heat pipes are more efficient to spread the heat more evenly across a radiator.

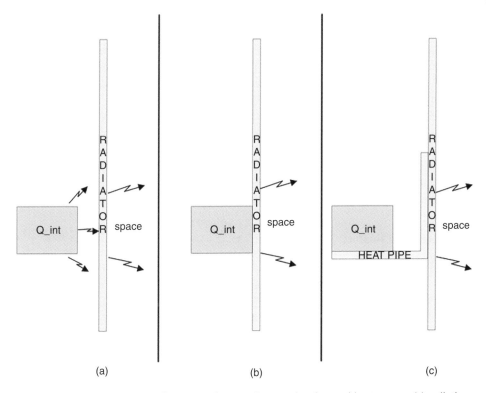

Figure 9.4 Different concepts for connecting a radiator and an internal heat source: (a) radiative connection, (b) direct conductive connection, and (c) conductive connection via interface.

9.3.4 Interface Connections and Heat Pipes

A major critical part in thermal design is the conductive connection between heat source and heat sink. In most cases, the heat from an electronic unit has to be transferred to a mounting structure that acts as the heat sink. The conduction between both parts depends mainly on the bolt pattern [19] as well as on the contact pressure and surface roughness of the connected surfaces. To overcome conductive resistance due to surface roughness, a thermal interface filler can be used. Fillers are available in different forms (e.g. grease, gasket) with different materials, whereas many of them have lower thermal conductivity than the parts they are connecting. Nevertheless, they can create a defined conductive link reducing the uncertainties in thermal modeling. It has to be considered that fillers also change other physical properties of the interface, such as the electrical conductivity.

Thermal straps are an option for a mechanical interface connection between two components that are not directly in touch. Thermal straps are usually made of copper, aluminum, or carbon fiber, and are available in different forms and variations, such as a stack of thin, woven threads that are fixated at the ends or thin-layered foils. An additional advantage of thermal straps is that they allow some mechanical flexibility between the parts that are connected. Thermal conductivity and flexibility depend on the materials used. Typical materials for thermal straps are copper, aluminum, or pyrolytic graphite.

Another solution for transporting large amounts of heat over a longer distance, from several centimeters up to meters, is the use of heat pipes. Heat pipes are closed pipes in which a two-phase liquid/gas heat transfer occurs [20]. At one end the heat pipe is connected to a heat source, where the contained liquid is evaporated (evaporator). The other end of the heat pipe is connected to the cold sink, where the contained gas condenses (condenser). Heat transfer between evaporator and condenser takes place via the gas phase. Capillary structures (e.g. grooves and composite wick materials) at the inner surface of the heat pipe allow the liquid to travel back from the condenser to evaporator. Depending on the gravity environment, the transfer of the liquid phase through the capillaries might be restricted. Furthermore, the liquid used in a heat pipe has to be chosen according to the required operating temperature range and freezing must be prevented to avoid bursting the heat pipe. Further options of heat transport are loop heat pipes or pumped fluid loops.

9.3.5 Electrical Heaters

Electrical heaters exist in a wide variety of forms, but for nanosatellites patch heaters are the most relevant. They consist of a thin resistance circuit embedded between two electrically insulating foils (e.g. Kapton polyimide). Patch heaters are mechanically flexible and available in different sizes and shapes to fit most satellite components. Often heaters need to be custom-made, in order to meet resistance, shape, or heat density requirements. The electrical resistance can be adjusted to the supply voltage and the desired amount of heater power by choosing different materials and lengths of the electrical circuit. The patch heater is either software-controlled in combination with a temperature sensor or by a thermostat. Whereas software control allows to adapt limits in flight, a thermostat solution is more robust and passive. Patch heaters are usually installed on the respective satellite component using adhesive. When applying adhesives for patch heaters, care must be taken to select adhesives with suitable temperature range and outgassing properties.

9.4 Design Approach and Tools

The thermal analysis of any space system becomes increasingly complex with the progress of a project. While an initial thermal model with 1–10 nodes yields a first rough estimate for nanosatellites of the expected thermal environment, a more detailed thermal analysis in later project phases requires increasingly more complex models with a much higher number of thermal nodes. The software packages used for thermal analysis utilize the fundamentals presented in Section 9.2. They allow the thermal engineer to efficiently implement the available means of thermal control discussed in Section 9.3 with the ultimate objective to meet the thermal requirements.

9.4.1 Numerical Methods

Numerical methods are widely used to solve physical problems that can be described with partial differential equations and are not solvable by analytical means [21]. A discretization of the real hardware is the basis for the computation of the heat transfer. Combined with boundary conditions of the system such as temperature and/or heat fluxes differential equations can be solved with numerical methods.

In the thermal analysis of satellites, a special form of the finite-difference methods (FDM) is typically used, called the lumped parameter method (LPM). In the LPM, a continuous medium is discretized into several nodes, each representing a concentrated or "lumped" node of its surrounding. The LPM has its benefits with respect to thin-walled structures and applications with a moderate level of integration [16, 22]. According to the fundamentals described in Section 9.2, heat is exchanged between the lumped nodes by conductive, convective, or radiative fluxes, whereas the convective aspect can be neglected for nanosatellite applications in most cases.

With regard to nanosatellites, which generally have a high level of integration and consist more of electrical components than structural parts, the use of the LPM should be scrutinized on a case-by-case basis. To accurately model the heat spread in small intricate structures (frame and stack of nanosatellites) or in PCB with numerous small heat sources, a finite-element method (FEM) approach can also be considered.

Independent of the applied numerical method, the radiative heat transfer must be incorporated. The view factors can be calculated in three ways: either by the use of empirical equations, by analytical equations, or by numerical ray-tracing calculations. Empirical and analytical equations are available for a wide variety of configurations [10]. Obviously, the analytical calculation of view factors becomes increasingly inefficient with increasing model complexity. When it comes to the interaction between solar and infrared rays between the Sun, reflecting or emitting surfaces on the outside of a spacecraft with different optical properties, ray-tracing approaches become attractive or might even be unavoidable. Whereas analytical view factor calculations become cumbersome for complex geometries, ray tracing produces statistic errors that diminish by increasing the number of rays [23].

The foundation for numerical solving is the so-called thermal mathematical model (TMM). The TMM is a finite network of thermal nodes and thermal connectors. The nodes represent thermal capacities and govern the transient behavior of the model. The connectors are conductive, radiative, or convective, and govern the temporal and spatial thermal gradients throughout a model. In order to compute temperatures, boundary conditions such as fixed temperatures or heat fluxes have to be applied. The TMM is often derived from a geometrical mathematical model (GMM), which is a geometrical representation of the real body, made up of primitive shapes, such as cylinders, spheres, or cuboids, discretized in triangles or quadrilaterals. The GMM is used to calculate view factors and the resulting radiative heat exchange between the nodes.

9.4.2 Modeling Approaches

A thermal design can be evaluated in a bottom-up or a top-down approach. The selection of approach depends on the requirements, the mission phase, the experience of the thermal engineer, and the available resources.

9.4.2.1 Top-Down Approach

Top-down analyses are fast, easy, and have low computational demand compared to the more detailed bottom-up approach. Top down models can be re-run often and fast.

Especially in the early project phases when the design changes regularly, a simple top-down model with few nodes is most efficient concerning outcome and effort. With a simple top-down model, the effect of optical surface properties can be assessed, and general thermal designs can be evaluated for passive thermal control methods. On the other hand, a top-down analysis has a low level of detail. The insight into the thermal behavior of the satellite is therefore limited. A top-down model is generally started from a single node or a simple shape (e.g. sphere, cylinder, or cube) with the overall dimensions and the approximate heat loads during the most likely operational modes. From there the model is further refined and subsystems or components are added as deemed necessary.

9.4.2.2 Bottom-Up Approach

Bottom-up analysis is time-consuming, requires a mature design and configuration status, and a database of material properties, optical surface properties, and contact conductances. Bottom-up analysis yields a deep insight in the thermal behavior of the nanosatellite and allows an accurate flight prediction to verify via analysis that the thermal design meets the thermal requirements. With a detailed model, design changes such as adding or removing insulation, heat paths, heaters, thermal washers, or filler material can be evaluated at all stages of the project. The bottom-up approach either begins from an already existing computer aided design (CAD), for example, for configuration aspects or structural model or is completely build-up from scratch. If a CAD or structural model exists, the thermal model can be derived from this model (Figure 9.5). The first step for the thermal analysis therefore is to translate this model into a meshed model for abstraction. It is not sensible to transport every detail of a CAD model into a thermal model. On the one hand, the thermal model should include only a few thermal nodes as possible for the sake of computational effort, postprocessing, and easy understanding of heat paths. On the other hand, the thermal model should contain as much details as necessary to simulate the thermal behavior in an appropriate way. While small cutouts or drill holes might have no big impact on

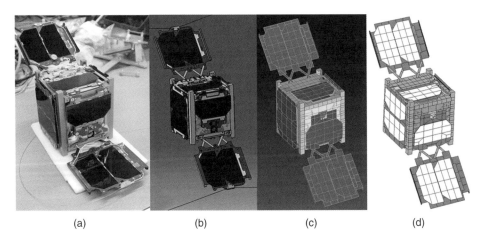

    (a)   (b)   (c)   (d)

Figure 9.5 Bottom-up thermal model development with the example of the CubeSat "First-MOVE" of the Technical University of Munich: (a) photograph after a test integration; (b) detailed CAD model; (c) simplified GMM in meshing tool; (d) simplified GMM/TMM in thermal analysis software. (*See color plate section for color representation of this figure*).

the thermal behavior of the satellite, other small components such as cables could create heat bridges that are not negligible, especially for small and highly integrated nanosatellites. An important step in the discretization of the model is the node numbering and the respective grouping of parts, assemblies, or subsystems. In a second step, the physical and thermooptical properties are added to the discretized model and the boundary conditions are assigned. Finally, the operational modes and the satellite environment or the orbital cases are defined. The most important scenarios to analyze are the "worst hot case" and the "worst cold case." Often these are also the cases that will be tested later on in the thermal qualification of the satellite.

9.4.3 Model Uncertainty and Margins

Modeling uncertainty and temperature margins are often confused although they describe different contingencies. Modeling uncertainty is acknowledging the fact that the mathematical models are always inaccurate. This is comparable to the understanding that a temperature sensor reading has a measurement uncertainty. The thermal margins on the other hand are safety margins applied in order to account for changes in the TCS (acceptance margins) or unforeseen design changes or events on system level (qualification). Modeling uncertainties are relevant for the thermal engineer. Acceptance and qualification margins are applied by the system engineers.

9.4.3.1 Modeling Uncertainty

Because models are only a simplified representation of reality, their predictions will always be wrong. The error created by a model is captured with the so-called modeling uncertainty. The so-called systematic modeling uncertainty is caused by simplifications in the thermal modeling process, such as for example the removal of drill holes, chamfers, or simplification of complex shapes. Apart from the systematic components there are many parameters that contribute to the overall modeling uncertainty, such as variations in material properties, contact conductances, external heat fluxes, and tolerances of thermal control hardware. An exhaustive list can be found in reference [24] along suggestions on how these parameters might be varied. The modeling uncertainty can be determined by a sensitivity analysis, which is a systematic variation of the abovementioned parameters. The differences between the parameter variation and a baseline case are combined in a root-sum-square manner because they are expected to be independent. The systematic modeling uncertainty is added linearly. Modeling uncertainty can only be reduced by correlating a thermal model with thermal balance test results. For uncorrelated thermal models, temperature deviations of 10–15 K are not uncommon. External components with low thermal inertia such as MLI can have higher uncertainties, whereas components inside a spacecraft mostly have lower uncertainties. Throughout the maturation of a design, the modeling uncertainties decrease, yet for nanosatellites uncorrelated models can still easily have modeling uncertainties of around 15 K even at critical design review (CDR). The temperatures output from a thermal model are called "calculated temperatures." As soon as the modeling uncertainty is added (hot case) or subtracted (cold case) the nomenclature changes to "predicted temperatures."

9.4.3.2 Temperature Margins

Acceptance and qualification temperature margins differ between agencies, companies, and even between projects. The acceptance and qualification margins are added to the predicted temperatures and define the test temperatures. In nanosatellite projects, acceptance and qualification temperature margins are often omitted. This is a result of using COTS components with predefined operational and nonoperational temperature ranges, and the fact that nanosatellite projects accept a higher risk.

9.4.4 Thermal Design Tools

Thermal design tools in aerospace engineering range from spreadsheet-based calculations to complex numerical simulation environments. Spreadsheet tools help in the early phase of a project for rough estimations of the overall conditions. They are relatively simple to use and have modest computational demands. The drawbacks are the restricted capabilities for a geometrical representation as well as limited numerical functionalities for thermal network solving and ray tracing.

Numerical thermal network solver tools specialized for thermal applications are often expensive, require a professional introduction, and a powerful workstation, or even a computer cluster. Furthermore, to set up models and interpret the results correctly, experience is required. The numerical thermal network tools provide the necessary means to either set up a TMM from scratch or import and adapt models generated in CAD software and meshing tools. They allow implementing thermal nodes based on different basic shapes (shells, spheres, cones, cubes, etc.) and nongeometric nodes. They also allow searching for, or manually implement, conductive and radiative interfaces. Tools from the aerospace domain also support the user with orbit simulation capabilities, taking into account different central bodies, distances, and orbit parameters for the investigated object. Examples for such aerospace-specific thermal software packages are ESATAN-TMS, Thermal Desktop, Thermica, Sinaps, FEMAP, ThermXL, NX Space Systems Thermal, and Thermal Synthesizer System.

9.5 Thermal Tests

The feasibility of a thermal design, the functionality of the used components, and the validity of the used numerical model have to be verified by thermal tests. Thermal tests can be done under ambient pressure or in a vacuum environment. Thermal-vacuum tests are the most realistic representation of the space environment, but they also require a high effort regarding preparation and execution and are therefore cost-intensive. In general, thermal tests are separated in Thermal Balance tests (TBal), Thermal Vacuum tests (TVac), and Thermal Cycle tests (TCyc). Further subdivision is possible in thermal parameter tests, bake-outs, thermal functional, and performance tests.

In nanosatellite projects, the necessity of thermal-vacuum tests is often underestimated. Long lead times for the test equipment might need to be taken into account and sometimes slots in the respective test facilities have to be organized many months in advance. The TVac test often is the first full system test in a space-like environment, and especially for nanosatellites it is often the first or sole fully integrated test where the operators are unable to access the device under test (DUT) for a prolonged period. Thermal test campaigns are projects of their own that need a schedule, allocation of personnel and facilities, design of ground support equipment (GSE), procurement and manufacturing of parts, test plan, test procedures, test predictions based on thermal analysis, test reports, and the subsequent model correlation with results of the thermal balance and parameter tests. Depending on the requirements, the overall schedule, and the plan for assembly, integration and test (AIT), thermal tests can be performed on different levels.

Since the thermal tests have a long lead time, it is useful to define as early as possible which tests shall be performed on which level (system, subsystem, unit) and at which point in time. For nanosatellites, subsystem level tests (e.g. a solar panel, or a PCB) and system level tests (integrated nanosatellite) are probably sufficient in most cases. For a payload, a separate sub-system level thermal test might be sensible. Planning the test includes the identification of the thermal, functional, and performance requirements that shall be addressed and possibly verified during the thermal tests. Significant reduction of test time and complexity is possible by a thorough test planning.

Vacuum chambers with a final pressure lower than 10^{-5} mbar are required to rule out an unwanted convective influence of residual gas. The required temperature range of the facility typically ranges between -100 and $100°C$ and depends on the space environment, the orbit, and the operational conditions of the mission that are simulated.

9.5.1 Types of Thermal Test

The main thermal tests are thermal balance, thermal-vacuum, and thermal cycle tests. Within these test types, several additional test modes can be conducted, such as bake-out, functional, and performance tests.

A bake-out is generally performed to accelerate or verify the outgassing levels of used components under vacuum and to meet contamination requirements of sensitive equipment. Typical requirements are to achieve a total mass loss (TML) of <1% and a collected volatile condensable material (CVCM) of <0.1% [25]. A bake-out must be performed under vacuum conditions. In nanosatellite projects, a bake-out is often required by the launch authority. Often bake-outs require to keep the DUT at or close to its maximum nonoperational temperature in vacuum for up to 72 h.

In a functional test, the functional requirements of the DUT are verified. Electrical and mechanical components are activated and deactivated according to the defined operational modes to verify their functionality at different temperatures, for example, on-board computer or deployment mechanisms. In general, thermal functional tests can be performed under ambient pressure, yet often it is a requirement to verify the functionality in a vacuum.

Performance tests are relevant for the satellite payload to determine their dependency on vacuum and temperature. Functional test with benchmarks or calibration standards are often conducted before and after the performance tests. Performance test data are particularly valuable for correlation with the in-orbit performance of the payload and to identify potential problems.

9.5.1.1 Thermal Balance Test

The main purpose of thermal balance tests is the correlation of the thermal model and the verification of the thermal design. The thermal balance test is the most relevant test for the thermal engineer. Usually one cold operational, one cold survival, and one hot operational thermal balance phase are conducted. On thermal balance phase during cold survival temperatures is used to check the functioning of heating lines. The DUT must dissipate heat during the hot and cold operational thermal balance phases. The dissipated heat must be known or measured with high accuracy for the correlation of the thermal model. The higher the temperature difference between the cold and hot balance phases, the more significant is the outcome and the clearer is the interpretation of results. The closer the test conditions are to the in-flight conditions, the more meaningful is the thermal model correlation. This means that the use of Sun simulators or Earth simulator plates is an asset. Thermal balance tests usually have a long duration depending on the thermal mass of the DUT, its conductive paths, its insulation, the dissipated heat, the environmental settings, and the applied temperature stabilization criteria. It has to be considered that the mere presence of test and measurement equipment, such as heaters and temperature sensors, alters the thermal behavior of the DUT compared to the thermal model. The steady state criterion for TBal tests is reached when the temperature rate of change falls below a certain threshold, for instance, $dT/dt < 1\,K/4\,h$ [15]. Either the DUT is switched on during the test or additional test heaters are integrated in the set-up to create a temperature gradient. In order to correlate test results with the thermal model, it is necessary to measure and monitor the temperatures and dissipated heat of the test configuration.

9.5.1.2 Thermal-Vacuum Test

A thermal-vacuum test serves to test the satellite in a space-relevant environment. Hot and cold cases that were predicted for the operation of the satellite in orbit shall be simulated as realistically as possible. This means that not only vacuum but also IR panels are used to simulate the central body (mostly Earth) and solar simulators for the Sun. Solar simulators are demanding in terms of power and thermal control. Furthermore, parallel rays and the reproduction of an appropriate solar light spectrum are required. Whether the use of solar simulators in a thermal vacuum test is necessary must be decided based on the thermal requirements. Hot and cold cases can also be simulated in vacuum by IR heat sources only. If sunlight is required for another reason, such as to test the performance of optical instruments, it should be evaluated whether such a test needs to be done with the entire nanosatellite under vacuum or if it can be done on subsystem or unit level under ambient pressure.

9.5.1.3 Thermal Cycle Test

The purpose of thermal cycle tests is to verify workmanship and reveal manufacturing errors, such as cold solder joints or material defects. The temperature of the DUT is cycled between nonoperational minimum and maximum temperatures. The DUT can be operated or switched off during the test. Before and after the thermal cycles, functional tests are performed. Thermal cycle tests are performed at very different levels of a product lifecycle. For part qualification thousands of cycles are performed. So-called accelerated life tests cycle components at high faster rate and with a wider temperature range in order to qualify the components' suitability for many cycles in orbit. On subsystem, instrument or satellite level usually of four to eight cycles is performed. Eight cycles are required for qualification tests and four cycles for acceptance tests. Thermal cycle tests can be combined with thermal-vacuum or thermal balance tests. Depending on the DUT, thermal cycle tests can be performed in a climate chamber under ambient pressure, for example, for purely structural components without internal heat dissipation. For nanosatellite system level thermal cycle tests, where often a proto-flight model approach is followed, it is common to conduct four cycles. The first cycle is usually a nonoperational cycle and three cycles are with the satellite in operational mode.

Figure 9.6 shows the exemplary test profile of a thermal-vacuum test of the CubeSat "First-MOVE," including one nonoperational temperature cycle, two thermal balance phases, a heater functional test, system functional tests at different temperatures, and two operational thermal cycles. The test profile shows an idealized qualitative procedure. The battery as the most temperature-sensitive component of the CubeSat is plotted individually. Note that the pressure curve is only indicative, because in reality the pressure changes with the temperature inside the thermal-vacuum chamber.

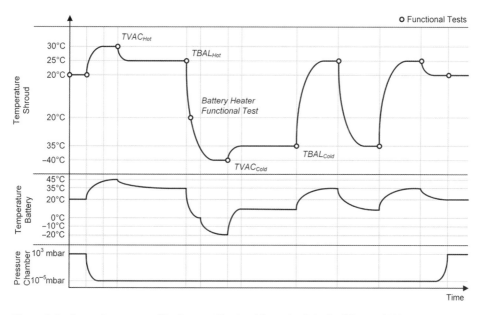

Figure 9.6 Exemplary test profile. Source: Obtained from the CubeSat "First-MOVE" test procedure.

9.5.2 Guidelines for Thermal-Vacuum Test Preparations

In a standard thermal-vacuum test, the DUT is placed inside a thermal-vacuum chamber, supported by mechanical ground support equipment (MGSE) and surrounded by the chamber thermal shroud. Electrical ground support equipment (EGSE) is connected to the DUT via connectors, feedthroughs, and cables. The test setup can easily become very complex due to a multitude of cables, connectors, adjunct computers, or data acquisition systems. The following considerations serve as guidelines for the preparation of thermal-vacuum tests:

- Thermal-vacuum test preparations for nanosatellite system level tests should be started several months in advance. If external test facilities are to be used, they should be contacted at least 6 months in advance. Ample time for the procurement of test hardware (MGSE, EGSE, thermal hardware, etc.) should be scheduled.

- The GSE as much as possible should be placed outside the vacuum chamber—otherwise the vacuum rating and temperature ranges of the flight hardware also apply to the GSE.
- The number of cables, harness length, connectors, pin allocation, feedthroughs, heaters, and temperature sensors should be documented before the test, as well as the setup of the test GSE with data acquisition, heater control, and data storage. Harness lengths can easily be underestimated depending on obstacles in the real test facility. On the other hand, long harness might cause additional errors to the data acquisition or transmission.
- MGSE that supports the DUT needs to be thermally isolated, or if not possible has to underlie thermal control in order to avoid thermal gradients, which would lead to artificial heat paths, leaks, and sources, and would potentially lead to thermally induced structural distortion, which might impact the functionality or performance of the DUT. Measurement and heater control data need to be saved for thermal model correlation.
- The use of vacuum-rated material is mandatory for thermal-vacuum tests. Many standard off-the-shelf components for test setups are not vacuum rated. Thus, the procurement of vacuum-rated cables, connectors, heat shrink tubes, tapes, cable ties, or resin needs to be taken into account. The acceptable outgassing rates for the utilized materials should be checked in advance [26, 27]. Note that professional test facilities for thermal vacuum test demand proof of the latter.
- The nanosatellite should be tested in a flight configuration, which includes both hardware and software. Malfunctions that are encountered during a thermal-vacuum test should be taken seriously and considered crucial for the flight mission.
- Nanosatellite mounting should minimize conductive links to the vacuum chamber. Ideally, the nanosatellite shall be suspended.
- The calibration status of all sensors should be recorded prior to the test. This includes internal sensors as well as facility sensors.

References

1 Harland, D.N. and Lorenz, R.D. (2005). *Space Systems Failures—Disasters and Rescues of Satellites, Rocket and Space Probes*. Springer.

2 Rotteveel J. and Bonnema A. (2005), Thermal control issues for nano- and picosatellites, *Proceedings of the International Astronautical Congress*, IAC-05-B5.6.07.

3 Baturkin, V. (2005). Micro-satellites thermal control – concepts and components. *Acta Astronautica* 56: 161–170.

4 GomSpace (2016), *NanoMind A712D Datasheet*, URL: https://gomspace.com/UserFiles/Subsystems/datasheet/gs-ds-nanomind-a712d-16.pdf, accessed 03/2019.

5 Innovative Solutions In Space (2016), *ISIS On board computer*, URL: https://www.isispace.nl/wp-content/uploads/2016/02/IOBC-Brochure-web-compressed.pdf, accessed 03/2019.

6 GomSpace (2018), *NanoPower Battery Datasheet*, URL: https://gomspace.com/UserFiles/Subsystems/datasheet/gs-ds-nanopower-battery-17.pdf, accessed 03/2019.

7 Clyde Space Ltd. (2010), *User Manual: CubeSat 1U Electronic Power System and Batteries: CS-1UEPS2-NB/-10/-20*.

8 Som, S.K. (2008). *Introduction to Heat Transfer*. PHI Learning Private Limited.

9 Prölss, G.W. (2004). *Physics of the Earth's Space Environment*. Springer https://doi.org/10.1007/978-3-642-97123-5.

10 Howell, J.R. (2015). *A Catalog of Radiation Heat Transfer Configuration Factors*, 3e, URL: http://www.thermalradiation.net/indexCat.html, accessed 03/2015.

11 Walter, U. (2012). *Astronautics – The Physics of Space Flight*, 2e. Wiley.

12 Messerschmidt, E. and Fasoulas, S. (2011). *Raumfahrtsysteme*, 4e. Springer.

13 Fixsen, D.J. (2009). The temperature of the cosmic microwave background. *The Astrophysical Journal* 707: 916–920. https://doi.org/10.1088/0004-637X/707/2/916.

14 Fröhlich, C. and Lean, J. (2004). Solar radiative output and its variability: evidence and mechanisms. *The Astronomy and Astrophysics Review* 12 (4): 273–320.

15 Gilmore, D.G. (2002). *Spacecraft Thermal Control Handbook, Volume I: Fundamental Technologies*, 2nd revised edition. AIAA.

16 Peyrou-Lauga R. (2017), Using real Earth albedo and Earth IR flux for spacecraft thermal analysis, Proceedings of the 47th International Conference on Environmental Systems, ICES-2017-142. doi: https://doi.org/10.1046/j.1474-919X.2003.00223.x.

17 Doenecke J. (1993), Survey and Evaluation of Multilayer Insulation Heat Transfer Measurements, SAE paper 932117.

18 Finckenor M. M. and Dooling D. (1999), Multilayer Insulation Material Guidelines, NASA/TP-1999-209263.

19 Bevans J. T., Ishimoto T., Loya B. R., and Luedke E. E. (1965), Prediction of Space Vehicle Thermal Characteristics, Air Force Flight Dynamic Laboratory Technical Report AFFDL-TR-65-139.

20 Reay, D. (2013). *Heat Pipes Theory, Design and Applications*, 6e. Butterworth Heinemann.

21 Özişik, M.N. (1994). *Finite Difference Methods in Heat Transfer*. CRC Press.

22 ITP Engines UK Ltd. (2014), ESATAN-TMS Thermal Engineering Manual.

23 Modest, M.F. (1993). *Radiative Heat Transfer*. McGraw-Hill.

24 ECSS-E-HB-31-03A (2016), Thermal analysis handbook.

25 NASA (1976), Thermal Vacuum Bake-out Specification for Contamination Sensitive Hardware, MSFC-SPEC-1238.

26 NASA (2014), *Outgassing Data for Selecting Spacecraft Materials Online*, URL: http://outgassing.nasa.gov, accessed 03/2015.

27 ESA (2015), *Outgassing Data*, URL: http://esmat.esa.int/services/outgassing_data/outgassing_data.html, accessed 03/2015.

10 I-2i

Systems Engineering and Quality Assessment

Lucas Lopes Costa[1], Geilson Loureiro[1], Eduardo Escobar Bürger[2], and Franciele Carlesso[1]

[1]National Institute for Space Research, São José dos Campos, Brazil
[2]Department of Mechanical Engineering, Federal University of Santa Maria, Brazil

10.1 Introduction

The number of nanosatellite space missions has sharply increased since the standardization of platforms and growth of launch opportunities. Nanosatellite projects arise as a new branch in the domain of spacecraft programs. The traditional engineering processes and practices may not be fully appropriate for smaller spacecraft development mainly due to large differences regarding schedule and cost limitations. The traditional systems engineering (SE) process has been successfully applied on larger satellites, however, it does not mean that implementing it on nanosatellite development is effective. The remarkable complexity difference between the concepts of large and small satellites identified since 2000s is becoming unclear. Recent nanosatellite developments are complex with the use of new technology and have the same engineering problems related to project development, space environment, and operations in a smaller scale.

The complexity and size attributes suggest that applying a traditional engineering approach to nanosatellite development may overwhelm project developers, and negatively affect two important features of small satellites: low cost and fast delivery.

The first nanosatellite development relates back to a variety of purposes. Their first uses were in university environment motivated by the need of students' involvement in real space projects and need of improvement in space education methods. The most common primary goal of nanosatellite development in university environment is conceiving, designing, and building a nanosatellite as an educational tool, and in second place is the mission assignment derived from organizational interest. Given this characteristic and the cost and schedule constraints related to nanosatellite development, a tailored SE process is required.

The absence of a specific process for small satellites made many projects to be developed without a systems engineering approach, being considered time consuming. A new set of practices and engineering processes are needed to achieve appropriate schedule, budget, quality, and risk management (RM) to develop a balanced solution that satisfies stakeholders' needs for this class of satellites. This chapter describes concepts of

Nanosatellites: Space and Ground Technologies, Operations and Economics, First Edition.
Edited by Rogerio Atem de Carvalho, Jaime Estela, and Martin Langer.
© 2020 John Wiley & Sons Ltd. Published 2020 by John Wiley & Sons Ltd.

systems engineering discipline and quality assessment trends and challenges applied to nanosatellite development.

10.2 Systems Engineering Definition and Process

SE is an engineering development philosophy that can be used for any system development: product, process (and service), or organization. This wide scope of application is due to the holistic way of view and broad thinking approaches used in the discipline. To see from a holistic point of view is challenging and it becomes even harder for professionals trained as "component engineers," who are used to solve specific engineering problems and do not see a complex product from a system and interface perspective. The SE thinking approach is needed due to the functional risks and concerns of stakeholders and architecture that must be considered during the entire development process and in parallel traded off with design and performance goals.

SE has different formal definitions, but the basic concept remains the same with the goal of obtaining a balanced solution to satisfy the stakeholders. Two main definitions used by the authors in their nanosatellite experience are shown as follows:

- Systems engineering is a multidisciplinary and collaborative engineering approach to derive, develop, and verify a balanced solution along its life cycle and attend stakeholders' expectations [1].
- Systems engineering is a transdisciplinary and integrative approach to enable the successful realization, use, and retirement of engineered systems, using systems principles and concepts, and scientific, technological, and management methods [2].

The SE definition states that a multidisciplinary approach is recommended to successfully develop complex systems. This approach shall be part of nanosatellite development due its importance to obtain a balanced solution avoiding rework, schedule overruns, and unplanned costs.

The decision of a SE process adoption in a satellite project is usually done through the tailoring of traditional space SE processes (e.g. National Aeronautics and Space Administration [NASA], European Space Agency [ESA], and Department of Defense [DoD]). In order to do the tailoring effort for nanosatellite development, the process structure shall be changed so that activities are clustered and adapted, modifying it into new processes or even excluding tasks to the point that the tailored process gets ideally fit for such application. The tailoring process shall be done consciously to extract the main purpose of each task from the original source by remodeling it to be performed in other way or clustered with other task and do not simply exclude it.

The tailored SE process must consider the specific characteristics of each project; for example, the development environment and team's technical experience in addition to the organization's culture and lessons learned. When these aspects are considered for the SE process definition, the chances for successful application and obtaining good results to the project development are higher.

Communication and knowledge of the principles and goals of the SE process are important for the entire project team. The expected results shall be shown to motivate engineers to

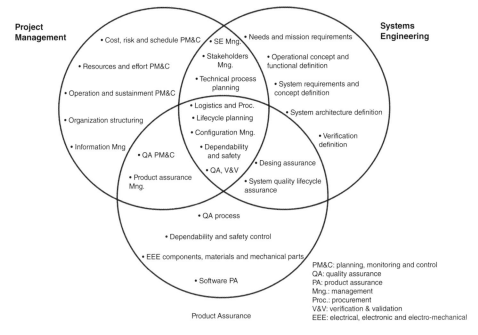

Figure 10.1 Systems engineering integration and interfaces.

engage and contribute to the development of the SE process. Larson [3] highlights the need for all project staff to agree in a common SE understanding in such a way that, no matter how roles and responsibilities are distributed, all members are on the same page regarding the development of the system in a clear and objective way.

SE discipline is integrated and interfaces with other high-level disciplines are needed in the space systems development. The deep integration attribute results in overlap of tasks to be performed in conjunction with other disciplines, and this characteristic demands effective management for communication and configuration control. Figure 10.1 shows one example of the interfaces and integration of SE with other disciplines.

The SE process framework shall be created according to the specific needs of each project. The activities can be divided into groups according to their goals or functions. The structure of functional tasks shall be flexible, and each organization or project uses the approach that best fits its environment. The functional division of the SE activities is essential for a nanosatellite development mainly when participants are frequently changed. The example shown in Figure 10.2 is a typical functional division for a nanosatellite project-tailored SE process. This approach allows the separation of activities into three logical work packages and facilitation for interfaces and interaction between definitions of activities.

Technical management (TM) consists of the project technical integration effort with the objective to comply with the schedule, costs, risks, and performance requirements. The TM performs the planning and control of the SE process, and aims to obtain a system optimal technical solution concerning the management constraints. Other activities to be developed by the TM include the definition and control of communication approach, control of technical activities and resources, technical risk management, configuration management

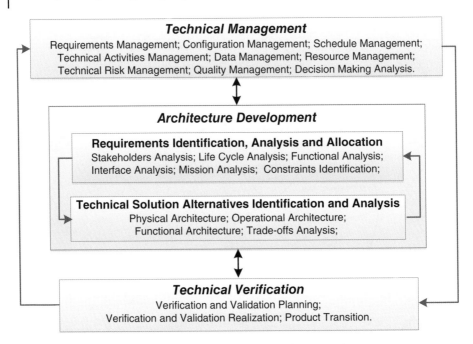

Figure 10.2 SE functional division for a nanosatellite project example.

support, requirements management, and decision-making analysis assistance. The following elements form the TM function for the nanosatellite's SE process example:

- *Requirements management*: Configuration, control, traceability, correctness, and requirements of conflict solving.
- *Configuration management*: Physical, functional, interfaces, documentation, and control of configuration changes.
- *Schedule management*: Planning, controlling, and schedule communication.
- *Management of technical activities*: Planning of activities, process control, and technical progress evaluation.
- *Data management*: Communication, information planning, and control.
- *Management of resources*: Budget, planning and control of workforce.
- *Management of technical risks*: Policy definition, planning, controlling, and assessment of risks.
- *Quality management*: Planning, controlling, and assessment of quality.
- *Decision-making analysis*: Strategy, methods, and criteria for technical decision making.

Architecture development (AD) function consists of identifying stakeholders' needs, translating stakeholders' needs into technical requirements, and developing and designing balanced alternative solutions to comply with the requirements. The system architecture characterizes the system elements, their organization, interfaces/interconnections, functional structure, and behavior in a structured way, representing the fundamental concept of the technical solution. The AD function is composed of two activity blocks in the example shown in Figure 10.2:

- *Requirements identification, analysis, and allocation*: Stakeholders' analysis, lifecycle analysis, functional analysis, interface analysis, mission analysis, constraints identification.
- *Technical solution development and analysis*: Physical architecture, operational architecture, functional architecture, and analysis of trade-offs.

Technical verification (TV) consists of planning of verification methods, definition and execution, definition of requirements verification levels, and model philosophy. The following elements form the TV functional block for the nanosatellite SE process example (Figure 10.2):

- *Verification and validation (V&V) planning*: V&V methods, levels, and strategy definition.
- *Verification and validation realization*: V&V realization and evaluation of results.
- *Product transition*: Planning and assessment of product transition from developers to users.

The functional division of SE activities allows a smooth assignment for SE tasks among project staff, which can comply with the organization's functional structure. Besides that, it is important to develop the functional division of SE activities according to the current organization structure, avoiding misunderstanding, providing technical knowledge maintenance, and utility of the lessons learned.

The SE process definition should be concurrently built with the functional division of SE activities and should be logically consistent. The SE process is the inter-related or iterative set of activities to be performed and intends to create a balanced system solution (output) to satisfy the stakeholders' needs (input). For the SE discipline, the process is highly iterative and concurrent activities are performed. The highly iterative characteristic brings the need for an effective configuration and communication management.

10.2.1 Architecture Development Process

This subsection presents an example of the AD process (part of the complete SE process) tailored for nanosatellite projects, and also some important aspects to be considered during their planning and development. Figure 10.3 shows the highest view of an SE process example for a nanosatellite mission and the specific AD process. The nanosatellite AD process example's main activity blocks and respective subactivity objectives are organized as follows:

- *Identification of needs*: Identify stakeholders' needs and define top level mission requirements.
- *Mission definition*: Define the operational concept of the mission, mission requirements, and mission architecture elements' requirements.
- *System definition*: Define the system of interest requirements (segment to be developed), identify technical alternative solutions to the system of interest, decompose the system requirements into its parts (lower hierarchy level), and define the physical solution, functional architecture, and its allocation (architecture).
- *Development of subsystems*: Define the requirements to the lower hierarchy level to be developed by specialists of each knowledge area (e.g. power subsystem developed by electrical/electronic engineer).

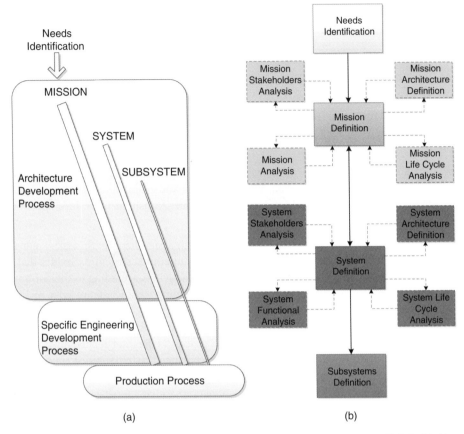

Figure 10.3 Nanosatellite architecture SE process example. (a) High view of a SE left side V model, (b) AD process.

Some points related to the process model example are highlighted in Table 10.1. These are important premises used for the process example that facilitate the common understanding and should be known for all process executioners.

The SE architecture process example can be modified for each project application in the way to achieve the objectives of developing an optimal solution to the system of interest. The activities of TM, AD, and TV are developed along the process and are iteratively updated following the development of dynamic nature.

Some important aspects of nanosatellite SE shall be defined to form the basis for the SE process. Typical characteristics of nanosatellite development are highlighted:

- *Simplicity*: Nanosatellite is known as a smart solution for the mission objective with the potential of innovation development. The simplification (but not poor) solution is used in all aspects of development. Excessive documents, processes, management, and too complex technical solutions can be avoided to keep the low cost and fast delivery attributes, unless to maintain those characteristics is not mandatory for the mission.
- *Modularity*: Nanosatellite architecture uses modularity approach to facilitate the multiple integration and disassembly activities during Integration & Testing (I&T) process.

Table 10.1 Premises and important aspects of the SE process example.

Subject	Description
Scope	The SE AD process comprises mission analysis, life cycle analysis, stakeholders' analysis, functional analysis, and implementation analysis. These processes consider, from the outset, the product not only in the context of its operations but also in the context of all other life cycle process scenarios, which are identified and explored at the life cycle analysis activity and are used as reference for the other activities.
Phasing correlation	The SE process is not presented in a space project phase condition due to the different phasing structure adopted for nanosatellite projects.
Analysis of life cycle scenarios	The SE process considers the analysis with the main life cycle scenarios early in the conceptual study phase of the system. This approach allows the identification of elements to be developed and processes that the nanosatellite must be submitted to.
Documentation planning	The SE process foresees the development of few documents, making the configuration and documentation management easier.
Modeling and tools	The SE process activities are suitable for modeling and tools. This is an important role of nanosatellites due to the new SE paradigm from document-centric to model-centric. Such freedom is important to contribute through evaluating the application of such new approaches.
Simplification	SE process simplifies the traditional standards, without losing the results and objective of core activities (Figure 10.3).
Hierarchy levels	The SE process considers three hierarchy levels, beginning for the mission level (space mission architecture), converging along the same process core to the development of a specific system segment (e.g. space segment), and the specific technical process for subsystem development (out of scope of SE process). It is important to note that activities are repeated for different hierarchy levels.
Iterative process	The SE process is highly iterative and nonlinear. The gates and maturity level to establish each baseline shall be according to project management and SE agreement.

The architecture often consists of a plug-and-play assembly and minimizes the use of wires and individual connectors, which provides less complex design and seeks reliability growth.

- *Accessibility*: This characteristic is related to the modular approach and mainly affects the mechanical structure and mechanical assembly. The CubeSat standard is an example of that; the satellites usually use a modular stack structure with bus interface that can reduce the need for wires and connections, providing easy accessibility to all satellite elements.
- *Commercial off-the-shelf (COTS)*: Mission stakeholders and budget availability guide the criticality level of the mission, and this leads to the quality level selection of the components. There is a trend for using COTS components and parts in nanosatellites that leads to a global study investigation of COTS use in space environment.

- *Open source*: Communication protocols, code libraries, tools, and other "open sources" are used by nanosatellite developers due to the availability of widespread information and lessons learned, which also leads to savings of cost and time.
- *Model-based systems engineering (MBSE)*: Agile, lean, effective, resilient, and adaptive MBSE tools are boosted for SE discipline in nanosatellite development. More discussion about MBSE will be addressed later in this chapter.

The top-level SE process is broken down into a more detailed view to clarify the relationship of activities and for management purpose. Figure 10.4 shows the example of breakdown activities and iterations of a tailored SE architecture development process for identification of needs and mission definition subprocess. The highly iterative characteristic should be managed, and the maturity point established by the SE authority.

10.3   Space Project Management: Role of Systems Engineers

The TM (SE management) is the set of activities of planning, controlling, and evaluating the project's course, integrating specific activities of various disciplines in the project execution, and measuring its progress. SE management acts from the concept development through retirement in order to maintain the system integrity according to stakeholders' needs.

SE planning scope considers all programmatic and technical parts of the project to guarantee a unique and integrated plan for all technical aspects including risk management, configuration management, information management, quality management, system definition, design, manufacturing, verification, integration, testing, and validation. The SE functions are integrated to ensure consistency and feasibility of plans, operational concepts, requirements, and architectures.

The planning effort of SE management provides important project elements:

- SE process definition (tailoring), methods, tools, and plans for the system's technical development.
- Technical organization, team functions, and definition of responsibilities.
- Technical knowledge, disciplines necessary for system development, and definition of its interfaces.
- System life cycle model definition.
- Technical review and definition of product assessment schedule.
- Success criteria of control gates and definition of management control approaches.
- Estimation of resources and management approach.
- Identification of critical risks and mitigation.
- Configuration and information planning.

The systems engineering management plan (SEMP) is the main document that guides the project's technical development, assigns responsibilities, establishes the management strategy, and identifies needed resources.

The SEMP has the function of coordinating the technical schedule and assuring the compatibility with other technical efforts. In order to do that, it is synchronized with the other technical plans of the project (project management plan, risk management plan, software

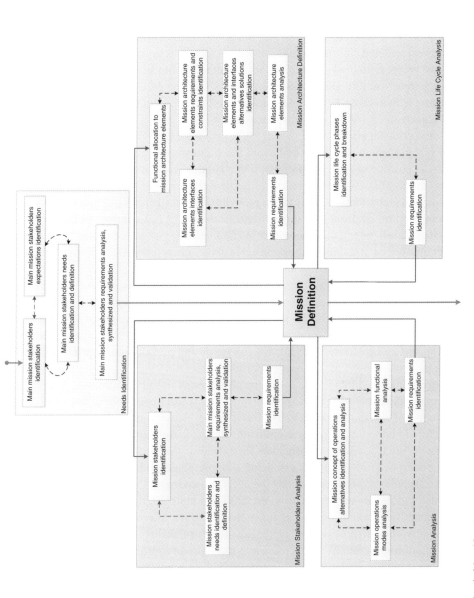

Figure 10.4 Needs identification and mission definition sub-process breakdown for a nanosatellite SE architecture development process.

management plan, dependability plan, and others). To be compatible with other plans, it is important to acknowledge that SE management has interfaces with all technical areas and this should be well established in the SEMP document.

The detailed effort, deliverables, and SE acceptance criteria are clearly identified, as well as planning of activities in collaboration with project management discipline. The technical planning is documented in the SEMP, which is frequently updated along the development to maintain such planning coherent with the project's resources and constraints.

SE assessment and control is a pillar activity that intends to ensure a real visibility of the technical progress and risks according to the planned technical activities. The importance of an accurate visibility allows that preventive actions be performed in a more effective way. Some activities of SE assessment and control can be highlighted: to monitor project performance and project risk, manage use of resources, project reviews, activities, action items, and critical path schedule.

RM is a process to anticipate and avoid the occurrence of negative risks' impact and encourage positive ones in the project using a continuous and collaborative approach to reduce (for negative risks) or increase (for positive risks) risk levels, and is part of SE management. RM is considered a cross-cutting activity through all disciplines involved in project development and controlled by project management. RM is a continuous process to aid the system development, otherwise it would be ineffective when performed just to satisfy project milestones. RM implies the definitions of risk management strategy, identifies and analyzes risks, and manages and monitors risk treatment.

Monitoring of technical activities is fundamental for SE management and provides information related to the system development, process implementation, and usage of resources to support the technical process development and the system quality. Adequate indicators and measures are fundamental inputs to trade off analyses, to balance schedule, performance, and cost. It helps to identify risks early and provides the basis for models and methods.

SE decisions are difficult due to the involvement of multiple objectives, uncertainty, risks, many stakeholders, and constraints of technical and managerial aspects. Due to the importance of decision making, it requires a formal decision-making process to provide a structured and logical basis for identifying, characterizing, and evaluating the different alternatives and providing information for selecting the optimal one. The most used method is trade study, known as trade-offs, whose objective is to define, measure, and assess stakeholders' needs to facilitate the decision making.

Configuration management aims to establish and maintain the integrity of all products (documents, requirements, physical, functional, software, interfaces, and so on) generated in the development process and make sure they are available and accessible. A great challenge of configuration management is to manage and control changes of configuration elements to maintain the updated baseline. This activity is essential for the success of nanosatellite development.

Data management is the process of gathering, collecting, and managing data from their source to the necessary receivers in a safe and reliable way. The choice of the data management system should be according to the organization which holds the project in order to maintain compatibility and access to both specific development project information and organization information. The information management for small satellite development can take advantage from the modern freely available communication and database systems; however, it should be noted that easy communications may make records and traceability difficult.

Quality management, as part of SE management, is concerned with attendance of customer needs and fit for use. It includes the planning effort for quality discipline, establishes a baseline for quality aspects during the system development process, and its attributes can be used as decision-making factors.

Management of technical activities consists in the planning, management, and support of the technical effort during the overall life cycle of the system. This includes the planning of engineering effort and other disciplines, the technical outputs, and required capabilities to structure these elements in a logical framework over the life cycle of the system.

10.4 ECSS and Other Standards

The SE processes, especially its breakdown and visualization, are the main differing aspects between standards and handbooks, while the approach and expected results are philosophically similar. The process generally describes the system development approach with a predefined set of activities and their iterations and relationship used to accomplish the SE tasks and obtaining the results. The standards are usually in a "what needs to be done" form and rarely show how to apply the process. As the overall SE concept should be, mostly available SE standards for development of space systems are of high level and generalist (to be used for different scope applications) and should be tailored to a specific project if necessary by adding, removing, and modifying the activities and their iterations.

In the space domain, some traditional and generally accepted SE process standards include DoD organizations, European Cooperation for Space Standardization (ECSS), and NASA. These SE standards will be briefly shown with their main characteristics.

The SE process adopted by DoD organizations, originated in the canceled MIL-STD-499A (and replaced by the IEEE 15288.1 in March 2017), presented in Figure 10.5, is composed of the main elements: requirements analysis, functional analysis, and allocation, synthesis, control, and systems analysis. There are continuous iterations and feedback between these activities and result refinements during the evolution of development.

The stakeholders' needs, objectives, and requirements in terms of capabilities, measure of effectiveness, environments, and constraints are the starting point to the process development as inputs. The SE process is then composed of two main loops, controlled and analyzed by a management activity. Some points related to the DoD SE process can be highlighted:

- The requirements loop is the initial iteration where the mission and environmental analysis results combined with the identification of functional requirements are inputs to the functional decomposition to the lower level and allocation of requirements to the lower-level functions. The feedback occurs to the requirements analysis to verify the compliance with the mission.
- The design loop is developed in parallel with requirements loop, where the functional architectures and interfaces are established allowing the development of physical architecture alternative solutions. As the architecture solution is developed, the design parameters are analyzed against the allocated requirements.

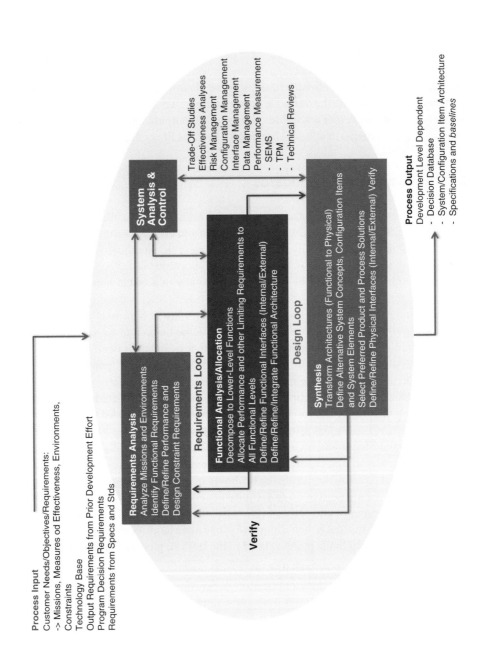

Process Input

Customer Needs/Objectives/Requirements:
-> Missions, Measures od Effectiveness, Environments,
Constraints
Technology Base
Output Requirements from Prior Development Effort
Program Decision Requirements
Requirements from Specs and Stds

Requirements Loop

Requirements Analysis
Analyze Missions and Environments
Identify Functional Requirements
Define/Refine Performance and
Design Constraint Requirements

Functional Analysis/Allocation
Decompose to Lower-Level Functions
Allocate Performance and other Limiting Requirements to
All Functional Levels
Define/Refine Functional Interfaces (Internal/External)
Define/Refine/Integrate Functional Architecture

Design Loop

Synthesis
Transform Architectures (Functional to Physical)
Define Alternative System Concepts, Configuration Items
and System Elements
Select Preferred Product and Process Solutions
Define/Refine Physical Interfaces (Internal/External) Verify

Verify

System Analysis & Control

Trade-Off Studies
Effectiveness Analyses
Risk Management
Configuration Management
Interface Management
Data Management
Performance Measurement
 - SEMS
 - TPM
 - Technical Reviews

Process Output

Development Level Dependent
 - Decision Database
 - System/Configuration Item Architecture
 - Specifications and *baselines*

Figure 10.5 DoD SE process. Source: Adapted from reference [4].

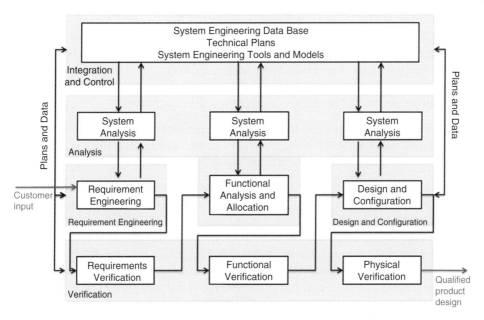

Figure 10.6 ECSS systems engineering process. Source: Adapted from reference [5].

- The system analysis and control is the activity responsible for planning, management, decision analysis, communication, process control, schedule development, and cost estimates.
- The SE process output includes a decision analysis database and the solution to the system architecture with the decomposition of top-level requirements toward the immediate lower hierarchy level.

The ECSS standard is a well-accepted SE reference for development of space systems, its SE process is shown in Figure 10.6. The ECSS SE process is described at ECSS-E-ST-10C through the SE discipline requirements and does not make a comprehensive description of the process itself and techniques to be used (as it is not the document objective). The process consists of five main elements: integration and control, analysis, requirements engineering, configuration and design, and verification.

Some aspects can be highlighted by analyzing the ECSS SE process:

- At the initial phases of the project, the SE develops design solutions using as input the stakeholders' requirements. This is reached due to an iterative top-down process that analyzes alternatives of solutions in an increasing detail level. Through this process, the SE performs multidisciplinary functional decomposition to obtain products for the lower hierarchy levels (hardware and software), and, in the meantime, the SE verifies optimal allocation for the entire system.
- The functional decomposition defines, for each system level, the technical requirements for the subsystems or lower level products to be developed, as well as the verification requirements for each product.

- The SE uses integration and verification activities (right side of the V model) from the lowest level through a bottom-up approach until to reach the validation of stakeholders' requirements. The main output is a system that satisfies the stakeholders' technical requirements within the established cost, time, and quality.
- The 2017 version of the standard presents a pretailoring table, which aims to help users on tailoring process and emphasizes the tailoring need evaluation.

The NASA SE process shown in Figure 10.7 is composed of three subprocess blocks: system design processes, technical management processes, and product realization processes. Some characteristics are highlighted about the NASA SE process:

- The systems design process is used to establish stakeholders' expectations, develop and define technical requirements, and create design solutions that satisfy the requirements and stakeholders' expectations for all system hierarchy levels.
- The product realization process is used for each product in the system hierarchy structure from the lowest-level product to the highest-level integrated product, defining a design solution for each system product through implementing or integration, verifying, validating, and transmitting to the next higher hierarchy level that satisfies the designed solution and stakeholders' expectations.
- The technical management process is applied to establish and control the technical planning of the project, manage communication between interfaces, evaluate the progress in accordance to plan and requirements, control technical execution of the project, and support the decision making.
- The technical processes are recursively and iteratively applied to decompose the initial system concepts in a sufficient detail level to permit a solution implementation by specialists. Thus, the process is applied recursively to integrate from the product's lowest detailed level to the integrated, verified, validated, and delivered system.

Other SE processes, not specific from space domain, are well accepted in the space systems engineering discipline, and a combination or tailoring of them may be useful for the development of a SE process for small satellite projects. Standardization organizations such as ISO recently developed specific aerospace standards on SE, and this aspect supports our argument of high degree of difficult-to-tailor SE program for space applications. The trends in SE process are linked to procurement and business opening with the aim of unifying a common understanding through an independent organization SE standard.

10.5 Document, Risk Control, and Resources

The SE process documents for a small satellite project shall concentrate the essential information that fits in a small configuration management effort. The baseline configuration of the evolving solution is critical for project success, and each control gate or project review shall freeze such baseline through documents. The number of documents generated during a nanosatellite project development is quite short and does not require a complex management approach, but should have a formal strategy, rules, and format for content changes, traceability, and availability.

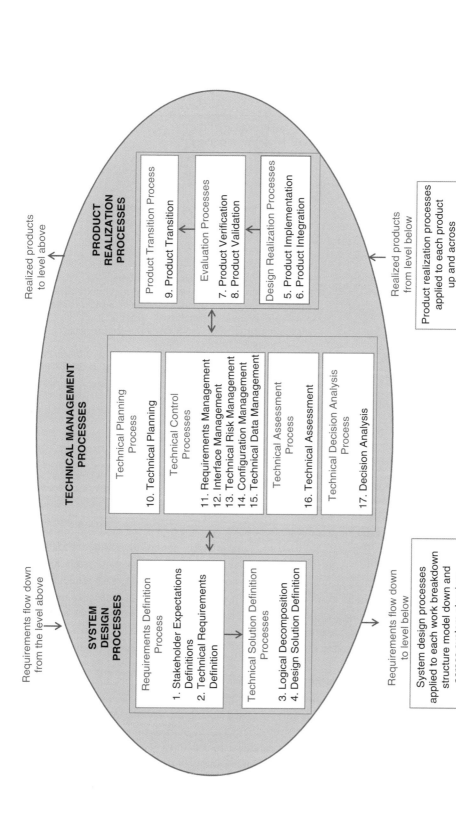

Figure 10.7 NASA SE process. Source: Adapted from reference [1].

Table 10.2 Example documentation descriptions for the AD process.

Document	Content
Systems engineering management plan	• Project overview • SE development strategy • SE activities and responsibilities • SE interfaces with other project disciplines • SE process description • SE activities per project phase • Methods, tools, and models for development of SE activities
Stakeholders' analysis and mission requirements	• Mission stakeholders and identification of expectations • Identification of needs • Definition of stakeholders' requirements • Mission requirements and definition of success criteria
Mission concept of definition and analysis of operations	• Functional analysis of mission • Identification and analysis of mission's architecture elements • Identification and analysis of mission's concept of operation • Mission life cycle and analysis of operation modes • Definition of mission architecture and requirements derivation for architecture elements
Analysis of system requirements	• Analysis of system stakeholders • System functional analysis • Identification and analysis of system's concept of operations (operational modes analysis) • Life cycle analysis (life cycle process identification and flow down of life cycle scenarios)
System architecture and subsystem requirements	• Definition of system's functional architecture • Definition of system's physical architecture • Analysis of technical risks • Flow down of requirements of subsystems

Some activities in the SE process are frequently evolving and being modified during the solution concept development, such as functional analysis, context diagrams, and state diagrams. This frequently changed baseline has great impact on the overall system and the most recent version shall be available through an online system for the SE group, at least during this concept phase, or ideally be developed in a concurrent engineering approach. Table 10.2 shows an example of AD process documentation and its content for a tailored SE process for a nanosatellite project. It is possible to note that there is a clustering of contents that are closely related in the same document, facilitating in case of small changes of content.

Risk is an event or uncertainty condition that can affect the project goals in a positive or negative way. It is usually described as a combination of the probability of occurrence and its impact and should be carefully managed during the overall project development. Any project development has the intrinsic characteristic of uncertainty in its occurrence or in generating the proper outputs. Nanosatellite projects often involve several risks by the extensive use of non-space-qualified components and new technologies. Thus, the project manager is responsible for planning and implementing a management system to eliminate or minimize the probabilities of undesired risk effects and maximize the probabilities of

Figure 10.8 Risk range related to uncertainty.

positive risk effects. Risks are considered when there is any level of uncertainty about an event, while facts and completely unknown events are out of the scope of risk definition and should not be treated as risk. Figure 10.8 shows the event characteristic related to uncertainty that can be considered as risk.

Risk management is a continuous and permanent activity during entire life cycle of the project and it is the responsibility of all persons involved in the development process, making it multidisciplinary. Some examples of risk sources can be technical (e.g. technology, performance, reliability, quality, etc.) or program related (e.g. schedule, costs, supplier, resources, etc.). The typical risk process consists of six main cycling steps: identification, assessment, acceptance decision, action (reduce the risk), monitor, and communicate.

A commonly used method for risk assessment is to assign a qualitative score for the probability of occurrence (likelihood linked to risk scenario) and severity of consequences, and through a scale categorization previously defined, the risks are classified and presented in a color scale 5×5 matrix, see Figure 10.9. Approaches with statistical techniques such as NASA's PRA (probabilistic risk assessment) are not suitable for situations with less available data, as most nanosatellite projects are new development at the organizations, there is low experience with these missions, becoming ineffective to use statistical methods for risk assessment.

Score	Severity[a]	Score	Probability[b]
1	negligible	1	minimal
2	significant	2	low
3	large	3	medium
4	critical	4	high
5	catastrophic	5	maximum

[a] Severity can be scored by schedule, cost and technical attributes.
[b] Probability can be scored by frequency of occurrence.

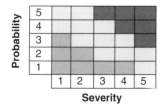

Risk level
High-implement new process or change baseline
Moderate-close monitoring, consider alternative process
Acceptable-monitor

Figure 10.9 Traditional risk assessment approach.

Project management in collaboration with SE management defines the risk management approach and these definitions are documented in a risk management plan. It is necessary to be aware of the possible changes during the project and report the risks as soon as they are identified. The high-level risks may decrease throughout the project to acceptable levels by performing actions and monitoring the results. The frequency of reviewing and monitoring risks shall be reasonable and depends on each project. For nanosatellite projects, a review in every 15 days is recommendable. Some strategies and recommendations for managing risks are mentioned in the following text:

- The project manager is responsible for the risk management plan implementation and communication.
- To define who is responsible for managing the risks in their respective field, and what communication, information and reports, and responsibilities shall be related to risk management issues.
- Each project area (such as engineering, software, verification, and schedule control) manages the risks arising in its domain, under the supervision of project manager.
- Risks are formally accepted for mitigation by higher levels of selection criteria.

In summary, the purpose of developing such a plan is to determine the best cost-benefit approach and time-performing risk management in the project. The risk management plan should be mature and consolidated for project design stages.

Nanosatellite project participants are essential for the mission success. Projects developed in university environment tend to have a high rotation of participants mainly due to the limited time they are in graduation courses and the concurrency between other projects. The engagement of human resource is one of the most important issues in nanosatellite projects. A good strategy to assure certain level of workforce consistency is to form a core team with some experience and available time to guide the development until the end of the project. Additionally, it is important to maintain a structured division of work breakdown structure (WBS) with each participant responsible and also a strategy to absorb the knowledge of the exchanging participants. Recruitment tools can then be used to attract students from different disciplines to join the project. However, the human resource is not a controllable attribute.

The risk of not having the proper funding in nanosatellite projects is also usually present. In this context, some good practices allow saving project budget:

- Use of prototype and simulation hardware.
- Use of free tools and open sources.
- Use of organizational standards and access to external bases.
- Serial manufacturing for reusable parts.
- Early planning of effort and resources.

Control of resources is one of the main responsibilities of managers, but the conscious use of resources shall be treated as a rule for every project participant together with organization's culture.

10.6 Changing Trends in SE and Quality Assessment for Nanosatellites

The quality discipline and its attributes make use of standards and traditional indicators as the main baseline for the development of space projects, but the traditional quality processes approach should not be the same for nanosatellites. The development approach using SE concepts together with the computational capability improvement brought the intense use of models to support requirements of management and development, simulation models for behavior analysis, safety, engineering performance, reliability, cost, and schedule. However, SE uses a document-based approach that brings complexity to the configuration management and decreases reliability.

The MBSE is the application of modeling to support system requirements, design analysis, verification, and validation since the conceptual design phase through all life cycle phases. The main focus is developing, managing, and controlling a representative model of the system instead of documents. This provides the potential to better quality, improved communications, reduced time, and reduced cost. MBSE is a new approach already in successful use for nanosatellite development.

The quality of Electronic, Electrical and Electromechanical (EEE) components to be used in space systems became a paradigm break with nanosatellite growth and can be considered a challenging trend for quality assessment of nanosatellites, since this satellite class design approach makes extensive use of commercially selected components. Many researches have been conducted to understand and establish procurement of components and quality assessment approach by classifying the criticality of use and the quality of available suppliers. Techniques such as screening, accelerated tests, and statistics are being explored when not-space qualified components shall be used in space systems.

The strict quality process and controls used for traditional satellites are too much for a nanosatellite development to afford with, but the balance between what and how much should be done in quality effort allowed the development of new approaches. A test-based quality assurance is being the trend for nanosatellite projects. This approach uses the same basic quality concerns, but with more flexibility, where the main quality assessment is made through testing the system and its components.

Nanosatellite quality assessment trends are always related to the effort of balancing between doing what is traditionally performed for a space system according to standards (ESA, NASA, DoD) and what involves the lowest resource usage to accomplish a successful mission.

References

1 Loureiro, G. 1999. A systems engineering and concurrent engineering framework for the integrated development of complex products. 391 p. Tese (Doctorate at Manufacturing Engineering) – Loughborough University, England.

2 INCOSE. 2019. Systems Engineering – Transforming needs to solutions. http://www .incose.org/systems-engineering (accessed 31 January 2019).

3 Larson, W.J., Kirkpatric, D., Sellers, J.J. et al. (2009). *Applied Space Systems Engineering*, 1e, 920. Learning Solutions ISBN-10: 0073408867; ISBN-13: 978-0073408866.

4 Pennell, L. W. and Knight, F. L. 2005. Systems Engineering. Military Standard: MIL-STD-499C. Aerospace Report no. TOR-2005(8583)-3. El Segundo, California. April 2005, 113 p.

5 European Cooperation for Space Standardization (ECSS) (2017). *Space Engineering: System Engineering General Requirements* v.3, Rev. 1, 116. Noordwijk: ESA Requirements and Standards Division (ECSS-E-ST-10C Rev.1).

11 I-2j

Integration and Testing

Eduardo Escobar Bürger[1], Geilson Loureiro[2], and Lucas Lopes Costa[2]

[1]Department of Mechanical Engineering, Federal University of Santa Maria – UFSM, Brazil
[2]National Institute for Space Research – INPE, Brazil

11.1   Introduction

The term "integration and testing" (I&T) relates to the sequential process of joining parts, confirming that the parts of this assembly are working together and testing to verify fulfillment of requirements. Even though this process is recursively repeated from lower-level configuration items (right after manufacturing) up to the complete system of interest, the I&T acronym is widely (and herein) used referring to the last stage of satellite integration and testing, in which the subsystems are integrated and tested to form the space element.

Nanosatellites are products that have several parts, equipment, and subsystems, which interact among themselves to provide output to the system functions. These systems are designed to survive very harsh conditions imposed by launch and space environments in which they are exposed. Beyond that, nanosatellite developers generally make extensive use of commercial off-the-shelf (COTS) components, which have been investigated about their performance and survivability in such environments. Therefore, coupled with the impossibility of repairing a failure after launch (with few exceptions) and limited resources, rigorous satellite integration and testing campaign is mandatory to ensure that this satellite class performs as designed reducing the risk of on-orbit failures.

I&T is performed during phase D of a typical space project life cycle. Though this process starts in the early phases within verification planning, well before all subsystems are integrated and tested, it ends at the moment the launch vehicle begins the lift-off. In I&T stage, system tests such as vibration, thermal-vacuum, and electromagnetic tests are performed to detect materials or integrated workmanship defects to ensure that the nanosatellite will survive upcoming environments, will not interfere with the launch vehicle nor neighboring satellites, and will perform according to designed parameters. By the mantra of "low cost and fast delivery," the I&T of this class of satellite is often underestimated. The focus is usually given simply on meeting launch requirements, keeping reliability at low priority. But the rapid and great advances on small satellite applications are demanding this situation to change.

The following items will introduce the reader to some basic concepts regarding I&T needed for a better understanding of this chapter.

Nanosatellites: Space and Ground Technologies, Operations and Economics, First Edition.
Edited by Rogerio Atem de Carvalho, Jaime Estela, and Martin Langer.
© 2020 John Wiley & Sons Ltd. Published 2020 by John Wiley & Sons Ltd.

11.1.1 Integration

Literature usually distinguishes the terms integration and assembly. Assembly activities are related to mechanical operations performed to physically position, secure, and interconnect all of the satellite parts [1]. In other words, it may be seen as integration from a mechanical point of view. Integration is the process of assembling elements with the following confirmation that each individual unit works properly when interconnected. Integration relates to functionally combining lower-level assemblies (hardware or software) so they operate together to constitute a higher-level functional entity. The integration of nanosatellites can be defined straightforwardly as the process of combining several subsystems into one system due to their small proportions. Each of these subsystems individually contributes to the whole system functionality, but only the combination of all subsystems provides the complete system functionality.

11.1.2 Testing

Testing is the process that demonstrates the expected behavior of a system under specific conditions. Tests are designed to reproduce the estimated operational environment and scenarios with the best accuracy whenever it is possible. It may be defined as the measurement of product characteristics, performance, or functions under representative environments [2]. The tests are part of a set of alternatives available to verification of requirements, and are considered an expensive, but effective method. There are different types of tests that can be performed at different levels of integration, development stages, and models. The test has generally auxiliary equipment that simulates environments, conditions, or restrictions to which the system will be exposed, in order to indicate performance, functionality, absence of failures, or resistance to harsh environment for a certain period [3].

System level testing can be divided into two types, environmental and functional tests. Environmental tests simulate various conditions (simultaneously or separately) on which an item is subjected during its operational life. They include natural and induced environments [2], for example, the vibration and thermal-vacuum tests, which are designed to replicate the launch and on-orbit environments, respectively. Functional tests are electrical tests that verify if the satellite functions are performing properly and the designed performance is achieved and maintained after exposure to a particular testing, transport, or to the space environment [1].

There are good reasons to perform as many tests as possible with nanosatellites, but in fact, this is not possible mainly due to programmatic constraints. Although there is the extensive use of COTS, there is absence of well-established testing standards specific for such class, and sometimes inexperienced workmanship (educational projects); the attribute of low cost is usually imperative for this class, and tests are expensive. Nanosatellite developers shall then find the proper balance between performing tests or other verification method, in order to have a cost-effective way to have a reliable system.

11.2 Overall Tasks

The I&T is part of the verification process. Verification is the right side (uprising) of the "V" development process model (Figure 11.1). It comprehends the confirmation that the

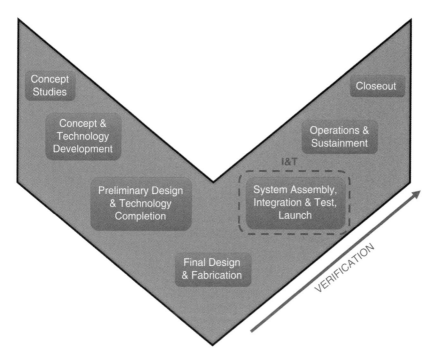

Figure 11.1 "V" development process model showing the I&T position. Source: Adapted from NASA [3].

product meets the designed requirements using verification methods, such as analysis, test, inspection, and demonstration. The I&T is a top-level verification activity of the system. It is composed by logical and sequenced activities in order to have subsystems assembled, integrated (working together), and tested to form a functional system.

Even though the I&T operations are executed during the phase D of a typical project life cycle, several activities are performed since early project phases. The important tasks of defining test requirements and creating integration and test procedures are examples of early activities. Table 11.1 shows an example of a nanosatellite's major I&T tasks along the project phases.

The nanosatellite's I&T tasks are subdivided according to Figure 11.2. Activities are broken down to organize and allow an effective management. This results in a work breakdown structure (WBS) to be incorporated by the overall project WBS. The next items address nanosatellite integration and tasks definitions and objectives of tests.

11.2.1 Integration Tasks

The integration and verification tasks are very interrelated and interactive to each other. Integration and verification tasks follow one another sequentially until forming a complete functional system. Müller [5] states that the main goal of integration is to reduce the project risk of being late and the risk of creating an ill-performing system by performing systematically and logically organized activities. These integration activities are performed throughout the system life cycle, they start when project activities are defined and end with launch integration.

Table 11.1 Example of a nanosatellite's major I&T tasks along the project phases (adapted from [4]).

Phases	Major I&T activities/products
Phase 0 Phases A and B	I&T requirements
	I&T plan—1st version
	I&T quality assurance plan—1st version
	GSEs specifications—1st version
Phase C	I&T plan—final
	I&T quality assurance plan—final
	GSEs specifications—final
	Tests procedures development—1st version
	Perform tests in satellite models (EM, EQM, etc.)
Phase D	Tests procedures—final
	Perform I&T of flight models (PFM or FM)
	Perform launch integration
Phase E	Support on-orbit tests

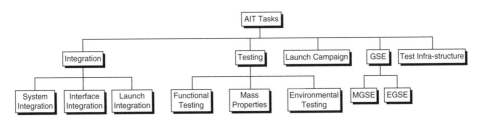

Figure 11.2 Nanosatellite I&T WBS.

Integration planning begins in the concept formulation phase and goes up to the actual start of system integration. It comprises the identification and preparation of all integration inputs, restrictions, and resources. The main activities are the definition of internal and external interfaces, requirements, set-up and evaluation of integration enabling products (support equipment), integration strategy and sequence definition, facility and tools preparation, and development of all support documentation, such as plans, procedures, instructions, and drawings.

The most important document of this activity is the integration plan. This plan should identify a sequence to receive, assemble, and activate the various components that make up the system. To avoid the occurrence of nonconformances during nanosatellite integration (typically high), it is wise to have well-detailed procedures (step-by-step) of assembly and integration, as well workmanship practicing with nonflight models. Establishing the integration strategy is an important part of such planning, and it shall include at least the following considerations: integration sequence (flow), description of the work to be done, responsibilities for each activity, resources required, schedule, procedures, tools, environment, and personnel skills.

The integration execution takes place after planning acceptance, which is generally done in the critical design review (CDR). Typical nanosatellite integration is composed of the following three tasks (adapted from [6]):

- *System integration*: Relates to the process of integrating components to form subsystems and then integrating subsystems to form a complete system ready to be tested.
- *Interface integration*: Relates to the process of integrating the nanosatellite on the launch interface. This activity may be the responsibility of the nanosatellite provider, interface provider, or a third-party integrator.
- *Launch integration*: Relates to the process of integrating the interface with satellite on launch vehicle. This activity is the responsibility of the mission integrator.

11.2.2   Testing Tasks

Testing is considered an expensive but effective method to verify requirements. Because of the need of special (and continuously calibrated) test equipment, specialized test operators, and appropriate facilities to test satellites, several nanosatellite developers outsource testing, involving another organization in the process, increasing the need for a well-planned and documented testing phase. This task of planning comprises the identification and preparation of all testing inputs, restrictions, and resources. The main activities define the "why," "what," "how," and "when" of testing, that are, respectively, the definitions of test requirements; test strategy and models; test procedure, levels, duration, and support equipment; and test schedule.

System level tests can be categorized into two main types: environmental and functional. Environmental tests use equipment to simulate several conditions (simultaneously or individually) that a system will be subjected during its operational life. Functional tests are mechanical or electrical operations performed to verify satellite's functional and performance requirements. Whether environmental or functional, each test is performed through the same basic set of tasks that are: confirmation of readiness of specimen (satellite), test equipment, support equipment (ground support equipment [GSE]), documentation and facility; performing test, document results, analyze results, and decide whether the nanosatellite passed or failed testing. Depending on the failure, corrective actions are taken and the test is redone.

11.2.2.1   Functional Tests

Functional tests are used to assess if the satellite parts were properly integrated, before and after exposure to testing, transport, or space environment. It is also usual to run functional tests during environmental tests, such as thermal-vacuum, and to evaluate compatibility with GSE.

Nanosatellite developers commonly use three or four levels of functional tests. They differ from the purpose and complexity. An example of this division is given in the following text (adapted from [7]):

- *Functional checks*: Conveniently abbreviated functional checks to verify if satellite is still working.
- *Baseline functional test*: Complete functional test demonstrating major satellite functions, prior and after environmental tests.

- *Flight simulation*: Simulates all satellite functions during operation.
- *Deployment test*: Demonstrates all satellite deployments, such as sensors, antennas, and solar panels function according to requirements.

11.2.2.2 Mass Properties

Measurements of mass properties are performed to demonstrate compliance with the requirements of satellite's mass properties, which generally are derived from constraints of launch vehicle, satellite attitude, and orbit control subsystem requirements. There can be four types of measurements:

- *Mass*: Demonstrates the mass of the system, meeting the mass requirements.
- *Center of gravity*: Demonstrates that the center of gravity is located according to requirements.
- *Momentum of inertia*: Demonstrates that the component values of moment of inertia (x, y, and z) are according to the requirements.
- *Alignment tests*: Demonstrate that the orientation of subsystems remains within the specified limits related to satellite axes and their stability before and after tests.

11.2.2.3 Environmental Tests

This section defines the most traditional tests performed in nanosatellite environmental test campaigns. It is important to note that there are other tests not mentioned herein, such as radiation, pressure, leakage, and other tests that come from specific missions or launch requirements. Nanosatellite launches involving manned spacecraft, such as deployments from the International Space Station, implicate the need to perform further tests that may be required to prevent astronauts from harm. An example is the off-gassing, a variation of the out-gassing test that differs by having specified limits for each compound detached from the satellite. Another example is the sharp-edge inspection, performed in order to prevent astronauts or hardware from damages. For further detailed information regarding thermal tests, please see Chapter 09 I-2h.

Electromagnetic tests: Electromagnetic compatibility (EMC) tests evaluate a possible interference of the nanosatellite's electronics with its own subsystems. This compatibility is also assessed regarding a potential interference with launcher emissions (even though most nanosatellites are cold launched).

Resonance test: Demonstrates first that satellite resonance modes are above values specified in requirements, in order to prevent dynamic coupling between nanosatellite and launcher. This test is also used to verify system structural integrity before and after all dynamic tests.

Shock test: Demonstrates the capability to bear the mechanical shock environment. This high-frequency transient acceleration results from the sudden application or release of loads associated with deployment, separation, or impact.

Random vibration test: Demonstrates the ability to withstand the random vibration environment expected during launch liftoff and ascent.

Sinusoidal test: The sinusoidal vibration test demonstrates the ability to withstand periodic excitations stemming from an instability (such as pogo, flutter, combustion) or to those due to rotating machinery.

Quasi-static test: Demonstrates the ability to withstand the highest acceleration generated by launch vehicle. Nanosatellite test engineers may conveniently run this test using the same shaker through a sine-burst test, a method to apply a quasi-static load using a vibration shaker.

Thermal-vacuum cycling test: Demonstrates the ability to withstand the thermal-vacuum stressing environment, and that the system is free of materials, process, and workmanship latent defects. The test consists of hot and cold temperature cycles under high vacuum with functional verifications on determined points. Depending on the nanosatellite's power level, this test may be replaced by thermal cycling test (at ambient pressure).

Thermal balance test: Demonstrates the proper functioning of the thermal control system in order to maintain components within the operational temperature limits (according to requirements). This test is also used to validate analytical thermal models.

Bake-out (out-gassing): It may be seen more as a procedure than a test. It accelerates out-gassing rates through increasing the hardware temperature under high vacuum in order to reduce the content of molecular contaminants within the hardware. This test is usually combined with thermal-vacuum test.

11.3 Typical Flow

The I&T sequence of nanosatellites follows the same rule of thumb of bigger satellites: to reproduce the order of environments that the system will be exposed through during its life cycle with the goal of detecting failures and latent defects as early as possible. Figure 11.3 shows the typical macro sequence flow of nanosatellite's I&T. It is important to note that this

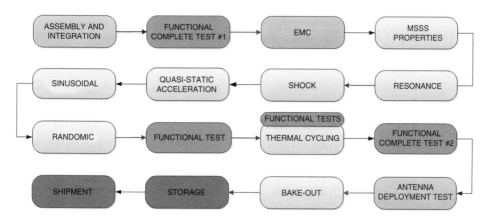

Figure 11.3 Typical macro sequence flow of nanosatellite I&T [4]. (*See color plate section for color representation of this figure*).

is a macro sequence, therefore (in practice) there are several other functional tests between each environmental test.

11.4 Test Philosophies

A wisely chosen test philosophy plays an important role in nanosatellite projects. Such decision directly affects their most significant characteristics of cost, schedule, and risks. The test philosophy decision shall be made with the proper balance of these attributes in order to have an effective solution for the test program and confidence in the product verification, while preserving the features of being a nanosatellite and not a larger-scale satellite. The term "test philosophy," also known as "test strategy," represents the type and characteristics of each physical model (model philosophy) and its associated verification objectives through tests. In other words, test philosophy stands for the definition of what will be tested, when, and with what purpose.

The verification process is mainly composed of development, qualification, and acceptance stages. In each stage, verification by test is performed on physical models to meet different verification objectives, such as design evaluation, verification of design margins, or workmanship verification. Therefore, tests can also be divided in the same stages of development, qualification, and acceptance.

11.4.1 Test Stages

This section defines the principal nanosatellite test stages. It shall be noticed that larger satellites deal with other test stages that for simplification purposes are not addressed herein, such as prelaunch and in-orbit stages. Since the vast majority of nanosatellite launches are secondary or tertiary payloads, there are few or no prelaunch verification activities. The on-orbit testing is an extensive subject, and it is beyond the scope of this chapter.

Development test: Conducted on new items to demonstrate that the design meets all applicable requirements.

Qualification test: Conducted to demonstrate that the design meets all applicable requirements and includes proper margins.

Acceptance test: Conducted to demonstrate that the flight item is free of workmanship, materials or latent defects of processes, integration errors, and is ready for subsequent operational use.

Protoflight test: Is the combination of qualification and objectives of acceptance testing. Conducted when no qualification hardware is available. Used to prove that the design on the first article produced has a predetermined margin above expected operating conditions but typically less then qualification levels.

11.4.2 Test Models

Given the lower complexity, low cost, and fast delivery approach of most nanosatellite projects, the number of system models is reduced as compared to larger satellites. Generally, the same model is used for different purposes in I&T campaigns. An example is the

Table 11.2 Common nanosatellite models and their respective objectives.

Model	Objective
Qualification model (QM)	Dedicated model to prove design on the first article produced. Subjected to qualification tests, environmental levels of which are above expected operating conditions; therefore, shall not be used for flight due to overtesting. Ideally, this model shall be fully similar to the flight model, but due to some constraints slight differences may be acceptable.
Protoflight model (PFM)	A combination of the qualification and acceptance testing objectives on the flight model. Usually the test levels are the same as qualification and test durations are the same as acceptance.
Flight model (FM)	Model to be launched. Tests are performed in acceptance levels to detect workmanship and latent defects of materials.

frequent use of an engineering and qualification model (EQM) both to engineering and qualification purposes [4]. The most common nanosatellite models and their respective objectives and characteristics are described in Table 11.2.

11.4.3 Test Philosophies

The test philosophy is chosen early in the project during verification planning and shall be established as soon as possible due to its impact on project implementation. The imperative aspects to define a nanosatellite test philosophy are programmatic constraints, technology maturity, acceptable risks, and number of satellites (in case of a fleet). The two most traditional nanosatellite test philosophies are described as follows:

Qualification and acceptance philosophy: A qualification model (QM) or an EQM is subjected to qualification tests and a flight model (FM) is subjected to acceptance tests. Also known as prototype philosophy, this is the reference philosophy for having the lower risks. The disadvantage is the high cost that involves the production and testing of two models. The QM (or EQM) shall be flight representative in terms of design, materials, tooling, and methods; such requirement may sometimes be difficult to meet by nanosatellite developers. This philosophy is usually taken by projects with new or complex design.
Protoflight philosophy: Only a protoflight model (PFM) is subjected to a combination of qualification and acceptance testing. This philosophy implicates more risks than the previous one; however, it involves a cost saving by having only one model to produce and test. This philosophy is recommended when design involves qualified or with flight heritage products and there are no new or complex technologies.

Alternative philosophies may be defined for nanosatellite programs with multiple elements, such as constellations projects. In these cases, it is common to skip part of testing program in order to save time and costs. Decisions such that shall be taken with considerations regarding cost, schedule, and mainly risks.

11.5 Typical System Integration Process

Nanosatellite integration may be a very tricky phase; it comprehends one of the most common sources of nanosatellite failures. The high presence of defects relates to poor systems engineering, documentation, and limited knowledge and experienced workmanship (human reliability). Nanosatellite projects were widely adopted by the educational community; this significant percentage of educational projects influences the whole nanosatellite class reliability. This is caused because most of nanosatellite reliability indicators have a broad generalization, mixing educational, commercial, scientific, and technological missions in the same statistics [8, 9].

The system integration activities begin with the preparation of lower-level assemblies through harness electrical testing, including signal, impedance, and other parameters. After lower level products are validated against requirements, the integration process is repeatedly performed until forming a complete functioning system. During the integration, the documentation is updated in the form of integration reports. When the complete assembly takes place, a complete functional test is performed to verify that all satellite functions and parameters are according to specifications.

During integration, the project team tries to find unforeseen problems as early as possible in order to solve them in time. Since imperfect fitting parts to more severe functional incompatibilities, the team must be mobilized to act efficiently at this stage. This phase often takes much more time and effort than planned. The order of integration of subsystems usually relates to function, beginning with power subsystem, following on-board computer, communications subsystem, attitude determination and control subsystem, and then payload. During testing, the problem takes another dimension with the difficulty to identify if the failure root cause lays in the integration or it is due to the environmental testing itself. A good approach to attenuate such problem can be through extensive use of functional testing in every integration phase.

Another important variable in the integration process is the integration facility. Despite being recommended, not all developers have access to a complete integration environment with contamination, temperature, and humidity controls. However, good engineering practices are usually inexpensive (such as the use of gloves and antistatic tools and equipment, such as electrostatic discharge [ESD] wrist straps and mats) and avoid catastrophic outcomes.

11.6 Typical Test Parameters and Facilities

11.6.1 Typical Test Parameters

Prior to running a nanosatellite's environmental test, test engineers need to have access to detailed specifications of the test, which describe every condition on how the specimen shall be tested to verify its requirements. The test specification gathers characteristics, such as test requirements, test approach, GSE and tools, test sequence, test parameters (or test conditions), pass/fail criteria, and schedule. This section defines the basic nanosatellite environmental test parameters for the major test areas: thermal-vacuum tests, dynamic tests, electromagnetic tests, and general parameters.

Table 11.3 Typical general parameters.

Tolerances	It is the range of allowable variation in specified parameter values. For example: ±2°C for temperature levels.
Functional tests	It consists of the functional tests performed to verify nanosatellite behavior before, during, or after environmental tests. For example: to verify if power is being supplied and signal transmission is being transmitted, both with the correct values.
Pass/fail/abort criteria	Pass and fail criteria address the conditions to approve or reprove the tested specimen and detail the means by which such conditions shall be verified. Abort criteria states the conditions followed by test operator to stop testing given abnormal situations.
Test interfaces (mechanical and electrical)	Mechanical interfaces comprehend all equipment that will interface the test equipment and nanosatellites. A wise choice of such interfaces is important to prevent interference with test results. An example is the impact of an interface on conductive thermal isolation of thermal-vacuum tests. Nanosatellites have a small heat budget, and the heat leakage through mechanical supports influences thermal balance. Electrical interface is essentially all cabling used for functional testing during environmental tests.

The following sections give a brief description of each parameter (Tables 11.3–11.6) featuring, when appropriate, some examples.

11.6.2   Typical Test Facilities

A satellite test facility (laboratory or center) is developed to meet the needs of assembly, integration, and testing of space products. For most of nanosatellites, the major test equipment required to perform an I&T program is shown in Table 11.7.

It is important to note that Table 11.7 shows just the major test equipment of each test area. Test centers shall have other supporting areas that were not described herein but are also important for testing.

11.7   Burden of Integration and Testing

11.7.1   I&T Costs

The I&T process is very time-consuming, laborious, and consumes a major chunk of the cost of any space program. Such fact is not different for nanosatellites, but I&T cost percentages of the total project may be even more substantial. Unlike larger satellites, there is not enough category cost data to make precise cost inferences to this small class of I&T campaigns. Furthermore, it is a fact that the wide range of applications and different mission objectives of satellites from 1 to 10 kg turns the expenses with I&T very different from one project to another. The discrepancy of I&T costs among nanosatellite projects relies on two main reasons: depending on project's nature, there are differences on what an I&T campaign

Table 11.4 Typical thermal-vacuum test parameters.

Temperature soaks	Refers to the maximum and minimum temperatures that shall be experienced by nanosatellites. They are usually in Celsius. For example: −15 and +50°C.
Temperature transition (ramp) rate	Refers to the temperature transition rate to which nanosatellites shall be heated or cooled. For example: ≤5°C min^{-1}.
Temperature stabilization	Also referred to as stabilization criteria, it defines the criterion to consider that nanosatellite temperature has reached a stable (or nearly stable) temperature within test tolerance. For example: temperature change is less than 1°C per 10 min.
Soak time	It is also called as "dwell at plateaus" by some organizations. It is the total time at plateau temperatures (hot or cold temperature soaks). The time starts to count after the temperature soaks are reached by control thermocouples. For example: 2 h.
Pressure level	It is the pressure level in which the test shall run. The vacuum measuring unit is usually expressed in mbar, Torr, or Pa. For example: 10^{-4} Torr.
Pressure drop rate	It is the maximum pressure drop rate that shall be imposed to the nanosatellite in order to avoid undesired effects, such as corona discharge. It may be expressed in the units similar to that of pressure level. For example: 3.92 kPa s^{-1}.
Number of cycles	It is the number of "loops" between hot and cold temperatures on which the nanosatellite shall be tested. For example: four cycles.
Test profile	The test profile is defined by graphs showing temperature over time and pressure over time.
Instrumentation information	It can be divided into the positioning of thermocouples and definition of the control thermocouple. In order to control the temperature of specific areas, the thermocouples shall be strategically positioned. Also known as temperature reference point (TRP), the control thermocouple is settled as reference for temperature soaks.

will verify (beyond launch provider's requirements); and there is a variation among launch providers' test requirements to secondary or tertiary payloads in terms of testing.

Costly or not, this amount shall be calculated early in the project budget planning. Unfortunately, this expense is often unforeseen or underestimated by project managers causing several impacts on project verification and then to reliability. Nanosatellite developers shall, case by case, find the proper balance between the extremes of assuming higher risks with a reduced and less expensive I&T, or lower risks with a costly and more complete I&T program.

System integration costs refer to person per hour costs and integrated clean room rental. The integration facility rental costs rarely exist because most developers use their own facilities, however, such expense should be considered in a particular case of need because of its high associated costs. Environmental testing can be performed in-house if developer team has all suitable test equipment and sensors, or it can be outsourced (most common). Other than saving costs, the in-house testing makes test troubleshooting a more flexible and less time-consuming activity. On the other hand, the facility equipment and sensors may be less

Table 11.5 Typical dynamic test parameters.

Sweep rate	Defined by the rate of sweeping from a lower frequency to an upper frequency. Usually specified in octaves per minute.
Test profile	Usually specified as a log–log plot and/or a table. Depending on the test, it is composed by amplitude (g) and frequency (Hz) or power spectral density (PSD), also referred to as amplitude spectral density (ASD), given in g^2 Hz^{-1} and frequency. PSD is the distribution of average vibration energy.
Test duration	Depending on the test, the duration is given in seconds per axis or in number of applications per axis.
Root mean square acceleration	Referred to as G_{rms}, it is the square root of the integral of the acceleration PSD over frequency. In other words, it is the square root of the area under the PSD/Hz curve.
Instrumentation information	Refers to the positioning of accelerometer sensors.
Quality factor	Referred to as Q-factor, it is usually used in shock test specifications to demonstrate the level of damping.
Direction	Refers to the nanosatellite direction during test. It is denoted by the X, Y, and Z axes.
Amplitude	The amplitude (G) varies with frequency for most dynamic tests, however, this parameter assumes a constant value for quasi-static tests. The amplitude method used may be 0-to-peak (0–p) or peak-to-peak (p–p). For example: $12G_{0-p}$.

Table 11.6 Typical EMC test parameters.

Maximum level of electromagnetic emissions	Usually specified as a graph or a table indicating the frequency band (Hz) and the respective electromagnetic field strength (V m^{-1}). This parameter is usually given by launch provider through a launch manual.
Standard	Refers to the standard that must be complied with by test operators, indicating test requirements and methods.
Critical parameters	Refers to important satellite parameters that shall be assessed during test. In some cases, inert electro explosive devices (EEDs) may be assembled in the satellite to monitor if appendages deployment activities (such as antenna deployment) are not influenced by electromagnetic radiation.

reliable, poorly calibrated, and with no official certifications as compared to an established testing center. The environmental test costs are composed by the use of equipment (usually calculated by equipment and depreciation of sensors), test supplies, workmanship hours, taxes, and other rates that vary at different test centers, such as profit rate. The most significant of these cost shares is assigned to thermal-vacuum tests due to the long test durations, which leads to large amounts of liquid nitrogen (N_2) and workmanship hours.

Table 11.8 shows an example of a nanosatellite's major environmental testing costs at a specific space I&T laboratory to give the reader a broad order of magnitude reference. It is

Table 11.7 Major test equipment to perform nanosatellite I&T.

Type of testing	Test equipment
Integration	• *Clean room*: A minimum clean room class of 100 000 is recommended (FED STD 209E or ISO 8) with temperature and humidity control. • ESD workbench
Mass properties	• Alignment system • Precision balance • Moment of inertia instrument • Center of gravity instrument
Thermal-vacuum	• Thermal-vacuum chamber • Thermal cycling chamber • Sun simulator or skin heaters
Dynamic	• Electrodynamic vibration shaker • Shock test equipment
Electromagnetic	• Anechoic chamber • Antennas

Table 11.8 Example cost of a nanosatellite's major environmental tests in a space I&T laboratory.

Test area	Environmental tests	Average price
Dynamic tests	Random vibration, sinusoidal, and resonance tests Specifications according to launch provider	US$2000
Thermal-vacuum tests	Thermal-vacuum cycling test and bake-out Four cycles from −15 to 50°C in high vacuum.	US$8000
Mass properties tests	Mass, center of gravity, and moment of inertia	US$1000
	Total cost	US$11 000

important to note that the price of testing varies depending on several factors, such as the type of institution it is performed, test specifications, local currency, workmanship cost per hour, and supply costs.

11.7.2 I&T Schedule

In contrast with cost percentages, the time spent with nanosatellite's I&T is considerably short. The average I&T time of larger satellites corresponds to 23% of the total development life cycle [10]. This same value would represent several months to integrate and test nanosatellites. However, the actual timeframes range from weeks to 2 months. The short time period relates to the lower system complexity, with much less interfaces, assembly activities, and shorter functional tests. Besides that, nanosatellites also do not perform long duration environmental tests, such as thermal-vacuum cycling test, which in larger satellites may take weeks.

Integration of nanosatellite is usually a potential source of several schedule delays. The lack of systems engineering and establishment and implementation of quality assurance

Table 11.9 Example of a CubeSat I&T timeframe.

I&T activity	Timeframe (d)
System integration	23
Mass properties	1
EMC	1
Thermal-vacuum cycling	3
Vibration tests	1
Total	29

processes in nanosatellite projects often leads to interface problems or malfunctions late discovered in the I&T. Troubleshooting when testing is outsourced is also a potential reason of schedule delays; the arduous task of rework usually takes more time than predicted. When a problem is found during an environmental test, the system may need to be disassembled, the problem shall be identified, corrected, and then the satellite shall be reintegrated and retested.

A rough cost estimate of a CubeSat's I&T performed in an integration and test laboratory is given as an example in Table 11.9.

11.8 Changing Trends in Nanosatellite Testing

Each nanosatellite project performs I&T using as guidance different references and standards, from tailoring large satellite standards to performing only the minimum to meet launch requirements. This difference is (in part) a consequence of the absence of well-established and internationally accepted nanosatellite testing standards. Fortunately, this situation is about to change due to international initiatives committed to fill-in this gap with standards that describe test requirements and methods specifically for this satellite class, such as ISO [11]. The broad scope of small satellite applications demands improvements to achieve satellites that are more reliable. A nanosatellite testing standard has much to contribute with that. Trend analysis suggests that the commercial sector will account for the majority of nanosatellites in the upcoming future [12]. The continuous seeking of better results and higher profits in such market indicates a tendency of high adherence to these standards.

Launch is another changing trend in the nanosatellite sector that directly impacts I&T. Nanosatellites are usually launched as secondary or tertiary payloads. But the inconstancy of these launch opportunities holds several projects down, making several nanosatellite initiatives not have a defined launch provider until late project phases. This setback is likely to be mitigated with the perspective of low-cost dedicated small launch vehicles entering the market, offering regular and low-cost space access for small satellites. This changing scenario of launch vehicles will also bring different launch requirements, affecting the whole I&T campaign, probably with more flexible and less severe test requirements. Another changing trend in the launch segment is the multispacecraft missions. Nanosatellites are becoming a worthy alternative to substitute larger and expensive missions. Estimates in space markets depict a point where quantity will replace quality, giving opportunity for multinanosatellite missions, formation flying, and constellation initiatives. This change shall push I&T to evolve for better cost-effective ways to qualify several systems in a row, expanding the use of verification by analysis and similarity.

Even though the I&T is imperative for the development of all kinds of space systems, its methods, results, and lessons learned are seldom shared in the nanosatellite community. The information sharing plays a major role in the nanosatellite world, but it is still limited to systems engineering, design, and manufacturing aspects. Little information exists about nanosatellite integration and testing. However, the increasing awareness of I&T and its significant impact on mission success tend to change this situation. The exchange of I&T experiences promotes access of information in a field that generally needs several years of expertise and also encourages new researches on the topic.

References

1 Silva, A. C. 2011. "Desenvolvimento integrado de sistemas espaciais – Design for AIT – Projeto para montagem, integração e testes de satélites – D4AIT." Doctorate Thesis – Aeronautics Technological Institute, São José dos Campos. 455 p.

2 ECSS (2012). *Space Engineering Testing (ECSS-E-ST-10-03C)*. Noordwijk: European Space Agency.

3 National Aeronautics and Space Administration (2007). *NASA Systems Engineering Handbook*. Washington, DC: NASA.

4 Burger, E. E. 2014. "Reference method to AIT pico and nanosatellites." Master Thesis – National Institute for Space Research, São José dos Campos. 308 p.

5 Muller, G. 2007. "Coping With System Integration Challenges in Large Complex Environments" INCOSE. INCOSE International Symposium, San Diego, CA. June 24-28

6 Kanter, M. H and Sedam, K. J. 2012. "Integration and Management Challenges for Multi-Nanosatellite Missions." IEEE. Aerospace Conference, Big Sky, MT. March 3–10, 2012.

7 Ruddy, M. A. 2012. "Pico-satellite integrated system level test program." Master Thesis. Faculty of California Polytechnic State University, San Luis Obispo.

8 Bouwmeester, J.G. (2010). Survey of worldwide pico- and nanosatellite missions, distributions and subsystem technology. *Acta Astronautica* 67: 854–862.

9 Pariente, M. 2013. "Nanosatellite Myths versus Facts." 5th European CubeSat Symposium. Brussels. June 3–5, 2013.

10 Anderson, N., Robinson, G.G., and Newman, D.R. (2005). Standardization to optimize integration and testing. In: *AIAA Responsive Space Conference, 2005, Los Angeles*. Proceedings of 3rd AIAA. Los Angeles: AIAA, (LRS3-2005-4005).

11 ISO. 2017. "Space Systems – Design qualification and acceptance tests of small spacecraft units" ISO 19683:2017 (E). Switzerland. July.

12 Spaceworks (2019). *2019 Nano/Microsatellite Market Forecast*, 9e, 36. Atlanta, GA: SpaceWorks Enterprises, Inc. (SEI).

12 I-3a

Scientific Payloads

Anna Gregorio

Department of Physics, University of Trieste, Italy
PICOSATS SRL, Trieste, Italy
INFN – Istituto Nazionale di Fisica Nucleare, Trieste, Italy
INAF – Istituto Nazionale di Astrofisica – Osservatorio Astronomico di Trieste, Trieste, Italy,

12.1 Introduction

One major question for many nanosatellite developers is to choose right kind of payloads to be accommodated on board those very small satellites. This question does not simply relate to the feasibility of the measurement, but rather its quality, intended as the performance of the scientific results small satellites can produce. Some solutions are inhibited by strict technical limitations that will be hardly overcome in the coming future, for example, power and thermal control, and lidars, or sophisticated instruments for astrophysics measurements.

If one does not consider the payloads excluded by these hard, technical limitations, the main point of nanosatellite is the size, as payload performance often depends on its dimension especially in case of scientific payloads. Here, a very small satellite as the CubeSat is taken as a reference—a 3U CubeSat with a box of $10\,\text{cm}^3$ base and 30 cm length. Although many of the instruments that will be described later better fit within a 6U CubeSat, the general discussion does not change.

To start with, the performance of optical imaging systems for Earth observations (nadir pointing) is linked to the physical limitations on the size of aperture area. If D is the aperture, h the altitude, Δx the ground spatial resolution, λ the reference wavelength, and $\Delta\theta$ the angular resolution, then for a diffraction-limited optical system:

$$\Delta x \approx h \cdot \Delta\theta \approx h \cdot 2.44 \cdot \lambda/D \tag{12.1}$$

By considering an average value of $\lambda = 550\,\text{m}$, an altitude of 600 km, and an aperture of the order of 10 cm, the maximum achievable resolution is of the order of 5 m.

Optical systems required to perform high-quality imaging of distant astronomical objects and other sophisticated cosmological studies are surely another example, since their sensitivity and signal-to-noise ratio (SNR), defined as the ratio between the detected signal S_{el}

Nanosatellites: Space and Ground Technologies, Operations and Economics, First Edition.
Edited by Rogerio Atem de Carvalho, Jaime Estela, and Martin Langer.
© 2020 John Wiley & Sons Ltd. Published 2020 by John Wiley & Sons Ltd.

and the total noise N_{tot} in the measured wavelength band $\Delta\lambda$, depend on physical limitations of the collecting and aperture area D. There is not a unique recipe to estimate the SNR, a reasonable one is the following:

$$\text{SNR} = S_{el}/N_{tot} = S \cdot D \cdot \eta \cdot T_{exp} / \sqrt{(S + 2B\text{FOV}^2) \cdot D \cdot \eta \cdot T_{exp} + 2N \cdot T_{exp} + 2R^2}$$

$$(12.2)$$

where T_{exp} is exposure time, FOV the system's field of view, η the total receiver efficiency, B the contribution coming from the background (the light coming from the sky itself), N the thermal noise, and R the read-out noise.

In an analogous way, the physical size of small satellites prevents the implementation of high-performance radio frequency (RF) based systems (imaging radars, radar altimeters, lidars, or even synthetic aperture radars (SARs)) limiting the gain G of the antenna and as a consequence the SNR:

$$G \approx \eta \cdot (\pi D/\lambda)^2 \qquad (12.3)$$

where η is the antenna efficiency, of the order of 0.55. To design a high-gain antenna, the antenna size shall be of the same order of the wavelength, for D being 10 cm, this corresponds to a 3 GHz radar system working in the C-band, while scientifically the most interesting bands stay at higher frequencies, for example, in the 94 GHz band for weather forecasting or at multiple frequencies.

These types of scientific programs are mostly out of the scope of small satellites, although some efforts are being carried on and interesting results are being achieved.

12.2 Categorization

Before starting with the proper payload description and analysis, it is worth analyzing the type and purpose of science missions. For small satellites, the mission categorization is not that different from the most general case [1]:

- *Earth observation*: These missions are dedicated to the Earth observation to understand how the global Earth system is changing and what are the sources of change in the Earth system and their magnitude and trends.
- *Solar and space weather*: Sun and near-Earth environment observation: corona, magnetic field, solar wind, Earth atmosphere, magnetosphere, etc.
- *Planetary*: These missions may resemble the Earth observation type missions, but they relate to the other solar system planets and might include landers and eventually rovers.
- *Astronomy*: These missions concern with measuring far-away bodies, such as galaxies, stars, and exo-planets.
- *Astrophysics and fundamental physics and biology*: Astrophysics and fundamental physics missions concern mostly with the observation of deep space, a branch of astronomy that studies the physical laws, the properties and dynamic processes of celestial bodies and of the Universe, and their evolution. Biology missions concern with the observation of the effect of the harsh space environment on biological systems.

Table 12.1 Summary of CubeSat statistics for NASA and NSF. Source: Photo courtesy of NASA.

Launch dates	Technology	Astronomy and astrophysics	Biological and physical sciences	Earth science	Solar and space physics	Planetary science	Total
2006–2015	28	0	4	1	14	0	47
Planned 2016–2018	32	1	4	7	10	3	57
Total	60	1	8	8	24	3	104

Source: NAP report [1].

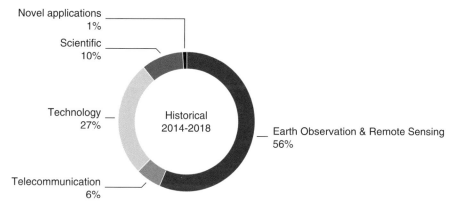

Figure 12.1 Small satellite (1–50 kg) application trend; 2019 market forecast by SpaceWorks Enterprises, Inc.

Table 12.1 gives a clear representation of the number and applications of CubeSat missions launched or under development by NASA and National Science Foundation (NSF) in these scientific fields [1]. Figure 12.1 shows the application trend.

The following sections will briefly discuss the typical scientific objectives for each of these categories and the payloads that have been conceived to meet these requirements by using small satellites. The number of CubeSat missions is rapidly increasing and is not feasible to discuss all the scientific payloads. Only the most recent payloads, grouped by type, or most representative concepts will be proposed.

It has to be stressed that in order to perform this review, five sources of information are considered crucial and have been used along this document, namely:

1. The National Academies Press (NAP) report, "Achieving Science with CubeSats: Thinking Inside the Box" [1].
2. The Earth Observation (EO) Portal Satellite Missions Database, available on https://directory.eoportal.org [2].
3. IDA—Global Trends in Small Satellites [3].

4. National Science Foundation (NSF), Cubesat-based Science Missions for Geospace and Atmospheric Research, October 2013 [4].
5. Gunter's Space Page, URL: http://space.skyrocket.de [5].

12.3 Imagers

An imager detects and conveys the information that constitutes an image, in a defined wavelength band. Optical imagers provide images, typically of Earth surface and atmosphere, and represent the most common instruments used for Earth observation and the most common payload for CubeSat small satellites. According to reference [1], in 2014, 71% of US CubeSat launches were Earth-imaging CubeSats. They are generally nadir-viewing instruments.

12.3.1 MCubed-2/COVE

As a representative optical imagining payload for Earth observation, the COVE (CubeSat On-board processing Validation Experiment) instrument on board the MCubed-2 developed by the UM (University of Michigan) aims at producing color images of the Earth by using an OmniVision 2 Mpixel CMOS camera chip (OV2655) from a low Earth orbit (LEO) at a resolution better than 200 m [2, 6, 7].

The camera, also referred to as μEye, has a 9.6 mm effective focal length (EFL) plano-convex lens, images are saved to a Colibri PXA270 microprocessor at a resolution of 1280 × 1024 pixels, with a pixel size of 3.6 μm × 3.6 μm. This allows for moderate- to high-resolution images of the Earth after postprocessing. The whole camera payload subsystem is fairly small with only 55 cm^3 of volume [8].

12.3.2 SwissCube

The objective of SwissCube is to perform spaceborne observations of the airglow occurring in the upper atmosphere at an altitude of about 100 km. The telescope consists of a printed circuit board (PCB), including a detector and its control electronics, and an optical system, comprising a baffle, the band-pass filter which selects the desired wavelength of the oxygen emissions and focusing optics [2, 9, 10].

The optical system has a size of 30 mm × 30 mm × 65 mm, the payload board 80 mm (length) × 35 mm (width) × 15 mm (depth), and the total mass is 50 g. A CMOS detector is used for photon detection, with a FOV of 18.8° × 25° and a resolution of 0.16° per pixel. The payload consumes 450 mW during 30 s for capturing each image. In order to avoid straylight, the payload is mounted to one of the sides of the satellite via a baffle, as shown in Figure 12.2.

In first phase of 3 months, airglow emissions are observed at different regions and under different angles of observation. These measurements are required to evaluate minimum, maximum, and mean intensities of airglow emissions during both day and night, and to analyze the background radiation due to scattered sun or moonlight. In second phase only, observations of airglow emissions at limb between 50 and 120 km are carried out to study more carefully their variation in intensity.

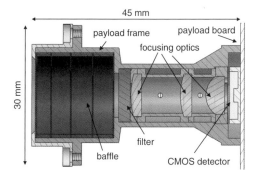

Element(s)	Material	Mass
Closing cap	Aluminium	3 g
Payload frame	Titanium	20 g
Vanes	Stainless steel	2 g
Spacers for the baffle	Aluminium	2 g
Spacers for the optics	Titanium	4 g
Lenses	Glass (BK 7)	1 g
Filter	Glass (RG 9)	1 g
Payload board	PCB	15 g
Screws	Stainless steel	2 g
Total mass		**50 g**

Figure 12.2   Cross-section of the telescope, its elements and mass budget. Source: Reproduced with permission from EPFL.

12.3.3   AAReST

The goal of Autonomous Assembly of a Reconfigurable Space Telescope (AAReST) is to demonstrate all key technological aspects of autonomous assembly and reconfiguration of a space telescope based on an optical focal plane assembly (camera) and multiple mirror elements synthesizing a single coherent aperture [2, 11].

The stowed volume of the telescope is $0.5\,m \times 0.5\,m \times 0.6\,m$. After separation from the primary payload, the telescope deploys its sensor package to the focus of the mirror array using a mast. Figure 12.3 depicts the components of AAReST: "CoreSat," two separate "MirrorCraft," a camera package, and a deployable boom [12].

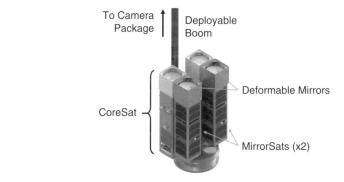

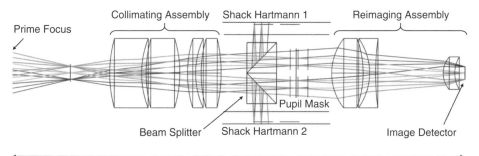

Figure 12.3   Left: AAReST spacecraft components. Right: Geometric ray-trace of the optical path within the camera package. Source: Reproduced with permission from JPL and Caltech/SSL.

The camera package, located at the focus of the telescope, contains various elements, such as corrective optics, wave front sensors, and an imaging detector. Light originating from the prime focus of the telescope enters the camera and is passed through a series of collimating lenses. A pair of beam splitters redirects a portion of this light onto two Shack–Hartmann wave front sensors providing the shape of the deformable mirrors. The remaining beam travels through a pupil mask and a set of lenses to reimage the light onto the imaging detector. Images are captured using this after a calibration process.

The MirrorCraft are independent spacecraft used to house the deformable mirrors, sitting atop these spacecraft. Each MirrorCraft is equipped with its own propulsion system to perform the autonomous reconfiguration and docking maneuvers by using electromagnets and a computer vision system. The deformable mirrors are made using thin glass wafers and a layer of piezoelectric polymer. A custom electrode pattern is incorporated onto the backside of the piezoelectric polymer allowing for shape control to be performed. The mirrors have a total stroke of ~40 µm, allowing for the correction of large shape errors [13].

12.4 X-ray Detectors

X-ray detection cannot be performed by using standard optical system but on small satellites, due to size limitations, nor even by using grazing incident systems. Rather particle detection systems typically based on silicon sensors are employed.

Quite often small satellites use silicon drift detectors (SDDs), able to detect both hard and soft X-ray (SXR) photons. SDDs are small and work at low voltage, making them perfectly fit to miniaturized spacecraft. Electrons are generated by the photon impinging on the SDD, drifted toward and finally collected by the anodes; SDDs can measure the energy of the incoming photon by counting the amount of electrons it produces. At room temperature, SDDs offer an extremely low noise level.

12.4.1 MinXSS

Miniature X-ray Solar Spectrometer (MinXSS) [14, 15] is a 3U CubeSat solar physics mission of the UC (University of Colorado), designed to better understand the energy distribution of solar flare SXR emissions and their impact on the Earth ITM (ionosphere, thermosphere, and mesosphere). MinXSS measurements of the solar SXR irradiance provide a more complete understanding of flare variability in conjunction with measurements from RHESSI and SDO EUV Variability Experiment (EVE).

MinXSS science measurements will be achieved using the commercially available X-123SDD advanced X-ray spectrometer of Amptek [16], with an active area of 25 mm^2, an effective Si thickness of 0.5 mm, an 8 µm thick Be filter on the detector vacuum housing, an active two-stage thermoelectric cooler (TEC) on the detector, and sophisticated multi-channel analyzer (MCA) detector electronics. Key to achieving MinXSS science goals is accurate pointing control and knowledge: with the wide field of view of ±4°, the pointing requirements are 2° (3σ) accuracy and 0.05° (3σ) knowledge. Finally, MinXSS observes the SXR solar spectrum from about 1–5 keV, with a resolution better than 0.5 keV and accuracy better than 30%.

12.4.2 HaloSat

HaloSat (SXR Surveyor) is a NASA 6U CubeSat astronomical science mission aimed at measuring SXR emissions from the halo of the Milky Way galaxy and contribute to solve the "missing baryon" problem looking at hot halos surrounding galaxies [2, 17, 18].

Also, HaloSat uses SDDs from Amptek Inc. [16] with an active area of 25 mm^2 behind a Si$_3$N$_4$ window, testing shows ΔE ~80 eV and Full Width at Half Maximum (FWHM) at 451 eV. The SDDs view the sky through a 13.3 mm diameter hole that is 135 mm away (9.2°–13.4°). The detector is enclosed in a scintillator readout with avalanche photodiodes (APDs) to reject charged particle background (see Figure 12.4).

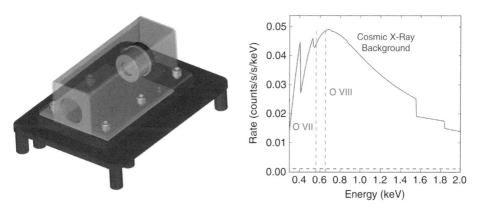

Figure 12.4 Sensor assembly: SDD detector surrounded by a copper–tungsten alloy shielding. Source: Reproduced with permission from University of Iowa. (*See color plate section for color representation of this figure*).

12.4.3 HERMES

Another interesting example is the new European mission HERMES (High Energy Rapid Modular Ensemble of Satellites)—a space-based instrument for high energy (keV–MeV) astrophysics, a science domain previously limited to large space missions. HERMES consists of a constellation of modular nanosatellites in LEO, equipped with X-ray detectors, again an SDD designed on purpose, with at least 50 cm^2 of active collecting area in the broadband between a few keV and ~1 MeV each, and very high time resolution (μs). The main science goal is to study and accurately localize high-energy astrophysical phenomena, such as gamma-ray bursts, possible electromagnetic counterparts of gravitational waves, and the potential high-energy counterparts of fast radio bursts. The transient localization is achieved through triangulations, by measuring the delay of signal arrival time on different nanosatellites, due to the very high temporal resolution of individual detectors.

12.4.4 CXBN

The goal of the Cosmic X-ray Background Nanosatellite (CXBN) science missions is to significantly increase the X-ray measurement precision in the 30–50 keV range, where the cosmic X-ray background peaks diffuse in order to constrain models that attempt to explain the relative contribution of X-ray sources. Results from the mission will lend insight into the underlying physics of this phenomenon [2, 19, 20].

The cadmium zinc telluride (CZT) array is the CXBN science instrument (or science array), developed by Redlen Technologies. The manufacturing process known as the traveling heater method (THM) allows an extremely uniform crystalline structure in the semiconductor material so that charge is evenly distributed, finally improving the energy resolution. The imaging module is formed by 2×2 array of 64-pixel CZT detectors with thicknesses of 5 and 2.46 mm pixel pitch, bonded to a common cathode plate.

12.4.5 MiSolFA

The goal of Micro Solar-Flare Apparatus (MiSolFA) [21] is to provide key insights into the evolution of the solar magnetic field by using a scaled-down version of the Helioseismic and Magnetic Imager (HMI) developed for the European Space Agency's (ESA's) Solar Orbiter mission, to be installed into a six-unit platform.

MiSolFA exploits indirect imaging techniques to provide independent spectra for different sources in the field of view, or to provide images in different energy ranges, achieving 10 arcsec = 0.003° angular resolution. With longer distance between front and rear grids and/or by adopting smaller slit periods, the angular resolution can become better than 1 arcsec. This will make it possible to measure the size of the chromospheric sources and compare them against the EUV "ribbons," and to study the source height dependence on the photon energy.

Viewed from the Sun, the first instrument unit is the imager, sitting just behind a passive thermal shield, which transmits only X-rays above about 5 keV, and hosting the aspect system, which precisely determines the location of the Sun inside the FOV of the imager. A pair of parallel grids is placed in front of a photon detector, the grids and the detector form a "sub-collimator," which provides information about one Fourier component of the image (two real values). The imager produces moiré patterns, see Figure 12.5, on photon detectors, enclosed in the detector box. The moiré pattern is sensitive to the source location along one direction (orthogonal to the slits); two point-like sources displaced along this direction will produce similar modulation patterns, but with different phase. Data from them and from the aspect system are collected by a system logically subdivided into two units, the detector interface unit (NASA) and the data processing unit.

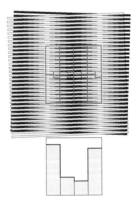

Figure 12.5
Sub-collimator response to point-like source. A minimum of four pixels spanning one moiré period are required to reconstruct amplitude and phase of the modulation [21].

The amplitude of the modulation, or equivalently the contrast of the moiré image, changes as a function of the source size, the maximum amplitude is obtained with a point-like source. A set of sub-collimators sampling different angular scales along independent directions is required to mathematically reconstruct an image of the X-ray emitting source.

12.5 Spectrometers

Spectrometers measure a spectrum, typically the flux intensity as a function of wavelength, of frequency, of energy, of momentum, or of mass.

12.5.1 SOLSTICE

The SOlar Stellar Irradiance Comparison Experiment (SOLSTICE) [2, 22, 23] is one of four solar irradiance measurement experiments launched as part of the Solar Radiation and Climate Experiment (SORCE) in 2003. The mission aims at producing daily solar UV irradiance measurements, comparing them to the irradiance from an ensemble of 18 stable early-type bright blue stars serving as monitor of instrument calibration.

SOLSTICE is a two-channel grating spectrometer aimed at measuring UV radiation in the spectral coverage of 115–320 nm with a spectral resolution between 0.1 and 0.2 nm. The instrument is made of two identical spectrometers, mounted at right angles to facilitate verification of stellar pointing. A mirror reflects the beam on one of two diffraction gratings mounted on a gimbal; rotating the gimbal causes the diffraction of a small wavelength band toward a second plane mirror. The grating disperses the radiation and directs a specific wavelength onto one of the two photomultiplier detectors (Figure 12.6).

The Compact SOLSTICE (CSOL) inherits the SOLSTICE experience for future developments of FUV/MUV CubeSat-sized instruments, still in the full 110–300 nm spectral range on a 10 s cadence, for routine monitoring of solar irradiance. CSOL occupies about 10% the volume of the SORCE SOLSTICE and provides the same wavelength coverage.

Laboratory for Atmospheric and Space Physics (LASP) is also making a miniature version of the SORCE solar irradiance monitor for nanosatellites.

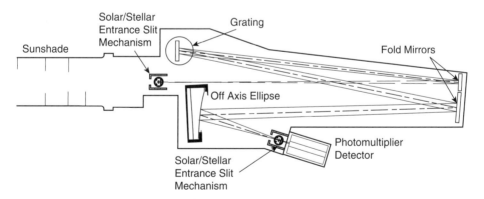

Figure 12.6 Optical-mechanical configuration of a single SOLSTICE I channel. Source: Reproduced with permission from Laboratory for Atmospheric and Space Physics–LASP [22].

12.5.2 OPAL

The goal of Oxygen Photometry of the Atmospheric Limb (OPAL) [2, 24, 25] is to analyze the thermospheric temperature signals of the dynamic solar, geomagnetic, and internal atmospheric forcing by remote sensing of the altitude temperature from the atmospheric limb from a mid- to high-inclination orbit (>50°).

The OPAL instrument is a high-resolution imaging spectrometer that simultaneously collects spatially resolved A-band spectra in multiple azimuthal directions and across the full altitude range of A-band emission. The neutral temperature can be estimated from the qualitative shape of the A-band spectrum over the range 150–1500 K [26]: as the temperature increases, the relative intensity of the R branch (759–762 nm) increases and shifts to shorter wavelength, while the P branch (762–770 nm) widens and shifts to longer wavelength.

OPAL is a grating-based imaging spectrometer with refractive optics and a high-efficiency volume holographic grating (VHG). The optical path is folded into three legs, each about 80 mm long. The scene is sampled by seven parallel slits that form nonoverlapping spectral profiles at the focal plane with a resolution of 0.5 nm (spectral), 1.5 km (limb profiling), and 60 km (horizontal sampling) (Figure 12.7).

The OPAL spectrometer is entirely comprised of commercial off-the-shelf (COTS) components, except for the band-pass filter and VHG, which are inexpensive semi-custom elements. The low-noise and high-sensitivity charge-coupled device (CCD) camera is model Trius-SX9 from Starlight Xpress, the sensor is a Sony ICX285AL monochrome CCD with 1392×1040 pixels at a 6.45 μm pitch. The photon detection efficiency in the A-band is 32%. The camera provides 16-bit readout over a USB interface with RMS readout noise of only $5\,e^-$. The ultralow dark current ($0.1\,e^-\,s^{-1}$ per pixel at 10°C) supports the extended exposure necessary to observe the A-band emissions.

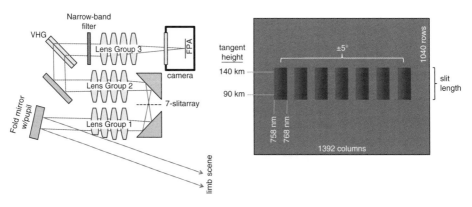

Figure 12.7 Left: Schematic view of the spectrometer optics and light path. Source: Image credit: USU/SDL. Light from the limb scene enters the instrument via a front fold mirror, coincident with the entrance pupil; the scene is imaged onto the slit array by a four-element lens assembly (group 1); light through the slits is recollimated by another four-element lens assembly (group 2, identical to the first); light is then dispersed along an axis perpendicular to the slits by the VHG aligned to the Littrow condition (incidence angle ~ diffraction angle); a narrow-band filter selects A-band wavelengths and finally the last four-element lens assembly (group 3) reimages the slits onto the FPA. Right: Focal plane utilization, dispersion is less than spacing between slit images to avoid spectra overlapping. Source: Image credit of USU/SDL.

12.5.3 Lunar IceCube/BIRCHES

The primary objective for the Lunar IceCube mission [2, 27, 28] is to prospect for water in solid, liquid, and vapor forms, while also detecting other lunar volatiles. The mission is designed to address existing strategic knowledge gaps related to lunar volatile distribution, focusing on the abundance, location, and transportation physics of water ice on the lunar surface at a variety of latitudes. The required scientific observations will be performed from a highly inclined, low-periapsis, elliptical lunar orbit using the Broadband InfraRed Compact High-Resolution Exploration Spectrometer (BIRCHES). A footprint of 10 km from a lunar altitude of 100 km is a key mission requirement, and a footprint of 10 km in track-direction regardless of altitude, larger in cross-track direction above 250 km.

BIRCHES is a miniaturized version of OVIRS on OSIRIS-Rex, a compact spectrometer (1.5U, 2.5 kg, 10–15 W) with a compact cryocooled HgCdTe focal plane array (Teledyne H1RG) for broadband (1–4 µm) measurements, achieving sufficient SNR (>400) and spectral resolution (10 nm) through the use of a linear variable filter (LVF) to characterize and distinguish important volatiles (water, H_2S, NH_3, CO_2, CH_4, OH, organics) and mineral bands. The instrument has built-in flexibility, using an adjustable four-sided iris, to maintain the same spot size regardless of variations in altitude (by up to a factor of 5) or to vary spot size at a given altitude, as the application requires. Also, electronics can be reconfigured to support the instrument in "imager" mode. Thermal design is critical for the instrument: a compact and efficient cryocooler is designed to maintain the detector temperature below 120 K. In order to maintain the optical system below 220 K, a special radiator is dedicated to optics alone, in addition to a smaller radiator to maintain a nominal environment for spacecraft electronics (Figure 12.8).

12.5.4 GRIFEX

The GEO-CAPE ROIC In-Flight Performance Experiment (GRIFEX) [2, 30–32] is a 3U CubeSat mission dedicated to measurements of rapidly changing atmospheric chemistry and pollution transport with the Panchromatic Fourier Transform Spectrometer (PanFTS) instrument. The aim is also to perform engineering assessment of a Jet Propulsion Laboratory (JPL)-developed all-digital in-pixel high frame rate read-out integrated circuit (ROIC), constituting the focal plane assembly, and to validate detector technology for the PanFTS.

In this context, the ROIC works with the PanFTS, a key component of the interferometer used for high-resolution measurements (temporal, spatial, and spectral) to capture rapidly evolving tropospheric chemistry from geostationary orbit on an hourly basis. The PanFTS design [29] combines a conventional Michelson interferometer with a number of features based on spectrometers such as TES (tropospheric emission spectrometer) to measure thermal emission, and such as OCO (Orbiting Carbon Observatory) and OMI (ozone monitoring instrument) to measure scattered solar radiation. The design has two parallel optical trains, see Figure 12.8 (right), one for infrared wavelengths and one for UV–Vis. The beam aperture of 5 cm is driven by the Rayleigh criterion to resolve the ground pixels. The instrument temperature is 180 K to minimize instrument's self-emission.

The working principle is schematically represented in Figure 12.8.

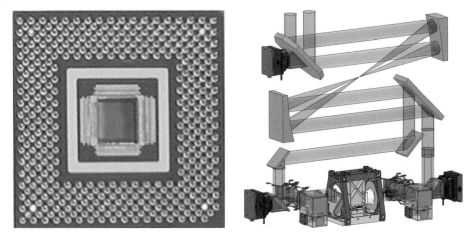

Figure 12.8 Left: ROIC/FPA design: 128 × 128 array, 60 μm pixel. Right: the PanFTS engineering model [29]. Source: Photo courtesy of NASA.

12.5.5 HyperCube

A similar concept of the one used by GRIFEX applies to HyperCube [2, 33, 34], a constellation of 6U CubeSats equipped with mid-wave infrared (MWIR) hyperspectral instruments for Earth observations to measure three-dimensional distributions of atmospheric water vapor. HyperCube wants to infer wind velocities at multiple vertical locations in the atmosphere by measuring the same region of the atmosphere at different times.

The instrument contains an FTS-based hyperspectral sounder, a cross-track step-stare scanner, a compact telescope, an infrared focal-plane array (IR FPA), and a calibration source. The scanner performs step-stares, the Earth radiance arrives to the scanner, enters the Michelson interferometer creating the interferogram that is finally acquired by the FPA.

12.6 Photometers

Photometry measures the flux, or intensity, of an object light.

12.6.1 XPS

XUV Photometer System (XPS) [2, 35, 36] is one of the payloads of the SORCE mission, heritage of SNOE and of the Solar EUV Experiment (SEE) instrument on the TIMED mission. The objective of XPS is to measure the extreme ultraviolet (XUV) solar irradiance from 1 to 35 nm. The instrument package is a set of filter photometers consisting of 12 silicon XUV photodiodes. Each photodiode has a thin-film filter to provide an approximately 5 nm spectral band pass. These thin film filters are deposited directly on the photodiode to avoid using delicate metal foil filters, which are difficult to handle, prone to develop pin holes, and degrade with time. The set of 12 XUV photometers is packaged together with a common filter wheel mechanism, which can rotate a closed aperture, a fused silica window, or an open aperture in front of any given photometer. The fused silica windows on this filter wheel allow accurate subtraction of the background signal from visible and near UV light. The 12 XUV photometers (XPs) are grouped into three sets. Each set of four XPs is arranged in a circle for use with the filter wheel mechanism. The filter wheel, which has three different rings of filters for the three sets of XPs, has eight positions: four blocked for dark measurements, two clear for solar XUV measurements, and two with fused silica windows for solar visible background measurements. An observation run is a sequence of measurements from five consecutive filter wheel positions, normally starting and ending with dark measurements. The electronics for each photodiode include only a current amplifier and a voltage-to-frequency (VTF) converter.

12.6.2 BRITE–Photometer

The objective of BRIght-star Target Explorer (BRITE) [2, 37, 38] is to examine the apparently brightest stars in the sky for variability using the technique of precise differential photometry in time scales of hours and more. The constellation of four nanosatellites is divided into two pairs, with each member of a pair having a different optical filter. The requirements call for observation of a region of interest by each nanosatellite in the constellation for up to 100 days or longer.

The science payload of each nanosatellite consists of a five-lens telescope with an aperture of 30 mm and the interline transfer progressive scan CCD detector KAI-11002-M from Kodak with 11 megapixels. A baffle is used to reduce stray light. The optical elements are housed inside the optical cell and are held in place by spacers. The photometer has a resolution of 26.52 arcsec per pixel and a FOV of 24°. The mechanical design for the blue and for the red instrument is nearly identical, only lens size is different (Figure 12.9).

The effective wavelength range of the instrument is limited in the red (550–700 nm) by the sensitivity of the detector and in the blue (390–460 nm) by the transmission properties of the glass used for the lenses. The filters were designed such that for a star of 10 000 K (average temperature of target stars) both filters would generate the same amount of signal on the detector.

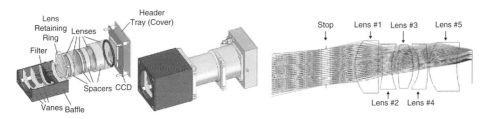

Figure 12.9 Left: Illustration of the BRITE telescope and baffle. Source: Image credit: UTIAS/SFL. Right: The optical design of the photometer. Source: Reproduced with permission from UTIAS/SFL, Ceravolo.

12.6.3 ExoPlanet and ASTERIA

The objective of ExoPlanet [2, 39–41] is to verify the capability of small satellites to search for unmapped planets and to complement existing planet-hunters. During the diurnal phase, the spacecraft points the solar panels toward the Sun, while during the eclipse, it slews to reacquire the target star. The operation concept is shown in Figure 12.10. An Earth-twin transit across the center of a star would last 13 h from a low inclination LEO, sufficient to detect such an event. The optimum integration time is a function of target star brightness and image size. The maximum integration time is on the order of 30 s.

The photon collection capability is a primary consideration given the photometric nature of the mission, this implies the lenses with large aperture (i.e. low f-number). The payload is a Zeiss 85 mm f/1.4 single lens reflex (SLR) camera with an aperture of 60 mm along with CCD (1 k × 1 k with 13.3 μm pixels, back-illuminated, 2 e^- rms read noise, 12.5 e^- per pixel s^{-1} dark current at 0°C, and 12.5° diagonal FOV) and comounted CMOS detectors for star tracking (2.6 k × 1.9 k with 1.3 μm pixels).

ExoPlanet has finally evolved into Arcsecond Space Telescope Enabling Research in Astrophysics (ASTERIA) [2, 42], capable to achieve arcsec-level line of sight-pointing error with highly stable focal plane temperature control. These technologies will enable careful measurement of stellar brightness over time.

The sensor suite includes an optics section, a dual-imager focal plane array realizing the two payload functions, and a piezoelectric nanopositioning stage capable of moving the focal plane to zero-out the platform jitter (Figure 12.11).

The optics system is a f 1.4/85 Zeiss lens with a 28.6° FOV, focusing an image 43 mm in diameter onto the focal plane. The focal plane array houses two active detector areas, one larger CMOS detector that fulfills the science function, and a smaller CMOS sensor with smaller pixel sizes to act as a rapid-cadence star camera to provide attitude information to the attitude control system. The science sensor operates at longer integration times and collects data on several pixel windows, one for the target star and several other bright stars for comparison plus dark areas of the sky for dark current correction.

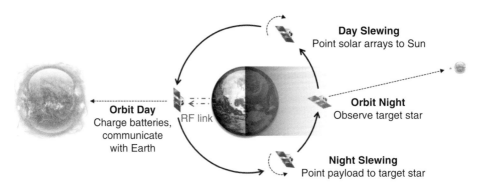

Figure 12.10 ExoPlanet operation concept [41]. Source: Photo courtesy of NASA.

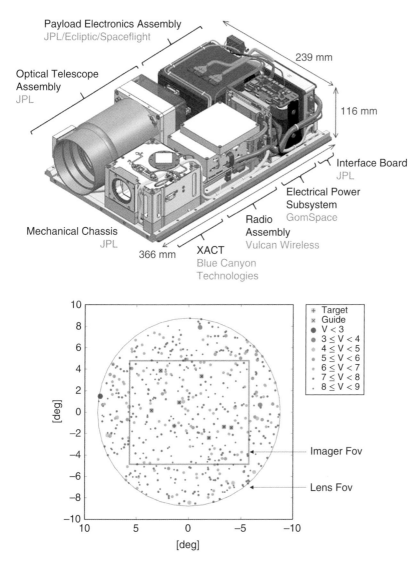

Figure 12.11 Left: Illustration of the ASTERIA spacecraft. Right: Its starfield [42]. Source: Photo courtesy of NASA. (*See color plate section for color representation of this figure*).

12.7 GNSS Receivers

Small satellites can provide atmospheric measurements on a global scale and data for weather and climate forecasting. One of the techniques is based on existing global navigation satellite system-radio occultation (GNSS-RO), a precise and cost-effective technique for measuring Earth atmosphere from space. The understanding of many Earth processes benefits from this kind of observation, including severe weather, cloud formation and evolutionary processes, aerosols or air quality related measurements, atmospheric photochemistry, vegetation, ocean color, solar irradiance, and Earth outgoing radiation. CubeSat-enabled constellations could deliver tens of thousands of occultations daily and make them available in near real time [43].

The cyclone global navigation satellite system (CYGNSS) receiver is chosen here as the reference instrument, but also Spire, GeoOptics, and PlanetIQ are developing CubeSat-based GNSS-RO constellations. Also, the Constellation Observing System for Meteorology, Ionosphere, and Climate (COSMIC) mission is another constellation of microsatellites making atmospheric soundings of temperature, moisture, and pressure in the troposphere and stratosphere using GNSS-RO.

12.7.1 CYGNSS

CYGNSS [2, 44, 45] Earth Venture mission is complementary to traditional spacecraft that measure winds. By going from active to passive technology using GPS, CYGNSS requires significantly less power than does a spacecraft using the traditional technique. CYGNSS is a constellation of eight low Earth-orbiting spacecraft that analyze reflected GPS signals from water surfaces shaped by hurricane-associated winds.

Each CYGNSS spacecraft carries a delay Doppler mapping (DDM) instrument consisting of a multichannel GPS receiver, low-gain zenith antennas, and high-gain nadir antenna. CYGNSS measures the ocean surface wind field associated with tropical cyclones by combining the all-weather performance of GPS-based bistatic scatterometry. The working principle is schematically shown in Figure 12.12.

Figure 12.12 illustrates the working principle. The direct GPS signal is transmitted from the orbiting GPS satellite and received by a right-hand circular polarization (RHCP) receive antenna on the zenith (i.e. top) side of the spacecraft that provides a coherent reference for the coded GPS transmit signal. The quasi-specular, forward-scattered signal that returns from the ocean surface is received by a nadir-(i.e. downward-)looking left-hand circular

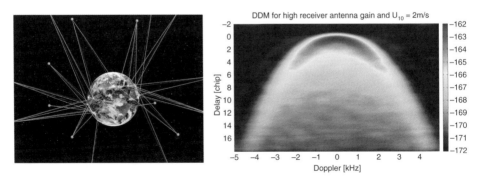

Figure 12.12 Left [2]: CYGNSS spacecraft (yellow dots) analyze reflected GPS signals (dark blue lines) from water surfaces shaped by hurricane-associated winds. Red lines indicate signals directly between the CYGNSS satellites and the GPS satellites (light blue dots further from Earth). Blue lines on the Earth surface represent measurements of the wind speed that CYGNSS derives from the reflected GPS signal. Source: Courtesy of NASA/University of Michigan. Right: Spatial distribution of the ocean surface scattering measured by the UK-DMC-1 demonstration spaceborne mission−referred to as the delay Doppler map. Source: Photo courtesy of NASA. (*See color plate section for color representation of this figure*).

polarization (LHCP) antenna on the nadir side of the spacecraft. The scattered signal contains detailed information about its roughness statistics, from which the local wind speed can be derived.

The image on the right of Figure 12.12 shows the scattering cross-section and demonstrates its ability to resolve the spatial distribution of ocean surface roughness. This type of scattering image is referred to as a DDM. There are two different ways to estimate ocean surface roughness and near-surface wind speed from a DDM. The maximum scattering cross-section (the darkest region) can be related to roughness and wind speed [47], but this requires absolute calibration of the DDM, not always available. Another technique estimates wind speed from a relatively calibrated DDM using the shape of the scattering arc (the lighter shades). The arc represents the departure of the bistatic scattering from the theoretical purely specular case, a perfectly flat ocean surface, which appears in the DDM as a single-point scatterer. This second approach relaxes requirements on instrument calibration and stability, however, it uses a wider region of the ocean surface; this implies a lower spatial resolution.

12.7.2 CADRE

The goal of CubeSat-investigating Atmospheric Density Response to Extreme driving (CADRE) [2, 48] is to study thermosphere properties. The secondary payload on CADRE is a dual frequency GPS receiver. The receiver will measure phase delays between two transmissions from the GPS constellation, L1 (1575.42 MHz) and L2 (1227.6 MHz). At transmission, L1 and L2 are in phase; the atmospheric media through which the signals pass introduces phase delays between the two signals. A measurement of the delay is used to measure total electron content (TEC) of the ionosphere and density measurements of the troposphere.

12.7.3 ³Cat 2

³Cat 2 [5, 49] is a 6U CubeSat developed at the Universidad Politécnica de Cataluña (UPC) to demonstrate a novel dual-frequency (L1 + L2) GNSS-R altimeter. The mission is designed to perform ocean altimetry by means of GNSS reflectometry (GNSS-R). The main payload is the novel dual-band altimeter P(Y) & C/A ReflectOmeter (PYCARO), designed and manufactured at UPC's Remote Sensing Lab and NanoSat Lab. Measuring GNSS signals reflected from the Earth allows performing simultaneous observations of the Earth on several points over a very wide swath, very promising for applications such as mesoscale altimetry.

It also integrates the Mirabilis star tracker, an experimental star tracker for attitude determination and the AMR ELISA magnetometer, designed and manufactured at IEEC for the future LISA mission of the ESA.

12.8 Microbolometers

The bolometer is a device for measuring the power of incident electromagnetic radiation via the heating of a material with a temperature-dependent electrical resistance. Bolometers have flat spectral response, are very stable and robust to on-orbit degradation. Bolometers can be efficiently used on board small satellites for measurements of solar spectral irradiance (SSI), a crucial measurement for climate modeling.

12.8.1 CSIM

The Compact Solar Spectral Irradiance Monitor (CSIM) [50, 51], developed by the LASP, University of Colorado and the Quantum Electronics and Photonics Division of the National Institute of Standards and Technology (NIST), Boulder, Colorado, represents a new generation of bolometric detectors based on a micromachined silicon substrate with a silicon nitride thermal link and using multiwall vertically aligned carbon nanotubes (VACNTs) as black absorber. A first prototype is shown in Figure 12.13 (left top panel).

The design provides a simple light path, see CSIM optical path in Figure 12.13 (right panel): entrance slit-prism–exit slit-detector, and wavelength is scanned by rotating the prism; the electrical substitution radiometer provides long-term stability to calibrate the photodiodes and carries the absolute calibration. CSIM incorporates two identical channels, stacked on top of each other, to permit tracking of exposure-induced degradation.

The bolometer (left bottom panel) is implemented as a paired electrical substitution radiometer, and thermistors on the two bolometers form one arm of a resistance bridge. A fixed heater power level is applied to the reference bolometer, the heater power applied to the active bolometer is adjusted to keep the bridge in balance ($V_B = 0$). To measure optical power, the active bolometer is illuminated; the measured optical power is the change in heater power applied to the active bolometer. The design is based on VACNTs

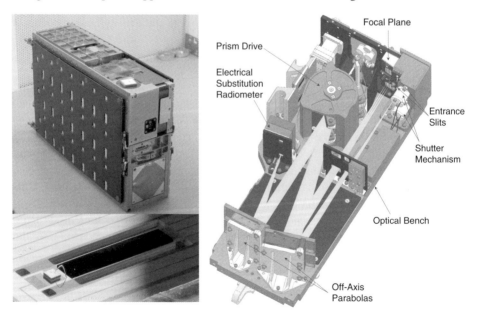

Figure 12.13 Left: (top) the CSIM prototype, (bottom) the bolometer prototype. Right: Optical overview [50, 51]. Source: Reproduced with permission from Laboratory for Atmospheric and Space Physics – LASP. (*See color plate section for color representation of this figure*).

(characterized by being extremely black, spectrally flat, and with large thermal conductivity), a silicon substrate, a SiN thermal link (low thermal conductivity and mass), patterned heaters and leads, and bonded thermistor.

These technologies both ease fabrication and provide higher performance: a continuous spectral irradiance in the 210–2400 nm, covering 96% of the total solar output, and a noise level of 260 pW for a 40 s measurement.

12.9 Radiometers

Three missions represent the status of the art of radiometers for small satellites and Cube-Sats in particular.

12.9.1 TEMPEST

The Temporal Experiment for Storms and Tropical Systems Demonstration (TEMPEST-D) [2, 52, 53] is a 6U CubeSat mission with a five-frequency millimeter-wave radiometer currently on orbit. The Earth Venture mission concept is to fly a constellation of identical TEMPEST CubeSats to observe the fine-scale temporal development of clouds and precipitation at 5 min intervals.

Another example is the Radiometer Assessment using Vertically Aligned Nanotubes (RAVAN) 3U CubeSat that will demonstrate a payload that could be incorporated into a CubeSat or hosted payload constellation for measuring Earth radiation budget.

It is also worth mentioning the Time-Resolved Observations of Precipitation Structure and Storm Intensity with a Constellation of Smallsats (TROPICS) mission that will allow to make rapid revisits, allowing its microwave radiometers to measure temperature, humidity, precipitation, and cloud properties frequently every 21 min.

The TEMPEST radiometer, chosen as the reference radiometer here, performs continuous measurements at five frequencies, 89, 165, 176, 180, and 182 GHz. The instrument design is based on a 165–182 GHz radiometer design inherited from RACE and an 89 GHz receiver developed under the ESTO ACT-08 and IIP-10 programs at Colorado State University (CSU) and JPL. The radiometer is based on the direct-detection architecture: the RF input to the feed horn is amplified, band-limited, and detected using Schottky diode detectors. The use of direct-detection receivers based on InP HEMT MMIC LNA (High Electron Mobility Transistor, Monolithic Microwave Integrated Circuit, Low Noise Amplifier) front ends substantially reduces the mass, volume, and power requirements of these radiometers. Input signals are band-limited using waveguide-based band-pass filters to meet the radiometer bandwidth requirements of 4 ± 1 GHz at center frequencies of 89 and 165 GHz, as well as 2 ± 0.5 GHz at 176, 180, and 182 GHz center frequencies [54].

The TEMPEST-D instrument occupies a volume of 3U and is mounted on a temperature-controlled bench that interfaces with the spacecraft structure using thermally isolating spacers.

The radiometer performs cross-track scanning, measuring the Earth scene between $\pm 45°$ nadir angles, providing an 825 km wide swath from a 400 km nominal orbit altitude. Each radiometer pixel is sampled for 5 ms. The radiometer performs end-to-end calibration during each rotation of the scanning reflector. The radiometer observes both cosmic background radiation at 2.7 K and an ambient blackbody calibration target (at approximately 300 K) every 2 s, for a scan rate of 30 rpm. A schematic representation of the TEMPEST-D observing profile over a 360° reflector scan and the resulting output data time series are shown in Figure 12.14.

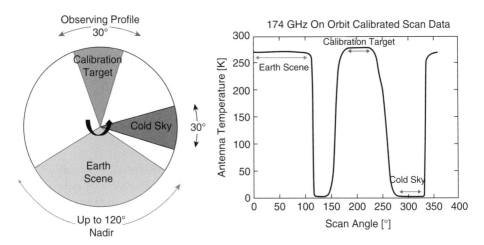

Figure 12.14 Schematic representation of TEMPEST-D observing profile (left) and output data time series (right) for each reflector scan [46]. Source: Padmanabhan, S., T. C. Gaier, S. C. Reising, B. H. Lim, R. Stachnik, R. Jarnot, W. Berg, C. D. Kummerow and V. Chandrasekar, "Radiometer Payload for the Temporal Experiment for Storms and Tropical systems Demonstration Mission (TEMPEST-D)," In: Proc. 2017 IEEE International Geoscience and Remote Sensing Symposium (IGARSS 2017), Fort Worth, Texas, Jul. 2017, pp. 1213-1215.

12.10 Radar Systems

Radar systems are active systems that transmit a beam of radiation in the microwave region of the electromagnetic spectrum. Unlike optical systems that rely on reflected solar radiation or thermal radiation emitted by Earth, imaging radar instruments work independently of light and heat.

12.10.1 RAX

The primary mission objective of Radio Aurora eXplorer (RAX) and RAX-2 [2, 55] is to study plasma instabilities that lead to magnetic field-aligned irregularities (FAI) of electron density in the lower polar thermosphere (80–300 km). These irregularities are known to disrupt communication and navigation signals. The RAX mission uses a network of existing ground radars that scatter signals off the FAI to be measured by a receiver on the RAX spacecraft (bistatic radar measurements) [5].

RAX mission measures the 3D k-spectrum (spatial Fourier transform) of ~1 m scale FAI as a function of altitude, with high angular resolution (~0.5°), in particular measuring the magnetic field alignment of the irregularities. The spacecraft will measure "radio aurora," the Bragg scattering from FAI that is illuminated with a narrow beam incoherent scatter radar (ISR) on the ground. The scattering locations are determined using a GPS-based synchronization between ISR transmissions and satellite receptions and the assumption that the scattering occurs only inside the narrow ISR beam. Figure 12.15 shows a drawing of the irregularities (red wiggles), the magnetic field lines, the radar beam, radio aurora (cones) and the satellite (cubes), and the satellite tracks. The irregularities inside the narrow radar beam and at a given altitude scatter the signals in a hollow cone shape. The thickness of

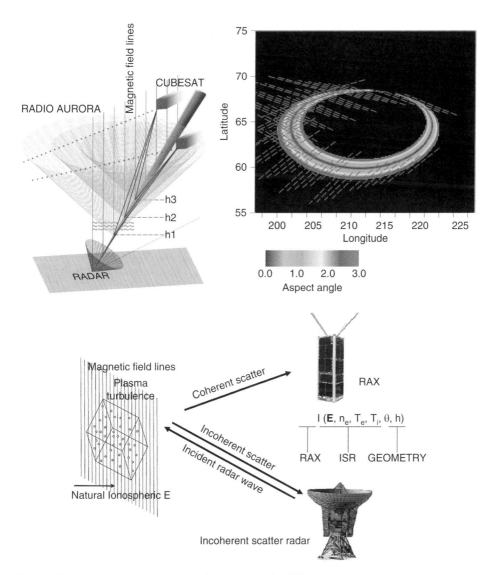

Figure 12.15 Top left and bottom: An illustration of the RAX experiment measurement concept. Top right: The loci of perpendicularity (<3°) and 1 min satellite tracks over the experimental zone over 30 days. Source: Reproduced/Adapted with permission from Hasan Bahcivan, SRI International, Menlo Park, CA, USA. (*See color plate section for color representation of this figure*).

the wall of each cone is a measure of magnetic aspect sensitivity, which is also a measure of plasma wave energy distribution in the parallel and perpendicular directions with respect to the geomagnetic field.

The radar receiver is capable of operation with five Ultra high frequency (UHF) ISRs: PFISR (Poker Flat Incoherent Scatter Radar), Poker Flat, AK; RISR (Resolute Bay Incoherent Scatter Radar), Alberta, Canada; ESR (EISCAT Svalbard Radar, in Longyearbyen, Spitsbergen, Norway); Millstone Hill ISR of MIT in Westford, MA, USA; and Arecibo ISR in Puerto Rico. The ground-based ISR emits high-power RF pulses into the ionosphere to be scattered off the FAI structures. Alternative ground radars are also being examined to achieve a greater number of experiments. The RAX payload receiver, developed at SRI, operates in a snapshot acquisition mode collecting raw samples at 1 MHz for 300 s over the experimental zone. Following each experiment, the raw data are postprocessed for range–time–intensity and Doppler spectrum. The snapshot raw data acquisition enables flexibility in forming different radar pulse shapes and patterns. In addition, the PFISR electronic beam steering capability can be utilized for simultaneous multiple beam position experiments. The receiver can also operate with the Modular UHF Ionospheric Radar (MUIR) radar located at the High Frequency Active Auroral Research Program (HAARP) facility in Gakona, AK, as part of active experiments.

12.10.2 Radar Altimeters and SAR (EO)

Radar altimeters [56] are active sensors that use the ranging capability of radar to measure the surface topography profile along the satellite track. They provide precise measurements of a satellite height above the ocean by measuring the time interval between the transmission and reception of very short electromagnetic pulses. Altimetry satellites provide important ocean surface topography, currents, and surface winds that are essential for all shipping services (scientific and commercial). The information generated by ocean sensing radars is crucial for weather predictions and global weather models. The working principle of a radar altimeter is given in Figure 12.16.

SAR [57] can provide day-and-night imagery of Earth. Being a radar system, clouds, fog, and precipitation do not have any significant effect, so that images can also be acquired independent of weather conditions. SAR profits of the motion of the radar antenna, mounted on board a satellite, above a target region to provide high spatial resolution images; its working principle is schematically shown in Figure 12.17.

For small satellites, SAR applications are physically limited by a number of factors [58]: the size (3U), the power/thermal capacity preventing the global coverage, a low orbit, and no station-keeping implying additional absolute height errors, using a Ku transmitter-receiver. In order to meet these requirements, the working principle of radar altimetry as the one implemented in TOPEX/Poseidon has been reduced: typically a single frequency is used and an ionosphere model is implemented, a single pulse (SNR of ~7 dB is achievable), no radiometer and implementing a troposphere model, no DORIS and use GPS and retro-reflectors to derive the satellite velocity.

Many science applications require sub-centimeter-level deformation measurements, but each individual SAR measurement is corrupted by up to several centimeters of atmospheric noise. Multiple acquisitions need to be averaged together to reduce atmospheric artifacts.

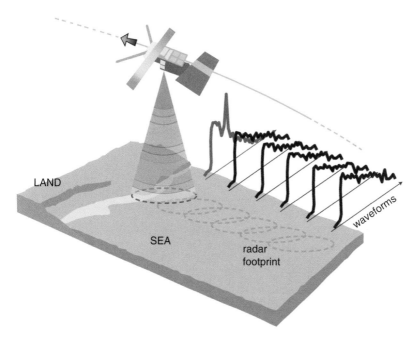

Figure 12.16 Working principle of radar altimetry: altimeter received waveforms in sea-to-land transition illustration. Source: COASTALT portal and http://www.altimetry.info.

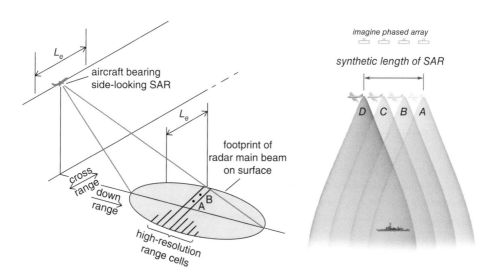

Figure 12.17 Schematic working principle of a SAR: range resolution achieved through pulse compression of broad-bandwidth pulse; cross-range resolution achieved with Doppler processing. Source: (a) Image Credit: McGraw-Hill and (b) Image Credit: Christian Wolff. Photo: ESA – https://www.esa.int/OurActivities/ObservingtheEarth/Copernicus/SARmissions.

12.10.3 SRI-Cooperative Institute for Research in Environmental Sciences (CIRES)

SRI International [59] has designed and developed a miniaturized SAR payload for Cube-Sats at 500 km altitudes (see Figure 12.18). SRI leverages expertise in airborne SAR and interferometric-synthetic aperture radar (InSAR) to create an S-Band radar subsystem capable of InSAR operations for the CubeSat platform. The radar subsystem has a volume less than 1.5U (10 cm x 10 cm x 15 cm), mass of 1.5 kg, and low phase noise (e.g. accurate, stable reference clock). The goal is to achieve 20 m spatial resolution and high-quality imaging (SNR > 13 dB) using a 5 m^2 supporting antenna and to obtain sub-centimeter level ground deformation accuracy. The payload subsystem is made of three modules: power amplifier (PA), transmit/receive (TX/RX), and high speed TX/RX processor for real-time processing and storage of raw I/Q data and on-board processing of stored data.

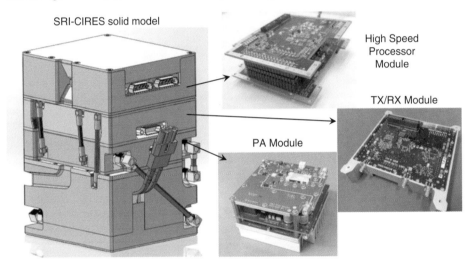

Figure 12.18 SRI-CIRES illustration. Source: Reproduced with permission from SRI International.

12.11 Particle Detectors

It is a wide range category of detectors, spanning from detectors measuring trapped electrons and protons, solar particles, or even cosmic rays. In the following sections, the most representative experiments are summarized.

12.11.1 REPTile

REPTile [60] is a small (6.05 cm in length and 6 cm in diameter), low-mass, and low-power article detector on board CSSWE (Colorado Student Space Weather Experiment) capable of measuring relativistic outer radiation belt electrons in the energy range from 0.5 to 3 MeV and solar energetic protons from 10 to 40 MeV [5]. The instrument is a scaled down version of the instrument built at LASP (Laboratory for Atmospheric and Space Physics) for NASA's RBSP (Radiation Belts Storm Probes) mission.

The REPTile instrument measures the Earth's outer radiation belt electrons, both trapped and precipitating, to study how their low rate and energy spectrum evolve in four energy bands (0.5–1.5, 1.5–2.2, 2.2–2.9, and > 2.9 MeV), and monitors the SEP protons associated with solar flares to study how flare location, magnitude, and frequency relate to the timing, duration, and energy spectrum of these protons that reach Earth in four energy channels (8.5–18.5, 18.5–25, 25–30.5, and 30.5–40 MeV).

The instrument is a loaded-disc collimated telescope; it consists of a stack of four solid-state doped silicon detectors (by Micron Semiconductor). The front detector has a diameter of 20 mm, while the following three are 40 mm across. The stack is contained in a tungsten chamber, encased in an aluminum outer shield. The choice of materials is optimized for their ability to shield energetic particles and minimize secondary electron generation within the housing. Tantalum, a good compromise between stopping power and relatively low secondary particle generation, is used for the baffles within the collimator preventing electrons from scattering into the detector stack from outside the instrument's 52° FOV and gives the instrument a geometric factor of 0.52 sr cm^2.

Particles impinge on the detector stack producing electron–hole pairs in the doped semiconductor; higher energy particles penetrate deeper into the stack. A bias voltage is applied to each detector to accelerate generated electrons to an anode where they are collected. Using coincidence logic, the electronics can determine the energy range of the particle. The total instrument mass is 1.25 kg, with a cylindrical envelope of 4.6 m (diameter) × 6 cm (length) (Figure 12.19).

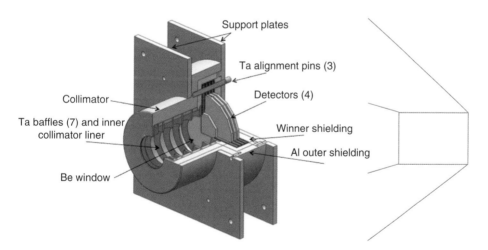

Figure 12.19 Cutaway view of the REPTile instrument [60]. Source: Reproduced with permission from CU- Boulder.

12.11.2 EPISEM

The Edison Demonstration of SmallSat Networks (EDSN) mission [2, 61, 62] wants to demonstrate the feasibility of multipoint space weather observations with a swarm of eight identical satellites. EDSN, equipped with energetic particle integrating space environment monitor (EPISEM) to simultaneously monitor spatial and temporal variations in penetrating radiation above the atmosphere, is important for understanding both the near-Earth radiation environment and as input for developing more accurate space weather models.

The EPISEM card [63] consists of a Geiger–Müller tube mounted on a PCB and is designed to consume approximately 100 mW during operation. The GPS card enables time- and location-correlation of the EPISEM science data. By combining data and position/time information, it is possible to characterize the variability of penetrating high-energy protons at a much smaller scale than previously accomplished. EPISEM employs a thin-walled Geiger–Müller tube located inside the spacecraft structure that detects penetrating beta/gamma radiation from energetic particles above a certain energy threshold, the threshold being different for electrons and protons. Incoming radiation knocks electron off the neon fill gas, neon becomes Ne$^+$ and frees electron avalanches toward anode (Figure 12.20).

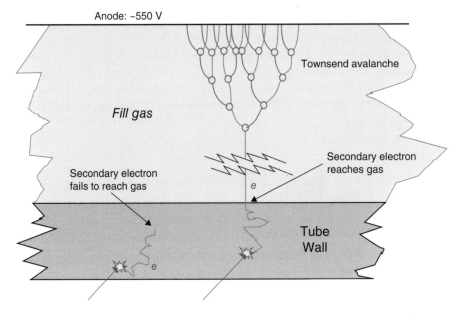

Figure 12.20 Schematic view of interactions in the Geiger–Müller tube of the EPISEM. Source: Photo courtesy of NASA.

12.11.3 FIRE

The goal of Focused Investigations of Relativistic Electron Burst Intensity, Range, and Dynamics (FIREBIRD) mission [2, 64, 65] is to study the spatial scale and spatial temporal ambiguity of magnetosphere microbursts in the Van Allen radiation belts, and to provide spectral properties at medium energies. The mission concept is made of two spacecraft. Each satellite carries two solid-state detector charged particle sensors with different geometric factors optimized to cover electron measurements over the energy range from 0.25 to ~1 MeV in six differential energy channels.

Focused Investigations of Relativistic Electrons (FIRE), developed by UNH, employs a heritage sensor design based on a single large-geometry-factor, solid-state detector set on each of the two spacecraft, and is sensitive to electrons precipitating from the radiation belts. The detectors are read out by a dual amplifier pulse peak energy rundown, a customized application-specific integrated circuit (ASIC) by The Aerospace Corporation. Fast sample observations are stored by the on-board memory; survey observations identify times of interest to retrieve the highest temporal resolution data.

12.12 Plasma Wave Analyzers

Plasma wave analyzers measure the electric and magnetic fields, electron density, and temperature in the interplanetary medium and planetary magnetospheres usually, thanks to the use of Langmuir probes. Indeed, one of the most fundamental parameters to observe the space environment is the electric field that drives the motion of the plasma in the Earth's ionosphere and magnetosphere.

In this category also, magnetometer sensors are included, encompassing several types of detectors for magnetic field variations, such as fluxgate magnetometers, search-coil magnetometers, which detect variations of the magnetosphere of very large-scale objects (planets, moons, or the Sun).

12.12.1 CADRE/WINCS

The overall objective of CADRE [2, 66] is the study of thermosphere properties. Earth's ionosphere and thermosphere are driven in a variety of regions on multiple scales; changes in the upper atmosphere can affect our society with the increasing usage of space-based infrastructure.

Wind Ion Neutral Composition Suite (WINCS) is the primary payload of the CADRE mission, developed by NASA/ Goddard Space Flight Cente (GSFC) and the Naval Research Laboratory (NRL). The size is 7.6 cm × 7.6 cm × 7.1 cm with a mass of ∼0.850 kg, and the power consumption is 1.3 W. The WINCS is made of four instruments [66], see Figure 12.21:

- WTS (Wind and Temperature Spectrometer)
- IDS (Ion Drift and temperature Spectrometer)
- NMS (Neutral Mass Spectrometer)
- IMS (Ion Mass Spectrometer)

Small-deflection energy analyzer (SDEA) is the core instrument of WTS/IDS. Ions enter the SDEA horizontally and deflect as they continue toward the exit slit; the exit slit plane is a circular cylinder section with vertical axis passing through the entrance slit—this defines the angle imaging function of SDEA. The upper and exit plates are biased at voltage V_{SDEA} (<10 V), while the entrance and lower plates are at ground potential. The inhomogeneous field produced in the upper left region of the cavity produces the focusing in the incident ions, this reduces the size of exit slit and helps reducing unwanted UV photons that may reach the micro channel plate (MCP) detector placed beyond the exit slit. The SDEA instrument has two identical mirrored sensors; one half of the SDEA is used for ions while the other half for neutrals.

The measurements are obtained from the angular and energy distributions of the particle flux. Two separate and orthogonal analyzers (WTS1 and WTS2 in Figure 12.21) measure the

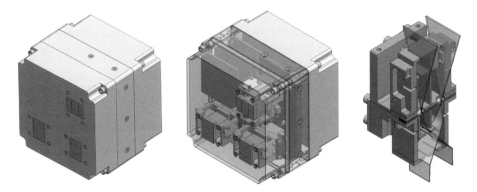

Figure 12.21 Illustration of the WINCS package with individual instrument apertures. Source: Photo courtesy of NASA. Left: Layout of the four WINCS spectrometers: WTS and IDS consist of two spectrometer modules with mutually perpendicular FOVs, as shown by the two pairs of long slits on the right side of the figure. The two round apertures on the other module show the parallel ion and neutral paths through the instrument. The multichamber WTS/IDS pair provides four separate chambers for photon rejection.

angular energy distributions, both pointing within a few degrees of the ram direction. Each analyzer spans 30° × 2° FOV in 15 pixels, each 2° × 2° pixel scanning in 20 energy steps with the energy analyzer voltage to determine particle velocity. The position of maximum flux in each analyzer determines the horizontal and vertical angles in which the particles enter the detector.

The IMS/NMS uses two gated electrostatic mass spectrometer (GEMS) time-of-flight mass spectrometers developed at GSFC. GEMS uses a SDEA, turning the potential on and off. When the potential is off, particles impinging on the detector at the entrance slit continue forward in a straight line along the SDEA and do not reach the exit slit; once the potential is turned on, charged particles are accelerated and deflected. The acceleration is a function of particle mass, their initial velocity being the same (orbital velocity): short after the potential turn on, low mass particles exit the slit (hydrogen needs to reach a kinetic energy of about 0.3 eV), followed after a while by heavier particles (iron needs a kinetic energy of about 18 eV). The final mass resolution $\Delta m/m$ is equal to the energy resolution $\Delta E/E$. The working principle is the same for IMS and NMS, hence, the same potential levels; in both cases the acceleration is achieved by two-aperture ion optic lenses mounted above the respective entrance apertures of the GEMS. In the NMS, ionization is produced by electron impact in its ion source.

12.12.2 Dynamic Ionosphere CubeSat Experiment (DICE)

The goal of DICE [2, 67] is to map the geomagnetic storm-enhanced density (SED) plasma bulge and plume formations in Earth's ionosphere. This objective is achieved via in situ ionosphere measurements of electric field and plasma density. The measurements are made by two instruments; the electric field probe (EFP) for electric field measurements and two separate spherical fixed-bias DC Langmuir probe (DCP) for absolute ion density measurements.

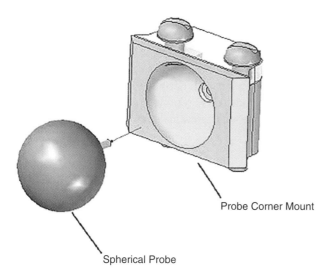

Probe Corner Mount

Spherical Probe

Figure 12.22 Corner mount geometry of the spherical probe. Source: Reproduced with permission from Dice Consortium.

The EFP is based on the well-known electric field double-probe technique that monitors the potential of pairs of sensors, deployed on a boom several meters from the spacecraft, at a well-known distance. The electric field in the direction of the boom is measured by the ratio between the potential differences. The EFP is made of four sensors: a 1 cm diameter sphere, gold plated, mounted at the end of each of four wire booms. Each sensor is mounted in a corner mount, also gold plated, shown in Figure 12.22 that has a semi-sphere that cups the EFP sensor.

The EFP makes DC and AC electric field measurements, both at equal time steps. The DC measurements are made at an 8 Hz rate, providing 45° rotational resolution per spacecraft spin to maximize the science return; DC measurements are a subset of AC measurements performed at 4096 Hz.

The DCP sensors operate in the ion saturation region providing measurements of ion density in the range of 2×10^9 to $2 \times 10^{13}\,\text{m}^{-3}$ and a resolution of $3 \times 10^8\,\text{m}^{-3}$. The two sensors are deployed on separate fiberglass cylindrical booms, 8 cm long, from the top and bottom of the spacecraft along its spin axis.

In addition to these two devices, a science-grade magnetometer, three-axis magnetometer (TAM), is aimed at measuring field aligned currents. The TAM measurements allow accurate identification of storm-time features, with simultaneous colocated electric field measurements, previously missing, provided by EFP.

12.12.3 INSPIRE/CVHM

Compact vector-helium magnetometer (CVHM) on board Interplanetary NanoSpacecraft Pathfinder In a Relevant Environment (INSPIRE) [2, 68, 69] is a JPL/University of California, Los Angeles (UCLA) instrument based on previous mission heritage (from Mariner, through Cassini missions), here significantly reducing the overall size: CVHM size is ~5 cm × 10 cm × 10 cm with a mass of 0.5 kg, still providing a very good stability, less than 10 pT, thus representing a significant improvement on previously flown CubeSat magnetometers.

The INSPIRE mission consists of two spacecraft that separate after deployment. Spacecraft make synchronous measurements of the solar-wind magnetic field, as they slowly separate after launch for 90 days out to 1.5 million km from Earth. Depending upon trajectory (Earth-leading or Earth-trailing), the magnetometers may provide real-time measurements across the bow shock of Earth magnetosphere, and a new look into turbulence effects.

12.13 Biological Detectors

A new discipline, called astrobiology, is recently entering the space field. As it is stated by European Astrobiology Network Association (EANA) on http://eana-net.eu/index.php?page=Discover/esaexo, astrobiology is a life sciences activity that covers every aspect of space life sciences, including biology, physiology, biotechnology, biomedical applications, biological life-support systems, exobiology, animal research, and access to ground facilities; thus, a large amount of detectors can be included in this category. On the side of small satellite, here only one main mission is given: ORganics Exposure to Orbital Stresses (OREOS), since this represents the state of the art in this field.

12.13.1 OREOS

The overall objective of O/OREOS [2, 70–72], heritage of PharmaSat and GeneSat missions, is to test how life and the components of life respond to the radiation environment in space. The experiment carries up organisms and organic compounds into space and monitors the changes induced by the harsh radiation environment and microgravity. There are two on-board experiments, both including specimens: Space Environment Survivability of Living Organisms (SESLO) and Space Environment Viability of Organics (SEVO). Each of them is equipped with its own electronics, microcontroller, and data storage requiring only a standard power and data interface.

The objective of the SESLO experiment is to characterize the growth, activity, health, and ability of microorganisms to adapt to the stresses of the space environment. The experiment contains two types of living biological specimens, that is, microbes that are commonly found in salt ponds and soil, *Halorubrum chaoviatoris* and *Bacillus subtilis*; the experiment provides active optical measurement of growth and/or metabolic activity. SESLO is sealed in a vessel at one atmosphere, and microbes are stored in a dried and dormant state.

Once OREOS is deployed in orbit, the experiment starts. It begins to rehydrate and grow three sets of the microbes at three different times, each test occurs in its own module, thermally isolated from the other two: (i) first test within 1–2 weeks after launch, (ii) in the following 3 months, and (iii) the final test in 6 months after launch. The payload uses time-resolved optical absorbance and density measurements to track the growth and metabolism of both the organisms, a technique inherited by the PharmaSat mission. A dedicated tricolor light-emitting diode (LED) illuminates along the axis of each system and a detector at the opposite side reads the intensity for each color, with center wavelengths at 470, 525, and 615 nm.

The objective of the SEVO experiment is to monitor the stability, modification, and degradation of organic molecules in specific space environments including interplanetary/interstellar space, the lunar surface, wet/salty environments, and the Martian atmosphere. SEVO consists of four experimental "microenvironments" containing miniature reaction cells of organic molecules, in thin-film form, representing chemical building blocks of life in our solar system and beyond.

The chemical behavior of molecule exposure to solar UV and visible light and space ionizing radiation is monitored by a spectrometer over the mission lifetime of 6 months. The spectrometer profits of the heritage of previous Lunar CRater Observation and Sensing Satellite (LCROSS) and Lunar Atmosphere and Dust Environment Explorer (LADEE) missions of NASA. The instrument covers the UV–Visible wavelength range (200–1000 nm) with a 1–2 nm spectral resolution and better than 0.1 nm spectral band shift measurement capability; it provides 0.03 absorbance unit resolution. The detector acquires a single spectrum in less than 100 ms.

A pair of baffled diffuser assemblies ensures that the sample film is uniformly illuminated over a 3 mm diameter spot. Light intensity varies by less than ±25% for Sun angles within ±35° from normal, and solar intensity sensors guarantee synchronous measurements for each spectral acquisition. The reaction cells are derived from those used for the EXPOSE experiment on the International Space Station (ISS). Each of the four organic materials is vacuum-sublimed onto several MgF_2 windows, to allow UV–Visible irradiation as short as 124 nm; some of the windows have either SiO_2 or Al_2O_3 thin coating layers between the organic and the MgF_2 to tune the band-pass and provide a chemical substrate relevant to the corresponding microenvironment.

A sapphire window is chosen for its high optical transparency along the spectrometer wavelength band and is placed on the far side, where light enters the collection optic and is routed to the spectrometer. Both windows are welded to a stainless steel spacer providing a hermetically sealed microenvironment with the desired simulated atmosphere. The sealed cells are then assembled in 11 mm diameter housings and installed into the rotatable payload carousel. The system is schematically shown in Figure 12.23. This assembly also helps in protecting the fragile windows during the launch phase. Samples are staggered and

Figure 12.23 Reaction cell cross-section. Source: Photo courtesy of NASA.

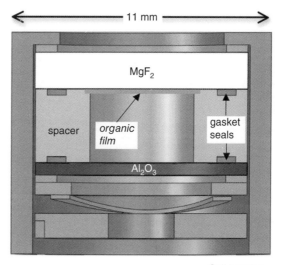

Cell housing

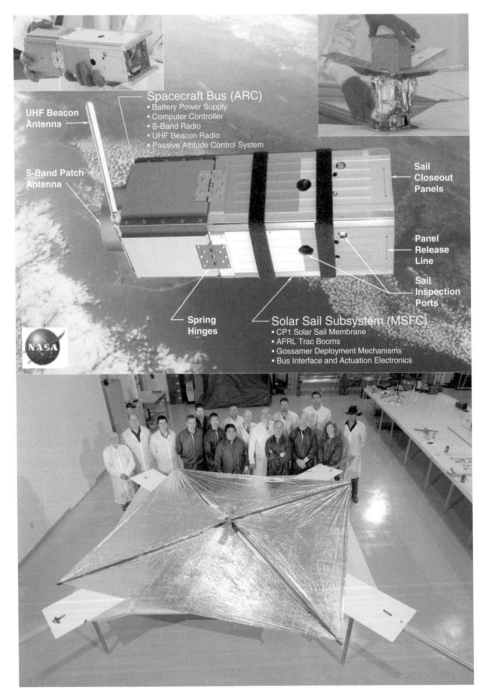

Figure 12.24 Top: On-orbit view of the NanoSail-D2 spacecraft before sail deployment. Source: Image credit: NASA/ARC. Bottom: Panel deployed configuration of NanoSail-D. Source: Photo courtesy of NASA.

divided into two concentric rings. Two optical fibers are placed under the carousel but, due to the cell spacing, only one fiber at a time is illuminated; their outputs are then combined for input to the spectrometer.

12.14 Solar Sails

Solar sails use sunlight to propel vehicles through space much like sailboats rely on wind to push through the water. NanoSail-D2 [2, 73], depicted in Figure 12.24 (top), is a technology and science experiment designed to validate solar sail capabilities. NanoSail-D2 is a 10 m^2 solar sail payload (Figure 12.24, bottom) packaged within a 3U nanosatellite, designed and developed at NASA/MSFC (Marshall Space Flight Center) in partnership with NASA/ARC (Ames Research Center) and several industry and academic partners. The NanoSail-D2 solar sail payload section occupies 2U or 2/3 volume of the spacecraft. Sail payload close-out panels provide protection for the sail and deployment booms during the launch phase of the mission, the panels have spring-loaded hinges that were released on-orbit, using the auto timer command of the spacecraft bus.

One of NanoSail-D's several mission objectives is to demonstrate the capability to deploy a large sail structure from a highly compacted volume without recontact with the space-craft. This demonstration can be applied to deploy future communication antennas, sensor arrays, or thin film solar arrays to power spacecraft. The mission will also demonstrate and test the deorbiting capabilities of solar sails. NASA hopes to use thin membranes to deorbit satellites and large space debris one day.

12.15 Conclusions

In this chapter, it has been demonstrated that small satellites have great capabilities. It is a sincere thought that the potential of small satellites stays in the satellite system configuration meant as a swarm of satellites, located in different points in the near-Earth environment or even in the solar system, that are working together to produce a combined result. These types of multipoint or constellation-type measurements provide much greater temporal coverage than that achievable with a single, large spacecraft.

Given this, at least a few more uses of small satellites shall be kept in mind. CubeSats can prove the usage of a payload to demonstrate technology for future use on larger missions. CubeSats also have additional advantages indeed; they can provide observations that can mitigate the data gap between critical satellite systems in different scientific fields, for example, Earth observation or space weather to guarantee a prompt warning, because of their lower cost and shorter development time.

References

1 National Academies of Sciences, Engineering, and Medicine (2016). *Achieving Science with CubeSats: Thinking inside the Box*. Washington, DC: The National Academies Press https://doi.org/10.17226/23503 NSF report.

2 The EO Portal Satellite Missions Database https://directory.eoportal.org/web/eoportal/satellite-missions

3 B. Lal, E. de la Rosa Blanco, J. R. Behrens, B. A. Corbin, E. K. Green, A. J. Picard, A. Balakrishnan, 2017. "Global Trends in Small Satellites," Science & Technology Policy Institute, Institute for Defense Analyses—IDA Paper P-8638, July 2017.

4 National Science Foundation (NSF), 2013. Cubesat-Based Science Missions for Geospace and Atmospheric Research, October 2013.

5 Gunter's Space Page, URL: http://space.skyrocket.de

6 Charles D. Norton, Steve A. Chien, Paula J. Pingree, David M. Rider, John Bellardo, James W. Cutler, Michael P. Pasciuto, 2013. "NASA's Earth Science Technology Office CubeSats for Technology Maturation," Proceedings of the 27th AIAA/USU Conference, Small Satellite Constellations, Logan, Utah, USA, August 10–15, 2013, paper: SSC13-XI-4, URL of paper: http://digitalcommons.usu.edu/cgi/viewcontent.cgi?article=2987&context=smallsat

7 Paula J. Pingree, 2014. "Looking Up: The MCubed/COVE Mission," Proceedings of the 11th Annual CubeSat Developers' Workshop—The Edge of Exploration," San Luis Obispo, CA, USA, April 23–25, 2014, URL: http://www.cubesat.org/images/.../DevelopersWorkshop2014/Pingree_MCubed_COVE_Mission.pdf

8 Dmitriy L. Bekker, Thomas A. Werne, Thor O. Wilson, Paula J. Pingree, Kiril Dontchev, Michael Heywood, Rafael Ramos, Brad Freyberg, Fernando Saca, Brian Gilchrist, Alec Gallimore, James Cutler, 2010. "A CubeSat Design to Validate the Virtex-5 FPGA for Spaceborne Image Processing," Proceedings of the 2010 IEEE Aerospace Conference, Big Sky, MT, USA, March 6–13, 2010.

9 Maurice Borgeaud, Noémy Scheidegger, Muriel Noca, Guillaume Roethlisberger, Fabien Jordan, Ted Choueiri, Nicolas Steiner, 2009. "SwissCube: The first entirely-built Swiss student satellite with an Earth observation payload," Proceedings of the 7th IAA Symposium on Small Satellites for Earth Observation, Berlin, Germany, May 4–7, 2009, IAA-B7-0205P, SwissCube URL: http://swisscube.epfl.ch

10 Guillaume Roethlisberger, Fabien Jordan, Anthony Servonet, Maurice Borgeaud, Renato Krpoun, Herbert R. Shea, 2008. "Advanced Methods for Structural Machining and Solar Cell Bonding Allowing High System Integration and their Demonstration on a Pico-satellite," Proceedings of the 22nd Annual AIAA/USU Conference on Small Satellites, Logan, UT, USA, August 11–14, 2008, SSC08-XI-4, URL: http://ctsgepc7.epfl.ch/12%20-%20SwissCube%20papers/G.Roethlisberger_SSC08-XI-4.pdf

11 Craig Underwood, Vaios J. Lappas, Chris Bridges, 2014. "AAReST Spacecraft DDR (Detailed Design Review): Spacecraft Bus, Propulsion, RDV/Docking and Precision ADCS," URL: http://pellegrino.caltech.edu/AAReST_Docs/AAReST_Surrey_DDR_2014.pdf

12 Sergio Pellegrino, "Autonomous assembly of a reconfigurable space telescope," URL: http://pellegrino.caltech.edu/aarest2.html

13 Keith Patterson, 2013. "Lightweight Deformable Mirrors for Future Space Telescopes," Thesis in Partial Fulfillment of the Requirements for the Degree of Doctor of Philosophy, California Institute of Technology, Pasadena, CA, December 2013, URL: http://thesis.library.caltech.edu/8043/1/patterson_keith_2013_thesis.pdf

14 S. Palo, T. Woods, X. Li, J. Mason, M. Carton, A. Caspi, A. Jones, R. Kohnert, S. Solomon, 2013. "MinXSS: A Three-Axis Stabilized CubeSat for Conducting Solar Physics," 5th European CubeSat Symposium, Royal Military Academy, VKI (Von Karman Institute), Brussels, Belgium, June 3–5, 2013.

15 Thomas N. Woods, Amir Caspi, Phil Chamberlin, Andrew Jones, Rich Konert, Xinli Li, Scott Palo, Stanley Solomon, 2013. "MinXSS—Miniature X-ray Solar Spectrometer (MinXSS) CubeSat Mission," Submitted to NASA on August 15, 2013, URL: http://www.pinheadinstitute.org/wp-content/uploads/2014/07/MinXSS_Proposal2013_NASA.pdf

16 "Amptek Complete X-Ray Spectrometer," Amptek, URL: http://www.amptek.com/x123.html

17 P. Kaaret, K. Jahoda, B. Dingwall, "HaloSat—A CubeSat to Study the Hot Galactic Halo," URL: https://files.aas.org/head14/116-21Philip_Kaaret.pdf

18 Philip Kaaret, 2016. "HaloSat Overview," The University of Iowa, August 17, 2016, URL: http://astro.physics.uiowa.edu/~kaaret/2016f_a6880/halosat_overview.pdf

19 Tyler G. Rose, Benjamin K. Malphrus, Kevin Z. Brown, 2011. "An Improved Measurement of the Diffuse X-Ray Background: The Cosmic X-Ray Background Nanosat (CXRN)," 8th Annual CubeSat Developers' Workshop, Cal Poly, San Luis Obispo, CA, USA, April 20–22, 2011, http://www.cubesat.org/images/2011_Spring_Workshop/thur_a9.30_rose_cxbn_cswkshp_sp2011.pdf

20 Kevin Z. Brown, Tyler G. Rose, Benjamin K. Malphrus, Jeffrey A. Kruth, Eric T. Thomas, Michael S. Combs, Roger McNeil, Robert T. Kroll, Benjamin J. Cahall, Tyler T. Burba, Brandon L. Molton, Margaret M. Powell, Jonathan F. Fitzpatrick, Daniel C. Graves, J. Garrett Jernigan, Lance Simms, John P. Doty, Matthew Wampler-Doty, Steve Anderson, Lynn R. Cominsky, Kamal S. Prasad, Stephen D. Gaalema, Shunming Sun, 2012. "The Cosmic X-Ray Background NanoSat (CXBN): Measuring the Cosmic X-Ray Background Using the CubeSat Form Factor," Proceedings of the 26th Annual AIAA/USU Conference on Small Satellites, Logan, Utah, USA, August 13–16, 2012, paper: SSC12-VII-6.

21 D. Casadei, N. Jeffrey, E.P. Kontar, 2017. "Measuring X-ray anisotropy in solar flares. Prospective stereoscopic capabilities of STIX and MiSolFA." A&A 606 (2017) A2, doi: https://doi.org/10.1051/0004-6361/201730629, MiSolFA URL: https://misolfa.i4ds.net/mission

22 Rottman, G. and George, V. (2002). An overview of the solar radiation and climate experiment. *The Earth Observer* 14 (3): 17–22; Rottman, Woods, and Sparn (1993) Solar-Stellar Irradiance Comparison Experiment I. 1. Instrument Design and Operation, JGR, vol. 98, D6, 10667, doi: 10.1029/93JD00462.

23 SORCE, URLs: http://lasp.colorado.edu/sorce, http://earthobservatory.nasa.gov/Library/SORCE, http://lasp.colorado.edu/home/sorce/mission

24 "National Scientific Foundation (NSF) CubeSat-based science missions for Geospace and Atmospheric Research," Annual Report, October 2013, pp. 36–37, URL: http://www.nsf.gov/geo/ags/uars/cubesat/nsf-nasa-annual-report-cubesat-2013.pdf

25 "OPAL (Oxygen Photometry of the Atmospheric Limb)," USU/SDL, URL: http://www.sdl.usu.edu/downloads/opal.pdf

26 Alan Marchant, Mike Taylor, Charles Swenson, Ludger Scherlies, 2014. "Hyperspectral Limb Scanner for the OPAL Mission," Proceedings of the AIAA/USU Conference on Small Satellites, Logan, Utah, USA, August 2–7, 2014, paper: SSC14-VII-5.

27 "Lunar IceCube to Take on Big Mission from Small Package," NASA, August 4, 2015 (updated on January 29, 2016), URL: https://www.nasa.gov/feature/goddard/lunar-icecube-to-take-on-big-mission-from-small-package

28 P. E. Clark, Ben Malphrus, Kevin Brown, Dennis Reuter, Robert MacDowall, David Folta, Avi Mandell, Terry Hurford, Cliff Brambora, Deepak Patel, Stuart Banks, William Farrell, Noah Petro, Michael Tsay, V. Hruby, Carl Brandon, Peter Chapin, 2016. "Lunar Ice Cube Mission: Determining Lunar Water Dynamics with a First Generation Deep Space CubeSat," 47th Lunar and Planetary Science Conference, The Woodlands, Texas, March 21–25, 2016, paper: 1043.pdf, URL: http://www.hou.usra.edu/meetings/lpsc2016/pdf/1043.pdf

29 S. Sander, D. Bekker, J-F. Blavier, M. Bryk, C. Donahue, R. Goullioud, B. Hancock, D. Johnson, R. Key, A. Lamborn, K. Manatt, J. Moore, J. Nastal, T. Neville, D. Preston, D. Rider, M. Ryan, J. Wincentsen, Y.-H. Wu 2015. "Panchromatic Fourier Transform Spectrometer Engineering Model (PanFTS-EM) for Geostationary Atmospheric Measurements," Fourier Transform Spectroscopy 2015, Lake Arrowhead, California, USA, 1–4 March 2015. ISBN: 978–1–55752-814-8.

30 Charles D. Norton, Michael P. Pasciuto, Paula Pingree, Steve Chien, David Rider, 2012. "Spaceborne Flight Validation of NASA ESTO Technologies," Proceedings of IGARSS (International Geoscience and Remote Sensing Symposium), Munich, Germany, July 22–27, 2012, URL: http://tinyurl.com/qgbzr9b

31 GRIFEX, URL: http://cubesat.jpl.nasa.gov/projects/grifex/overview.html

32 Panchromatic Fourier Transform Spectrometer (PanFTS). 2011. 2nd GEO-CAPE Community Workshop, May 12, 2011.

33 "HyperCubeTM3D Wind Measurement," Harris Corporation, 2017, URL: https://www.harris.com/sites/default/files/downloads/solutions/55493_hypercube_data-sheet_v2_2_final_.pdf

34 Ronald Glumb, Michael Lapsley, Peter Mantica, Anna Glumb, 2017. "TRL6 Testing of Hyperspectral Fourier Transform Spectrometer Instrument for CubeSat Applications," Proceedings of the 31st Annual AIAA/USU Conference on Small Satellites, Logan, UT, USA, August 5–10, 2017, paper: SSC17-VI-08, URL: http://digitalcommons.usu.edu/cgi/viewcontent.cgi?article=3641&context=smallsat

35 Woods, T.N., Rottman, G., and Vest, R. (2005). XUV photometer system (XPS): overview and calibration. *Solar Physics* 230 (1–2 (special issue of SORCE)): 345–374.

36 XPS, URL: http://lasp.colorado.edu/sorce/instruments/xps/xps_instrument_design.htm

37 N. C. Deschamps, C. C. Grant, D. G. Foisy, R. E. Zee, A. F. J. Moffat, W. W. Weiss, 2006. "The BRITE Space Telescope: A Nanosatellite Constellation for High-Precision Photometry of Bright Stars," Proceedings of the 20th Annual AIAA/USU Conference on Small Satellites, Logan, UT, August 14–17, 2006, paper: SSC06-X-1, URL: http://www.utias-sfl.net/docs/brite-ssc-2006.pdf

38 N. C. Deschamps, C. C. Grant, D. G. Foisy, R. E. Zee, A. F. J. Moffat, W. W. Weiss, 2006. "The BRITE Space Telescope: Using a Nanosatellite Constellation to Measure Stellar

Variability in the Most Luminous Stars," Proceedings of the 57th IAC/IAF/IAA (International Astronautical Congress), Valencia, Spain, October 2–6, 2006, IAC-06-B5.2.7, URL: http://www.utias-sfl.net/docs/brite-iac-2006.pdf

39 Arcsecond Space Telescope Enabling Research in Astrophysics (ASTERIA), NASA/JPL, URL: http://www.jpl.nasa.gov/cubesat/missions/asteria.php

40 Matthew W. Smith, Sara Seager, Christopher M. Pong, Sungyung Lim, Matthew W. Knutson, Timothy C. Henderson, Joel N. Villaseñor, Nicholas K. Borer, David W. Miller, Shawn Murphy, 2011. "ExoplanetSat: A Nanosatellite Space Telescope for Detecting Transiting Exoplanets," 8th Annual CubeSat Developers' Workshop, Cal Poly, San Luis Obispo, CA, USA, April 20–22, 2011.

41 Smith, M.W., Seager, S., Pong, C.M. et al. (2010). ExoplanetSat: Detecting transiting exoplanets using a low-cost CubeSat platform. In: *Proceedings of SPIE, 'Space Telescopes and Instrumentation 2010: Optical, Infrared, and Millimeter Wave,'*, vol. 7731 (eds. J.M. Oschmann, M.C. Clampin and H.A. MacEwen). San Diego, CA, USA: SPIE https://doi .org/10.1117/12.856559, http://dspace.mit.edu/openaccess-disseminate/1721.1/61644

42 Pong, C.M. and Smith, M.W. (2019). Camera modeling, centroiding performance, and geometric camera calibration on ASTERIA. In: *2019 IEEE Aerospace Conference, Big Sky, MT*, 1–16.

43 Hand (2012). Microsatellites aim to fill weather-data gap. *Nature* 491 (7426): 650–651.

44 Christopher Ruf, 2012. "The NASA EV-2 Cyclone Global Navigation Satellite System (CYGNSS) Mission," August 27, 2012, URL of presentation: http://svcp.jpl.nasa.gov/ meetings/2012/es/082701/CYGNSS_27Aug2012_Ruf_CYGNSS_JPL_Seminar.pdf

45 Christopher S. Ruf, Scott Gleason, Zorana Jelenak, Stephen Katzberg, Aaron Ridley, Randall Rose, John Scherrer, Valery Zavorotny, 2012. "The CYGNSS Nanosatellite Constellation Hurricane Mission," Proceedings of IGARSS (International Geoscience and Remote Sensing Symposium), Munich, Germany, July 22–27, 2012.

46 Padmanabhan, S., Gaier, T.C., Reising, S.C. et al. (2017). Radiometer payload for the temporal experiment for storms and tropical systems demonstration Mission (TEMPEST-D). In: *Proc. 2017 IEEE International Geoscience and Remote Sensing Symposium (IGARSS 2017), Fort Worth, Texas*, 1213–1215.

47 www.sstl.co.uk/getattachment/8ea1074b-b37a-4829-8345-b1bb02656d02/SGR-ReSI, http://ktb.engin.umich.edu/RSG/pubsfiles/AeroConf-2013Unwin-etalSGR-ReSI.pdf

48 James W. Cutler, Aaron Ridley, Andrew Nicholas, 2011. "CubeSat Investigating Atmospheric Density Response to Extreme Driving (CADRE)," Proceedings of the 25th Annual AIAA/USU Conference on Small Satellites, Logan, UT, USA, August 8–11, 2011, paper: SSC11-IV-7, URL: http://rax.engin.umich.edu/wp-content/uploads/files/ CADRE-SmallSat-2011.pdf

49 3Cat 2, URL: https://nanosatlab.upc.edu/en/missions-and-projects/3cat-2

50 A Compact Solar Spectral Irradiance Monitor for Future Small Satellite and CubeSat Science Opportunities, Challenges & Opportunities in Solar Observations, 12 November 2015.

51 New Generation Bolometric Detector for Measurement of Solar Spectral Irradiance, January 2017. 2017. AMS, American Meteorological Society.

52 Anne Ju Manning, 2015. "Small satellites to pave way for future spaceborne weather observations," CSU, December 2015, URL: http://source.colostate.edu/small-satellites-to-pave-the-way-for-future-space-borne-weather-observations

53 Reising, S. C., T. C. Gaier, S. Padmanabhan, B. H. Lim, C. D. Kummerow, V. Chandrasekar, W. Berg, C. Heneghan, R. Schulte, C. Radhakrishnan, S. T. Brown, M. Pallas, 2018. "Temporal Experiment for Storms and Tropical Systems Technology Demonstration (TEMPEST-D) to Enable Temporally Resolved Observations of Clouds and Precipitation on a Global Basis Using 6U-Class Satellite Constellations," In: Proc. NASA Earth Science Technology Forum (ESTF 2018), Silver Spring, Maryland, June 2018, A3P6.

54 Steven C. Reising, Christian D. Kummerow, V. Chandrasekar, Wesley Berg, Jonathan P. Olson, Todd C. Gaier, Sharmila Padmanabhan, Boon H. Lim, Cate Heneghan, Shannon T. Brown, John Carvo, Matthew Pallas, 2017. "Temporal Experiment for Storms and Tropical Systems Technology Demonstration (TEMPEST-D) Mission: Enabling Time-Resolved Cloud and Precipitation Observations from 6U-Class Satellite Constellations," Proceedings of the 31st Annual AIAA/USU Conference on Small Satellites, Logan, UT, USA, August 5–10, 2017, paper: SSC17-III-01, http://digitalcommons.usu.edu/cgi/viewcontent.cgi?article=3609&context=smallsat

55 Sara C. Spangelo, James Cutler, Louise Anderson, Elyse Fosse, Leo Cheng, Rose Yntema, Manas Bajaj, Chris Delp, Bjorn Cole, Grant Soremekum, David Kaslow, 2013. "Model Based Systems Engineering (MBSE) Applied to Radio Aurora Explorer (RAX) CubeSat Mission Operational Scenarios, IEEEAC Paper #2170, Version 1, Updated 29 January 2013," URL: http://www.omgsysml.org/mbse_cubesat_v1-2013_ieee_aero_conf.pdf

56 ESA Altimetry missions, URL: https://www.esa.int/OurActivities/ObservingtheEarth/Copernicus/Altimetrymissions

57 ESA SAR missions, URL: https://www.esa.int/OurActivities/ObservingtheEarth/Copernicus/SARmissions

58 Stacy N., 2012. 6U Radar Altimeter Concept, 6U Cubesat Low Cost Space Missions Workshop, Mt. Stromlo Observatory, Canberra, Australia, 17–18 July 2012.

59 Wye L., 2016. SRI CubeSat Imaging Radar for Earth Science (SRI-CIRES), Earth Science Technology Forum.

60 Schiller, Q., Abhishek Mahendrakumar, Xinlin Li (2010), REPTile: A Miniaturized Detector for a CubeSat Mission to Measure Relativistic Particles in Near-Earth Space, 24th Annual AIAA/USU Conference on Small Satellite.

61 Jim Cockrell, Richard Alena, David Mayer, Hugo Sanchez, Tom Luzod, Bruce Yost, D. M. Klumpar, 2012. "EDSN: A Large Swarm of Advanced Yet Very Affordable, COTS-based NanoSats that Enable Multipoint Physics and Open Source Apps," Proceedings of the 26th Annual AIAA/USU Conference on Small Satellites, Logan, Utah, USA, August 13–16, 2012, paper: SSC12-I-5, URL of paper: http://digitalcommons.usu.edu/cgi/viewcontent.cgi?article=1095&context=smallsat

62 "Edison Demonstration of Smallsat Networks Mission—A Swarm of Advanced, Affordable, COTS-based Nanosatellites that Enable Cross-Link Communication and Multipoint Physics," NASA Factsheet, URL: http://www.nasa.gov/sites/default/files/files/EDSN_FactSheet_031014.pdf

63 Adam Gunderson, David Klumpar, Matthew Handley, Andrew Crawford, Keith Mashburn, Ehson Mosleh, Larry Springer, James Cockrell, Hugo Sanchez, Harrison Smith, 2013. "Simultaneous Multi-Point Space Weather Measurements using the Low Cost EDSN CubeSat Constellation," Proceedings of the 27th Annual AIAA/USU Small-Sat Conference, August 29, 2013, paper: SSC13-WK-41, URL: http://digitalcommons.usu .edu/cgi/viewcontent.cgi?filename=0&article=2904&context=smallsat&type=additional

64 Brian Larsen, Harlan Spencer, David Klumpar, Larry Springer, J. Bernard Blake, 2009. "Focused Investigations of Relativistic Electron Burst Intensity, Range, and Dynamics (FIREBIRD)," CubeSat Developers' Workshop, Cal Poly, April 22–25, 2009, URL: http:// mstl.atl.calpoly.edu/~bklofas/Presentations/DevelopersWorkshop2009/2_Science/2_ Larsen-FIREBIRD.pdf

65 David M. Klumpar, Harlan E. Spence, Bernie Blake, 2009. "Overview: The Dual-CubeSat FIREBIRD Mission (Focused Investigations of Relativistic Electron Burst Intensity, Range, and Dynamics)," November 30, 2009, URL: http://mstl.atl.calpoly.edu/~bklofas/ NSF_comm/20091130_telecon/FIREBIRD _Overview_NSF_Telecon_113009.pdf

66 "The WINCS 'Factory' First of Many Miniaturized Helio Instruments to be Delivered this Fall," NASA, 2011. July 16, 2011, URL: https://www.nasa.gov/mission_pages/ sunearth/news/wincs.html

67 DICE, URL: http://www.sdl.usu.edu/programs/dice

68 Brent Sherwood, Sara Spangelo, Andreas Frick, Julie Castillo-Rogez, Andrew Klesh, E. Jay Wyatt, Kim Reh, John Baker, 2015. "Planetary CubeSats come of age," Proceedings of the 66th International Astronautical Congress (IAC 2015), Jerusalem, Israel, October 12–16, 2015, paper: IAC-15-A3.5.8.

69 "Interplanetary Nano-Spacecraft Pathfinder in Relevant Environment (INSPIRE)," NASA/JPL, URL: http://www.jpl.nasa.gov/cubesat/missions/inspire.php

70 NASA Factsheet Factsheet, URL: http://www.nasa.gov/pdf/467036main_OOREOS_ FactSheet_FINAL_2010-10-29.pdf

71 Michael Schirber, 2009 "Outer Space Oreos," Space Daily for Astrobiology Magazine, Moffett Field, CA (SPX), May 08, 2009, URL: http://www.spacedaily.com/reports/Outer_ Space_Oreos_999.html

72 Bramall, N.E., Quinn, R., Mattioda, A. et al. (2012). The development of the space environment viability of organics (SEVO) experiment aboard the organism/organic exposure to orbital stresses (O/OREOS) satellite. *Planetary and Space Science* 60: 121–130. URL: http://www.gwu.edu/~cistp/assets/docs/research/articles/Ehrenfreund_SEVO_O-OREOS .pdf

73 NANOSAIL. URL: https://www.nasa.gov/centers/marshall/pdf/484314main_ NASAfactsNanoSail-D.pdf

13 I-3b

In-orbit Technology Demonstration

Jaime Estela

Spectrum Aerospace Group, Munich, Germany

13.1 Introduction

The market of small satellites continues to grow, and more and more nanosatellites will be built and launched. New nanosatellite hardware manufacturers develop hardware for such satellite platforms. Most of these new hardware products are tested only on the ground, with no flight heritage. For a commercial product, a flight heritage is a very important issue and potential customers ask always for such characteristic. Nowadays, very few flight opportunities exist for technology demonstration. Almost all these flights will be granted by space agencies. The problem is that the demand is much bigger than the available flights. Furthermore, such flights involve a long process, and it takes at least 2 years until the experiment is launched into space. In-Orbit Demonstration (IOD) and In-Orbit Validation (IOV) are at priority for most space agencies and suitable programs have already started. Such programs have long selection and implementation processes and the competition is hard. The technology evolves very fast and the validation of new products in a real space environment is necessary. The absence of efficient IOD/IOV programs hinders the introduction of new products in the space market. Small satellites will be used for more ambitious missions, such as high-resolution remote sensing, communications, IoT services, deep space missions, and others. These kinds of missions require high-performance electronics. Small satellites are still using mainly commercial components and the validation of such hardware in space is an important topic.

The products to be tested vary from complete systems to single components. The major group comprises single components which were designed and tested for space missions and require a validation in space. Single components can be integrated as part of a payload in a big satellite. In other cases, complete modules or systems have to be validated. The bigger the experiment or product, the difficult is the organization of a flight into space, and also the higher costs. Generated data, power consumption, size, weight, and materials play an important role in the project plan for an IOD/IOV campaign.

Several projects are ongoing and each project follows a different strategy. Each one has advantages and disadvantages. In this chapter, IOD/IOV programs will be described and analyzed.

Nanosatellites: Space and Ground Technologies, Operations and Economics, First Edition.
Edited by Rogerio Atem de Carvalho, Jaime Estela, and Martin Langer.

13.2 Activities of Space Agencies

Space agencies are the major financier of IOD/IOV activities. National agencies support the local industry and prioritize the introduction of new technologies. Regional agencies such as the European Space Agency (ESA) promote programs for the development of strategic technologies relevant for the member states and also the cooperation among companies of different countries. Some of the activities are mentioned in the following text.

13.2.1 NASA

The NASA, the Air Force Office of Scientific Research (AFOSR), the Air Force Research Laboratory (AFRL), the American Institute of Aeronautics and Astronautics (AIAA), and the Space Development and Test Wing organized the University Nanosatellite Program. The program started in 1999 and six competitions were conducted. The aim of the program was to train young space engineers to design and build a satellite in a period of 2 years. The launch of the satellites succeeded as piggyback in one American satellite launcher, and the launch service was sponsored. The winner got the funding for the launch. The last competition was in 2011 [1].

A part of the cargo capacity of American launchers is reserved for such government initiatives. These piggyback slots benefit a lot to the educational American space program. The validation of the satellite and payload succeed constantly. The motivation of the American students, thanks to such initiatives, increases the interest in the space industry and allows America to get graduate professionals already with practical experience.

It is important to mention that American deep space missions give important technical advantages to national companies due to the fact that the development of high-performance hardware for real missions is the best praxis for designers. The validation in space confirms the expected quality and performance.

13.2.2 ESA

The ESA started its Technology Demonstration Program (TDP) in 1987 with the aim of ensuring the availability of technology for future space missions. The TDP was superseded by the General Support Technology Program (GSTP). The Project for On-Board Autonomy (PROBA) satellites were used as platform for the IOD missions. The IOD tradition of the ESA can be summarized in the following programs [2, 3]:

- TDP (1992–1997)
- GSTP2 (1996–2000): Mission PROBA-1
- GSTP3 (2001–2004): Mission PROBA-2
- GSTP4/5 (2004–2013): Mission PROBA-V
- GSTP6 (2014–ongoing): Mission PROBA-3

Figure 13.1 PROBA-2 satellite. Source: Reproduced with permission from ESA.

In the GSTP6, CubeSats will be also used for technology demonstration (Figure 13.1).

Since years, the ESA promotes the development of new hardware for small satellites, the use of commercial off-the-shelf (COTS) components, and the use of nanosatellites for remote sensing, communication, and deep space missions. The use of COTS in space systems will also be supported from the European Union (EU) in the Horizon 2020 program. Following examples can be mentioned:

Laser communication

Payloads such as high resolution or hyperspectral cameras generate big amount of data. Nanosatellites mainly use UHF or S-band for communication. X-band transceivers are now in development and data rates above 20 Mbps will be necessary for future missions. The ESA Call AO-9095 requested the development of a laser communication module for small satellites. The module should allow the communication from low Earth orbit (LEO) to LEO and from LEO to ground.

High-tech camera

In the ESA EMITS Calls AO-8220 [4] and AO-8643 [5], the use of commercial high-resolution cameras and multispectral or hyperspectral cameras was a requirement to comply with remote sensing activities. For the AO-8220, two satellites, one from the NASA and one from the ESA, will investigate the impact of a satellite in the binary asteroid Didymos. The event should be observed and recorded with cameras in order to analyze the impact in detail.

The AO-8643 call handles a mission to the Moon, and remote sensing was also important for the detection of minerals in areas selected for future Moon bases (Figure 13.2).

Figure 13.2 PROBA-3 satellite. Source: Reproduced with permission from ESA.

Use of COTS

For this topic, the ESA opened several calls in the same way for the JUICE mission.

JUICE (Jupiter mission)

The use of silicon carbide–based power devices is a young technology that increases the performance of electronics, especially in power modules. Power MOSFET will be often used in cars or industrial machines. By comparing the silicon carbide technology with the traditional silicon technology, the most important advantages are high break-down voltage, higher switching frequency, high operating electric field, operation in high-temperature environments, and low energy losses. For space systems, the use of silicon carbide can contribute with significant improvements. The major challenge is the radiation robustness, and in space this is the most important environmental factor that degrades electronics. First test indicated a good tolerance of total ionizing dose (TID), but high sensitivity to single event effects (SEEs). Furthermore, for deep space missions such as the JUICE, a slow degradation of the components must be simulated for the proper qualification. It means the TID test must be achieved with a low dose rate. This specific long TID test is called enhanced low dose rate sensitivity (ELDRS). A low dose rate requires longer irradiation times and in consequences higher facility costs. For the JUICE mission, strong shielding is considered; however, high dose level will be reached during the mission around Jupiter. The accumulated dose is calculated between 50 and 100 krad. In comparison to a LEO orbit, the accumulated dose of 1 year is around 1 krad [6] (Figure 13.3).

Figure 13.3 JUICE mission. Source: Reproduced with permission from ESA.

ESA academy

The ESA encourages the participation of universities in the development of the European nanosatellite industry. The launch of the VEGA launcher will be used to carry satellites and experiments from universities in a regular manner. This IOD/IOV strategy already benefits many universities in Europe [7].

13.2.3 DLR

The German Space Agency has IOD/IOV activities as priority. German space companies have a close contact with Deutschen Zentrum für Luft- und Raumfahrt (DLR) and the exact demand for such activities is well known (over 300 companies). The satisfaction of such demand will be made using microsatellites or in cooperation with other space agencies. Here, the biggest part of the demand will not find a flight opportunity. DLR is always open for new projects that can support the IOD/IOV activities.

13.3 Nanosatellites

In space missions, an important issue is the reduction of costs. The same situation is for IOD/IOV programs and cost efficient solutions are preferred. Nanosatellites are cost-effective test platforms. Nanosatellites are not expensive and single components can be tested in such platforms. The implementation of a payload based on single components keeps an easy design of the satellite and is flexible for the configuration of the satellite. One important demerit of this technology is the use of COTS components without qualification. This characteristic makes the platform unstable and is not reliable enough for commercial projects. In fact, more than a half of the nanosatellite missions fail. This lack of quality should be solved in the future in order to have nanosatellites as reliable test platforms. Nowadays, around 50 nanosatellite hardware manufacturers exist worldwide, and the components are not qualified for space. This uncertainty is accepted and aims to keep prices low [8].

An important advantage of nanosatellites is that they can be built in few months. An upgrade of the hardware can also succeed in a short time.

13.3.1 IOV/IOD Providers

For IOD/IOV missions, a nanosatellite can be configured with the adequate resources and can be set up to the exact mission requirements. Currently, the following hardware is available for nanosatellites:

- *On-board computer (OBC)*: For this subsystem, many possible configurations exist. The computing power can be selected with the adequate capacity and also with the needed memory size. Solid-state disks (SSDs) play an important role for reliable storage, short access time, and low power consumption.
- *Power module*: The configuration of this module also varies a lot. It can be kept small for low power experiments or can be dimensioned for high power consumption when required.
- *Communication module*: This is one of the modules that evolves constantly and tries to get the best reliability and the highest data rate. The higher the data rate, the higher the power consumption and the pointing accuracy of the attitude system.

13.3.2 SSTL

Surrey Satellite Technology Ltd. (SSTL) coordinates the UK IOD program and it uses Cube-Sats which will be launched from the International Space Station (ISS). The platforms are Clyde Space CubeSats and SSTL supports the assembly, integration, and testing. The launch service is from NanoRacks. This program is limited to four missions and is funded by the Innovate UK program.

13.3.3 Alba Orbital

This Scottish company offers launch services for the new CubeSat standard called Pock-etQube. PocketQube is a satellite with the form of a cube with the size of 5 cm. The launch of an eighth (1/8) 1U CubeSat allows the reduction of the service price. The market of Cube-Sat launch services is offered already from several companies, but the launch services for PocketQubes is a niche market that Alba Orbital is developing for its own business model. The reduced launch cost is the strongest argument for potential customers. The launch service of 1U (8 PocketQubes) by Alba Orbital will be around €25 000 per 1p compared with €80 000 for a 1U CubeSat. 1p is the unit used for PocketQubes and has following relation with the standard CubeSat: $1p = 1/8U$ or $1U = 8p$. Comparing the launch cost between both standards, the launch service of 1U CubeSat by Alba Orbital would cost €200 000. The first PocketQubes will be launched in 2019. Alba Orbital developed the PocketQube platform "Unicorn" in cooperation with the ESA, and this 2p PocketQube will be used for future IOD missions [9].

13.3.4 GAUSS Srl

Based in Rome, this Italian company was founded in 2012 and is a spin-off from the Scuola di Ingegneria Aerospaziale. The main activities of the company are engineering support and launch services. The launch services are available for all nanosatellite sizes including CubeSats, PocketQubes, and others. GAUSS Srl also has its own IOD program that started with EU funding and the project is called IODISPLay, where with other partners the IOD market is analyzed. GAUSS Srl has the UniSat platform for IOD missions. The satellite has a weight above 20 kg and has different form factors [10].

13.3.5 Open Cosmos

This is a UK start-up company that got important early-stage investors and has a partnership with the ESA. The cooperation with the ESA is related to the support of flight opportunities for IOD missions. Open Cosmos offers an end-to-end service, which includes a CubeSat, mission simulation, integration, testing, launch service, frequency allocation, insurance, and operations. Open Cosmos also developed the Qbee platform, which is a CubeSat with the sizes 3U, 6U, and 12U and based on COTS components. A complete packet configured with a 3U CubeSat can be acquired for a price of €560 000. In the space market, the typical launch price for a 3U CubeSat is €240 000, and a fully equipped 3U CubeSat from the shop costs maximum €100 000. The launch of a customer satellite should succeed 2 years at the earliest after project start [11].

13.3.6 Deep Space ESA Calls

The ESA motivates the European small and medium enterprises (SMEs) and educational community with serious deep space missions. The participation in such calls is big and the proposed solutions demonstrate great creativity of the participants. The following calls were published:

Didymos: Mission to an asteroid
Didymos is a Greek word and means "twin." This asteroid system consists of two components. The official name of the primary component is 65803 Didymos and of the secondary component is S/2003(65803)1. The primary component has a size of around 800 m and the secondary has a size of 170 m. The NASA–ESA mission Asteroid Impact and Deflection Assessment (AIDA) is a cooperation of several institutions, such as the ESA, NASA, DLR, Observatoire de la Côte d'Azur (OCA), and the John Hopkins University Applied Physics Laboratory (JHU/APL). The Asteroid Impact Mission (AIM) is the European contribution to the AIDA mission. The American contribution consists of the NASA's Double Asteroid Redirection Test (DART) satellite. The DART satellite will impact the Didymos secondary component at very high velocity, with the intention to vary the orbit of the asteroid. The ESA satellite AIDA has the mission to observe the impact and measure the variation of the orbit. The satellite

AIM will have two support probes. These probes consist in two CubeSats and this constellation got the name CubeSat Opportunity Payload Intersatellite Networking Sensors (COPINS). The CubeSats will observe the impact from a closer position than the AIM and in high resolution and with multispectral cameras. It is also planned to land a probe in the small asteroid. The mission was canceled in 2017 for lack of funding [12] (Figure 13.4).

LUCE: Mission to the Moon

In 2017, the ESA started the program "Moon Village" which promotes the development of technologies needed for future mining activities on the Moon. The first missions to the Moon are in planning and first ideas for experiments were collected. The Lunar CubeSats for Exploration (LUCE) call was searching for proposals, such as experiments for Moon remote sensing, Lagrange points missions, lunar landing experiments, and Moon ground analysis. A feasibility study will define next steps for the ESA's Moon exploration program.

Figure 13.4 AIM satellite and Didymos. Source: Reproduced with permission from ESA.

Both these calls represent a good example for IOD/IOV initiatives and also define future demand and projects for technology validation.

13.4 Microsatellites

Microsatellites are platforms with a good capacity for experiments and the mission costs can be managed by space agencies. For IOD/IOV missions, microsatellite platforms allow the installation of several small experiments but also complete modules. Microsatellites can also be launched as piggyback payload. One disadvantage is that each satellite platform has its own interface and requirements. It increases the complexity and cost for the integration of new experiments. Space agencies such as DLR and the UK Space Agency have their own microsatellite platforms and they use these for technology demonstration. Both institutions also have available ground stations to support microsatellite missions.

Microsatellites are still being expensive platforms for technology validation due to the size and complexity.

13.4.1 BIRD and TET

Bi-Spectral Infrared Detection (BIRD) was a microsatellite from the German Space Agency DLR. It had the form of a cube with the size of 62 cm and a weight of 94 kg. It was developed in Germany from different DLR institutes. The satellite was based on commercial components and modules. The payload was an infrared camera to detect fire on the ground. Further other technologies were tested in the mission BIRD, such as:

- Failure tolerance on-board computer (OBC)
- Bird Operating System Sergio (BOSS)
- High-precision reaction wheels
- Star camera
- Global Navigation Satellite System (GNSS) receiver

The OBC consisted of four identical computers: two in hot redundancy and two in cold redundancy. The main computer (worker) executed the commands and the backup (supervisor) monitored the main computer activity.

The satellite was operational for more than 10 years. One of the protection strategies for the electronics was to position the electronic boards' plane perpendicular to the sunshine plane so that the irradiated area of electronics was reduced to a minimum value. After some years, the electronics was degrading constantly and at the end of the mission it was not possible to control the attitude because only one reaction wheel was still operational [13].

Technology Experiment Carrier or Technologie Erprobungsträger (TET) is part of the On-Orbit Verification (OOV) program from DLR. The satellite is based on the BIRD technology, but it was necessary to completely redesign the electronics because the components used in BIRD were not anymore available (10 years old technology). The aim of TET is to have a low-cost platform for experiments [14]. The platform consists of following parts:

- Service segment includes batteries, reaction wheels, power unit, and laser gyro.
- Electronics segment includes the OBC of the satellite.
- Payload segment includes star sensors, Hall sensors, antennas, and all experiments.

The payload support system (PSS) controls all the experiments and is the interface with the satellite bus. The PSS consists of:

- *Power supply*: Supplies the experiments with energy.
- *Processor board*: Controls all PSS activities.
- *I/O board*: Provides the interfaces to the experiments.

The PSS is a very important system for the mission success. TET-1 has installed 11 different experiments and not all experiments can be activated at the same time. The management of the experiment activities was a big challenge to be solved. A bad coordination of the experiment activities risks the power resources. The PSS has following functions:

- Control of the experiments
- Housekeeping collection
- Science data collection
- Temporary storage of data

- Encapsulation of data
- Execution of commands
- Control of PSS devices, such as switches, circuit breakers, etc.
- Thermal control management
- Software update and self-test
- Redundancy management
- Executing Fault-Detection, Fault-Isolation and Recovery (FDIR) tasks

Following experiments are installed in the TET-1 satellite:

- Lithium polymer batteries
- Flexible thin-layer solar cells
- New generation of highly efficient solar cells (2×)
- Sensor bus system
- Miniature propulsion system
- Infrared camera
- GNSS receiver
- Ceramic microwave circuits for radio communication
- Memory with recovery mechanisms
- Experimental OBC (Figure 13.5)

Figure 13.5 TET satellite. Source: Reproduced with permission from DLR.

13.4.2 TDS

The TechDemoSat (TDS) was manufactured by SSTL. The UK Space Agency ordered the design and construction of the TDS as national technology demonstration satellite. The TDS is based on the SSTL-150 platform and has the size of $77 \times 50 \times 90 \, cm^3$ and a weight of 157 kg including the experiments. SSTL has the tradition to use commercial components in its satellites, whose qualification for space will be done in-house [15].

Following experiments were installed:

- GNSS receiver
- Altimeter
- Radiation environment monitor
- Charged particle spectrometer
- Cosmic ray intensity detector
- Deorbit sail CubeSat attitude determination system
- Hollow cathode thruster

13.4.3 Euro IOD

The Euro IOD program is a cooperation between Germany and the UK. DLR and the UK Space Agency with ESA support plan to launch an IOD/IOV satellite every 2 years. For this, the TET and the TDS platforms will be used. The characteristics of both are similar. First, a German satellite will be launched, and after 2 years a British satellite will be started. This sequence will be repeated again and again. Each mission will carry European experiments. The main objective of the Euro IOD is long-term and sustainable European IOD missions [16].

13.5 ISS

Nowadays, the ISS is a platform for technology validation. From the different space agencies and in several ISS modules, it is possible to install private experiments. Advantages and technical details vary from one program to another. In this chapter, few examples will be mentioned (Table 13.1). The experiments installed on the ISS can be operational for at least 3 months. One important advantage is that the cargo ships take flights to the ISS every 3 months. It means new experiments can be installed every 3 months and old experiments can be brought back to the ground. Further, it is possible to transport payloads in pressurized cargo ships. The advantage is that the payloads can be easily unloaded direct to the habitable areas of the ISS. Another advantage is that the payloads can be packed in foam and thus reduce the vibrational stress during the transport. It is also important to mention that the vehicle manifest is more flexible in order to easily support the high frequency of the cargo flight to the ISS (Figures 13.6 and 13.7).

13.5.1 NanoRacks

The best-known company with services on the ISS is NanoRacks. NanoRacks has a long story in the space field. NanoRacks CEO was involved in the first space tourism flights on the MIR (former Russian space station). NanoRacks offers launch services from the ISS for nanosatellites and microsatellites. Furthermore, they can use the external platform Japanese Experiment Module–Exposed Facility (JEM-EF) where the NanoRacks has a container installed, called NanoRacks External Platform (NREP). Experiments can also be achieved inside the ISS in the pressurized area. The experiments are installed in containers or racks and these containers follow the CubeSat standard. The advantage inside the ISS is that astronauts can support the experiment if necessary. In both cases, the experiments get power from the ISS and also a data link through a Universal Serial Bus (USB) interface. These advantages are very comfortable because the experiment design can be kept simple.

Table 13.1 Available experiment platforms on the ISS.

Facility	View	Size (cm³)	Mass (kg)	Data (Mbps)	Interface	Power	Transport	Return
EXPRESS Logistics Carriers (ELCs)	Obstructed	86 × 116 × 124	600	10	Ethernet	500–750 W	Unpressurized	No
COL-EPF	Good	86 × 116 × 124	370	100	Ethernet		Unpressurized	No
JEM-EF	Obstructed	80 × 100 × 185	500–2500	100	Ethernet/ Wireless Local Area Network (WLAN)	3 kW	Unpressurized	No
NREP	Obstructed	10 × 10 × 40	40	10	USB/WLAN	300 W	Unpressurized	Yes
Bartolomeo JEM Camera, Light, Pan and Tilt Assembly Adapter Plate (JCAP)	Very good	64 × 83 × 100	100	100	WLAN	100 W	Unpressurized/ pressurized	Yes
Bartolomeo Flight Releasable Attachment Mechanism (FRAM)	Very good	86 × 116 × 124	300–600	100	WLAN	120–800 W	Unpressurized	No

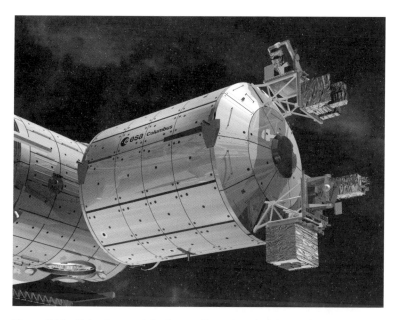

Figure 13.6 Columbus module. Source: Reproduced with permission from ESA.

Figure 13.7 Biolab supports biological experiments. Source: Reproduced with permission from ESA.

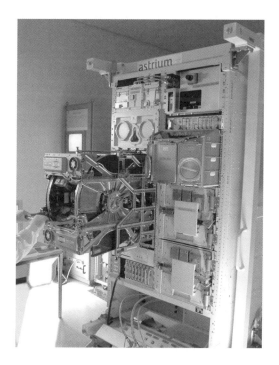

The telemetry is received in almost real-time and the science data are distributed through Internet where the customers can download the data through a secure portal.

NanoRacks has many services such as suborbital flights, etc. [17–19] (Figures 13.8 and 13.9).

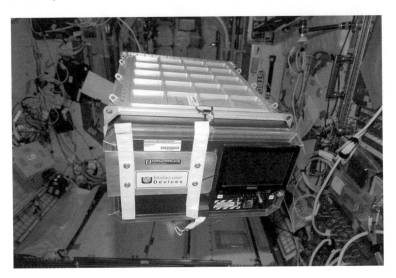

Figure 13.8   NanoRacks platform. Source: Reproduced with permission from NanoRacks.

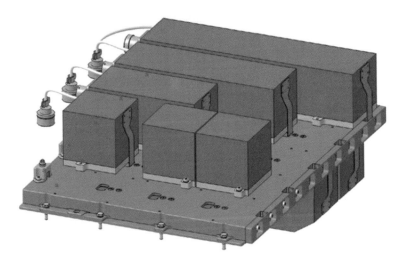

Figure 13.9   Payload configuration example. Source: Reproduced with permission from NanoRacks.

13.5.2   Bartolomeo

Airbus and Teledyne Brown Engineering (TBE) prepared a new external platform for private payloads. The new platform will be installed outside of the European Columbus module and has the name Columbus Laboratory Module – Exposed Payload Facility (COL-EPF) or Bartolomeo. The ESA tries to maximize the use of the Columbus module and Bartolomeo is the best example of that. The Bartolomeo external facility will be launched in the Q3 2019. The first customer for Bartolomeo is the Australian company Neumann Space, which signed a contract in September 2016. Neumann Space has rented a volume of 50 l where the Facility for Australian Space Testing (FAST) will be installed [20].

13.5.3 ICE Cubes

The Belgian company Space Applications Services in partnership with the ESA has a container installed in the Columbus module where experiments in CubeSat format can be installed. The container has the name ICE Cube Facility (ICF) and up to 20 different experiments can be installed. This initiative supports the European industrial and educational IOD/IOV program. In this case also power and data link will be available for all experiments. The service price is €50 000 per kilo and universities get a discount of €15 000 [21] (Figure 13.10).

Figure 13.10 ICE Cubes facility. Source: Reproduced with permission from ESA.

13.5.4 Starlab

The German company Spectrum Aerospace Group also has a new concept for IOV/IOD solution. In this case, a container with the size of 4U will be used and it will be installed on the ISS. In one container, many experiments from different sizes can be installed. In the container, power and data link (USB) is available for all experiments. Nanosatellites and microsatellites can also be used as platform for this IOV/IOD concept. The main idea of this program is the strong reduction of the service cost and also the availability of a comfortable

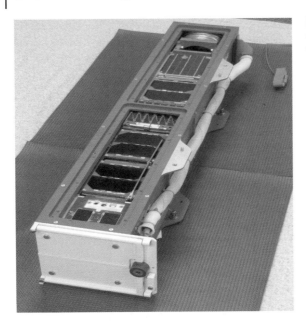

Figure 13.11 4U container for experiments. Source: Reproduced with permission from NanoRacks.

service where the customer gets all necessary resources and can concentrate on its own experiment (Figure 13.11) [22].

References

1 University Nanosatellite Program, University Nanosatellite Program website, https:// universitynanosat.org

2 IOD/IOV. 2016. Context and Orientations ESA, Frederic Teston, September 2016.

3 In-Orbit Demonstration Activities – ESA, Frederic Teston, 2015. Horizon 2020 IOD Workshop, 17 November 2015.

4 SysNova, 2015. R&D Studies Competition for Innovation – No. 3, Asteroid Impact Mission (AIM) Cubesat Opportunity Payloads (COPINS), April 2015.

5 SysNova, 2016. R&D Studies Competition for Innovation – No. 4, Lunar Cubesats for Exploration (LUCE), October 2016.

6 Survey of total ionising dose tolerance of power bipolar transistors and Silicon Carbide devices for JUICE, December 2014.

7 ESA Academy, https://www.esa.int/Education/ESA_Academy, ESA Academy website.

8 Nanosatellites & Cubesat companies, Nanosats.eu, Nanosatellite Database website.

9 Alba Orbital, Alba Orbital website, http://www.albaorbital.com.

10 GAUSS Srl, GAUSS Srl website, https://www.gaussteam.com.

11 Open Cosmos, Open Cosmos website https://www.open-cosmos.com.

12 SysNova, 2015. R&D Studies Competition for Innovation, Statement of Work, ESA, April 2015.

13 BIRD (Bi-Spectral Infrared Detection), https://directory.eoportal.org/web/eoportal/ satellite-missions/b/bird, eoPortal website.

14 TET (Technology Expermient Carrier or Technologie ErprobungsTräger), https://directory.eoportal.org/web/eoportal/satellite-missions/t/tet-1, eoPortal website.

15 TechDemoSat-1 (Technology Demonstration Satellite-1) TDS-1, https://directory.eoportal.org/web/eoportal/satellite-missions/t/techdemosat-1, eoPortal website.

16 Euro-IOD, 2015. A Sustainable TET/TDS-Based European IOD Concept, Norbert Lemke (OHB), Shahin Kazeminejad (DLR), Norbert Püttmann (DLR), Workshop on IOD Opportunities, ESTEC, 17–18 November 2015.

17 Current edition of Interface Document for Internal Platform 1A/2A NanoLab Customers, NanoRacks Documents, http://nanoracks.com/resources/documents, NanoRacks website.

18 Current edition of Interface Document for NanoRacks External Platform (NREP) Customers, NanoRacks Documents, http://nanoracks.com/resources/documents, NanoRacks website.

19 Learn How to Build a Great NanoLab Payload, NanoRacks Documents, http://nanoracks.com/resources/documents, NanoRacks website.

20 ISS: Bartolomeo, ISS-Utilization: Bartolomeo—External Payload Hosting Platform, https://directory.eoportal.org/web/eoportal/satellite-missions/i/iss-bartolomeo, eoPortal website.

21 ICE Cubes Service, ICE Cubes website, http://www.icecubesservice.com.

22 Starlab Project, Starlab Project website, http://starlab-project.com.

14 I-3c

Nanosatellites as Educational Projects

Merlin F. Barschke

Institute of Aeronautics and Astronautics, Technische Universität Berlin, Germany

14.1 Introduction

In the early 1980s, universities started to investigate the capabilities of very small satellites to exploit financial advantages that originated from their reduced complexity and the lower associated launch costs. Gradually, such projects were discovered to be excellently suited to provide hands-on education to engineering students. Nowadays, a large number of universities conduct educational satellite projects, which significantly changed the role of nanosatellites in the space sector. Hence, the present understanding of nanosatellites' capability to fulfill tasks that required much bigger spacecraft in the past, or even to open the way for entirely new application areas, owes much to the educational satellite projects of the past decades.

14.2 Satellites and Project-based Learning

Problem-based learning as a method for engineering education is widely regarded as both successful and innovative, as students generally show greater motivation with this learning model. Project organized learning as a specific form of problem-based learning is especially useful in engineering education. Here, the students work in teams to apply the theoretical knowledge gained in the lectures on practical engineering problems. Besides the higher motivation of students, such projects are ideally suited to support the development of design, teamwork, and project management capabilities, which can only be taught insufficiently using traditional teaching methods [1, 2]. Furthermore, students are more likely to graduate in time with the problem-based learning approach. At Aalborg University, where this approach is firmly integrated in the curriculum, 75% of the students manage to complete their studies within the designated duration of study [3]. As a result of these advantages, project work is an integral part of the engineering curricula of many universities around the world.

The more realistic a project is, the more effective is the learning process [4]. In this context, nanosatellites have proven to be especially well suited as subject for project-based

Nanosatellites: Space and Ground Technologies, Operations and Economics, First Edition.
Edited by Rogerio Atem de Carvalho, Jaime Estela, and Martin Langer.
© 2020 John Wiley & Sons Ltd. Published 2020 by John Wiley & Sons Ltd.

learning. Kitts summarized designing a multidisciplinary system, experiencing the entire design cycle and executing systems engineering as well as project management tasks as key benefits of nanosatellite projects for educational purposes [5].

14.2.1 A Brief History of Educational Satellite Projects

The launch of UoSAT-1 in 1981 marks the beginning of the university-class satellite era. UoSAT-1 was a 52 kg satellite developed and build by the University of Surrey to pursue science and educational objectives [6]. In the following two decades, 35 university satellites were developed and launched [7]. Most of these satellites had a mass below 50 kg, with the smallest even going down to 200 g.

The introduction of the CubeSat reference design by Stanford University and California Polytechnic State University in 1999 has radically changed the landscape of educational satellite missions. The aim was to develop a new class of extremely small standardized spacecraft for education and research. Originally, a CubeSat was a cubic-shaped spacecraft with an edge length of 10 cm and a mass of 1 kg. The form factor was later expanded to 2 U (10 cm × 10 cm × 20 cm) and 3U designs, and even CubeSats of 6U and larger were developed. A CubeSat has to be ejected from a containerized separation system, which could accommodate three 1U CubeSats [8]. The idea of using an ejection container originates from Stanford University's Orbiting Picosatellite Automatic Launcher (OPAL) microsatellite project [9]. The first CubeSats were launched in 2003. Since then, the number of launched university-class spacecraft vastly increased, with the majority being CubeSats.

Today, hundreds of satellites based on the CubeSat form factor have been developed and launched, not only in educational projects but also by industry and research institutions [10]. Launches of educational CubeSats are supported by programs, such as NASA's CubeSat Launch Initiative (CSLI) or the "Fly your Satellite!" program of European Space Agency (ESA) [11]. Furthermore, some national programs, such as the Norwegian student satellite program (ANSAT), provide funding for both, the development and the launch of CubeSats [12]. Other programs, such as the University Nanosat Program (UNP) or EduSAT from Italy, are not limited to CubeSats. The UNP is a joint effort between different entities of the US Air Force and the American Institute of Aeronautics and Astronautics (AIAA), and provides launch capabilities as well as guidance during the development process for university projects. Within this program, about 10 universities develop a satellite design and compete for 2 years for a launch slot, while being professionally supported by design workshops and reviews during this period. The program was initiated in 1999, with approximately 5000 students participated since then and eight satellites with a mass between 15 and 50 kg brought to orbit [13]. Similarly, the EduSAT program of the Italian Space Agency currently funds the development and the launch of an educational nanosatellite [14].

Between 2006 and 2018, more than 275 satellites below 10 kg and CubeSats up to 12U developed by universities were brought to orbit. Several developing nations managed to establish their first space project in an educational environment, some of which even resulted in the launch of a country's first satellite [13]. Figure 14.1 gives an overview of satellites with significant student involvement in developing the spacecraft, launched

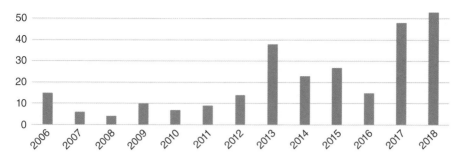

Figure 14.1 Overview of small satellites with a mass below 10 kg (CubeSats up to 12 U) that were launched by universities between 2006 and 2018. Source: Reproduced with permission from ESA.

between 2006 and 2018 [15]. Notwithstanding the classification presented in Section 14.2.2, the figure lists all satellites launched by universities.

14.2.2 Project Classification

While all educational satellite projects share the objective to improve aerospace education, the approach followed may differ significantly among different missions. This ranges from projects where students develop a satellite largely independently to developments mainly driven by university staff.

Swartwout defined university-class missions such that "untrained personnel (students) performed a significant fraction of key design decisions, integration and testing activities, and flight operations" [16]. Furthermore, he states that for a university-class mission, "the training of these people was as important as (if not more important than) the nominal 'mission' of the spacecraft itself" [16]. However, this definition does not allow a distinction among different types of educational projects and might even exclude some research-driven missions with nevertheless strong student participation. Therefore, this chapter defines an educational project, somewhat more broadly, as an effort with significant educational impact. Additionally, educational satellite missions are further divided into education-focused and research-focused projects. Here, education-focused missions are defined as projects, where education is the primary mission objective, similarly to the definition provided by Swartwout. By contrast, the mission is prioritized over the educational impact for research-focused missions.

According to this, an education-focused satellite project might have the following characteristics: An average number of around 80 undergraduate and graduate students will participate in the work throughout the project duration with a highly varying degree of commitment. The average time a student will be involved is 1 year, during which he contributes to the project as part of his unpaid course or thesis work. High-level decisions on mission, satellite design, and payload will be made by the students, which will also be responsible for systems engineering. Students will furthermore be involved in the project management and may even be, at least to a certain extent, responsible for fund raising. While education is clearly stated as primary objective, the mission may also include a secondary technology demonstration or science payload.

A research-focused satellite mission, on the other hand, will mainly be driven by its technology or science objective, while education will be secondary. Employed researchers and Ph.D. students will form the core team and undergraduate and graduate students will mainly be involved as employed assistants or during their thesis work. High-level decisions are taken by the core team, which is also responsible for systems engineering and project management. Overall, the developed satellite will be considerably more sophisticated when compared to an education-focused project.

While many projects combine characteristics from both types of missions, a clear distinction is sensible at this point, as certain characteristics are very different from one another.

14.3 University Satellite Programs

Universities around the world established educational nanosatellite programs to improve their engineering education by providing the students with practical design and teamwork experience. Other university missions are more driven by their science and technology objectives but will nevertheless have a large educational impact. The following three programs are introduced to serve as examples for what can be achieved.

14.3.1 Aalborg University

Aalborg University started its satellite development activities by contributing the attitude control system to the first Danish satellite Ørsted, which was launched in 1999. Although Ørsted itself was not an educational project, several Ph.D. candidates at Aalborg University conducted research on this mission [3].

As a next step concerning satellite development, Aalborg University initiated its own CubeSat series. The first satellite developed was AAU CubeSat that carried a CMOS imager as main payload. AAU CubeSat was launched in mid-2003 and was therefore one of the first CubeSats ever brought to orbit [17]. Launched in 2008, AAUSAT-II was the second CubeSat developed at AAU Student Space laboratory. It carried an experimental attitude determination and control system, as well as a gamma ray detector. Its successor AAUSAT3 was launched in 2013 and carried an Automatic Identification System (AIS) receiver as payload [18].

Aalborg University also contributed subsystems and payloads to the several satellite missions of other institutions. This was for example the case for the on-board computer, the attitude determination and control system, and a camera payload of the Student Space Exploration and Technology Initiative (SSETI) Express student mission of ESA, launched in 2005 [19]. Similarly, the on-board computer of Bauman Moscow State Technical University's Baumanetz satellite originated from Aalborg University.

More recently, AAU Student Space laboratory launched AAUSAT4 and AAUSAT5. While the first one was brought to orbit as part of the "Fly your satellite!" program of ESA in 2016, the second one was released by the first Danish astronaut Andreas Mogensen during his stay on the International Space Station (ISS) in 2015. Figure 14.2 shows AAUSAT4 in the clean room and while being removed from the deployer after vibration testing. Currently, students at AAU Student Space laboratory are defining the mission for AAUSAT6.

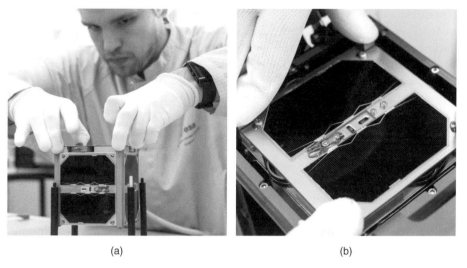

(a)   (b)

Figure 14.2 Impressions of the educational single-unit CubeSat AAUSAT4 of AAU Student Space laboratory that was launched within ESA's "Fly your satellite!" program in 2016. (a) AAUSAT4 in the clean room. (b) AAUSAT4 being removed from the deployer. Source: Reproduced with permission from ESA.

The AAUSAT program is a very successful example of an education-focused satellite program. Satellites built at AAU Student Space laboratory are entirely developed by the students and all subsystems are designed in-house.

14.3.2 Technische Universität Berlin

Technische Universität Berlin started developing small satellites in the early 1980s and launched its first satellite in 1991. TUBSAT-A was the first member of the TUBSAT family, which was developed as a low-cost test facility for on-orbit experiments [20]. The mission of TUBSAT-A was to provide store-and-forward communication services and perform several technology experiments. Among many other applications, the satellite was used for animal tracking and provided communication capabilities to Arctic and Antarctic expeditions. The satellite was successfully operated for more than 15 years.

Within the TUBSAT series, seven more satellites with launch masses ranging from 3 to 56 kg were developed and launched, with a special focus laid on demonstrating Earth observation capabilities [21]. As a novel feature, a later satellite of the TUBSAT-family, DLR-TUBSAT, incorporated adaptive attitude control capabilities, which allowed to visually follow a target via video life-feed using one of the satellite cameras. DLR-TUBSAT was a cooperation of the German Aerospace Center (DLR) and Technische Universität Berlin, which developed the satellite bus, while DLR was responsible for the payload [22].

Technische Universität Berlin also established a CubeSat series called Berlin Educational and Experimental SATellite (BEESAT). In 2009, the single unit CubeSat BEESAT-1 was launched to demonstrate miniaturized reaction wheels on orbit. In 2013, BEESAT-2 and BEESAT-3, two further single-unit CubeSats, were brought to orbit together on one launcher. While BEESAT-2 was to demonstrate three-axis attitude control, the primary mission objective of BEESAT-3 was to provide hands-on education to students of Technische Universität Berlin [23, 24]. Both satellites are still operated successfully to date [25]. Figure 14.3 shows the flight model of BEESAT-3 during vibration testing and while being integrated onto the Fregat upper stage at the launch site. BEESAT-4, which was launched in 2016, is based on the design of BEESAT-1 and BEESAT-2 but supplemented with a Global Positioning System (GPS) receiver. Currently, Technische Universität Berlin is developing four 0.25U CubeSats for a technology demonstration mission that is scheduled for launch in 2019 [26].

Besides the BEESAT program, Technische Universität Berlin is also currently establishing the TUBiX platform series, comprising two satellite platforms, namely, TUBiX10 and TUBiX20 [27]. While TUBiX10 supports missions of up to 10 kg, TUBiX20 missions have a launch mass of roughly 20 kg. The TUBiX10 platform is implemented for the first time within the S-Net project, which was launched in 2018 and currently demonstrates intersatellite communication in the S-band with four 8 kg spacecraft [28]. Further, three TUBiX20-based missions are currently worked on at Technische Universität Berlin. TechnoSat was launched on a Soyuz-Fregat from Baikonur in 2017 and currently demonstrates a number of newly developed nanosatellite components on orbit [29, 30]. Technische Universität Berlin Infrared Nanosatellite (TUBIN) will carry two uncooled infrared imagers to evaluate their orbital performance regarding wildfire detection [31]. Within the third TUBiX20 mission, QUEEN, Humboldt-Universität zu Berlin, Technische Universität Berlin, and the Ferdinand-Braun-Institut, Leibniz-Institut für Höchstfrequenztechnik, Berlin are planning to demonstrate optical quantum technologies for compact rubidium vapor-cell frequency standards in space [32].

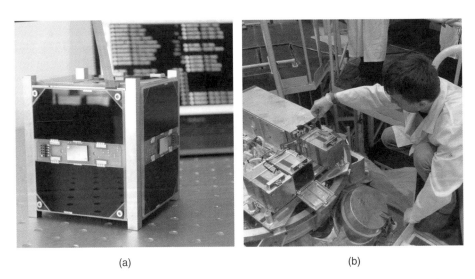

(a) (b)

Figure 14.3 Educational single-unit CubeSat BEESAT-3 of Technische Universität Berlin, Germany that was launched aboard a Soyuz rocket in 2013. (a) BEESAT-3 functional testing in the laboratory. (b) Remove Before Flight (RBF) removal from the BEESAT-3 deployer. Source: Reproduced with permission from ESA.

With the exception of BEESAT-3, all satellite projects of Technische Universität Berlin are research-focused missions with the majority of the development team members being university staff, complemented by employed students. However, many students take the opportunity to perform their thesis work on certain aspects of these missions and mission operations are firmly integrated into the lectures.

14.3.3   University of Tokyo

The Intelligent Space Systems Laboratory (ISSL) of the University of Tokyo was among the firsts to develop CubeSats. ISSL's XI-IV, a single unit CubeSat launched in 2003, was the first CubeSat to deliver images from Earth. Being still operational after more than a decade, XV-IV impressively demonstrated the durability of its design. In 2005, XI-V, the flight spare model of XI-IV, which was originally not intended for launch, was brought to orbit as well [33]. The XI missions have been education-focused projects, where not only the satellite development and production but also most management tasks were performed by students [34].

As a next project, the university was aiming at the development of a larger and more sophisticated satellite, based on the CubeSat technology, which was already proven on orbit by XI-IV and XI-V. As a result, Pico-satellite for Remote-sensing and Innovative Space Mission (PRISM), an 8 kg satellite that carries an extensible telescope boom was developed [35]. Despite its lager size, it was also ejected from a containerized dispenser that was specially designed to support the flexible extensible telescope boom. After its launch in 2009, the telescope boom was successfully deployed, and images were transmitted to Earth. Figure 14.4 shows the PRISM satellite in the clean room, as well as a picture that was taken by the spacecraft from orbit.

(a)   (b)

Figure 14.4   The 8 kg nanosatellite PRISM built by the Intelligent Space Systems Laboratory (ISSL) of the University of Tokyo that was launched in 2009, which carries a flexible extensible telescope boom. (a) PRISM spacecraft in the clean room. (b) Picture taken by PRISM on orbit. Source: Reproduced with permission of Tokyo University, Japan.

Together with the National Astronomical Observatory of Japan, Tokyo University developed Nano-Japan Astrometry Satellite Mission for Infrared Exploration (Nano-JASMINE), which is a technical demonstrator for Japan's space astrometry mission JASMINE [36]. Nano-JASMINE has a mass of 35 kg and a cubic shape with 50 cm edge length. The satellite was scheduled for launch in late 2018. Most recently, Tokyo University developed three satellites of the Hodoyoshi series, of which the first one was developed in cooperation with Axelspace Corporation [37]. All three Hodoyoshi satellites have a mass of about 60 kg, carry Earth observation imagers, and were launched in 2014.

Starting off with a student-type CubeSat mission, the University of Tokyo has now established a world-leading satellite design facility that is developing satellites for both science and Earth observation missions.

14.4   Outcome and Success Criteria

Outcome and success criteria of an educational satellite project strongly depend on the nature of the project. Within an education-focused mission, the success criteria strongly reflect the educational nature of the project. As such projects seek to maximize the training impact for the students, education is the primary mission objective and a success rate of more than 50% can typically be archived, even if the satellite fails to respond, once brought to orbit. Even though the satellite may carry an innovative payload, the generation of relevant science data, which is of interest outside the project itself, tends to be the exception rather than the rule. One reason for this is a lack of reliability of the spacecraft bus that often prevents payload operation. More than a quarter of the university CubeSat missions launched until 2014 never established a radio link to their satellite, launch failures not taken into account. Additionally, less than half of the remaining projects that could establish communication claim to achieve their primary science or technology objective [38]. While this makes about one third of the university CubeSat missions achieving their primary mission objective, this number also includes missions with no relevant payload. Basically, it can be observed that for student-type programs, the probability of mission success increases with the number of satellites launched. But unfortunately, some universities could not raise sufficient funds to establish a second mission, after the first one was launched [39].

As most of the satellite design is developed by students in such projects, the educational outcome is obvious. For a survey by Straub and Marsh, the majority of students indicated that the participation in an educational small satellite program improved their comfort in working collaboratively with others, as well as their understanding of what everyday research work is like [40]. Moreover, many students participated in education-focused projects hold important positions in industry and academia today, not least due to the experience gained in these missions. Many universities can directly relate a vast increase in students placed in space industry to the introduction of an educational satellite project [41].

However, although it might be acceptable for a university to peruse a high-risk and low-performance mission in order to enter the field, the capabilities and especially the reliability of education-focused nanosatellite missions need to be vastly increased to justify the launch. This is especially true considering the opportunities offered by CanSat projects, as well as sounding rocket and high-altitude balloon experiments that could be used for project-based education with similar impact [42]. The possibility to recover the system after its mission as well as the shorter project duration might even allow for a better learning outcome, as all students can experience the entire project circle including a comprehensive fault analysis, if needed. Satellite projects, on the other hand, provide an extra motivation for students with the prospect of actually launching the satellite to orbit and resemble the real industry environment more closely.

Research-focused satellite projects, on the other hand, have stronger emphasis on the performance of satellite bus and payload, which also reflects in the success criteria defined for such projects. While education is still an important factor, science and technology objectives are dominant.

One reason why a low success rate is still a large concern, especially for education-focused missions, is that inexperienced universities are continuously entering the field of satellite development. In addition, the problem originates, at least partially, from the demand for higher performance that in turn arises from the call for a higher science return. The growing complexity of today's educational satellites allows to meet the requirements of ever more demanding payloads, when being compared to early designs. Such payloads might come with high power requirements or demand accurate pointing, which will notably increase the complexity of the system. When designing a satellite to fulfill such demanding requirements, reaching a high level of reliability is an even greater challenge.

Successful programs show that the reliability and therefore the science outcome of educational nanosatellite missions can be improved by increasing the project continuity. For education-focused missions, this means continuity in design, as a continuity in staff cannot be achieved. When a program follows a long-term objective, the design can be improved and extended in smaller steps, once a reliable solution is established, which limits the risk introduced by any changes. For research-focused missions, the more demanding mission objectives may require major modifications of a given design between consecutive missions. Here, a lower turnover of staff can help to conserve experience and knowledge within the team to facilitate higher reliability. Another approach, which is nowadays followed by many educational programs, is to concentrate the development efforts on one specific part of the satellite, such as the payload, and purchase commercial prequalified components for the remainder of the system. While the complexity of systems engineering tasks and software development for such an approach should not be underestimated, it can support the development of a more reliable system in a shorter time frame.

The greatest achievement of educational satellite projects so far is the nanosatellite itself. While this class of satellite was long time regarded as not powerful enough for real applications, industry and agencies nowadays show a great interest and get involved, which is mainly a consequence of the capabilities demonstrated in educational missions. As a result, many of today's most impressive examples of on-orbit performances in the nanosatellite class have not been achieved by academia. Examples are NASA's O/OREOS mission [43], the AeroCube series of the Aerospace Corporation [44], and the American company Planet, which demonstrated that nanosatellites can be part of a successful business model [45]. As education-focused missions cannot compete with high profile missions of agencies and industry, they should instead seek to develop scenarios, which are not yet covered by such missions.

14.5   Teams and Organizational Structure

The organizational structure of an educational satellite project strongly depends on the academic environment in which it is set. This might be a university where teaching is based on project-based learning to a large extent or an institution than mainly makes use of traditional lectures.

An education-focused project is usually organized around a university employee or a steering group, which coordinates the students, gives technical advice, and manages the financial aspects of the project. While university staff is actively participating in designing the spacecraft in some projects, development and production of the satellite is left entirely to the students in others. The students are commonly organized in teams, each of them responsible for one subsystem of the spacecraft. One dedicated team member of each subsystem group is usually responsible for systems engineering tasks and presents the subsystem on management level [18]. Throughout the development of a student satellite, which usually takes 2–4 years, often up to 50 or 100 students are involved in the project through project work, as student staff or during their final thesis [46].

The organizational structure of a research-focused satellite projects is quite different: While the team is also divided in subsystem groups, those are usually grouped around a member of staff. While a comparable degree of responsibility might be transferred to students here, design decisions are ultimately left up to staff members. As a research-focused project is most probably larger and more complex than a student-driven project, it resembles the industry work environment more closely. Also, the work on a comparatively more complex system might be more challenging from a technical point of view. Furthermore, as more experienced staff is working on the project, it is more likely that experts from all relevant disciplines are present, which can share broader experience in a certain field.

In conclusion, it can be said that the educational benefits for students participating in research-focused missions are in part rather different but equally large. While they have the chance to work on a spacecraft that is more sophisticated, certain systems engineering and project management trades are not executed by the students. Furthermore, the number of involved students is usually lower, when compared to an education-focused mission.

14.6   Challenges and Practical Experiences

Due to the complexity and the demand for a high level of reliability, developing a satellite is a very challenging task in itself. While educational satellites are considerably less complex when compared to commercial spacecraft, there are a number of additional challenges that arise from their educational nature. These include, for example, high staff turnover rates and limited financial resources. Students, while often highly motivated and enthusiastic for the project, are still in the process of being trained and their experience is therefore limited. Hence, a mission that sorely relies on students as workforce will have a much different dynamic when compared to a research-focused mission, which might also have strong student involvement, but which builds on a more experienced workforce partly composed of researchers. In the following, challenges as well as practical experiences that evolve from educational satellite projects are described.

14.6.1 Staff Turnover

One of the major challenges faced by educational satellite projects results from the fast turnover rates of the students. As most students are involved in such projects as part of their thesis or course work, their involvement is usually limited to about 1 year in average. The drawbacks of continuous student fluctuation are obvious, especially in critical phases of the mission. New students that replace the ones that left are likely to be unfamiliar with both the general work status and the concept plan of a work package assigned to, and they require more supervision. This is especially a problem if large parts of the team are exchanged in the semester break, and new participants can therefore not be incorporated by more experienced team members. To minimize this effect, the duration of the student's time of involvement can be extended, if a further employment as student staff is possible. This will have significant impact, especially in critical project phases [47].

14.6.2 Development of Multidisciplinary Skills

A satellite is a multidisciplinary task, which requires expertise from all sorts of fields. Therefore, students from many different backgrounds, such as electrical engineering, aerospace, and software, should be incorporated in the team, each one working on their field of expertise. In this manner, students gain hands-on experience in a field during their studies and contribute important specialist knowledge to the project's success. Furthermore, working within a team that combines expertise from many different fields maximizes the learning outcome for all team members.

14.6.3 External Experts

Design reviews play an important role in satellite development. For educational projects, such reviews are a unique opportunity to have the design evaluated by external experts. Experience showed that there is a good chance to attract experienced reviewers from space industry to attend reviews of educational projects [48]. The gains for the project are twofold: On the one side, the experts are likely to reveal weaknesses in the design and can share their experience within such reviews. On the other hand, the gathered contacts can also be useful in later stages of the project, when external expertise is required.

14.6.4 Project Documentation

A signification portion of the manpower of an educational satellite project is usually spent on documentation. This includes designing documents along with configuration management and quality assurance. Documentation is vital for distributing information and data within the team, to preserve it for later project phases, and it serves as preparatory reading for new team members to familiarize with the project. Furthermore, it might be used by experts to evaluate the developed design in reviews.

However, documentation is useless, if not properly maintained throughout the project. Often it is seen that a project starts with a very high level of documentation, which is outdated quickly and cannot be continued due to the limited resources in workforce. Therefore, it is very important to find the appropriate level of documentation effort right from the start of the project. In recent years, modern and effective tools for documentation purposes, such as ticket systems' bug-trackers and project Wikis, have emerged, especially from the field of software development. Such tools are especially useful for educational projects, as they can replace many paper-based documents, which increases the accessibility for all team members.

Paper-based design documents are still required for reviews and have proved effective for preserving the knowledge about the overall design of the spacecraft within the project or for future reference. For many other purposes, server-based databases and tools are often more effective. However, experience shows that establishing and maintaining such tools in large teams is a challenge. Therefore, strictly, only necessary tools should be introduced that can realistically be maintained throughout the project. In the following, a few tools that can be used effectively for educational satellite projects are introduced.

- A project Wiki that contains instructions about how to use hardware and software can significantly reduce the enquiries about such procedures within the team. Furthermore, a Wiki may be used to distribute systems engineering level information, such as interface definitions. Another useful application is the documentation of integration procedures.
- Bug trackers are of course useful to streamline the software development process but can also be effectively applied for other purposes, such as configuration management and replacing paper-based solutions.
- A ticket system, if consequently used, makes the distribution of tasks transparent to all team members.

14.6.5 Testing

A well-tested satellite is a prerequisite for continuous and reliable operations once brought to orbit. Especially in educational missions, extensive testing is a key factor for mission success. This applies to mechanical and environmental qualification and acceptance tests, but may be even more important for functional testing. It is widely believed that for many missions that failed on orbit, the problem could have been eliminated before launch, if the spacecraft would have been tested more extensively on the ground beforehand [49].

14.6.6 Software

Software development and testing takes up a significant part of any space project. In fact, many completed educational satellite projects state that they have greatly underestimated the software part of the project [47]. Furthermore, not only the software of the satellite itself is required, also ground support and ground station software need to be developed. Another drawback of the commonly applied hardware first approach is that the hardware cannot effectively be tested, if no software is available. Software shall be considered from the very beginning of the project and sufficient project resources need to be allocated.

14.6.7 Ground Station

In the final stage of the project, all resources are usually focused on delivering the flight model in time. For projects that do not have a well-tested ground station available, this has often proved problematic. More than one project managed to deliver a working flight model, only to find that they could not communicate with their satellite once in orbit due to ground station issues.

14.7 From Pure Education to Powerful Research Tools

In conclusion, it can be said that hundreds of educational missions conducted in the past have demonstrated that educational satellite projects are a very successful tool for the training of young space engineers. Today, the prerequisites for successful student-driven nanosatellite missions are better than ever before. Ten years ago, students had to develop most of the satellite in-house and only very few relevant publications were available. Today, components are widely available, thanks to the CubeSat standard and hundreds of educational nanosatellite projects. Companies such as Innovative Solutions in Space (ISIS), Tyvak Nano-Satellite Systems, GomSpace, and ClydeSpace offer a variety of components and even entire satellite buses and associated services. Furthermore, a large community generated a wealth of useful publications. However, measured by the number of projects conducted in recent years, the science return, especially for education-focused missions, is still comparatively small. In order to increase the science impact of future educational space missions, two criteria must be met. First, the mission must carry a payload with substantial scientific value, and second, the mission success rate needs to be vastly increased, to justify the investment of such payload. While current projects indicate that the share of missions without a valuable payload will drop in the near future, the current generation of educational satellites still suffers from relatively high failure rates.

However, different approaches to increase the reliability of educational space missions emerged from past projects. One strategy, which was successfully applied by several projects, is to start with a reliable and comparatively simple satellite design, whose capabilities are gradually increased during subsequent missions. Another way to address this challenge is to purchase flight-proven hardware and to concentrate the development efforts on a certain subsystem or on the payload, as well as on system integration, software, and testing. Finally, certain missions that require a comparatively complex spacecraft or a high percentage of newly developed components might be more suitable for a research-focused mission, conducted by a team formed by experienced staff and students.

When following these approaches, future educational nanosatellite missions can serve as powerful tools for university space research. However, this can only be achieved when the mission is designed on the basis of a realistic assessment of available resources both in staff and funding and lessons learnt from past missions are considered.

References

1 de Graaf, E. and Kolomos, A. (2003). Characteristics of problem-based learning. *International Journal of Engineering Education* 19: 657–662.
2 Dym, C., Agogino, A., Eris, O. et al. (2005). Engineering design thinking, teaching, and learning. *Journal of Engineering Education* 94: 103–120.

3 Bhanderi, D., Bisgaard, M., Alminde, L., and Nielsen, J.F.D. (2006). A Danish perspective on problem based learning in space education. *IEEE Aerospace and Electronic Systems Magazine* 21: 19–22.

4 Lenschow, R.J. (1998). From teaching to learning: a paradigm shift in engineering education and lifelong learning. *European Journal of Engineering Education* 23: 155–161.

5 Christopher, A. (1999). Kitts. Three project-based approaches to spacecraft design education. In: *Proceedings of the Aerospace Conference*, vol. 5, 359–366. Snowmass at Aspen, USA.

6 Sweeting, M.N. (1982). UoSAT-1: an investigation into cost-effective spacecraft engineering. *Radio and Electronic Engineer* 52: 363–378.

7 Michael Swartwout. 2004. University-class satellites: From marginal utility to "disruptive" research platforms. In *Proceedings of the 18th Annual AIAA/USU Conference on Small Satellites*, Logan, USA.

8 Michael Swartwout, 2008. Standardization promotes flexibility: A review of CubeSats success. In *Proceedings of the 6th Responsive Space Conference*, Los Angeles, USA.

9 Hank Heidt, Jordi Puig-Suari, Augustus S. Moore, Shinichi Nakasuka, and Robert J. Twiggs. CubeSat, 2000. A new generation of picosatellite for education and industry low-cost space experimentation. In *Proceedings of the 14th Annual AIAA/USU Conference on Small Satellites*, Logan, USA.

10 Michael Swartwout 2016. Secondary spacecraft in 2016: Why some succeed (and too many do not). In *Proceedings of the IEEE Aerospace Conference*, Big Sky, USA.

11 Elizabeth Buchen. 2015. Small satellite market observations. In *Proceedings of the 27th Annual AIAA/USU Conference on Small Satellite*, Logan, USA.

12 Joran Grande. 2013. Educational benefits and challenges for the Norwegian student satellite program. In *Proceedings of the 64th International Astronautical Congress*, Beijing, China.

13 Woellert, K., Ehrenfreund, P., Ricco, A.J., and Hertzfeld, H. (2011). CubeSats: cost-effective science and technology platforms for emerging and developing nations. *Advances in Space Research* 47: 663–684.

14 Filippo Graziani, Giuseppina Pulcrano, Fabio Santoni, Massimo Perelli, and Maria Libera 2013. Battagliere. EduSAT: An Italian Space Agency outreach program. In *Proceedings of the 64th International Astronautical Congress*, Beijing, China.

15 Erik Kulu. Nanosats Database. www.nanosats.eu. [Accessed: 2019-03-19].

16 Swartwout, M. (2009). The first one hundred university-class spacecraft, 1981–2008. *Aerospace and Electronic Systems Magazine* 24: 1–25.

17 Alminde, L., Bisgaard, M., Vinther, D. et al. (2003). Educational value and lessons learned from the AAU-CubeSat project. In: *Proceedings of the International Conference on Recent Advances in Space Technologies*, 57–62. Istanbul, Turkey.

18 Larsen, J.A. and Nielsen, J.D. (2011). Development of CubeSats in an educational context. In: *Proceedings of the International Conference on Recent Advances in Space Technologies*, 777–782. Istanbul, Turkey.

19 Alminde, L., Bisgaard, M., Melville, N., and Schaefer, J. (2005). The SSETI-express mission: from idea to launch in one and a half year. In: *Proceedings of the 2nd International Conference on Recent Advances in Space Technologies*, 100–105. Istanbul, Turkey.

20 Udo Renner, Bernard Lübke-Ossenbeck, and Pius Butz. 1993. TUBSAT, low cost access to space technology. In *Proceedings of the 7th Annual AIAA/USU Conference on Small Satellites*, Logan, USA.

21 Merlin F. Barschke, Klaus Brieß, and Udo Renner. 2016. Twenty-five years of satellite development at Technische Universität Berlin. In *Proceedings of the 6th Small Satellites Systems and Services Symposium*, Valletta, Malta.

22 Roemer, S. and Renner, U. (2003). Flight experiences with DLR-TUBSAT. *Acta Astronautica* 52: 733–737.

23 Merlin F. Barschke, Frank Baumann, and Klaus Brieß. 2012. BEESAT-3: A picosatellite developed by students. In *Proceedings of the 61st German Aerospace Congress*, Berlin, Germany.

24 Sebastion Trowitzsch, Frank Baumann, and Klaus Brieß. 2014. BEESAT-2: A picosatellite demonstrating three-axis attitude control using reaction wheels. In *Proceedings of the 65th International Astronautical Congress, Toronto*, Canada.

25 Merlin F. Barschke, Philipp Werner, Sascha Kapitola, and Marc Lehmann. 2018. BEESAT-3 commissioning—better late than never. In *Proceedings of the 69th International Astronautical Congress*, Bremen, Germany.

26 Frank Baumann, Nicholas Korn, Kjell Pirschel, Ronny Wolf, and Klaus Brieß. 2017. Distributed picosatellites for technology demonstration. In *Proceedings of the 66th German Aerospace Congress*, München, Germany, September.

27 Merlin F. Barschke, Zizung Yoon, and Klaus Brieß. 2013. TUBiX—The TU Berlin Innovative Next Generation Nanosatellite Bus. In *Proceedings of the 64th International Astronautical Congress*, Beijing, China.

28 Yoon, Z., Frese, W., Bukmaier, A., and Briess, K. (2014). System design of an S-band network of distributed nanosatellites. *CEAS Space Journal* 6: 61–71.

29 Merlin F. Barschke, Philipp Werner, Karsten Gordon, Marc Lehmann, Walter Frese, Daniel Noack, Ludwig Grunwaldt, Georg Kirchner, Peiyuan Wang and Benjamin Schlepp, 2018. Initial results from the TechnoSat in-orbit demonstration mission. In *Proceedings of the 32nd AIAA/USU Conference on Small Satellites*, Logan, USA.

30 Merlin F. Barschke, Karsten Gordon, Philip von Keiser, Marc Lehmann, Mario Starke, and Philipp Werner. 2018. Initial orbit results from the TUBiX20 platform. In *Proceedings of the 69th International Astronautical Congress*, Bremen, Germany.

31 Barschke, M.F., Bartholomäus, J., Gordon, K. et al. (2017). The TUBIN mission for wildfire detection using nanosatellites. *CEAS Space Journal* 9: 183–194.

32 Aline N. Dinkelaker, Akash Kaparthy, Sven Reher, Ahmad Bawamia, Heike Christopher, Andreas Wicht, Philipp Werner, Julian Bartholomus, Sven Rotter, Robert Jrdens, Merlin F. Barschke, and Markus Krutzik. 2018. Optical quantum technologies for compact rubidium vapor-cell frequency standards in space using small satellites. In *Proceedings of the 16th Reinventing Space Conference*, London, UK.

33 Rye Funase, Ernesto Takei, Yuya Nakamura, Masaki Nagai, Akito Enokuchi, Cheng Yuliang, Kenji Nakada, Yuta Nujiri, Fumiki Sasaki, Tsukasa Funane, Takeshi Eishima, and Shinichi Nakasuka. 2005. Technology demonstration on University of Tokyo's picosatellite XI-V and its effective operation result using ground station network. In *Proceedings of the 56th International Astronautical Congress*, Fukuoka, Japan.

34 Yuichi Tsuda, Nobutada Sako, Takashi Eishima, Takahiro Ito, Yoshihisa Arikawa, Norihide Miyamura, Kazutaka Kanairo, Shinichi Ukawa, Shiro Ogasawara, Sanae Ishikawa, and Shinichi Nakasuka. 2002. University of Tokyo's CubeSat "XI" as a student-build educational picosatellite—Final design and operations plan. In *Proceedings of the 23rd International Symposium on Space Technology and Science*, Matsue, Japan.

35 Toshiki Tanaka, Yuki Sato, Yasuhiro Kusakawa, Kensuke Shimizu, Takashi Tanaka, Sang Kyun Kim, Mitsuhito Komatsu, Yoo Il-Yun, Casey Lambert, and Shinichi Nakasuka. 2009. The operation results of Earth image acquisition using extensible flexible optical telescope of PRISM. In *Proceedings of the International Symposium on Space Technology and Science*, Tsukuba, Japan.

36 Nobutada Sako, Yoichi Hatsutori, Takashi Tanaka, Takaya Inamori, and Shinichi Nakasuka. 2001. Nano-JASMINE: A small infrared astrometry satellite. In *Proceedings of the 21st Annual AIAA/USU Conference on Small Satellites*, Logan, USA,

37 Korehiro Maeda and Shinichi Nakasuka. 2014. Overview of Hodoyoshi microsatellites for remote sensing and its future prospect. In *Proceedings of the Geoscience and Remote Sensing Symposium*, Quebec, Canada.

38 Michael Swartwout and Clay Jayne 2016. University-class spacecraft by numbers: success, failure, debris, (but mostly success). In *Proceedings of the 30th Annual AIAA/USU Conference on Small Satellites*, Logan, USA.

39 Michael Swartwout. 2007. Beyond the beep: Student-built satellites with educational and "real" missions. In *Proceedings of the 21st Annual AIAA/USU Conference on Small Satellites*, Logan, USA.

40 Jeremy Straub and Ronald Marsh. 2014. Assessment of educational expectations, outcomes and benefits from small satellite program participation. In *Proceedings of the 26th Annual AIAA/USU Conference on Small Satellite*, Logan, USA.

41 Michael Swartwout. 2009 The promise of innovation from university space systems: Are we meeting it? In *Proceedings of the 23rd Annual AIAA/USU Conference on Small Satellites*, Logan, USA.

42 David Voss, Jared Clements, Kelly Cole, Melody Ford, Christopher Handy, and Abbie Stovall. 2011. Real science, real education: The University Nanosat Program. In *Proceedings of the 25th Annual AIAA/USU Conference on Small Satellites*, Logan, USA.

43 Christopher Kitts, Mike Rasay, Laura Bica, Ignacio Mas, Michael Neumann, Anthony Young, Giovanni Minelli, Antonio Ricco, Eric Stackpole, Elwood Agasid, Christopher Beasley, Charlie Friedericks, David Squires, Pascale Ehrenfreund, Wayne Nicholson, Rocco Mancinelli, Orlando Santos, Richard Quinn, Nathan Bramall, Andrew Mattioda, Amanda Cook, Julie Chittenden, Katie Bryson, Matthew Piccini, and Macarena Parra. 2011. Initial on-orbit engineering results from the O/OREOS nanosatellite. In *Proceedings of the 25th Annual AIAA/USU Conference on Small Satellites*, Logan, USA.

44 Joseph W. Gangestad, Darren W. Rowen, Brian S. Hardy, and David A. Hinkley. 2014. Flying in a cloud of CubeSats: Lessons learnt from early orbit operations of AeroCube-4, -5, and -6. In *Proceedings of the 65th International Astronautical Congress*, Toronto, Canada.

45 Christopher Boshuizen, James Mason, Pete Klupar, and Shannon Spanhake. 2014. Results from the Planet Labs Flock constellation. In *Proceedings of the 28th Annual AIAA/USU Conference on Small Satellites*, Logan, USA.

46 Elstak, J., Amini, R., and Hamann, R. (2009). A comparative analysis of project management and systems engineering techniques in CubeSat projects. In: *Proceedings of the 19th Annual International Symposium of the International Council on Systems Engineering*, vol. 19, 545–559. Singapore.

47 Sebastian Trowitzsch, Frank Baumann, Merlin Barschke, and Klaus Brieß. 2013. Lessons learned from picosatellite development at TU Berlin. In *Proceedings of the 2nd IAA Conference on University Satellites Missions*, Rome, Italy.

48 Muriel Noca, Fabien Jordan, Nicolas Steiner, Ted Choueiri, Florian George, Guillaume Roeth-lisberger, Nomy Scheidegger, Herv Peter-Contesse, Maurice Borgeaud, Renato Krpoun, and Herbert Shea. 2009. Lessons learned from the first Swiss pico-satellite: SwissCube. In *Proceedings of the 23rd Annual AIAA/USU Conference on Small Satellites*, Logan, USA.

49 Swartwout, M. (2013). The first one hundred CubeSats: a statistical look. *Journal of Small Satellites* 2: 213–233.

15 I-3d

Formations of Small Satellites

Klaus Schilling

[1] *Chair for Informatics VII, University of Würzburg, Germany*
[2] *Zentrum für Telematik, Würzburg, Germany*

15.1   Introduction

Compared to traditional satellites, miniaturized satellites exhibit limited capabilities of each satellite, but enable multisatellite systems at the same costs. Realized as a distributed self-organizing formation of picosatellites or nanosatellites, they result in significant perspectives for innovative scientific approaches and functionalities [1–9]. There is significant application potential in the field of telecommunication in the context of Internet of Things [10], as well as in Earth observation for sensor networks with higher temporal and spatial resolutions [4, 11, 12]. Today, multisatellite systems are mainly organized as constellations (such as GPS, Iridium, OneWeb, A-train), where each satellite is individually controlled from ground stations. Further, improved performance is expected from formations, where the satellites self-organize on basis of direct information exchange and distributed control approaches (such as the two-satellite missions of GRACE (http://grace.jpl.nasa.gov), TanDEM-X [13], PRISMA [14], CanX-4/CanX-5 [15]). On the hardware side, miniaturized attitude and orbit control systems as well as on-board data-handling capacities support realization of formations even at picosatellite level [6, 16–18], [8]. The goal of this chapter is to elaborate the theoretical and technology background for small satellite formations, as well as to provide some related examples.

15.2   Constellations and Formations

Distributed networks of satellites receive significant interest because of the following advantages:

- Higher temporal and spatial resolutions in observation data.
- Higher availability.
- Graceful degradation capabilities of the sensor network in case of defects.

Of course, it is counterbalanced by the complexity of operations of a multisatellite system. Therefore, in a formation, various levels of autonomous reaction capabilities are

Nanosatellites: Space and Ground Technologies, Operations and Economics, First Edition.
Edited by Rogerio Atem de Carvalho, Jaime Estela, and Martin Langer.

implemented in the network. Special features to be addressed include intersatellite links, relative navigation toward the other satellites, and distributed control approaches [17]. There are significant similarities to robotics, where coordination of several vehicles based on relative navigation sensor data is also intensively analyzed.

15.2.1 Definitions for Multivehicle Systems

A multivehicle system is described as:

- A constellation, when each vehicle is individually controlled from a ground center.
- A formation, when the vehicles directly exchange information and relative navigation measurements to realize a closed-loop control on-board in order to preserve the topology and to control relative distances.

So far, mainly constellations of classical satellites are realized in various application fields, such as:

- Telecommunication: Iridium (66, LEO), Globalstar (64, LEO), TDRSS (3, GEO), Orbcomm (24, LEO), and Eutelsat (38, GEO).
- Earth observation: RapidEye (5, LEO), A-train (6, LEO), Dove (150, LEO)
- Navigation: BeiDou (4 GEO, 12 IGSO, 9 MEO), GPS (24 MEO), GLONASS (24 MEO), and Galileo (30 MEO).
- Science: Cluster (4, elliptic orbit), Swarm (3 LEO), and MMS (4, elliptic orbit).

Currently, large constellations like OneWeb (about 650, LEO) are in implementation and first six satellites were launched in February 2019. Despite formations are discussed for years, so far only the missions Gravity Recovery And Climate Experiment (GRACE) (launched in 2004) [3] (https://grace.jpl.nasa.gov), TerraSAR-X-Add-on for Digital Elevation Measurements (TanDEM-X) (launched in 2010) [13], Hyperspectral PRecursor of the Application Mission (PRISMA) (launched in 2010) [14], and CanX-4/CanX-5 (launched in 2014) [15] had been realized, all just composed of two satellites. Nevertheless, algorithms and methods for formations are well developed, some high-fidelity computer and hardware-in-the-loop simulations [2, 3, 12] are available.

15.3 Orbit Dynamics

When satellites fly in a formation on almost similar circular orbits, it is of interest to derive the relative motion to each other with respect to one moving reference satellite, described by the Euler–Hill equations. With r_1 and r_2 being the vectors from Earth to the reference satellite and the second satellite, the relative distance between them is (Figure 15.1):

$$\rho = r_2 - r_1 \tag{15.1}$$

The related equations of motion of the two satellites are:

$$\ddot{r}_1 = \frac{-\mu r_1}{r_1^3}, \ddot{r}_2 = \frac{-\mu r_2}{r_2^3} + f \tag{15.2}$$

with driving force f.

Figure 15.1 Scenario of the Euler–Hill equations with relative distance ρ between two spacecraft on circular orbits.

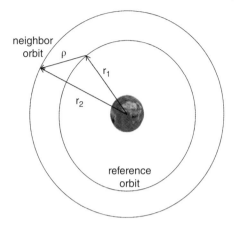

Replacing r_2 by r_1 and ρ, then for small ρ the relative acceleration results as:

$$\ddot{\rho} = -\rho + 3\,(r_1/r_1 \bullet \rho)\,r_1/r_1 + f + O(\rho^2) \tag{15.3}$$

with the higher order terms $O(\rho^2)$ in the power series expansion.

For an almost circular orbit (with small eccentricity ε), with neglecting higher-order terms (ε^2 and ρ^2, products of ε and ρ) in the components of relative motion, the second-order differential equation system is resulted by representation of ρ in Cartesian coordinates, named Clohessy–Wiltshire or Hill's equations:

$$\ddot{x} - 2n\dot{y} - 3n^2x = f_x$$
$$\ddot{y} + 2n\dot{x} = f_y \tag{15.4}$$
$$\ddot{z} + n2z = f_z$$

with $n = (\mu/r_1^3)^{1/2}$ being the angular velocity of the reference orbit. Key assumptions for this linearization approach are a spherical Earth and a circular reference orbit (Figure 15.2).

Here, the motions in the orbit plane (x- and y-coordinates) and out-of-plane (z-coordinate) are decoupled. The analytical solution of this second-order system is:

Figure 15.2 Local vertical local horizontal (LVLH) coordinate frame for x and y, while z is perpendicular to the orbit plane.

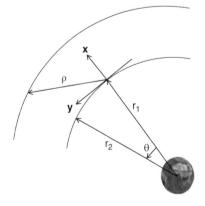

Out-of-plane:

$$z = c \cos(n t + \text{ß})$$
$$\acute{z} = -c\, n\,(n t + \text{ß})$$

(15.5)

with the parameters c and β, which can be derived from the initial values at $t = 0$.
In-orbit plane:

$$x = c_1 \cos(n t + \alpha) - 3/2\, n\, c_2$$
$$y = -2\, c_1 \sin(n t + \alpha) - 3/2\, n\, c_2 t + c_3$$
$$\acute{x} = -c_1\, n \sin(n t + \alpha)$$
$$\acute{y} = -2\, c_1 \cos(n t + \text{a}) - 3/2\, n\, c_2$$

(15.6)

Here, the parameters c_1, c_2, c_3, and α correspond to integration constants and can be derived from the initial values at $t = 0$.

The infinite-dimensional satellite trajectories, varying over time, can be approximated this way by an approach parameterized by just six real numbers c, c_1, c_2, c_3, α, and β, resulting in a significant simplification.

Extensions to elliptical orbits are addressed in the Tschauner–Hempel Equations [19], while Alfriend and Yan [20] also include J_2 terms to approximate the Earth's gravity field (Figure 15.3).

In order to minimize efforts and fuel consumption for orbit corrections within the formation, orbits are designed such that average disturbances during one revolution are almost

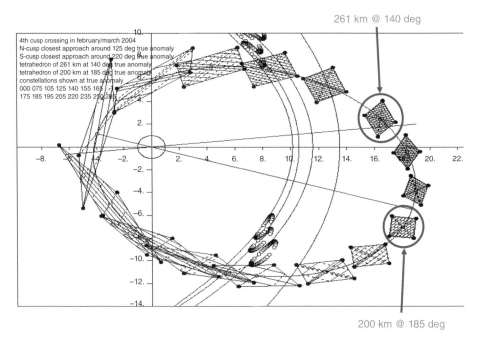

Figure 15.3 The relative positions of the four CLUSTER satellites during one highly elliptic orbit. Source: Reproduced with permission from ESA.

identical for all satellites. In case of two satellite systems, this leads to Helix orbits (like at TanDEM-X [13]). These are generalized to Cartwheel orbits for the case of n satellites [21].

15.4 Satellite Configurations

When a constellation is to be established, relevant analyses concern the best way to distribute the satellites around different orbit planes and within one orbit plane in order to achieve a good coverage. This topic has been first addressed for low Earth orbits (LEOs) in telecommunication systems. LEO-satellites benefit from the shorter distance to the Earth's surface, in particular, from shorter signal propagation periods and lower power needs for data transmission and instrumentation. Disadvantages related to LEO are high velocities relative to the surface, having short contact periods to ground stations and short observation periods of specific surface areas as consequence. Therefore, several satellites in appropriate complementary orbits are placed to increase coverage. Dimensioning the number of satellites and selection of orbits is based on the following main criteria:

- Temporal and spatial coverage requirements.
- Minimization of the number of necessary satellites by selecting appropriate orbit constellations.

A frequently used solution approach based on symmetry arguments is the "Walker Delta constellation" [22] for a continuous coverage of the Earth's surface by a minimum number of spacecraft.

15.4.1 Definition of Walker Delta Pattern Constellation

A Walker constellation with the parameters

- i: inclination
- t: total number of satellites
- p: number of equally spaced orbit planes
- f: relative phase difference between satellites in adjacent planes

is presented in standard notation for a constellation:

$$i: t/p/f.$$

Here t/p equally spaced satellites are placed in each plane. $0 \leq f \leq p^{-1}$ is measured in the direction of motion from the ascending node to the closest satellite in units of $360°/t$.

Example 15.1   The Galileo Navigation Satellites
The Galileo navigation satellites are placed as a $56°{:}27/3/1$ constellation, having 27 satellites in orbit, inserted into three orbit planes separated by $\Delta\Omega = 120°$. Each of the three orbit planes with an inclination $i = 56°$ hosts nine satellites at angular distances of $40°$ between subsequent satellites. The phase shift between adjacent orbits is $f = 40°/3 = 13\,^1/_3°$. In addition, in each orbit plane, one spare satellite is parked for fast replacement capability in case of a failure.

Figure 15.4 The Galileo formation. Source: Reproduced with permission from ESA.

The Galileo satellite constellation (Figure 15.4) has been optimized to the following nominal constellation specifications:

- Circular orbits (satellite altitude of 23 222 km).
- Nine operational satellites equally spaced in each plane.
- One spare satellite (also transmitting) in each plane.

15.5 Relevant Specific Small Satellite Technologies to Enable Formations

The capability for intersatellite links as well as relative navigation are essential to realize a formation, for example, the direct measurement of relative distances and directions among the different satellites [17]. On that basis, the control activities among the different formation members can be coordinated. This way, the attitude and orbit control loops between the satellites can be closed in orbit.

15.5.1 Intersatellite Communication

In order to exchange information for planning joint control action of several satellites for observations or for obstacle avoidance, an intersatellite link needs to be established at the level of small satellites. Initial trade-offs concern approaches by omnidirectional broadcasting, requiring significant energy, and targeted radio beams, requiring higher-precision attitude control. Both options are so far implemented at picosatellite level in missions in UHF/VHF and S-band communication, even higher-bandwidth approaches using X-band and optical links are in preparation. Broadcast solutions are by example used in the missions S-Net and NetSat. In particular, missions requiring high precision pointing for observation with their instrumentation also use this capability for the intersatellite link, such as Telematics Earth Observation Mission (TOM) [11].

15.5.2 Relative Navigation

Today, Global Navigation Satellite System (GNSS) data (such as from GPS, Galileo, GLONASS, BeiDou) provide a very convenient source to determine relative range and direction information [23]. These are complemented by ranging data determined from time-of-flight measurements in the communication link or by optical scanners. Using triangulation methods in a multisatellite formation of relative positions, the three-dimensional space can be derived form range measurements.

While in larger satellites the difference between GNSS antennae distributed with maximum distances on the satellite surface can be used to determine attitude, for picosatellites, the possible distances for antennae placements limit accuracies; thus, Sun and star sensors provide precise absolute attitude data, from which relative attitudes are derived. Also, magnetometers (with lower accuracies) and gyros (characterizing change rates to an initial reference) are typically included for attitude determination [24].

15.5.3 Attitude and Orbit Control

The orbit control requirements face two challenges: formation initialization after deployment from the launcher with subsequent formation, keeping and adapting the formation topology for optimal performance during the operational lifetime. For picosatellites and nanosatellites, many technology approaches are developed. For a survey on US technology, see (https://sst-soa.arc.nasa.gov/04-propulsion). Here, cold gas and electric propulsion systems are most advanced. Electric propulsion has the advantage that there is no danger to cosharing launcher customers, and it is available even as a picosatellite subsystem. At picosatellite level, a field emission electric propulsion (FEEP) system [25] was demonstrated in the UWE-4 mission launched on 27th December 2018 (see Figure 15.5).

Figure 15.5 The field emission electric propulsion system (FEEP) for UWE-4, integrated in the guiderail of this 1U CubeSat.

Figure 15.6 The brushless motor in the background is complemented by a mass (at the left of the figure) and placed in the protection cover (at the right of the figure) to be integrated into a 2 cm^3.

For joint observations by the formation, as well as for directed communication links, precision in attitude control is very demanding for small satellites. As actuators, typically the magnetorquers in combination with the Earth's magnetic field are applied [24]. Higher accuracy at picosatellite level is achieved by recently developed miniature reaction wheel in a cube of 2 cm side length and with the very modest power requirement of 150 mW. These reaction wheels are orthogonally mounted in the three axes providing a three-axes attitude control system, frequently complemented by a fourth wheel for redundancy purposes. Desaturation of the wheels is handled via the magnetorquers (Figure 15.6).

Classical PID control algorithms combine the sensors and actuators to a three-axis control system at each satellite. Those local satellite control systems are coordinated by a distributed control system at the formation level. Performance of current low-energy consumption microprocessors is quite sufficient to implement this at picosatellite level.

15.6 Application Examples

Formation flying technologies for nanosatellites were prepared by the technology demonstration missions CanX-4 and CanX-5 [15]. Picosatellite UWE-4 demonstrated an electric propulsion system [25] for a 1U CubeSat (launched in December 2018). In the next step, in the end of 2019, NetSat will employ the same orbit control technology to fly first time a four-satellite formation. This way, control of a three-dimensional formation topology for future innovative observations will be demonstrated in orbit (Figure 15.7).

In the context of Earth observation within the TOM [11], three small satellites are coordinated for joint imaging of the same target area from different positions. The major challenge is appropriate pointing of the cameras by the attitude control system. By photogrammetric methods, three-dimensional images of the target area are generated. This is in particular used to characterize the ash clouds from volcano eruptions [26].

As an additional technology experiment, the optical link Osiris4CubesSats is on board to test the capabilities for a high-performance downlink capacity up to 100 Mbits s^{-1} onboard a CubeSat [11] (Figures 15.8 and 15.9).

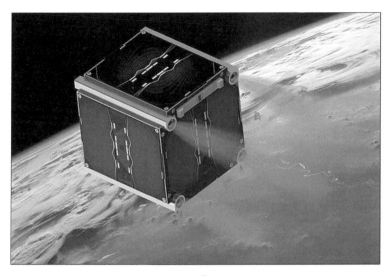

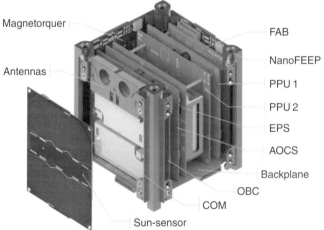

Figure 15.7 UWE-4 will demonstrate a FEEP electric propulsion system in orbit at 1U CubeSat level.

A further increase in complexity will be realized in the CloudCT mission, composing by 10 nanosatellites a formation to characterize the interior of clouds by a computed tomography approach. This is expected to provide relevant information to cloud models in order to improve longer term climate predictions [12].

In particular, for Earth observation of applications such as disaster monitoring, formations of picosatellites in LEO get enabled due to high temporal resolution of the monitoring of very dynamic processes. Nevertheless, the high relative velocity to reference points on ground results in short observation and communication contact periods in the target areas. Higher temporal resolution is therefore provided by satellite constellations with several satellites in the same orbit. The achievable temporal and spatial resolutions of such a formation open new application areas in biomonitoring and surveillance.

Figure 15.8 The three TOM picosatellites coordinate joint observations to generate 3D images by photogrammetry.

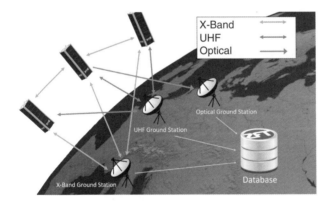

Figure 15.9 Intersatellite and ground station links in the TOM scenario also including an optical link as payload. (*See color plate section for color representation of this figure*).

15.7 Test Environment for Multisatellite Systems

In order to perform hardware in the loop tests of formations, a test facility was established at the Zentrum für Telematik (ZfT) in Würzburg. For simulation of changing topologies for formation motion in the orbit plane, several mobile robots able to carry small satellites floating on air cushions are employed (cf. Figure 15.10). With respect to crucial attitude coordination tests, high precision and high dynamics turntables of relevance are used for intersatellite link and coordinated Earth observation aspects (cf. Figure 15.11).

Figure 15.10 Mobile robot systems carrying the small satellites in a ball floating on an air cushion.

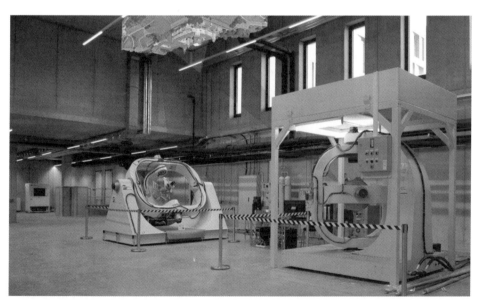

Figure 15.11 Two turntables to simulate attitude changes for the camera system during photogrammetric observation of 3D surface features mounted on the ceiling of the test hall.

15.8 Conclusions for Distributed Nanosatellite Systems

Currently, a paradigm shift from traditional spacecraft with multiple payloads toward decentralized, distributed small satellite systems can be observed. Here, particular advantages for Earth observation and surveillance are higher fault tolerance and robustness of the overall system, as well as higher temporal and spatial resolutions. In addition, multisatellite systems are scalable: according to application needs, additional satellites can be added to increase resolution and coverage. This opens new application perspectives for combinations of satellite systems, composed of few large and many small satellites to provide the required data quality in short time intervals, as well as flexibility and robustness.

Acknowledgments

The author acknowledges the financial contributions and the inspiring framework of the ERC Advanced Grant "NetSat" and the ERC Synergy Grant "CloudCT," as well as the advanced host environment of the Zentrum für Telematik, allowing to realize inspiring small satellite formation missions.

References

1 Alfriend, K.T., Vadali, S.R., Gurfil, P. et al. (2010). *Spacecraft Formation Flying. Dynamics, Control and Navigation*. Elsevier Astrodynamics.

2 Ankersen, F. (ed.), 2008. Proceedings 3rd International Symposium on Formation Flying, Missions and Technologies. *ESA SP-654*.

3 D'Errico, M. (ed.) (2012). *Distributed Missions for Earth System Monitoring*. Springer.

4 Jones, N. *Mini satellites prove their scientific power. Nature* 508 (2014): 300–301. http://www.nature.com/news/mini-satellites-prove-their-scientific-power-1.15051.

5 Sandau, R., Nakasuka, S., Kawashima, R., Sellers, J., 2012. *Novel Ideas for Nanosatellite Constellation Missions*, IAA Book Series.

6 Schilling, K. 2015. *Winzlinge im Orbit*, Spektrum der Wissenschaft, Mai, S. 48–51.

7 Schilling, K., Pranajaya, F., Gill, E., Tsourdos, A., 2011. *Small Satellite Formations for Distributed Surveillance: System Design and Optimal Control Considerations*. NATO RTO Lecture Series SCI-231.

8 Schilling, K. (2017). Perspectives for miniaturized, distributed, networked Systems for Space Exploration. *Robotics and Autonomous Systems* 90: 118–124.

9 Zurbuchen, T. H., von Steiger R., Bartalev S., Dong X., Falanga M., Fléron R., Gregorio A., Horbury T. S., Klumpar D., Küppers M., Macdonald M., Millan R., Petrukovich A., Schilling K., Wu J., Yan J., Performing high-quality science on CubeSats, *Space Research Today*, Vol. 196 (2016), pp. 10–30.

10 Schilling, K., 2018. Machine-to-Machine Communication by Networks of Small Satellites, invited lecture at topical Workshop on Internet in Space (TWIoS), Proceedings IEEE Radio & Wireless Week, Anaheim.

11 Schilling, K. Tzschichholz, T., Motroniuk, I. Aumann, A. Mammadov, I. Ruf, O. Schmidt, C. Appel, N. Kleinschrodt, A. Montenegro, S. Nuechter, A. 2017. TOM: A Formation for Photogrammetric Earth Observation by Three CubeSats. *4th IAA Conference on University Satellite Missions, Roma*, IAA-AAS-CU-17-08-02.

12 Schilling, K., Schechner, Y., Koren, I., 2019. CloudCT—Computed Tomography of Clouds by a Small Satellite Formation. Proceedings "12th Symposium on Small Satellites for Earth Observation"; IAA-B12–1502.

13 Krieger, G., Zink, M., Moreira, A., 2012. *TANDEM-X: A Radar Interferometer with Two Formation Flying Satellites*. Proceedings International Astronautical Congress 2012, IAC-12-B4.7B.3.

14 Persson, S., Harr J., Gill E., Bodin, P., 2006. PRISMA: An Autonomous Formation Flying Mission. *Proceedings of ESA Small Satellite Systems and Services Symposium (4S)*.

15 Bonin, G., Roth N., Armitage S., Newman J., Risi B., Zee R. E., 2015. CanX-4 and CanX-5 Precision Formation Flight: Mission Accomplished!, Proceedings Small Satellite Conference, Logan, SSC15-I –4.

16 NASA Ames Research Center, 2014. Mission Design Division Staff; *Small Spacecraft Technology State of the Art*, NASA/TP–2014–216648.

17 Schilling, K., Schmidt, M., Busch, S., 2012. *Crucial Technologies for Distributed Systems of Pico-Satellites*. Proceedings 63rd International Astronautical Congress, IAC-12-D1.2.4, Naples, Italy.

18 Schilling, K. and Schmidt, M. (2012). *Communication in distributed satellite systems*. In: *Distributed Space Missions for Earth System Monitoring* (ed. M. D'Errico). Springer.

19 Tschauner, J. and Hempel, P. (1964). Optimale Beschleunigungsprogramme für das Rendezvous-Manöver. *Astronautica Acta* 10: 296–307.

20 Alfriend, K. T., Yan, H., 2002. "An Orbital Elements Approach to the Nonlinear Formation Flying Problem," International Formation Flying Conference: Missions and Technologies, Toulouse, France, October 2002.

21 Jochim, E. F., Fiedler H., Krieger G., *Fuel consumption and collision avoidance strategy in multi-static orbit formations*, Acta Astronautica 68 (2011) 1002–1014.

22 Walker, R. 2013. *CubeSat Nano-Satellite Systems*, Technical Dossier ESA/ESTEC TEC-SY/145/2013/TNT/RW.

23 Montenbruck, O. and D'Amico, S. (2013). *GPS based relative navigation*. In: *Distributed Space Missions for Earth System Monitoring* (ed. M. D'Errico), 185–223. Springer.

24 Busch, S., Bangert P., Schilling K., 2014. *Attitude Control Demonstration for Pico-Satellite Formation Flying by UWE-3*, Proceedings 4S-Symposium, Mallorca.

25 Bock, D.; Kramer, A.; Bangert, P.; Schilling, K., 2015. NanoFEEP on UWE platform—Formation Flying of CubeSats Using Miniaturized Field Emission Electric Propulsion Thrusters. Proceedings *34th International Electric Propulsion Conference, Hyogo-Kobe*. IEPC-2015-121/ISTS-2015-b-121.

26 Zakšek, K., James, M.R., Hort, M. et al. (2018). Using picosatellites for 4-D imaging of volcanic clouds: proof of concept using ISS photography of the 2009 Sarychev peak eruption. *Journal Remote Sensing of Environment* 210: 519–530.

16 I-3e

Precise, Autonomous Formation Flight at Low Cost

Niels Roth, Ben Risi, Robert E. Zee, Grant Bonin, Scott Armitage, and Josh Newman

Space Flight Laboratory (SFL), UTIAS, Toronto, Canada

16.1 Introduction

The use of multiple autonomously coordinated spacecraft, often—though not necessarily—in close proximity to one another, is a critical capability to the future of spaceflight. Formation flight applications range from synthetic aperture radar and optical interferometry, to on-orbit servicing of other spacecraft, and to gravitational and magnetic field science. Groups of small, relatively simple spacecraft can also potentially replace single large and complex ones, reducing risk through distribution of instruments, and cost by leveraging nonrecurring engineering costs. Performance of the entire formation can be gradually built up over several launches, maintained over time with replacement units when others fail, or allowed to degrade gracefully.

The benefits of formation flight are best realized as the size of spacecraft decreases, nanosatellites being the foremost example. These spacecraft are cost-effective, easily mass-produced, and capable of being deployed en masse from a single launch. Nanosatellite technology has already matured to the point where this is possible. However, there had been no successful demonstrations of formation flight with spacecraft of this scale prior to CanX-4 and CanX-5. With their success, CanX-4 and CanX-5 have paved the way for these miniaturized technologies to be integrated on spacecraft of all scales at affordable cost amenable to commercial constellations and constrained budgets.

16.1.1 Formation Flight Background

Early in both the United States and the Soviet Union space programs, it was recognized that the ability to operate spacecraft in close proximity to one another would become increasingly important in order to facilitate the rendezvous of vehicles for the purposes of crew and material transfer. The first attempt at coordinated spacecraft operation was the Soviet Vostok 3 and 4 mission launched in 1962. These spacecraft were launched a day apart into nearly identical orbits, with an initial distance of about 6.5 km. Given their lack of maneuvering thrusters, this distance quickly grew to nearly 3000 km after a few days [1].

Nanosatellites: Space and Ground Technologies, Operations and Economics, First Edition.
Edited by Rogerio Atem de Carvalho, Jaime Estela, and Martin Langer.
© 2020 John Wiley & Sons Ltd. Published 2020 by John Wiley & Sons Ltd.

In 1965, in preparation for the Apollo missions where docking the lunar and command modules would be a critical mission step, US astronaut Wally Schirra successfully maneuvered his Gemini 6 spacecraft as close as 0.3 m from the Gemini 7 spacecraft and kept station around its target at ranges up to 90 m, including a 20 min period where no control thrusts were performed at all [2].

More recently, advances in on-board computing capability have allowed for automated spacecraft rendezvous and docking down to the small satellite scale. The Swedish-led Prototype Research Instruments and Space Mission technology Advancement (PRISMA) mission, launched on 15 June 2010, was designed to demonstrate autonomous homing, rendezvous, formation flight, and other proximity operations, among other things. The space segment is composed of a main and a target spacecraft, with masses of 145 and 50 kg, respectively [3]. The PRISMA mission cost an order of magnitude more than CanX-4 and CanX-5 to develop [4].

Work toward autonomous formation flight of nanosatellites has been ongoing at the Space Flight Laboratory (SFL) for several years. This work can be traced back to the CanX-2 spacecraft, launched in 2008, which demonstrated a number of technologies required for formation flight, including a cold-gas propulsion system and a GPS receiver with precision, in a 3U form factor [5]. CanX-4 and CanX-5 represent the latest efforts in the field and have set the bar for the state of the art in nanosatellite formation flying [6] with the completion of their primary mission in November 2014.

16.2 Mission Overview

The primary goal of the CanX-4 and CanX-5 mission was to demonstrate relative position control accuracy better than 1 m, 2σ, for a duration of at least 10 orbits per formation in four formations: a 1000 m along-track orbit (ATO), a 500 m ATO, a 100 m projected-circular orbit (PCO), and a 50 m PCO. The ATO can be thought of as a "leader–follower" configuration, whereby one spacecraft maintains a fixed relative separation from the other in the same orbital plane. The PCO is so named because, when viewed from Earth, one spacecraft appears to draw a circle around the other over the course of one orbit. Formation control was accomplished using one active orbit-controlled spacecraft, designated the Deputy, and one uncontrolled spacecraft, designated the Chief.

The reference trajectories were periodic solutions to the Hill–Clohessy–Wiltshire (HCW) equations, which describe relative satellite motion assuming a circular Chief orbit and close relative separation between the spacecraft as compared to the orbit radius. These reference trajectories are given by reference [7]:

$$x(t) = \frac{1}{2} d_1 \sin(nt + \alpha)$$
$$y(t) = d_1 \cos(nt + \alpha) + d_3 \tag{16.1}$$
$$z(t) = d_2 \sin(nt + \beta)$$

where n is the mean orbital angular velocity of the Chief spacecraft and d_1, d_2, d_3, α, and β are the formation design parameters. These solutions are expressed in the rotating local-vertical local-horizontal (LVLH) reference frame of the Chief. The x-axis of the

Table 16.1 Formation design parameters.

Formation	d_1 (m)	d_2 (m)	d_3 (m)	α (rad)	β (rad)	Duration (orbits)
ATO 1000	60	30	1000	0	$\pi/2$	11
ATO 500	60	30	500	0	$\pi/2$	11
PCO 100	100	100	0	0	0	11
PCO 50	50	50	0	$3\pi/2$	$3\pi/2$	11

LVLH frame is aligned with the position vector, while the z-axis is aligned with the orbital angular momentum vector, and the y-axis completes the orthonormal triad such that it is nominally aligned with the velocity vector. These directions are often referred to as radial, cross-track, and along-track, respectively. It is important to note that the cross-track motion is decoupled from the other components and its phase with respect to the radial motion can be adjusted to provide passively safe relative orbits, where at least one component is guaranteed to be non-zero.

The design parameters for the four target formations are given in Table 16.1. For the ATOs, a passively safe relative separation of 30 m in the radial and cross-track directions was selected to safeguard against collisions in the event of unexpected formation control loss. The PCOs could not be made passively safe since this formation required both phase angles to be equal. The phase angles for the PCOs were selected to minimize fuel during the formation reconfiguration maneuvers. The duration of 11 orbits was selected so that fine formation control could be maintained for full 10 orbits, allowing one orbit for convergence.

16.3 System Overview

Each of the CanX-4 and CanX-5 spacecraft are approximately 6 kg nanosatellites based on the SFL's Generic Nanosatellite Bus (GNB) architecture. The GNB structure is a 20 cm cube, designed to interface with the SFL's X-Picosatellite Orbital Deployer (XPOD) launch vehicle deployment system. The GNB platform (Figure 16.1) was designed with mission flexibility in mind. It is the basis for several successful missions currently on orbit. In particular, the GNB platform has been used for the BRIght Target Explorer (BRITE) constellation of stellar astronomy spacecraft [8], consisting of five operational satellites; the ship-tracking Automatic Identification System Satellite (AISSat) constellation [9], consisting of two operational satellites and a third slated for launch shortly; and ExactView-9, a ship-tracking mission.

Both CanX-4 and CanX-5 are identical to each other in design. Figure 16.2 illustrates the CanX-4 and CanX-5 spacecraft layout, while Figures 16.3 and 16.4, respectively, show the two spacecraft during the vibration and thermal vacuum (TVAC) portions of their acceptance testing.

For downlink, CanX-4 and CanX-5 use an S-band transmitter connected to two wide-beam S-band patch antennas, mounted on opposite faces to provide near-omnidirectional coverage, with downlink speeds between 32 and 256 kbps. Command uplink is

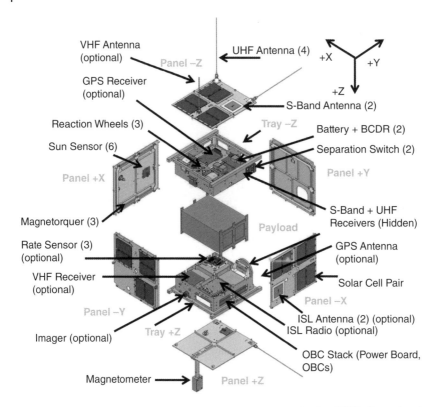

Figure 16.1 Exploded view of the SFL's Generic Nanosatellite Bus (GNB). (*See color plate section for color representation of this figure*).

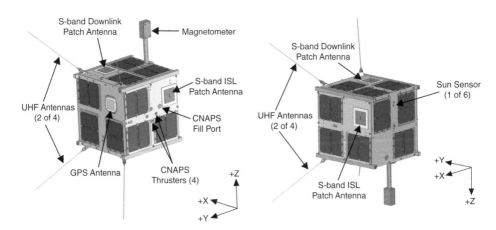

Figure 16.2 CanX-4 spacecraft (CanX-5 identical).

Figure 16.3 CanX-4 undergoing vibration testing.

Figure 16.4 CanX-5 undergoing TVAC testing.

implemented via a UHF receiver with a canted turnstile antenna system, also providing near-omnidirectional coverage. This overall communications approach avoids so-called "death modes" in the communications system, allowing spacecraft communications in all attitudes. During autonomous formation flight, data are passed between the spacecraft using an S-band intersatellite link (ISL), which has a demonstrated range exceeding 100 km with an omnidirectional antenna system.

The power system is a parallel-regulated direct energy transfer (DET) system, with dual parallel battery charge/discharge regulators (BCDRs) responsible for battery charging and enabling peak power tracking when required. All power is distributed centrally via a power board using solid-state switches for current consumption monitoring and overcurrent detection and fault isolation when needed.

As with all GNB spacecraft, CanX-4 and CanX-5 each carry a suite of attitude sensors and actuators for full three-axis attitude determination and control. These include six fine sun sensors, a three-axis rate sensor, a three-axis magnetometer mounted on an external pre-deployed boom, and three sets of orthogonally mounted magnetorquers and reaction wheels. A GPS receiver and antenna are used to collect high precision information on spacecraft position.

16.3.1 Propulsion

The Canadian Nanosatellite Advanced Propulsion System (CNAPS) provides orbital control for orbit acquisition and phasing (drift recovery), station keeping, and formation control and reconfiguration. CNAPS is equipped with four thrusters and fuelled with 260 g of liquid sulfur hexafluoride (SF_6) propellant, providing a specific impulse of 45 s and a total Δv capability of $18\,\mathrm{m\,s^{-1}}$.

SF_6 was selected for its high storage density and vapor pressure, making the system self-pressurizing, as well as its inert and nontoxic properties, making it safe to handle and compatible with most materials. Two filters are present in the system to remove contaminants that could damage the solenoid valves, and a pressure relief valve on the storage tank prevents the possibility of an overpressure event compromising safety on the ground or the launch vehicle (Figure 16.5).

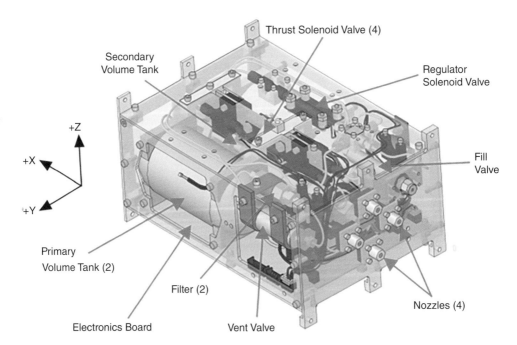

Figure 16.5 Interior view of CNAPS. (*See color plate section for color representation of this figure*).

Thrust levels range from 12.5 to 50 mN, depending on the chamber pressure and the number of selected thrusters. As the four nozzles are located on a single face of the spacecraft bus and offset from the center-of-mass, thruster selection also allows the system to be used for momentum management, with the nozzle set being autonomously selected to reduce momentum build-up on the spacecraft.

16.3.2 Intersatellite Link

The ISL radio enables autonomous on-orbit communications between the two spacecraft. The ISL is a compact, medium-range, low-data-rate S-band radio link. Each spacecraft is equipped with a radio module (Figure 16.6) and two dedicated patch antennas. The ISL provides the timely and bidirectional exchange of data messages between the two spacecraft at distances up to 5 km and data rates up to 10 kbps. The maximum distance is set by estimates of the worst-case spacecraft separation distance during reconfiguration maneuvers. In addition, this range allows for autonomous recovery of formation flight from a free-drift configuration, such as might occur if a fault interrupted nominal conditions.

The radio module is housed in a small enclosure, which provides electromagnetic shielding, crucial to avoiding mutual interference between the ISL and the spacecraft telemetry transmitter, and substantially simplifies handling during spacecraft assembly, integration, and testing. The radio uses an RF transceiver subassembly for the transmission, modulation, demodulation, and reception of wireless data. The output of the RF transceiver is routed to a power amplifier and via a power splitter to each of the antenna ports. A baseband processor provides an interface between the spacecraft bus and the RF cores,

Figure 16.6 ISL radio module.

communicating with the spacecraft payload computer using a serial link. This baseband processor also provides protocol translation and is firmware-upgradeable, allowing for the implementation of different protocol stacks.

The ISL consumes 400 mW of power when receiving and 600 mW of power during transmission. It provides 21.8 dBm of RF output power to the antennas, which emit 15.3 dBm of equivalent isotropically radiated power. With the measured antenna gains, simulations showed that a link availability of greater than 98% is achievable during formation flight, and better than 90% availability at the 5 km maximum design distance [10].

16.3.3 Algorithms

There are three pieces of navigation and control software that help fulfill the high-level formation control requirements—the formation flying integrated on-board nanosatellite algorithm (FIONA), the relative navigation algorithm (RelNav), and the on-board attitude system software (OASYS). FIONA and RelNav run at a 5 s period on the Deputy spacecraft, while OASYS runs asynchronously at a 2 s period on both spacecraft. The primary roles of OASYS are to reorient the spacecraft to commanded attitude targets, reverting to zenith tracking in the absence of an attitude target, and to select the thruster(s) to be used for upcoming maneuvers. RelNav is responsible for estimating the relative orbital state of the two spacecraft using differential GPS techniques. Finally, FIONA computes formation keeping and reconfiguration control maneuvers and performs absolute state estimation of the Chief and Deputy orbits.

16.3.4 OASYS

During fine formation control, OASYS's commanded attitudes are inertial quaternions computed by FIONA on the Deputy spacecraft. The +X face of the spacecraft is aligned with the target thrust direction, while the +Y face is constrained to be as close to zenith as possible. This attitude maximizes the number of GPS satellites in view. The target attitude computed on the Deputy is sent to the Chief via the ISL so that both spacecraft acquire the same attitude. Identical attitudes both maximize the number of common GPS satellites in view—improving relative navigation—and minimize the impact of differential perturbation forces.

Attitude determination is performed using an extended Kalman filter (EKF) operating on all available sensor data at each epoch to estimate the quaternion and angular velocity. Attitude propagation between epochs employs the quaternion kinematics and Euler's equation of rotational motion. The modeled disturbance torques include gravity gradient, magnetic control torque, wheel control torque, and thrust torque. The PID feedback control laws are formulated in terms of the Euler axis and angle error. That is, if the error quaternion is given by:

$$q_e = \begin{bmatrix} a_e \sin\left(\frac{\phi_e}{2}\right) \\ \cos\left(\frac{\phi_e}{2}\right) \end{bmatrix} \tag{16.2}$$

then here the proportional error term is $a_e\phi_e$. This formulation was found to have faster response and settling times than the typical formulation. During long thrust periods, the

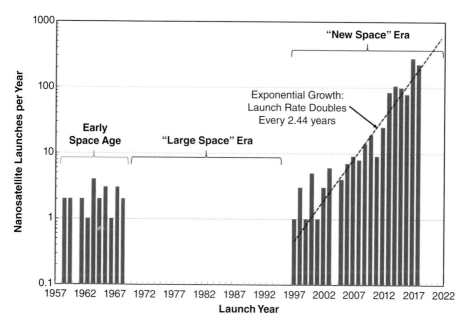

Figure 1.1 Yearly launch rates of nanosatellites from 1958 through 2017. Source: Data compiled from [2–4].

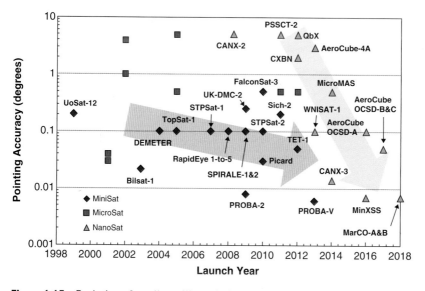

Figure 1.13 Evolution of small satellite pointing accuracy over the last 20 years.

Nanosatellites: Space and Ground Technologies, Operations and Economics, First Edition.
Edited by Rogerio Atem de Carvalho, Jaime Estela, and Martin Langer.
© 2020 John Wiley & Sons Ltd. Published 2020 by John Wiley & Sons Ltd.

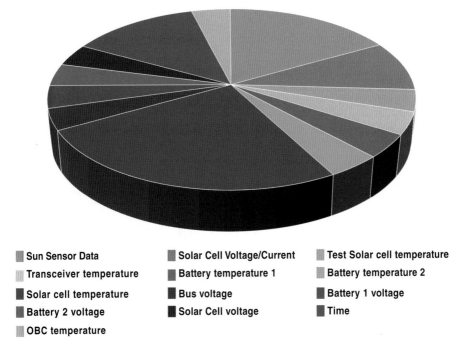

▨ Sun Sensor Data	▨ Solar Cell Voltage/Current	▨ Test Solar cell temperature
▨ Transceiver temperature	■ Battery temperature 1	▨ Battery temperature 2
■ Solar cell temperature	■ Bus voltage	■ Battery 1 voltage
■ Battery 2 voltage	■ Solar Cell voltage	▨ Time
▨ OBC temperature		

Figure 2.10 Nanosatellite data budget example. Source: Photo courtesy of NASA.

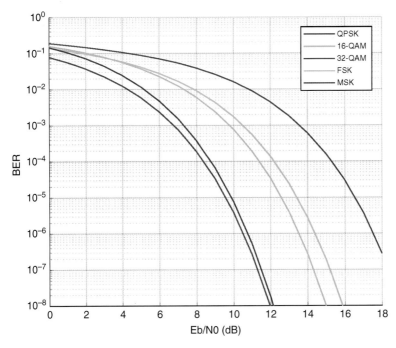

Figure 6.4 Expected bit error rate (BER) for common modulation schemes in an AWGN channel without additional channel coding.

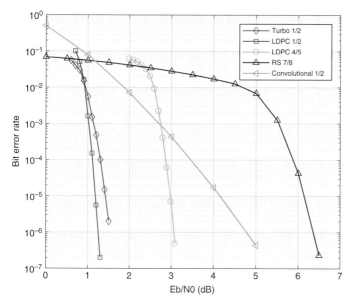

Figure 6.5 Bit error rate for CCSDS turbo with rate 1/2 and block length $k = 1784$, LDPC with block length $k = 4096$, the (255, 223) Reed–Solomon code, and the (7, 1/2) convolutional code.

Figure 7.7 First two vibration modes of a 3U structure alone.

(a)

(b)

(a)

(b)

Figure 7.8 First two vibration modes of a 6 kg 3U satellite assembly.

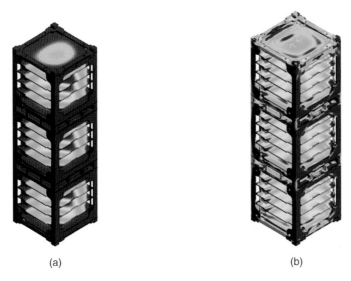

(a)   (b)

Figure 7.9 Examples of displacement and stress distributions of a satellite assembly when loaded along vertical direction.

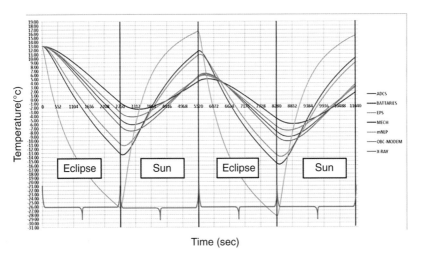

Time (sec)

Figure 7.12 Temperatures of some subsystems as a function orbital position for the BeEagleSat [18].

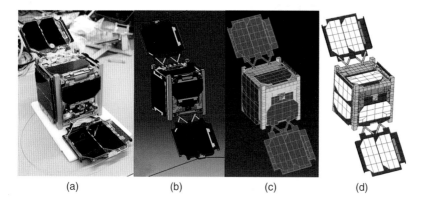

Figure 9.5 Bottom-up thermal model development with the example of the CubeSat "First-MOVE" of the Technical University of Munich: (a) photograph after a test integration; (b) detailed CAD model; (c) simplified GMM in meshing tool; (d) simplified GMM/TMM in thermal analysis software.

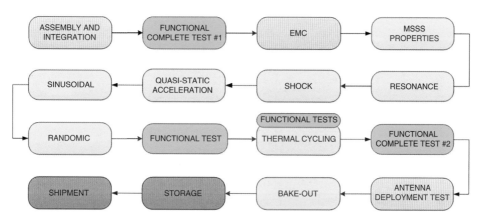

Figure 11.3 Typical macro sequence flow of nanosatellite I&T [4].

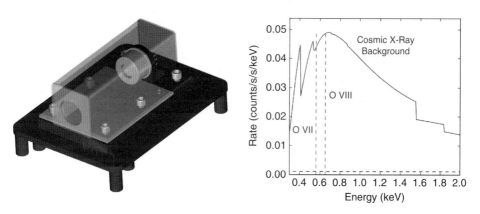

Figure 12.4 Sensor assembly: SDD detector surrounded by a copper–tungsten alloy shielding. Source: Reproduced with permission from University of Iowa.

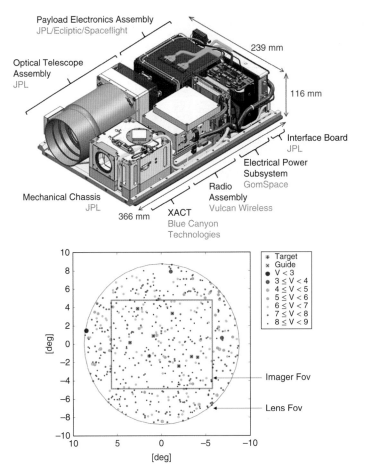

Figure 12.11 Left: Illustration of the ASTERIA spacecraft. Right: Its starfield [42]. Source: Photo courtesy of NASA.

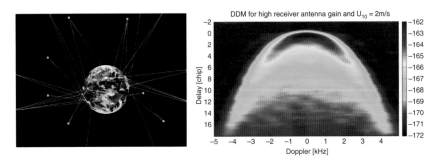

Figure 12.12 Left [2]: CYGNSS spacecraft (yellow dots) analyze reflected GPS signals (dark blue lines) from water surfaces shaped by hurricane-associated winds. Red lines indicate signals directly between the CYGNSS satellites and the GPS satellites (light blue dots further from Earth). Blue lines on the Earth surface represent measurements of the wind speed that CYGNSS derives from the reflected GPS signal. Source: Courtesy of NASA/University of Michigan. Right: Spatial distribution of the ocean surface scattering measured by the UK-DMC-1 demonstration spaceborne mission—referred to as the delay Doppler map. Source: Photo courtesy of NASA.

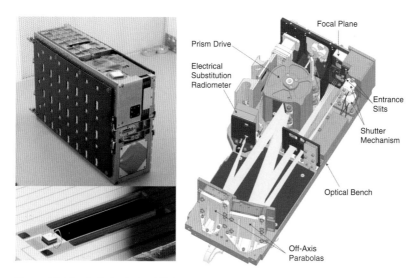

Figure 12.13 Left: (top) the CSIM prototype, (bottom) the bolometer prototype. Right: Optical overview [50, 51]. Source: Reproduced with permission from Laboratory for Atmospheric and Space Physics – LASP.

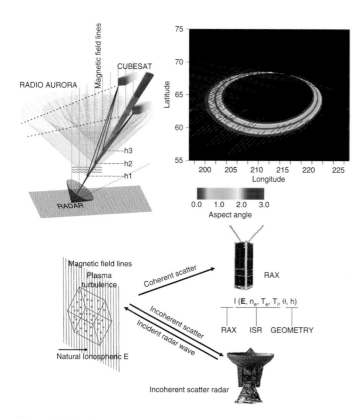

Figure 12.15 Top left and bottom: An illustration of the RAX experiment measurement concept. Top right: The loci of perpendicularity (<3°) and 1 min satellite tracks over the experimental zone over 30 days. Source: Reproduced/Adapted with permission from Hasan Bahcivan, SRI International, Menlo Park, CA, USA.

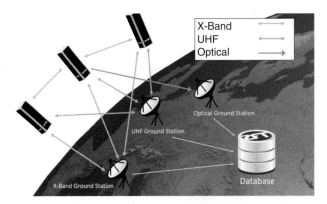

Figure 15.9 Intersatellite and ground station links in the TOM scenario also including an optical link as payload.

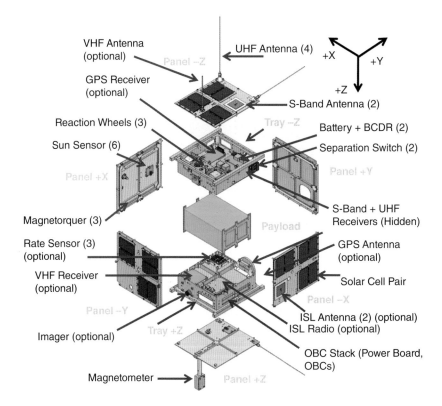

Figure 16.1 Exploded view of the SFL's Generic Nanosatellite Bus (GNB).

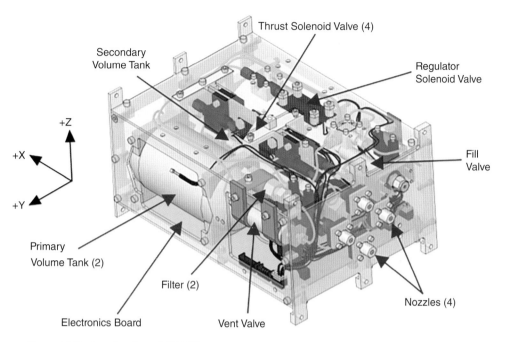

Figure 16.5 Interior view of CNAPS.

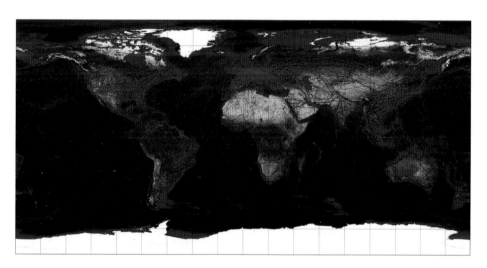

Figure 20.5 Global representation of the ADS-B data collected by GOMX-3 where each dot corresponds to a plane location captured and downloaded by the satellite.

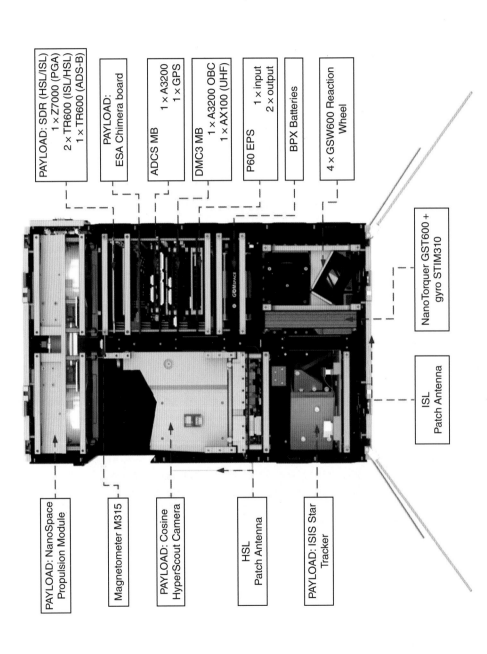

Figure 20.7 GOMX-4 internal layout: CAD system design (top) and flight model satellite (bottom).

PAYLOAD: SDR (HSL/ISL)
1 × Z7000 (PGA)
2 × TR600 (ISL/HSL)
1 × TR600 (ADS-B)

PAYLOAD:
ESA Chimera board

ADCS MB   1 × A3200
1 × GPS

DMC3 MB   1 × A3200 OBC
1 × AX100 (UHF)

P60 EPS   1 × input
2 × output

BPX Batteries

4 × GSW600 Reaction Wheel

NanoTorquer GST600 + gyro STIM310

ISL Patch Antenna

PAYLOAD: NanoSpace Propulsion Module

Magnetometer M315

PAYLOAD: Cosine HyperScout Camera

HSL Patch Antenna

PAYLOAD: ISIS Star Tracker

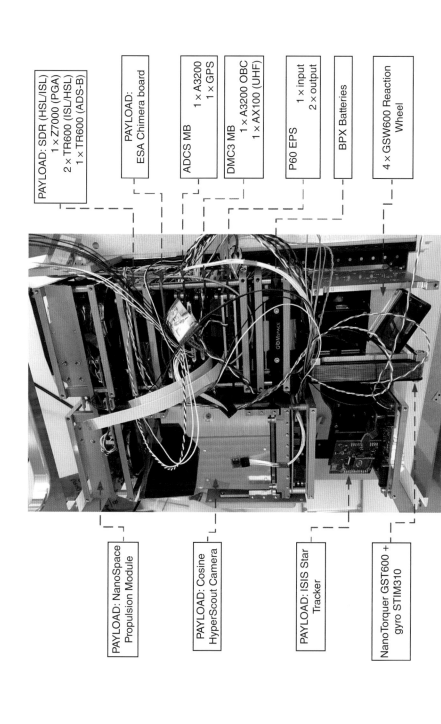

PAYLOAD: SDR (HSL/ISL)
1 × Z7000 (PGA)
2 × TR600 (ISL/HSL)
1 × TR600 (ADS-B)

PAYLOAD:
ESA Chimera board

ADCS MB   1 × A3200
               1 × GPS

DMC3 MB   1 × A3200 OBC
            1 × AX100 (UHF)

P60 EPS   1 × input
               2 × output

BPX Batteries

4 × GSW600 Reaction
Wheel

PAYLOAD: NanoSpace
Propulsion Module

PAYLOAD: Cosine
HyperScout Camera

PAYLOAD: ISIS Star
Tracker

NanoTorquer GST600 +
gyro STIM310

Figure 20.7 (Continued)

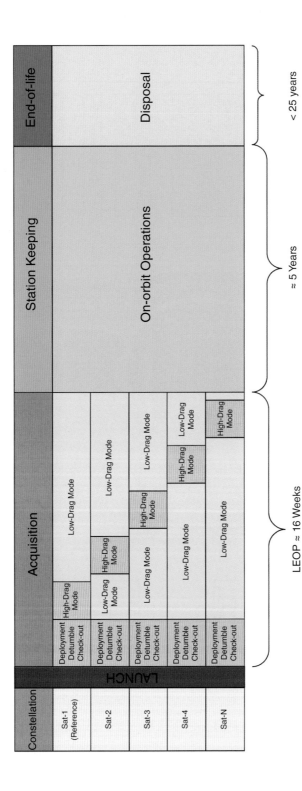

Figure 20.13 Relation between satellites' operation phases and constellation's phases.

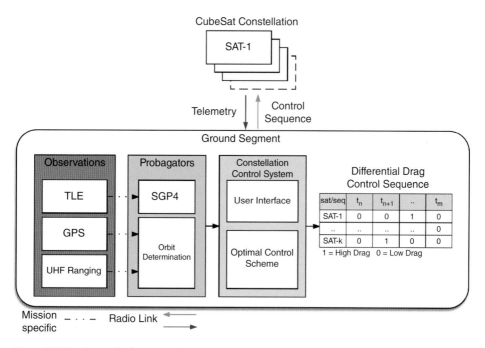

Figure 20.14 Constellation determination and control system diagram.

Figure 22.1 Clash problem of multiple satellites and stations.

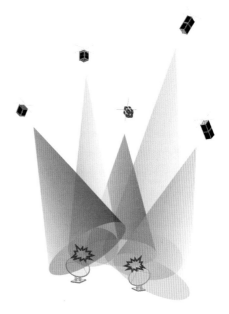

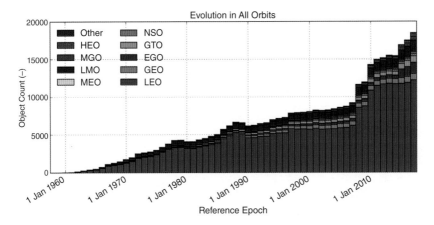

Figure 23.1 The evolution of number of objects in geocentric orbit-by-orbit class [1].

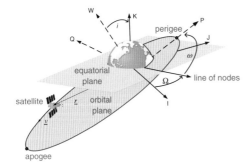

Figure 23.4 Earth-centered inertial (*IJK*) and perifocal system (*PQW*).

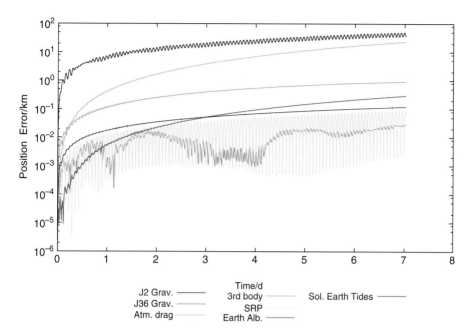

Figure 23.6 Comparison of the position error from a propagated state over 7 days with a numerical propagator for a sun-synchronous object on a $640 \times 619 \text{ km}^2$ orbit. The J_2 and J_{36} gravitational models are compared against a J_{70} model. The remaining lines are created by disregarding the respective force models in the propagation of the object.

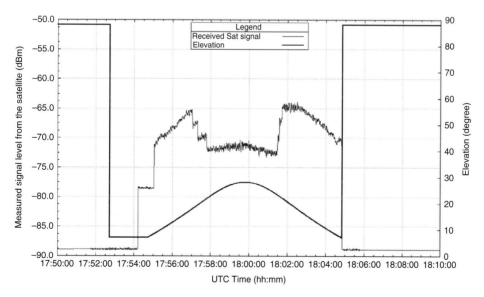

Figure 23.11 Measurement data from a satellite pass: received signal level measured at the ground station (red) and elevation of the ground station during the pass (blue).

Figure 26.1 Soyuz Fregat piggyback launch with 72 satellites in addition to the main passenger. Here, the majority of the piggyback satellites were 3U CubeSats, four of which were launched out of one 12U container. Source: reproduced with permission from Roscosmos, Russia.

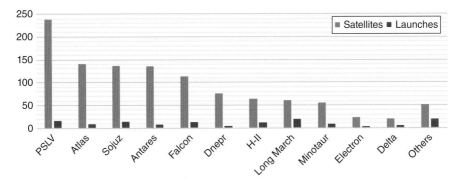

Figure 26.3 Overview of satellites below 10 kg and CubeSats up to 12U launched until March 2019 grouped by launch vehicle. Blue bars indicate the number of satellites that fall in the given mass category brought to orbit by the respective launch vehicle, while red bars illustrate the number of launches that carried such satellites performed by the launcher in question. Source: Reproduced with permission from ESA.

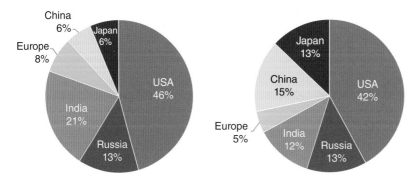

Figure 26.4 Launches of satellites below 10 kg and CubeSats up to 12U until March 2019 sorted by country. The fraction of the total number of satellites launched by each country is shown in panel (a), while panel (b) shows the countries' fraction of the total number of launches. Source: Reproduced with permission from ESA.

assumed thrust torque is used as a feedforward term to maintain pointing accuracy. On the Chief, spacecraft magnetorquers are used for wheel momentum regulation, while on the Deputy, OASYS selects the set of thrust nozzles that will result in the greatest reduction in the wheel angular momentum, or the smallest increase if no reduction is possible. Magnetorquers are not used on the Deputy during formation flying to improve pointing accuracy. To improve the navigation performance in the absence of commanded target attitudes, OASYS is programed to revert autonomously to a zenith-tracking attitude with the GPS antenna boresight.

16.3.5 RelNav

The relative navigation algorithm, or RelNav, is an EKF which uses carrier phase differential GPS techniques to estimate the relative state of the Deputy with respect to the Chief as an input to the formation control laws. The concepts in RelNav's design were adapted from numerous sources [11–16], whose contributions are gratefully acknowledged. The salient points in the filter design are summarized in what follows.

The RelNav state vector is given by:

$$x = [\Delta r^T, \Delta \dot{r}^T, \Delta b, \Delta N_1, \cdots, \Delta N_m]^T \tag{16.3}$$

Where Δr is the relative position expressed in the World Geodetic System (WGS) 84 Earth-centered Earth-fixed (ECEF) reference frame, $\Delta \dot{r}$ is the relative velocity, Δb is the differential clock error, and ΔN_i is the ith floating point single-difference carrier phase ambiguity. This is a dynamic state vector whose length changes with the number of GPS satellites commonly tracked by the two spacecraft. The maximum number of common satellites is 14, the number of independent channels on the GPS receiver. The relative orbit state is propagated using pseudo-relative dynamics [15, 16]. One step of a fourth order Runge–Kutta integration method is used to propagate between epochs; the nominal filter step is 5 s. Filter initialization is performed using the scheme proposed by Leung [16].

At each filter update step, RelNav processes single-difference pseudo-range and carrier phase measurements. During measurement evaluation, the time of signal transmission from each GPS satellite is solved iteratively [12]. The GPS satellite orbits are evaluated using the latest broadcast ephemeris parameters logged from the receiver. The measurements are assumed to be uncorrelated, thus, allowing the use of scalar measurement updates [16], which obviates the need for matrix inverses and greatly reduces computational cost.

16.3.6 FIONA

The FIONA is responsible for autonomously computing the formation reconfiguration and formation keeping control maneuvers. FIONA implements an EKF to estimate the absolute states of both the Chief and Deputy spacecraft used to compute auxiliary control parameters—for example, reference orbital elements—as well as to map the relative state estimated by RelNav into the LVLH frame required for the formation control laws. The EKF is necessary to smooth the single-point position and velocity estimates coming from the GPS receiver, which suffer from a high degree of noise, especially in the velocity terms.

The formation-keeping controller is a discrete-time linear quadratic regulator (LQR) designed using the error dynamics of the HCW equations [17]. This formulation is possible since the reference trajectories are solutions to the equations of relative motion. The output of the LQR is converted to a control impulse to be applied. The nominal control time step is 75 s, of which 15 s allotted for thrusting. In the worst case, this gives the attitude control system 60 s to perform a full 180° reorientation.

FIONA's reconfiguration algorithm identifies a set of impulsive maneuvers which minimize an energy-like cost function subject to the relative motion constraints as described by an arbitrary state transition matrix (STM) [18]. The reconfiguration algorithm requires a start time, end time, as well as a number of thrusts. The thrusts are spaced equally throughout the time domain. After the application of each thrust in the sequence, the remaining thrusts are recomputed, effectively leading to a closed-loop reconfiguration that is more robust to maneuvering errors. In practice, the Ankersen–Yamanaka STM [19] was used since it provided the best accuracy for the least computational complexity of the STMs considered.

FIONA will only command a thrust/attitude at the start of a control time step and if the RelNav solution has been marked as "reliable," otherwise the current attitude is held. The nominal values for the minimum and maximum impulses are 7.5 and 375 mN s, respectively. Commanded impulses greater than the maximum are set to the maximum and any values less than the minimum are set to zero. A commanded impulse of zero will not result in a new attitude command. Successive periods of no commanded thrust will thus result in an autonomous reorientation to a zenith-tracking attitude, which subsequently improves the relative navigation solution and ensures that the next commanded thrust is as accurate as possible.

16.4 Launch and Early Operations

CanX-4 and CanX-5 were launched on PSLV-C23 from Sriharikota, India on 30 June 2014. Commissioning of the two spacecraft was completed successfully in mid-July 2014, and all systems were found to be performing above expectations. During commissioning, the two satellites drifted apart from each other which required that drift recovery maneuvers be performed. This would mark the beginning of the main mission.

16.4.1 Drift Recovery and Station Keeping

The objective of the drift recovery and station keeping (DRASTK) system was to place one spacecraft directly behind the other, approximately 3 km apart, with as close to zero relative motion as possible. In mean orbital element terms, this means going from an initial state, with the spacecraft drifting under the effects of differential elements, to a final state where the elements of one spacecraft match those of the other, except for a small difference in the true anomaly. To do this, an impulsive control scheme was developed, based on Gauss' variation equations [20].

In a Chief–Deputy formation architecture, one spacecraft is designated the "Chief" since it is used to define the uncontrolled reference orbit, and the second spacecraft is designated

Table 16.2 Differential mean orbital elements of
CanX-5 to CanX-4 immediately after launch.

Differential mean element	Value
Semi-major axis	−708 m
Eccentricity	-1.75×10^{-4}
Inclination	$-2.32 \times 10^{-3\circ}$
RAAN	$-1.51 \times 10^{-3\circ}$
Argument of perigee	55.2°
Mean anomaly	−57.6°

the "Deputy" since its orbit is controlled relative to the Chief. CanX-4 was assigned to be the nominal Chief and CanX-5 the Deputy.

From GPS data postprocessed on the ground, the relative mean orbital elements immediately after launch vehicle kick-off were determined and are shown in Table 16.2. The most important relative element from a drift recovery standpoint is the relative semi-major axis (Δa), as it defines the secular drift rate between the spacecraft. With a Δa of −708 m, the spacecraft were drifting apart at about 95 km per day.

These relative states were input to the DRASTK algorithm. The algorithm accounts for fuel spent on maneuvers, propellant leakage over time, and the desire to maximize the number of thrusts that take place in sunlight where attitude control is more reliable than in eclipse. It also allows faster or slower trajectories to be chosen based on operational requirements.

The optimal trajectory required Δa to be changed to 306 m and Δi to be changed to 0.00129°. Inclination is changed along with semi-major axis because that allows the secular change in right ascension of the ascending node (RAAN), known as the precession of the node, to be controlled. Nodal precession is caused by the oblateness of the Earth, also known as J_2, and is a function of semi-major axis, eccentricity, and inclination [7]. Failing to correct the RAAN difference would create a large and undesirable out-of-plane motion between the spacecraft, and it is generally cheaper to correct it via a small inclination change propagated over time than to correct the RAAN alone impulsively.

Maneuvers to put the Deputy onto the return trajectory took place on 24–27 July 2014. During these maneuvers, it was discovered that the propulsion system was performing near its theoretical maximum specific impulse, exceeding expectations by ~20%. This, combined with knowledge that drift recovery could be completed for far less than the 5 m s^{-1} that was originally budgeted, meant that a considerable amount of margin was available to use. Therefore, the decision was made to increase the speed of drift recovery such that station keeping would be entered in early September, at an additional cost of about 29 cm s^{-1}. Thus, the return trajectory was altered on 29 July to have a Δa of 720 m and Δi of 0.0030°.

On 16 August, the spacecraft reached a relative range of 315 km, from a maximum of 2300 km on 25 July (Figure 16.7). At this point, deceleration thrusts began, such that the spacecraft maintained a minimum separation of 3 days for safety. Control thrusts were applied every 2 days, which was a compromise between thrusting every day, which would

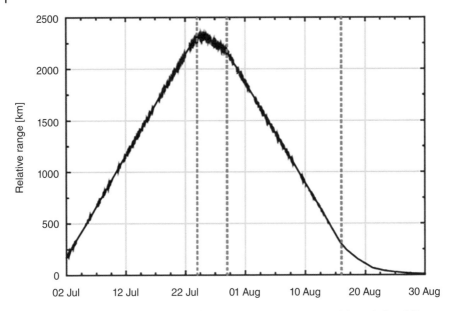

Figure 16.7 Relative range of the two spacecraft and trajectory transitions during drift recovery phase.

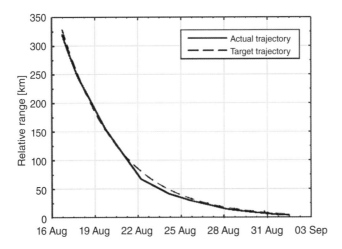

Figure 16.8 Actual and targeted return trajectory during deceleration phase.

allow slightly faster recovery, and thrusting less often which requires less operator time. Using this method, the Deputy stayed within 12 km of the reference trajectory. That error dropped to less than 2 km when the spacecraft were 15 km or closer (Figure 16.8). The process took about 17 days, ending on 2 September. When the final drift arresting thrust was sent on 3 September, the spacecraft were within 50 m of their nominal parking positions with nearly zero residual relative orbital elements (Table 16.3).

Total Δv expended in maneuvers during drift recovery is predicted to have been 2.032 m s^{-1}, based on the best estimates of on-orbit thruster performance. Based on

Table 16.3 Differential mean orbital elements of CanX-5
to CanX-4 after completing drift recovery.

Differential mean element	Value
Semi-major axis	$0.5\,\text{m}$
Eccentricity	8×10^{-7}
Inclination	$1 \times 10^{-6\circ}$
RAAN	$1.3 \times 10^{-6\circ}$
Argument of perigee	$0.014°$
Mean anomaly	$0.009°$
Range	$2.95\,\text{km}$

simulations done on the ground, assuming no attitude or navigational errors, the minimum cost to perform these maneuvers would be $1.922\,\text{m s}^{-1}$. The error, 5.71%, is well within expectations from simulations, where the mean error was found to be 5.8% with a standard deviation of 2.7%. An additional $0.813\,\text{m s}^{-1}$ was spent on station keeping maneuvers after and between each of the four formations. The majority of this fuel was used putting the spacecraft into passively safe relative states immediately upon exiting the PCO formations, when the risk of a collision was highest [20].

16.5 Formation Control Results

Before discussing the formation control results, it is worth summarizing the typical experiment planning methodology. The first step was to download GPS data and use DRASTK to compute a set of thrusts to bring the spacecraft to within roughly 1000 m of the target formation in a passively safe relative orbit. Next, offline formation control simulations were performed using the predicted relative orbit as the initial condition to establish reconfiguration start and end times, along with the expected fuel consumption. The reconfigurations were designed to begin near the start of a morning pass block—the two to four communications windows that occur for CanX-4 and CanX-5 each morning over the SFL ground station—with an overall duration of three or four orbits so that the early reconfiguration progress could be monitored during the remainder of the morning passes.

All formations were held for 11 orbits and the end of the 11th orbit was designed to take place before or during the next pass block. This was done so that, if required, maneuvers to keep the relative orbits safe could be performed as soon after the end of the formation as possible. Typically, at least 2 days were needed to download the large volume of payload data collected. During this time, the spacecraft were placed into safe relative orbits prior to commencing the next experiment.

Special considerations were required for the PCOs, since this formation can result in a collision within a few orbits following loss of formation control. During experiment planning, a set of contingency thrusts (one set for each ground contact during the experiment) were

Table 16.4 Timeline of formation flying experiments.

Date	Formation	Notes
01 Oct	1000 m ATO	Navigation errors larger than desired due to attitude targeting between maneuvers
15 Oct	500 m ATO	Formation control requirements met
21 Oct	100 m PCO	Formation control requirements met
02 Nov	50 m PCO	Formation control requirements met
06 Nov	1000 m ATO	Formation control requirements met

computed in DRASTK and prepared for upload in case it was found that the spacecraft had fallen out of formation. These thrusts were designed to restore the 90° phase offset between the radial and cross-track motion to ensure passive safety of the formation. Fortunately, the contingency thrusts were never required.

The timeline for the formation control experiments performed is shown in Table 16.4. The first formation attempt was the 1000 m ATO. In this attempt, the formation was established and maintained for the required 10-orbit period; however, the control error was sub-meter only 88% of the time, instead of the 95.45% requirement. The ultimate cause was found to be poor navigation performance due to FIONA commanding target attitudes even if the desired impulse was naught. This led to the GPS antennas often pointing away from zenith, resulting in fewer commonly tracked satellites with acceptable C/N_0 and thus less reliable solutions. The higher number of unreliable solutions resulted in less control thrusts and thus more excursions outside the desired control window. Even so, the maximum control error observed during this experiment was only 2.25 m, which was still an excellent result.

After analyzing the first experiment's results and identifying the cause for the degraded navigation performance, a new software upload was performed prior to attempting the 500 m ATO. With the improvements in the attitude targeting, the 500 m ATO was a complete success. As shown in Figure 16.9, after the initial convergence period following the

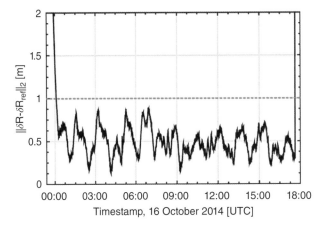

Figure 16.9 Position control error for the 500 m ATO.

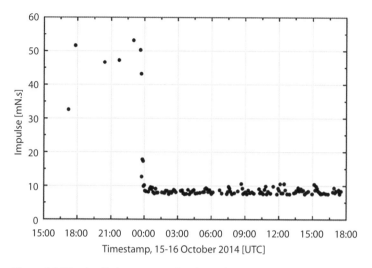

Figure 16.10 Applied maneuvers for the 500 m ATO.

end of the reconfiguration maneuver, the control error remained sub-meter for the duration of the experiment. The periods of time where the control error is increasing correspond to times where the commanded impulse was below the minimum impulse bit. The applied impulses for this experiment are shown in Figure 16.10. The first seven thrusts correspond to the reconfiguration maneuver. Following this, three more maneuvers are required before settling into a steady state of operation. The mean time between control thrusts in this experiment was 8 min and 40 s—far less frequent than the originally anticipated control period of 75 s. The actual fuel consumption was 1.6 cm s^{-1} per orbit—just under the expected value of 1.7 cm s^{-1} per orbit based on hardware-in-the-loop tests on the ground.

Having succeeded with the ATOs, the PCO experiments were started a few days later. The reconfiguration error was larger than expected, at 14 m. The fuel consumption was also 5 cm s^{-1} larger than expected based on pre- experiment planning. Although the exact cause is not known, the most likely cause of this is an error in the magnitude/direction of one of the computed thrusts due to a slightly degraded relative navigation solution. Despite this, the LQR was successful at reducing the control error from 14 to 1 m over the first 45 min. Once converged, the formation control error remained well below 1 m (Figure 16.11).

The sequence of applied impulses for this formation is shown in Figure 16.12, where the periods of convergence and steady state operation are evident. Not accounting for the additional fuel used during convergence, the steady state fuel consumption was 1.15 cm s^{-1} per orbit, roughly 0.15 cm s^{-1} per orbit above the predicted value.

The 50 m PCO was attempted 9 days following the 100 m PCO. Operationally, this was the most dangerous of all the formations due to the proximity of the spacecraft and the fact that loss of formation control could easily lead to collision. However, by this point in time, the team had a high degree of confidence in the system given the previous successes. As shown in Figure 16.13, the initial reconfiguration error was less than 2 m. At this point, the LQR took over and maintained a sub-meter control error for the duration of the formation. The mean fuel consumption was roughly 1.3 cm s^{-1} per orbit—less than half the expected

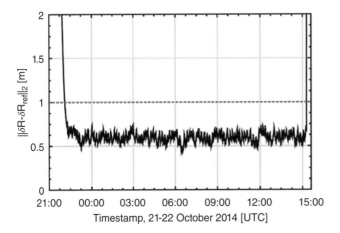

Figure 16.11 Position control error for the 100 m PCO.

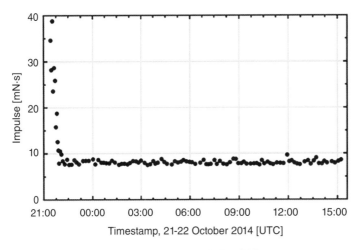

Figure 16.12 Applied maneuvers for the 100 m PCO.

value based on preflight simulations. This is attributed to the high accuracy of the relative navigation solution—in particular the relative velocity.

Following the successful completion of the 50 m PCO, it was decided to revisit the 1000 m ATO to demonstrate unequivocally that high-level mission requirements could be met in every formation. The control error following reconfiguration was the largest seen in the mission so far—35 m in relative position and 4 cm s^{-1} in relative velocity. This occurred despite no apparent relative navigation or thrust application issues. Fortunately, this was still inside the stability boundary for the LQR and the control error was successfully reduced to the required level over the course of one orbit.

As shown in Figure 16.14, there was a large fuel penalty associated with using the LQR to reduce the control error by such an extent—almost 26 cm s^{-1}. However, once the control converged the fuel consumption came to a steady state value of roughly 3.4 cm s^{-1} per orbit, just under the expected value of 3.65 cm s^{-1} per orbit.

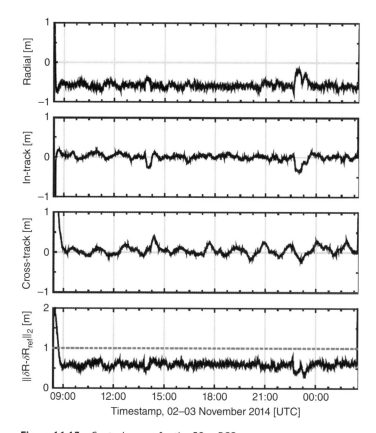

Figure 16.13 Control errors for the 50 m PCO.

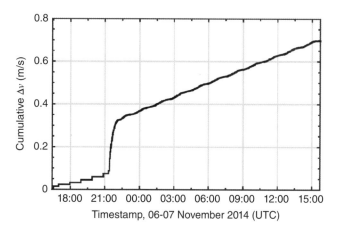

Figure 16.14 Fuel usage for the second 1000 m ATO.

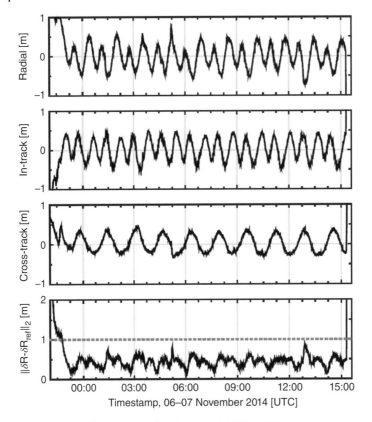

Figure 16.15 Control errors for the second 1000 m ATO.

As shown in Figure 16.15, the control error following the convergence remained sub-meter for the duration of the experiment. Note that the apparent sudden changes in relative position are in fact gradual—the data points in between are removed since they were flagged as unreliable. The control error grows during these navigation outages since no thrusts are performed.

A performance summary for the fuel consumption and formation control error for all formation flying experiments is shown in Table 16.5. As mentioned previously, the discrepancy between the actual and expected fuel consumption is due to the initial convergence period of the LQR. During steady state operation, the fuel consumption was close to the expected value in all cases. The 3D-RMS control error was well within the 1 m, 2σ requirement in all cases.

An example of the RelNav performance following commissioning is shown in Figure 16.16. Here, the measurement residuals during the 50 m PCO formation flying experiment are evaluated using the on-orbit solution. The GPS satellite orbits are computed using the broadcast ephemeris parameters since GPS orbit errors do not contribute significantly to the overall error as a result of the single-difference measurements.

For the most part, the residuals meet the expected result of a zero-mean Gaussian distribution, which indicates that the EKF is operating correctly. It can also be seen that there are

Table 16.5 Summary of formation control results.

Formation	$\Delta v_{\text{expected}}$ (cm s^{-1} per orbit)	Δv_{actual} (cm s^{-1} per orbit)	Δr_{actual} 3D-RMS (m)	Δr_{actual} 3D-RMS (m)
ATO 1000	3.65	5.55	0.590	0.453
ATO 500	1.71	1.62	0.345	0.513
PCO 100	0.99	1.63	0.517	0.602
PCO 50	3.07	1.27	0.554	0.594

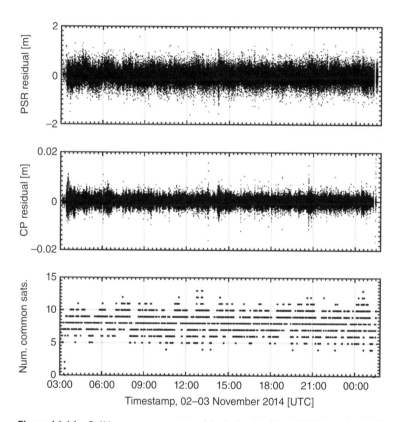

Figure 16.16 RelNav measurement residuals for the 50 m PCO formation flying experiment.

several periods where the residuals show a larger spread. These periods are typically due to dynamic events—rapid reorientations where many satellites are added and removed from the state vector, as well as thrusts where the change in spacecraft velocity is instantaneous.

It is also common to see a period of large residuals following a period of open-loop propagation caused by too few commonly tracked satellites, such as at the start and end of the time span. Examination of the solution at these epochs reveals that the relative position and velocity estimates remain consistent and smooth and it is the differential clock bias that affects the residuals most. This is likely due to the fact that the GPS receivers automatically steer their clocks to GPS time so that the differential clock bias cannot be predicted in the absence of EKF updates.

16.6 Conclusion

In only 4 months following launch, the CanX-4 and CanX-5 dual satellite formation flying mission was accomplished, ahead of schedule and with all mission objectives met. This exciting mission has broken new ground in the capabilities of nanosatellite formation flying performance—techniques that are entirely portable to larger satellites and will enable much higher-performance missions in turn. The pioneering contribution of the CanX-4/CanX-5 mission is to place high-performance formation flight within reach of cost-constrained programs, so as to efficiently support even the most challenging of business models, such as the HawkEye 360 Pathfinder mission currently under development at the time of writing.

CanX-4 and CanX-5 have pushed the boundary of what can be achieved with nanosatellites. The technology and algorithms demonstrated on CanX-4 and CanX-5 open a wide range of potential missions and applications, ranging from on-orbit inspection and repair, to sparse aperture sensing, interferometry, and ground moving target indication.

Both satellites continue to perform exceptionally well in orbit, with a large fraction of their propellant remaining, even several years after having completed their mission. Future experiments are possible with the remaining propellant, although the CanX-4 and CanX-5 mission has already exceeded the best of expectations and represents an outstanding foundation for future operational formation flying missions.

Acknowledgments

The authors wish to acknowledge the CanX-4/CanX-5 formation flying mission funding sponsors: Defence Research and Development Canada (Ottawa), the Natural Sciences and Engineering Research Council, the Canadian Space Agency, MacDonald Dettwiler and Associates Ltd., and the Ontario Centres of Excellence. Without their financial support, development of this mission would not have been possible. The authors would also like to thank Professor Susan Skone and Professor Elizabeth Cannon from the University of Calgary as well as Professor Christopher Damaren from UTIAS for contributions to the formation flying determination and control algorithms implemented on the satellites.

References

1 Petrov, G.I., 1971. Conquest of Outer Space in the USSR: Official Announcements by TASS and Material Published in the National Press from October 1967 to 1970, Moscow.
2 Godwin, R. (2002). *Gemini 7—The NASA Mission Reports*. Burlington, ON: Collector's Guide Publishing Inc.

3 "PRISMA, 2014. (Prototype Research Instruments and Space Mission Technology Advancement)," https://directory.eoportal.org/web/eoportal/satellite-missions/p/prisma-prototype, accessed 4 December 2014.

4 Clark, S., 2015. "French Sun Satellite and Swedish Experiment Blast Off on Russian Rocket," http://www.space.com/8608-french-sun-satellite-swedish-experiment-blast-russian-rocket.html, accessed 20 January 2015.

5 Sarda, K., Grant, C., Eagleson, S., Kekez, D. D., Zee, R. E. 2010. "Canadian Advanced Nanospace Experiment 2 Orbit Operations: Two Years of Pushing the Nanosatellite Performance Envelope," Proceedings of the European Space Agency Small Satellite Systems and Services Symposium, Funchal.

6 Bandyopadhyay, S., Subramanian, G. P., WFoust, R., Morgan, D., Chung, Soon-Jo, Hadaegh, F. Y. 2015. "A Review of Impending Small Satellite Formation Flying Missions," Proceedings of the 53rd AIAA Aerospace Sciences Meeting, Kissimmee, FL.

7 Schaub, H. and J unkins, J.W.L. (2009). *Analytical Mechanics of Space Systems*. American Institute of Aeronautics and Astronautics.

8 Grant, C., Sarda, K., Chaumont, M., Seung Yun Choi, Johnston-Lemke, B., Zee R. E. 2014."On-orbit performance of the BRITE nanosatellite astronomy constellation," Proceedings of the 65th International Astronautical Congress, Toronto, ON.

9 Helleren, Ø., Olsen, Ø., Narheim, B. T., Skauen, A. N., Olsen, R. B. 2012. "AISSat—1–2 Years of Service," Proceedings of the European Space Agency Small Satellite Systems and Services Symposium, Portorož, Slovenia.

10 Armitage, S., Stras, L., Bonin, G. R. R., Zee, R. E., 2013. "The CanX-4&5 nanosatellite mission and technologies enabling formation flight," Proceedings of the 7th International Workshop on Satellite Constellations and Formation Flight, Lisbon, Portugal.

11 Marji, Q., 2008. "Precise Relative Navigation for Satellite Formation Flying Using GPS," MASc thesis, University of Calgary, Calgary, AB.

12 Busse, F.D., 2003. "Precise formation-state estimation in low Earth orbit using carrier differential GPS," PhD thesis, Stanford University, Stanford, CA.

13 Ebinuma, T., 2001. "Precision spacecraft rendezvous using GPS: an integrated hardware approach," PhD thesis, University of Texas at Austin, Austin, TX.

14 Montenbruck, O., Ebinuma, T., Lightsey, E., Leung, S., 2002. A real-time kinematic GPS sensor for spacecraft relative navigation. *Journal of Aerospace Science and Technology* 6: 435–449.

15 Kroes, R., 2006. "Precise relative positioning of formation flying spacecraft using GPS," Netherlands Geodetic Commission, No. 61.

16 Leung, S. and Montenbruck, O. (2005). Real-time navigation of formation-flying spacecraft using GPS measurements. *Journal of Guidance, Control, and Dynamics* 28 (2): 226–235.

17 Pluym, J.P. and C.J. Damaren, 2006. "Dynamics and Control of Spacecraft Formation Flying: Reference Orbit Selection and Feedback Control," Proceedings of the 13th Canadian Astronautics Conference, Montreal, QC.

18 Roth, N.H. and C.J. Damaren, 2010. "Computationally Efficient State-Transition Matrix Based Multiple-Thrust Satellite Formation Reconfigurations," Proceedings of the 15th Canadian Astronautics Conference, Toronto, ON.

19 Yamanaka, K. and Ankersen, F. (2002). New state transition matrix for relative motion on an arbitrary elliptical orbit. *Journal of Guidance, Control, and Dynamics* 25 (1): 60–66.

20 Newman, J.Z., 2015. "Drift Recovery and Station Keeping for the CanX-4 & CanX-5 Nanosatellite Formation Flying Mission", MASc thesis, University of Toronto, Toronto, ON.

17 I-4a

Launch Vehicles—Challenges and Solutions

Kaitlyn Kelley

Spaceflight Industries, Seattle, USA

17.1 Introduction

Although nanosatellites have launched on many vehicles, they are deceptively difficult to manifest. One reason is that the current launch market is built around supporting large satellites. These large satellites purchase the entire launch vehicle and effectively own all the lift capacity. Occasionally, the large satellites allow nanosatellites and small satellites to be accommodated on the launch vehicle. Then the large satellites function as the "primary payload." While the primary satellite providers do not pay for the entire vehicle, they certainly have purchased the majority of the launch capacity, which allows them to determine orbital parameters (altitude and inclination of orbit) as well as the launch schedule. Nanosatellites and microsatellites also manifested on the vehicle are subject to the primary payload's orbit and schedule. The portion of the rocket that they purchased, while it does offset the total cost, is much less than the primary payload. Nanosatellites launching with a primary satellite are often deemed "secondary" or "rideshare" payloads. Although there are occasional options for nanosatellites to launch on sounding rockets or other small vehicles, the majority of them launch rideshare. This method is an excellent way of utilizing all of a launch vehicle's energy, but there are obstacles when adding nanosatellites to the launch manifest.

Launching rideshare may seem an excellent solution for nanosatellites, but it has its own challenges. Adding nanosatellites to the launch manifest requires approval by both the primary satellite and the launch services provider (LSP). ("Launch vehicle provider" and "LSP" are used interchangeably in this chapter.) While there are many satellite launches scheduled, primary satellites often prefer to launch with unused capacity, rather than incur the technical challenges of rideshare nanosatellites. First, there is a physical integration challenge. The nanosatellite must be integrated in a location where it does not pose a risk to the primary satellite. On many launch vehicles, this accommodation is on the exterior of the rocket or on the aft bulkhead. In the SpaceX launch shown in Figure 17.1, the nanosatellite deployers were placed around the inside wall of the second stage. The bright circle in the distance is the departing SpaceX Dragon capsule. (The black boxes around the deployers were added to show the location.) When nanosatellites do ride in the fairing with the

Nanosatellites: Space and Ground Technologies, Operations and Economics, First Edition.
Edited by Rogerio Atem de Carvalho, Jaime Estela, and Martin Langer.
© 2020 John Wiley & Sons Ltd. Published 2020 by John Wiley & Sons Ltd.

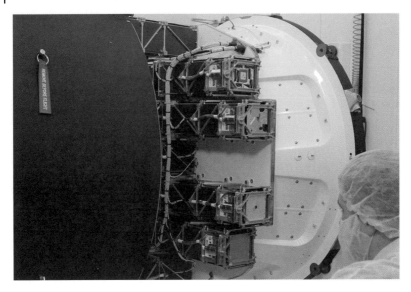

Figure 17.1 Nanosatellites launching "rideshare" on NASA's Educational Launch of Nanosatellites (ElaNa 19).

primary satellite, there are additional risks. The nanosatellite cannot contact the primary satellite or the rocket upon deployment. Additionally, if the primary satellite has strict outgassing or cleanliness requirements, the nanosatellite must comply. Rather than work with a nanosatellite developer, many primary satellites choose to launch alone.

Certifying that the nanosatellite can meet the primary payload's requirements and schedule is a logistical hurdle for the LSP. Due to this additional work, LSPs also do not always welcome nanosatellites onto their launch vehicle. Each satellite requires a certain amount of documentation, even a nanosatellite. Sorting through the extra paperwork for a nanosatellite is often not worth the financial gain. Launch vehicle providers often must meet stringent requirements to be certified for launch. They also must certify their payloads, the primary satellite, and secondary satellites. Assessing two satellites (a nanosatellite and a primary payload) is often twice as much work as assessing one, even if the nanosatellite is much smaller or simpler. A LSP is typically more interested in dedicating its time to primary satellite that occupies 90% or more of the vehicle than assessing the last 10% of launch capacity taken up by nanosatellites. The financial incentive to LSPs for launching a nanosatellite is fairly low, compared with the financial gain of launching a primary satellite. When paying by the kilogram, the financial benefit of launching a 5 kg nanosatellite is often not enough to motivate LSPs and primary satellites to include the small satellites on the manifest.

Despite these challenges, there are growing opportunities for nanosatellite launches, particularly CubeSats. CubeSats' standard form factor allows launch vehicles to allocate a standard-sized space within the structure. This can be filled by any nanosatellite that conforms to the standard volume, eliminating any unique integration work. Innovative launch platforms, like nanosatellite deployment from the International Space Station, are further expanding nanosatellites' space access.

17.2 Past Nanosatellite Launches

Since the 1960s, nanosatellites have launched on a wide variety of launch vehicles. As small payloads, nanosatellites can launch on a small sounding rocket, or on the aft bulkhead carrier of an Atlas V. The small mass of a nanosatellite may make finding a launch vehicle more difficult, but it also allows flexibility once accepted onto an Launch Vehicle (LV) manifest. Nanosatellites also have the option of launching in groups, since most rockets' launch capability is greater than the lift capability required to insert a single nanosatellite into orbit. This flexibility in launch configuration has enabled nanosatellites to launch on many launch vehicles since their inception in the dawn of the satellite era.

Although there is a current industry focus on the CubeSat nanosatellite form factor, nanosatellites were launched decades prior to the initiation of the Cal Poly standard. The fourth satellite ever launched into orbit, Vanguard 1, was a nanosatellite of less than 2 kg. An experimental satellite designed to test the effects of the space environment, Vanguard 1 set the standard for many nanosatellites to follow. For many years, nanosatellites were primarily launched as technology demonstrations. A recent study [1] notes that small satellite launch rates (including nanosatellites) were high in the 1960s, as scientists and engineers explored the new technology of the satellite. As a technology demonstration device, nanosatellites were already characterized as a good investment nearly half a century before the CubeSat emerged as a technology demonstration platform.

Yet, throughout the 1970s and 1980s, small satellite launches decreased. In an assessment of nanosatellite launches completed by Bouwmeester and Guo [2], the authors had no data points between the 1963 and 1996. This is not to say that there were no nanosatellites launched, but the lack of launches included in the study is indicative of the trend. As launch vehicles became powerful enough to loft larger satellites, the satellite technology also grew in mass and size. For many years, science payloads were massive and heavy satellites were needed to support them. A review of launches in the 1970s and 1980s shows a great number of interplanetary exploration satellites. One example was the Galileo Jupiter probe, which was a massive 2223 kg at launch, according to the NASA website. Pioneer 10, Pioneer 11, Voyager 1, and Voyager 2 were also launched in a similar time frame. All the programs listed are NASA programs. The Soviet Union also launched a number of orbiters. With an increase in launch vehicle power, satellites continued to increase in size. It was not until low power electronics became widespread that nanosatellites returned en masse to space.

As electronics grew cheaper and smaller in the late 1980s and 1990s, nanosatellites slowly began to return. This time however, the launch vehicles were large and could loft massive satellites. Instead of occupying the entire launch vehicle payload, as Vanguard 1 did, nanosatellites were a secondary payload. This provided new launch opportunities. In 1989, AMSAT and European Space Agency (ESA) partnered to launch six microsatellites under 12 kg. ESA needed test spacecraft for its new Ariane Structure for Auxiliary Payloads (ASAP), and AMSAT wanted to build small cheap satellites to serve the amateur radio community. According to Space Today Online [3], the spacecraft launched were UoSAT-OSCAR-14, UoSAT-OSCAR-15, AMSAT-OSCAR-16, Dove-OSCAR-17, WEBERSat-OSCAR-18, and LUSAT-OSCAR-19; they were developed in four different countries. New launch opportunities as secondary payloads were becoming available and satellites were shrinking to fit in the available spaces.

Then, in 1999, the CubeSat standard was introduced by Jordi Puig-Suari of the California Polytechnic State University (Cal Poly) and Bob Twiggs of Stanford University; the nanosatellite launch rate increased dramatically. The limitations on size and mass were a boon to universities. A 10 cm cube weighing about a kilogram could be launched [4] for a price affordable to many universities. With the CubeSat standard form factor guiding size and mass, graduates could do a project with an achievable scope and the opportunity to see their hardware in space. The first CubeSat-class nanosatellites were launched in 2000. Swartwout's statistical analysis of the first 100 CubeSats launched pinpoints six nanosatellites launched in 2000, PICOSATs 1, 2, 3, 4, 5, and 6. Built by students at Santa Clara University, the satellites were the first of many CubeSat-class nanosatellites built by universities in the next 10 years.

While the CubeSat standard was effective at garnering university support, the Poly Picosatellite Orbital Deployer (P-POD) deployment system is what made CubeSats palatable to launch vehicle providers. The P-POD is a separations system that allows three $10 \times 10 \times 10\,\text{cm}^3$ units of CubeSats to launch in a single standardized container. It is simple, nonpyrotechnic, and encapsulates the CubeSat. These traits are intended to minimize impact to the primary satellite and launch vehicle. In an industry in which every satellite and mission is unique, the P-POD was a revelation. CubeSats could be exchanged, one for another, without any change to the launch configuration. The P-POD was the interface for the launch vehicle and it was static. If the CubeSats followed the standard, they could fit in the P-POD. Any launch vehicle that had capacity for a P-POD could find a CubeSat (or a few) to fill it (Figure 17.2).

NASA quickly embraced the CubeSat platform and organized launches to support the growing number of university CubeSats. Called the Education Launch of Nanosatellites

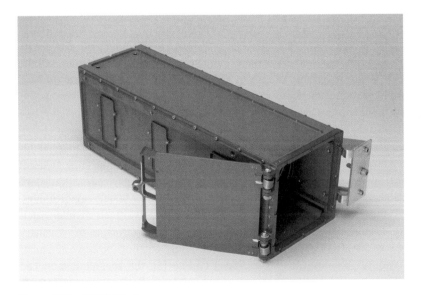

Figure 17.2 P-POD Mk. II.

(or ELaNa), the NASA program has been manifesting qualified CubeSats on government launches. To date, 59 CubeSats have been successfully launched through the ELaNa program, according to NASA's website (one launch with three CubeSats did not achieve orbit). NASA has tried to provide launch opportunities for CubeSats on multiple government launches. Each launch vehicle has completed design work to accommodate these NASA CubeSats. Now these NASA-specific interfaces can be offered as a commercial accommodation for other CubeSats riding on the same launch vehicle. For example, the interface developed to accommodate one P-POD on the Antares launch vehicle could be offered commercially, since the P-POD interface has already been developed. NASA's approach has opened more opportunities on US launch vehicles to CubeSats.

Outside of NASA's work in the USA, many CubeSats have launched on non-US vehicles. As explained in the following section, Russian and Indian launch vehicles have been quick to provide rideshare launch opportunities to CubeSats, other nanosatellite form factors, and microsatellites.

Realizing the potential of the CubeSat form factor, commercial companies also began offering CubeSats launch services. Innovative Solutions in Space, a Dutch company, offers integration services on a number of launch vehicles. Their first launch in 2009 successfully deployed four CubeSats from a Polar Satellite Launch Vehicle (PSLV). The company manufactures their own deployers. Another commercial company, Terran Orbital (formerly Tyvak Nanosatellite Systems Inc.), had completed 15 launch campaigns at the time of this composition. Terran Orbital also manufactures and tests CubeSats, in addition to providing launch opportunities. Spaceflight Industries also began to sell commercial launch services for CubeSats and larger microsatellites. Their first satellites were deployed in April 2013; three CubeSats were deployed from an Antares launch vehicle, and a single CubeSat was deployed from a Soyuz vehicle. NanoRacks, a third commercial company, began deploying CubeSats from the International Space Station in early 2014, releasing 33 CubeSats in total. As the launch demand grows, more companies and launch opportunities are emerging for nanosatellites.

With government and commercial entities working together to provide launch opportunities to CubeSats, the nanosatellites have been launched on a variety of different vehicles.

17.3 Launch Vehicles Commonly Used by Nanosatellites

With government support of CubeSats and nanosatellites, the United States launch vehicles would be the assumed leader in launches. Two studies in 2010 supported this estimation. Bouwmeester and Guo counted 41 US nanosatellite launch opportunities, which was closely followed by 39 nanosatellites launched on Russian vehicles. Swartwout [4] also completed a study in 2010 that analyzed the first 100 CubeSats launched. He concluded that the USA launched 46% of CubeSats and Russia launched 27%. These numbers correlated with the US government's demonstrated interest in CubeSats and nanosatellites. The interfaces driven by NASA's ELaNa program allow CubeSats to launch on almost every US launch vehicle.

Yet, an analysis of nanosatellite launches on individual launch vehicles completed by SpaceWorks Enterprises (SEI) in 2014 [5] showed that Russian vehicles have increasingly supported nanosatellite launches. As of 2014, two Russian launch vehicles held the top spots for nanosatellite launches. The first US launch vehicle to make the list was the Minotaur 1, the third most prolific nanosatellite launch vehicle. Summing the Russian vehicles (Kosmos-3M, Dnepr-1, Soyuz, and Rokot) and the American launch vehicles (Minotaur 1, Space Shuttle, Falcon 9, Delta II, Minotaur 4, Antares, and Atlas V) showed that the Russian launch vehicles supported more nanosatellite launches than the US launch vehicles. PSLV, the Indian Polar Satellite Launch Vehicle and the two Japanese launch vehicles (H-2 and M-5) have also enabled large numbers of nanosatellite launches. SEI concluded that "low cost piggyback opportunities attract nanosatellites and small satellites to [non-US] launch vehicles." This is interesting because the USA is the primary developer of nanosatellites (according to Bouwmeester and Guo), which means that US-developed CubeSats launched internationally must overcome the legal obstacles put in place by the United States International Traffic in Arms Regulations (ITAR).

One launch platform not included within SEI's study is the NanoRacks deployment system. Since no satellites were deployed by NanoRacks until 2014, they were not included within the 2000–2013 analysis. However, the amount of CubeSats that NanoRacks is able to deploy would quickly bump NanoRacks up into the top five. In the first deployment round, NanoRacks deployed 33 CubeSats. NanoRacks deployment platform would rank right after the Minotaur and Dnepr. But the Dnepr has also continued to launch CubeSats at a high rate, nearly doubling the amount of CubeSats launched on that launch vehicle in 2014. (The June 2014 launch of Dnepr deployed 37 satellites in total, including three main payloads.) While the 2014 Nano/Microsatellite Market Assessment is certainly a good indication of what vehicles are popular, NanoRacks and other commercial launch platforms are changing where nanosatellites go for launch.

A more recent report shows how the industry is realigning to support nanosatellite launches. The 2019 SpaceWorks Enterprises Nanosatellite Launch Vehicle Assessment [6] shows the emergence of Electron, a launch vehicle specific to small satellite launches. With only three launches in its inaugural year, Electron propelled 20 satellites into space. However, the majority of payloads still seem to launch rideshare. SpaceX's launch manifest included a large rideshare mission called Spaceflight SSO-A. The mission, organized by Spaceflight Industries, had 64 payloads. PSLV and Soyuz are also still active in rideshare launches. However, the change over 5 years demonstrates how the influx of nanosatellites is changing the launch vehicle industry.

17.4 Overview of a Typical Launch Campaign

The differences between nanosatellites and primary satellites can often create obstacles for sharing a launch vehicle. Three unique aspects of nanosatellites that affect the launch campaign are shown in Figure 17.3. Nanosatellites can be built in months, as compared to the years required for large satellites. CubeSats in particular also have little unique hardware. Entire satellite buses can be purchased at low cost. Finally, the launch costs for nanosatellites are orders of magnitude less than large satellites. Challenges caused by these differences are explored in the following sections.

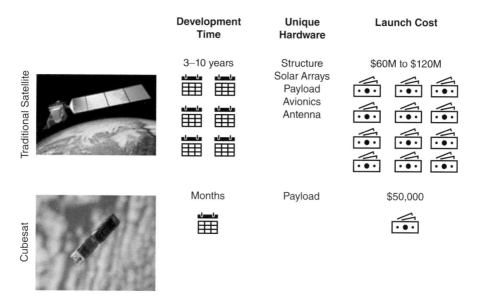

	Development Time	Unique Hardware	Launch Cost
Traditional Satellite	3–10 years	Structure Solar Arrays Payload Avionics Antenna	$60M to $120M
Cubesat	Months	Payload	$50,000

Figure 17.3 Unique aspects of nanosatellites that affect launch.

The short development time of nanosatellites, while a boon to developers, can be difficult for launch vehicle providers. Aligning nanosatellite milestones with the standard primary satellite milestones can be nearly impossible. A CubeSat or nanosatellite may begin searching for a launch with a year of the expected launch timeframe. If they want to ride with a primary satellite, this can cause issues. At 12 months to launch, much of the analysis has already been completed. Some launch vehicle providers run coupled loads analysis 2 years prior to launch. Nanosatellites joining a mission after this analysis must stay within a certain tolerance or the analysis has to be re-run at great cost.

Looking for launches earlier does not always help either. If it takes 1 year to develop a nanosatellite and they manifest on a launch 2 years in advance, the developers will not have hard data for another year, when the development actually begins. The developer could build up the nanosatellite earlier, but then it sits waiting for a launch instead.

How can nanosatellites have such a quick development timeline? Vendors such as Clyde Space and Pumpkin Inc. offer standard parts that can be integrated into a CubeSat, which cuts down drastically on time spent in development. As shown in the graphic, the payload may be the only unique component on a CubeSat; the rest of the parts can be purchased as commercial off-the-shelf parts. A prime example of the speed at which nanosatellites can be built is Planet Labs, developer of the 3U CubeSat "Dove." NanoRacks launched 28 Planet Labs satellites on both the ORB-1 and ORB-2 missions. Fifty six satellites were launched on the two consecutive launches. ORB-1 was integrated by NanoRacks in November 2013. ORB-2 reached the International Space Station in July 2014. During this 8-month period, Planet Labs created at least 28 satellites for the ORB-2 launch. This conservative estimate has Planet Labs churning out one satellite in little over a week (8.5 days). Planet Labs can build nearly 43 satellites per year at this calculated rate.

While CubeSat and other nanosatellite development can be completed within months, generating the documentation required for launch is not congruous with the nanosatellite development timeline. Nanosatellites must meet launch vehicle requirements by complying with safety and design standards. Although nanosatellite components may be smaller or lower power, these must still comply with requirements. In the USA, in particular, satellites must meet US Air Force Range Safety requirements to launch. The CubeSat standard provides many of these baseline safety requirements; therefore, if the CubeSat conforms to the CubeSat standard, it conforms to many of the safety requirements. Little additional analysis is needed. However, more safety documentation may need to be completed as CubeSats increase in complexity. The most recent CubeSat standard (Revision 13) specifies that propulsion is now allowable, but it has to meet the Air Force safety requirements. Now CubeSat developers must evaluate their conformance to the US Air Force or comparable agency for propulsion. Nanosatellites are often asked to submit this information at the same time as the primary payload. However, the nanosatellite design may not be completed at the time, so documents are submitted late or incomplete.

One method of addressing this timing difference between the primary and rideshare nanosatellites is standardization. The CubeSat standard and the standard integration hardware derived from it allow later entry of CubeSats onto the launch manifest. Launch vehicles can set aside unused portions of the launch vehicles for a CubeSat dispenser. Many US vehicles offer the capability to add CubeSat deployment. The NASA ELaNa program has opened up many opportunities for standard accommodations on government launches. This is a nice setup because any CubeSat that conforms to the standard can be integrated into the dispensers. One CubeSat could even be exchanged for another if the original occupant does not finalize the satellite before the launch date. The CubeSat design requirements are one of the reasons that this is possible; the design is set and CubeSats may simply adjust their environmental testing depending upon the selected launch vehicle. Additionally, the standard interface already built into the launch vehicle makes integration easier for the launch vehicle services provider.

Because most nanosatellites launch as "rideshare" payloads, the launch campaign standard is to make integration with the primary satellite as simple as possible. Auxiliary payloads must remain powered "off" until deployment. They are carefully evaluated for outgassing and offgassing. It is not permissible for them to impact the primary mission in a typical launch campaign. This is a challenge for all nanosatellites. The CubeSat standard has eased integration through standard interfaces and documentation requirements. Many of the nanosatellites launching today are CubeSats; uniquely shaped nanosatellites often have more challenges finding an appropriate interface. For many, the P-POD becomes the design interface. The odd shaped nanosatellites are deployed like CubeSats, but can use different adapter to conform to the standard P-POD interface.

Another option for launch vehicles to better support nanosatellites is to offer a "reflight" option. Nanosatellite developers can launch a CubeSat with 90% of the same hardware as a previous. Only the new hardware needs to submit to the onerous documentation process. This saves time for both the launch vehicle provider and the nanosatellite.

There are many options to address how to better integrate nanosatellite timelines with those of the primary payload. However, it is not particularly beneficial to the LSP to address the needs of nanosatellites. Another difference between the large satellites and the small satellites is the cost of launch. When the benefit of launching a nanosatellite is a small financial incentive, why should launch vehicle providers provide custom accommodations or adjust the launch timeline? NASA has engineered solutions through the ELaNa program. Microsatellites until the launch cost is a real benefit to the launch vehicle. India's PSLV also accommodates aggregated small satellites and nanosatellites. With incentives for both the launch vehicle and the developer, more solutions will be developed in the near future.

17.5 Launch Demand

As nanosatellites have proliferated, helped along by the miniaturization of payloads and other equipment, launch demand has skyrocketed. SpaceWorks Enterprises has evaluated how many satellites have launched since 2000. Their estimates of the future market for small satellites (less than 50 kg) seem to increase each year. In the 2019 report, the year 2023 was forecasted to have 513 small satellites launched. This is nearly 200 more satellites than the estimate from 2017, which was 320 launched in 2023. The numbers are growing even faster than forecasted. Interestingly, more nanosatellites and small satellites were likely developed than the number estimated. Some of them could have been grounded for different reasons—whether due to technical issues or difficulty in finding a launch.

Although CubeSats and nanosatellites have been technology demonstration vehicles in the past, commercial companies are beginning to use them as revenue-generating satellites. For companies such as Planet Labs, the effectiveness of the satellite platform is increased if the satellites are launched as a constellation. With hundreds of nanosatellites in the sky, Planet Labs photographs nearly the entire Earth daily. The number of satellites required to meet this goal is high. Instead of investing in a single satellite with a 5-year development plan, commercial companies can simply launch massive amount of nanosatellites. If one is lost, it is not catastrophic because there is a whole constellation to support the platform. This business model has certainly increased the launch demand over the recent years. (For reference, Planet Labs began launching satellites in 2013, a year in which the number of small satellite launches doubled.) Nanosatellites and small satellites are now an intriguing business opportunity, which is driving up launch demand.

Even before commercial companies began using nanosatellites as a revenue platform, launch demand was an issue identified by NASA. Many universities embraced the CubeSat standard and began their own programs to launch the CubeSats. Launch opportunities seemed too scarce to support all the university CubeSats. To increase launch opportunities, NASA made launch opportunities available on US government launch vehicles through the ELaNa program. The space agency even took the launch demand issue a step further and issued a Centennial Challenge for a nanosatellite launcher in 2010. A US$2 million prize was made available to any team that could deliver small satellites into orbit safely and at low cost. Although this challenge was canceled prior to any submission, it showed the government interest in the area.

With the high nanosatellite launch demand, commercial companies began to consider developing their own nanosatellite and small satellite launch vehicles. The companies determined that the nanosatellite market was enough to sustain a launch vehicle dedicated to small payloads. Three of the companies dedicated to small and nanosatellite launches are Generation Orbit, Firefly Space Systems, and Rocket Lab Ltd. Each has developed a unique launch vehicle to get nanosatellites into orbit and meet the launch demand. Rocket Lab's Electron vehicle had three launches in 2018, sending more than 20 satellites into space. Even Virgin Galactic has added a small satellite launch capability to address the growing launch demand. Its vehicle, LauncherOne, is slated for a captive carry test in 2019. As more commercial entities get involved in the development of small satellites and small launch vehicle, the launch demand is likely to continue to grow. Many have forecasted hundreds of nanosatellites and small satellites launching each year in the near future.

17.6 Future Launch Concepts

To serve the growing nanosatellite market, different launch concepts have been proposed. The first concept is maximizing the space available to small satellites and nanosatellites on existing launch vehicles. The second concept is the creation of launch vehicles specifically designed to serve small satellites and nanosatellites. Both concepts are discussed in this chapter.

Maximizing the space available to nanosatellites on launches is a concept that has already been demonstrated, but may be taken a step further in the future. NASA has slowly expanded the space available to CubeSats on US vehicles. The PSLV and H2A launch vehicles have portions of the fairing set aside for small satellite and nanosatellite integration. The Dnepr cluster missions have used the rocket's unique platforms to send a record number of nanosatellites to orbit. These missions have demonstrated successful deployments of nanosatellites. However, in these cases, the primary satellite is still a factor in determining orbit parameters and other requirements. The next step in this evolution would be to eliminate the primary payload and launch a large amount of nanosatellites and small satellites that would like to go to similar orbits. A launch vehicle with a large payload capacity could deploy hundreds of small satellites and nanosatellites in a single launch, using this method. Spaceflight Industries, which follows this model, used a SpaceX Falcon 9 to deliver 64 payloads in 2018.

Launch vehicles that are intended to support only nanosatellites and small satellites are also a potential launch concept. In this section, launch vehicles that are capable of delivering a nanosatellite to low Earth orbit (LEO) (<2000 km above Earth's surface) are described. Vehicles that are limited to suborbital trajectories are not covered. The four launch vehicles discussed in detail here were selected using these additional criteria: first launch scheduled prior to 2019, less than 500 kg to LEO payload capability, and built by commercial companies. Because nanosatellites are small enough to launch in sounding rockets, there are other options than the ones listed here. The intention of this chapter is to discuss new concepts, and commercial vehicles were selected due to the availability of information.

The first vehicle selected for discussion, the Electron, can support up to 250 kg to LEO. This capability is more consistent with a microsatellite launch. However, this does not mean that nanosatellites cannot take advantage of this launch vehicle. Multiple nanosatellites can launch on a single Electron. In 2018, the Electron sent over 20 satellites into space in three launches. The Electron launches vertically. This is the typical launch configuration. The unique technology demonstration for the Electron is the light composite materials used for the rocket's structure. The rocket is constructed of a carbon fiber composite to reduce weight and maximize the launch capability.

An aircraft-assisted launch vehicle that is in development is LauncherOne, Virgin Galactic's launch vehicle. With a larger payload capability than Electron, the LauncherOne is expected to send up to 300 kg to orbit [7]. Interestingly, the LauncherOne rocket is interchangeable with the anticipated man-rated SpaceShipTwo vehicle. It is launched from the same plane that provides the air launch capability (White Knight Two). Like the others, it purports that it is a good option for CubeSats and other nanosatellites, even though it can accommodate microsatellites. The smallest launch vehicle discussed here is the Vector-R, which has a payload of only 26 kg to sun-synchronous orbit and 50 kg to LEO. Due to its small launch payload, the Vector-R is designed for launching Cubesats. The company had two successful test flights in 2017. They are slated to launch multiple Cubesats in 2019.

Each of these vehicles has a unique method to address the nanosatellite launch demand. The vehicles have different payload capacities, fairing sizes, and reachable orbits. However, they all offer one common service, the ability for nanosatellites to select an appropriate orbit for their mission. With rideshare launch, nanosatellites are often injected into the same orbit as the primary satellite or the orbit available after a collision avoidance maneuver is completed by the launch vehicle. With small satellite launchers, nanosatellites can occupy enough of the launch capability to determine the final orbital characteristics. New orbits open new opportunities for innovative payloads and demonstrations.

Small launchers also benefit orbital debris mitigation. Nanosatellites can launch into lower orbits than primary satellites, which allow the spacecraft to meet orbital debris mitigation standards. Currently, many nonpropulsive nanosatellites launched into orbits higher than 600 km above the Earth's surface do not re-enter within 25 years of the end of mission life. However, rideshare launches often deploy at these altitudes. Control of the launch vehicle's orbit and inclination allows nanosatellites to more easily comply with orbital debris best practices.

The small launchers provide capabilities that are not currently available to nanosatellites. The larger small launchers may seem to have too much capability for nanosatellites, but the excess capability also enables nanosatellite missions. Constellations of nanosatellites should be able to use the extra mass capability to perform additional burns, allowing the nanosatellites to be spread along orbital planes. Imagery nanosatellites can particularly benefit from this capability. Imagery nanosatellites spread throughout a sun-synchronous orbital plane can almost continuously image a single location. This enables constant monitoring of certain areas. Precision injection by new small satellite launchers can increase the efficiency of current platforms.

Finally, small satellite launchers can address the issues with rideshare nanosatellite launch campaigns. Most small satellite launchers have aligned their development time with the amount of time that it takes to design and construct a nanosatellite or microsatellite. Rocket Labs even created an infographic to show how the Electron aligns with small satellite lifespans and the typical wait time for launch. The timing differences between the primary satellite and nanosatellites are no longer an issue. Additionally, nanosatellites occupying the entire vehicle are not restricted by the primary satellite's risk profile and cleanliness requirements. This provides additional flexibility to the nanosatellite developer.

The nanosatellite launch demand has helped create a new interest in small launch vehicles. For years, the only methods of getting to orbit were the massive launch vehicles hauling 4000 kg satellites to orbit. Nanosatellites could simply hope to "tag-along" for the ride. As the number of small satellites awaiting launch grows, the number of nanosatellites deployed from a single large launch vehicle may rise as commercial companies find new ways to launch nanosatellites to orbit. However, with small satellite launchers, nanosatellites could begin to choose their own orbit, instead of following the primary satellite's plan. The varying capabilities of the developing launchers will enable even more types of missions for nanosatellites, particularly constellations. With new innovations, launch campaigns can be less of an obstacle and allow more ingenious nanosatellite designs to successfully complete their missions.

References

1 Depasquale, J., Charania, A.C., Kanamaya, H., and Matsuda, S. (2010). Analysis of the earth-to-orbit launch market for nano and microsatellites. In: *AIAA SPACE 2010 Conference & Exposition*, 8602.

2 Bouwmeester, J. and Guo, J. (2010). Survey of worldwide pico- and nanosatellite mission, distributions and subsystem technology. *Acta Astronautica* 67: 854–862.

3 Space Today Online. (2005). "Microsatellites were Invented as Amateur Radio Satellites in the 1990s." Space Today Online. Retrieved 15 January 2015, http://www.spacetoday.org/Satellites/Hamsats/Hamsats1990s/Hamsats90sMicrosats.html

4 Swartwout, M. (2013). The first one hundred CubeSats: a statistical look. *Journal of Small Satellites* 2: 213–233. http://www.jossonline.com/downloads/0202%20The%20First%20One%20Hundred%20Cubesats.pdf.

5 Buchen, E., DePasquale, D., (2014). "2014 Nano/Microsatellite Market Assessment." SpaceWorks Enterprises. https://digitalcommons.usu.edu/cgi/viewcontent.cgi?filename=0&article=3018&context=smallsat&type=additional

6 Piplica, A.J. (2014). "GOLauncher 2: Fast, Flexible and Dedicated Space Transportation for Nanosatellites." Small Satellite Conference, Logan, UT.

7 Charania, A.C., Isakowitz, S., Pomerantz, W., Morse, B. Sagis, K. (2013). "LauncherOne: Revolutionary Orbital Transport for Small Satellites." Small Satellite Conference, Logan, UT.

18 I-4b

Deployment Systems

A. Rüstem Aslan[1], Cesar Bernal[2], and Jordi Puig-Suari[3]

[1] Astronautical Engineering Department, Istanbul Technical University, Turkey
[2] ISIS—Innovative Solutions In Space B.V., Delft, The Netherlands
[3] Cal Poly, Aerospace Engineering Department, San Luis Obispo, USA

18.1 Introduction

The present chapter introduces various nanosatellite deployment systems. Nanosatellites are generally designed and developed based on the CubeSat standard and usually launched using a picosatellite orbital deployer (POD), a special container originally called "poly picosatellite orbital deployer (P-POD)." The POD, which is also an integral part of the CubeSat design specification, is fully tested and qualified to survive the launch loads exerted by launch vehicles (LVs). Since nanosatellites are, till now, launched as secondary payloads together with a main spacecraft, they should not endanger the success of the main spacecraft and the rocket. The POD is the interface between the nanosatellite and the launch vehicle. It may be compared to the large containers used to carry various goods on ships. Currently, there are many new nanosatellite deployers (also called dispensers) that can accommodate various sizes (CubeSat form factors) of nanosatellites, multiple of them, or swarms of them. Most PODs fully enclose the nanosatellite while some do not. The PODs are the enabling technology for the nanosatellites, although they may impose certain constraints on their design, particularly for the attachments, such as booms, cameras, sensors, or antennas. Deployment from the International Space Station (ISS) or from a mother spacecraft are recent trends in nanosatellite deployment.

18.2 Definition and Tasks

The deployment system usually called "POD" provides a standard interface between a nanosatellite and a launch vehicle. Nanosatellites are usually in rectangular form, obey the CubeSat design specification [1], and are multiples of 1U form factor. The deployment system tasks are:

- To fully enclose the nanosatellite to provide a safer environment for it.
- To protect the main payload and the rocket from possible damages that may be caused by the nanosatellite (which may not have been fully tested).

Nanosatellites: Space and Ground Technologies, Operations and Economics, First Edition.
Edited by Rogerio Atem de Carvalho, Jaime Estela, and Martin Langer.
© 2020 John Wiley & Sons Ltd. Published 2020 by John Wiley & Sons Ltd.

- To act as an interface between the nanosatellite and the launch vehicle.
- To deploy the satellites to the target orbit with a suitable velocity of $1-2\,\mathrm{m\,s^{-1}}$ and tip-off rate.

18.3 Basics of Deployment Systems

The deployers are basically developed for holding a single (1U) or multiple unit (up to 16U) CubeSats. Larger PODs may house a combination of different-sized CubeSats as well. Very first examples are the P-POD of Cal Poly [2, 3] and the Tokyo Picosatellite Orbital Deployer (T-POD) of University of Tokyo [4], Japan. The XPOD of University of Toronto [5–7], Canada followed the first two. None of these were commercially available to other CubeSat developers. They could be used only if the launch contract was carried out by the POD producer. The PicoSatellite Launcher (PSL) family of Astro- und Feinwerktechnik Adlershof GmbH (Astrofein) [8–10] of Germany and the ISIPOD of Innovative Solutions In Space (ISIS) [11] of the Netherlands are PODs that are also commercially available without the need for a launch contract, in the market.

The square cross-sectioned tube-like design of the POD is intended to provide a linear trajectory for the CubeSats, thus resulting in a low spin rate following deployment. A CubeSat is deployed from the POD by the push of a compressed spring, thus gliding along the smooth corner rails toward exit to open space. The CubeSats are placed inside the POD, and then the lid is closed and locked using a mechanism. The release mechanism is activated by a signal sent from the LV computer, thus releasing the lock, opening the lid, and thus pushing out the CubeSats one by one with the energy of the compressed spring.

The POD material is of aluminum alloy, which protects the CubeSats mechanically and forming a Faraday cage and a grounding point. It is coated with Teflon or hard anodized, making it resistant to cold welding, and providing a smooth surface for an easy deployment. The CubeSat POD exit velocity is about $1-2\,\mathrm{m\,s^{-1}}$, which can be adjusted to meet different requirements of launch vehicle and orbit by simply replacing the ejector spring with a suitable one. Side panels of PODs provide access to the CubeSat after their placement in it. Thus, CubeSat batteries can be charged or a connection with CubeSat is established to run diagnostics. Before the integration to the LV, the ports are locked into positions.

The empty mass of the POD depends on the size and make, and given separately for each POD presented. Natural frequencies of vibration shall be all above 90 Hz. The POD, CubeSats enclosed within, is attached to the launch vehicle's upper stage usually below the main spacecraft, and usually the nanosatellites are deployed after the main payload. Overall, the basic deployers do not control the deployment conditions, only the release velocity is known.

18.3.1 POD Technical Requirements

POD development shall follow the following requirements [12]:

- PODs shall be designed and verified to the environments given in references [1, 12, 13].
- PODs shall be structurally qualified in accordance with strength qualification requirements given in references [1, 12].
- CubeSat size limitations are established as 1U, 2U, 3U, 6U, 8U, 12U, and 16U to occupy the full usable volume of a POD.
- PODs shall not violate the primary mission static and/or dynamic envelopes.

- PODs shall not affect LV avionics qualification status or architecture.
- POD shall incorporate a sensor for door position (open/closed).
- POD door release mechanism shall be designed to accept redundantly initiated signals.
- PODs shall be designed to accommodate ascent venting per ventable volume/area <2000 in. in accordance with accepted standards.
- POD shall deploy CubeSats at a velocity sufficient to prevent recontact with primary mission hardware.
- POD shall not deploy CubeSat mass simulator(s).
- PODs shall utilize industry standards for locking methodologies on all fasteners consistent with NASA-STD-6016 (https://standards.nasa.gov/training/nasa-std-6016/index.html).
- POD material shall be in accordance with NASA-STD-6016, standard materials and processes requirements for spacecraft.
- PODs shall conduct vehicle specific CubeSat separation analyses. The separation analysis shall determine the nominal and three sigma dispersion values of the impulse imparted to the LV for each CubeSat separation event to include consideration of uncertainties of separation system mechanism and CubeSat mass properties. The separation analysis shall confirm that deploying CubeSat(s) during the CubeSat separation event(s) remains within the allowable separation cone(s) as specified by the LV contractor.
- POD system shall be designed to provide a minimum of 20 dB EMI Safety Margin (EMISM) for nonexplosive actuator (NEA) circuits.
- POD system shall have a fixed base frequency greater than 120 Hz.

18.3.2 POD Testing Requirements

The POD shall be tested in a similar fashion to the CubeSat to ensure the safety and workmanship before integration with the CubeSats. Testing shall be performed to meet all launch provider requirements as well as any additional testing requirements deemed necessary to ensure the safety of the CubeSats and the POD. If the launch vehicle environment is not known, GSFC-STD-7000 may be used to derive testing requirements (http://standards.gsfc. http://nasa.gov/gsfc-std/gsfc-std-7000.pdf) [13]. The launch provider requirements supersede any derived requirements. Test pods may be employed for qualification testing of CubeSats.

18.4 State of the Art

There are a number of nanosatellite deployers [14, 15] in the market usually provided as a part of a launch contract. Some of them may be purchased as a standalone product. In this section, most of existing deployers and their properties are explained in some detail. The PODs considered are explained in detail in the following subsections.

18.4.1 P-POD

The P-POD, shown in Figure 18.1, is a standard CubeSat deployment system, which ensures all CubeSat developers conform to common physical requirements, which in turn reduces cost and development time. The P-POD plays a critical role as the interface between the launch vehicle and picosatellites. The P-POD is versatile, with a small profile and the ability to mount to different launch vehicles in a variety of configurations. The P-POD utilizes a tubular design and can hold up to 34 cm × 10 cm × 10 cm of hardware. The most common configuration is three CubeSats of equal size (typically 10 cm cubes with a mass of less than 1.33 kg); however, the capability exists to integrate picosatellites of different lengths. The tubular design creates a predictable linear trajectory for the picosatellites resulting in a low spin rate upon deployment. The satellites are deployed from the P-POD by means of a spring and glide along smooth flat rails as they exit the P-POD. After a signal is sent from the launch vehicle, a spring-loaded door is opened, and the picosatellites are deployed by the main spring.

The P-POD material is of Aluminum 7075-T73, which protects the CubeSats mechanically and forming a Faraday cage and a grounding point. It is coated with Teflon, making it resistant to cold welding, and providing a smooth surface for an easy deployment. The CubeSat P-POD exit velocity is about 1.6 m s^{-1}, which can be adjusted to meet different LV and orbit requirements by simply replacing the ejector spring with a suitable one. Side panels of P-PODs provide access to the CubeSat after their placement in it. Thus, CubeSat batteries can be charged or a connection with CubeSat is established to run diagnostics. Before the integration to the LV, the ports are locked into positions. The empty mass of the P-POD is 2.25 kg leading to a maximum mass of 5.25 kg just before deployment. Natural frequencies of vibration are all above 90 Hz, a value usually required by most LVs, at 180, 360, 700, and 920 Hz. The P-POD, CubeSats enclosed within, is attached to the launch vehicle's upper stage usually below the main spacecraft, and deployed after the main payload, Figure 18.1 [16]. New variants of P-POD hosting CubeSats of various sizes are currently being developed and used by Tyvak Inc. (https://www.tyvak.com/).

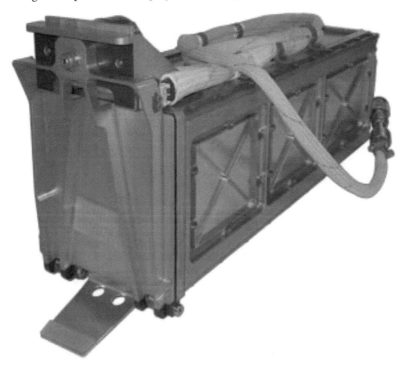

Figure 18.1 P-POD of Cal Poly, Mark III (Alodined, Rev.E) [16].Source: Reproduced with permission from ESA.

18.4.2 T-POD

T-POD is the University of Tokyo's CubeSat separation system. The T-POD v1.7, which holds just a 1U CubeSat, was used to successfully deploy the XI-IV picosatellite in June 2003. A version of the T-POD was also developed by University of Toronto, Institute for Aerospace Studies/Space Flight Laboratory (UTIAS/SFL) in collaboration with the University of Tokyo, which included an upgrade to the electronics design, an improved separation mechanism, and lower friction ejection rails. The T-POD v1.7 was tested thoroughly at UTIAS/SFL, European Space Research and Technology Centre (ESTEC), and the launch site prior to launch on 27 October 2005 aboard a Kosmos-3M rocket from Plesetsk, Russia. The current T-POD specifications as provided by Professor Nakasuka of University of Tokyo of Japan are given in Table 18.1. The T-POD is shown in Figure 18.2 [4].

Table 18.1 T-POD specifications.

Mass	3600 ± 50 g (including CubeSat UT)
Material	Aluminum alloy 5052 (primary)
Size (before separation)	250 mm(W) $\times 170$ mm(D) $\times 200$ mm(H)
Size (during separation)	250 mm(W) $\times 330$ mm(D) $\times 350$ mm(H) (maximum)
Separation speed of CubeSat UT	90 ± 10 cm s^{-1}
Waiting time after the separation signal	70 s
The error of separation direction	$\leq \pm 3.45°$
Dimension error	± 2 mm

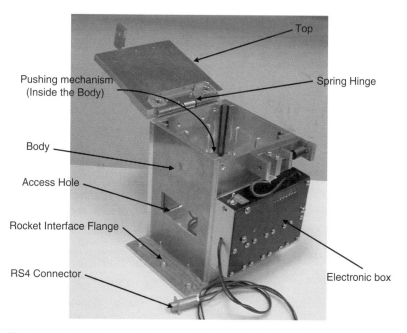

Figure 18.2 T-POD separation system. The CubeSat is released from T-POD into the space with the separation speed of 90 ± 10 cm s^{-1} [4].

18.4.3 XPOD Separation System

The XPOD family as shown in Figure 18.3 [5, 7] is a custom-made nanosatellite separation system designed and built at the Space Flight Laboratory of University of Toronto, UTIAS. The XPOD is an enclosed "jack-in-the-box" container for separating nanosatellite from virtually any launch vehicle. The XPOD separation system was developed through the SFL Can-X "nanosatellite program." Its purpose is to secure the satellite during the extreme conditions of the launch environment. In addition, it serves as the interface between the satellite and the launch vehicle and deploys the satellite upon reaching the target orbit [7]. Once a deployment signal is received from the launch vehicle, a power supply inside the XPOD activates a release mechanism causing a door to open and the spacecraft to be ejected. XPOD stands for either "eXoadaptable PyrOless Deployer" or "eXperimental Push-Out Deployer." Various models are available, including models compatible with the Cal Poly CubeSat standard. The "standard" XPODs include the XPOD Single (1U), XPOD Double (2U), and XPOD Triple (3U). Typical launch configuration is one spacecraft per XPOD. XPOD Generic Nanosatellite Bus (GNB) (for $20\,cm \times 20\,cm \times 20\,cm$, 7.5 kg, with fixed appendages) and XPOD Duo ($20\,cm \times 20\,cm \times 40\,cm$, 15 kg, with fixed appendages) are used by UTIAS for its own missions.

Figure 18.3 Various XPODs [5, 7].

Recently, an XPOD called H27, which accepts $27\,cm \times 27\,cm \times 27\,cm$, 10 kg spacecraft, with fixed appendages, is qualified. The XPOD technology implements a single-failure fail-operational design and is customizable for different spacecraft with size up to 15 kg and different launch vehicles.

Semi-enclosed (or "open concept") designs that accommodate fixed appendages are also available. Recently, six XPODs flew successfully aboard PSLV-C9 including the XPOD Single, XPOD Triple, and XPOD GNB models (Figures 18.4 [7] and 18.5 [17]). The XPOD family deployment systems have successfully ejected over 22 nanosatellites into low Earth orbit (LEO). Specifications of two XPOD models are summarized in Table 18.2 [5]. Additional models and customized units are also available.

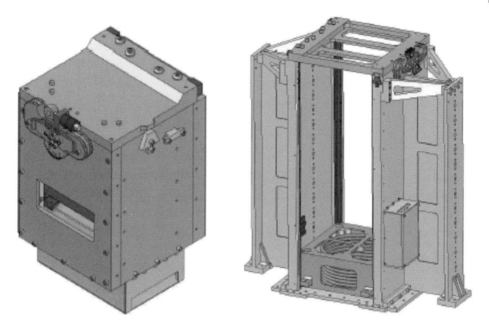

Figure 18.4 XPOD family of separation systems. Left: Single; right: DUO [7]. Source: Reproduced with permission of F.Pranajaya from CASI Astro 2012 conference, Quebec, Canada, 2012.

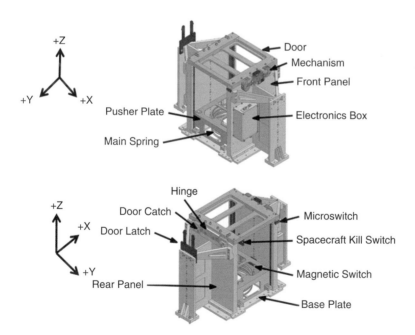

Figure 18.5 XPOD ($20 \times 20 \times 20$ cm^3) [17].

Table 18.2 XPOD Triple and XPOD DUO Specifications.

Dimensions and mass	XPOD Triple	$42 \times 17.5 \times 15 \, cm^3$, 3.5 kg w/o spacecraft
	XPOD DUO	$47 \times 47 \times 52 \, cm^3$, 10 kg w/o spacecraft
Accommodation	XPOD Triple	$10 \times 10 \times 34 \, cm^3$, 3.5 kg spacecraft
	XPOD DUO	$20 \times 20 \times 40 \, cm^3$, 14 kg spacecraft
Release mechanism		Clamp-type with Vectran tie-down
Deployment signal		28 V, 1.4 A, 500 ms
Telemetry		Door status, deployment status
Mechanical interface		Customizable per LV requirement
Dimensions and mass	XPOD Triple	$42 \times 17.5 \times 15 \, cm^3$, 3.5 kg w/o spacecraft
	XPOD DUO	$47 \times 47 \times 52 \, cm^3$, 10 kg w/o spacecraft
Temperature, operational		−35 to +65°C
Temperature, survival		−35 to +65°C
Vibration		12.9 gRMS
Natural frequency		>100 Hz
Miscellaneous		The XPOD power supply requires charge top off every 30 d

18.4.4 ISIPOD CubeSat Deployers

The ISIPOD is an affordable European launch adapter developed by ISIS [11] for use with its ISILaunch to accommodate CubeSats on board a large variety of launch vehicles. Yet, it can also be purchased separately in case a CubeSat developer has arranged a launch independently. By design, the ISIPOD provides simple, well-defined interfaces with the CubeSats internally and with the launch vehicle externally. During launch, the CubeSats are fully enclosed by the ISIPOD and are only dispensed upon signal by the launch vehicle. The ISIPOD is compatible with most launch vehicle pyro pulse signals to operate its electrical actuator. This is a nonexplosive device, which is testable, resettable, and reusable without refurbishments.

The ISIPOD is also compatible with the CubeSat standard and available at various form factors, up to 16U. It uses an electrical actuator and compatible with most LVs. Figure 18.11 presents the 1U, 2U, and 3U ISIPODs. Moreover, shared configuration is possible. Custom sizes from 0.5U to 16U in length are developed on request. The general features of ISIPOD are:

- Provides deployment status signal.
- No battery ("unlimited shelf life").
- No pyrotechnics.
- No export restrictions.
- Protects the CubeSats from external environment.
- Activated by launch vehicle.
- Extended inner envelope for deployables compared to other CubeSat launch adapters.

ISIPOD mechanically interfaces with CubeSats by means of guiderails, and with the launch vehicle by means of standard fasteners. Electrically, it interfaces with the launch vehicle for command and telemetry. ISIPOD adheres to latest CubeSat standards. The 3U ISIPOD CubeSat deployer is shown in Figure 18.6a,b. The ISIPOD details are depicted in Figure 18.7 and its specifications are presented in Table 18.3 [11].

(a)

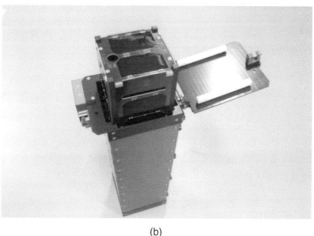

(b)

Figure 18.6 (a) 3U ISIPOD. (b) 3U ISIPOD with TurkSat-3USat [18].

The ISIPOD is qualified for multitude of launch vehicles, such as Long March, Dnepr, and PSLV. The 3U ISIPOD is qualified for heavier (6 kg) than standard CubeSat payloads (up to 4 kg). The qualification temperature range is between −40 and +90°C. The tests performed are functional, vibration, mechanical shock, thermal cycling, and thermal vacuum

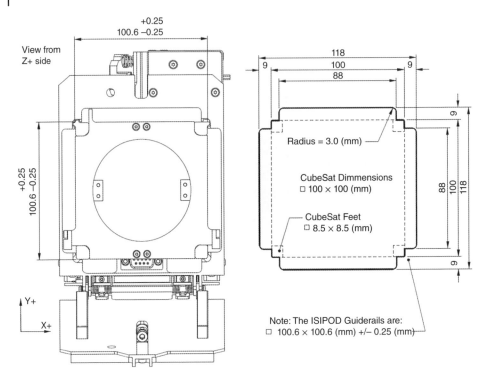

Figure 18.7 ISIPOD CubeSat deployer details.

Table 18.3 ISIPOD specifications.

	1U	2U	3U
ISIPOD mass	1.5 kg	1.75 kg	2 kg
Payload mass (typical)	1.0 kg	2.0 kg	3 kg
Payload mass (maximum)	2.0 kg	4.0 kg	6 kg
Envelope [H × W × L] (mm³)	~182 × 127 × 186	~182 × 127 × 300	~182 × 127 × 414

at qualification level on the qualification model, and functional, vibration, and thermal cycling at acceptance level on the unit to be flown.

18.4.5 QuadPack ISIS Deployer

Shown in Figure 18.8, the QuadPack [19] is a custom, reliable, nonhazardous, low shock, International Traffic in Arms Regulations (ITAR)-free, low-cost deployer. It is capable of providing flexible accommodation for a wide range of CubeSat sizes (from 1U up to 16U) and forms (not only squared shapes but also spherical). It also allows for larger extended volumes in the lateral sides and top and bottom sides (tuna-can) to accommodate extra payloads and deployable elements, such as booms and solar panels. In addition, the QuadPack is flexible with regard to the launcher vehicle's interface; it can be launched both in vertical and horizontal positions and interface through any of the lateral or back side. The Quad-Pack has an independent Hold-Down and Release Mechanism (HDRM) system for each

Figure 18.8 ISIS QuadPack deployer.

of its doors. One of the main features of the QuadPack is the so-called dynamic rail; it is able to remove the gap between the CubeSat's rails and the deployer's rails. The Quadpack also provides door hatches for data and power access to the CubeSats even after they are integrated into the QuadPack.

In addition, the QuadPack can be provided with a sequencer box, called iMDC, which is able to provide the power and signals for the deployment of each CubeSat according to a pre-established sequence, it can also retrieve all the telemetry information and download it to the ground. Figure 18.9a–h represents the different QuadPack types or configurations:

On 15 February 2017, 25 of ISIPOD QuadPack deployers were successfully used to deploy 101 CubeSats of various sizes (99 3U, 2U, and 1U) into LEO during record-breaking Indian Space Research Organisation (ISRO) PSLV-37 launch (www.isispace.nl/news/dutch-nanosatellite-company-gets-101-cubesats-launched-recordbreaking-pslv-launch).

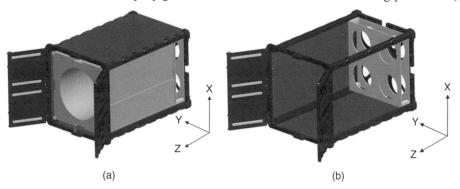

(a) (b)

Figure 18.9 (a) Type 1 with one 12U CubeSat dummy. (b) Type 1 empty internal view. (c) Type 2 with two 6U CubeSat dummies. (d) Type 2 empty internal view. (e) Type 3 with one 6U and two 3U CubeSat dummies. (f) Type 3 empty internal view. (g) Type 4 with four 3U CubeSat dummies. (h) Type 4 empty internal view.

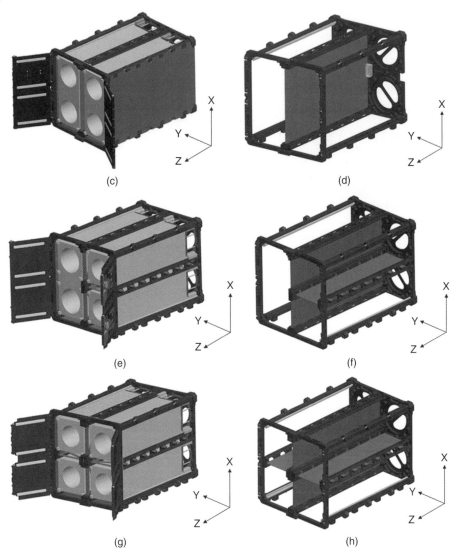

Figure 18.9 (*Continued*)

18.4.6 SPL/DPL/TPL/6U/12U of Astro- Und Feinwerktechnik Adlershof GmbH (Astrofein)

Shown in Figure 18.10, the family of PSL consists of the Single Picosatellite Launcher (SPL), the Double Picosatellite Launcher (DPL), and the Triple Picosatellite Launcher (TPL). The SPL, DPL, and TPL are designed to deploy picosatellites, which are built according to the CubeSat Design Specification (CDS). The SPL is used to deploy one 1U CubeSat. The DPL is used to deploy one 2U CubeSat or two 1U CubeSats. The TPL is used to deploy three 1U CubeSats, one 2U CubeSat, and one 1U CubeSat, or one 3U CubeSat with design specifications according to reference [1]. In Figure 18.11, main components of the PSL are shown.

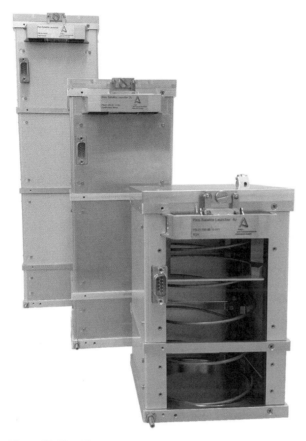

Figure 18.10 PSL of Astrofein, TPL for 3U, DPL for 2U, and SPL for 1U from back to front, respectively [8].

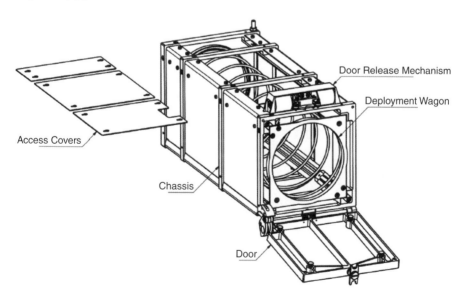

Figure 18.11 Main components of the PSL. Source: Reproduced with permission of Thomas, Hellwig from Astro- und Feinwerktechnik Adlershof GmbH, Germany, www.astrofein.com.

The 6U and 12U systems, developed recently, can deploy a combination of CubeSats of various sizes from 3U to 12U. All the information given in this section is based on the documents provided by the Astrofein [8–10].

The PSL consists of five main components, which are the chassis, the door release mechanism, the deployment wagon, the door, and the removable access covers. All components are firmly mounted onto the container in flight configuration. The chassis is the rigid framework of the PSL that serves as structure, mounting interface, and guide of the deployment wagon. The rails of the chassis are guiding the deployment wagon in a manner that no canting and twisting of this wagon is possible. This is violated, if the maximum difference between center of gravity of the CubeSat and its geometric center exceeds the values specified in reference [1]. The chassis is closed on all sides during launch to prevent any parts from leaving the PSL prior to the deployment process of the CubeSat.

The door opens approximately 100° upon the initiation of the deployment process, and is locked in this position immediately after reaching the end stop. Upon the complete opening of the door, the deployment wagon is released. Besides the opening function, the door is also used to hold the CubeSat in place during transportation and launch. Therefore, the door is equipped with four spring-loaded retainers, which are in contact to the contact faces of the

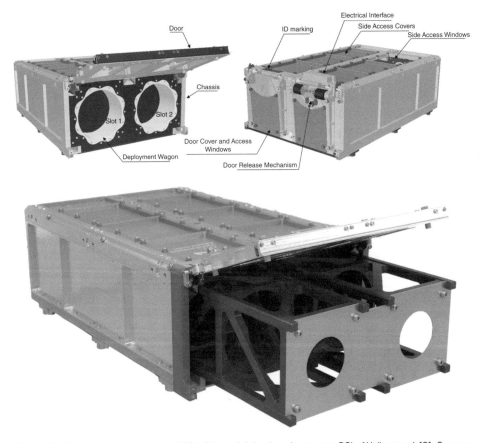

Figure 18.12 Main components of PSL-6U model (top) and two-way PSL-6U (bottom) [9]. Source: Reproduced with permission of R. Bretfeld from Astro- und Feinwerktechnik Adlershof GmbH, Germany, www.astrofein.com.

guide rails of the CubeSat at any specified environmental conditions. These spring force pins also contribute to the opening energy of the door. The door release mechanism maintains the door closed during any transportation and launch operation and initiates the deployment process upon the deployment signal. Two redundant electric permanent magnets are employed as actuators to release the door using a lever mechanism. The deployment wagon ejects the CubeSat into the open space, once the door is completely opened and locked. The wagon is guided in the chassis with roller bearings. Hence, a smooth and continuous motion with predictable deployment energy is accomplished. The deployment spring provides the force to accelerate the CubeSat and the deployment wagon during deployment process. The PSL deployment wagon is optimized for a low spin rate of the CubeSat. The access covers allow access to the CubeSat after its complete integration into the PSL for energy and data service. To provide a high reliability and ergonomic handling, the fasteners of the easy access panes are preinstalled, so that no small fasteners have to be handled on launch site.

Available 6U and 12U SPLs of Astrofein are shown in Figures 18.12 and 18.13.

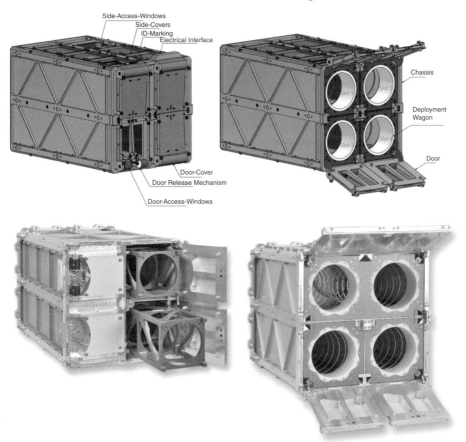

Figure 18.13 Main components of PSL-P-12U model (top) and three-way PSL-12U (bottom) [10].

18.4.7 Canisterized Satellite Dispenser (CSD)

The Canisterized Satellite Dispenser (3U, 6U, 12U, and 27U CSD) is developed by Planetary Systems Corporation [20, 21], Figure 18.14. While having similar internal

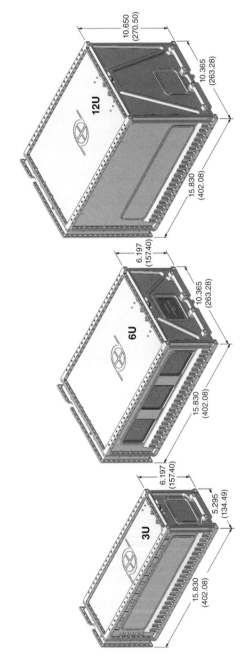

Figure 18.14 3U, 6U, and 12U CSD [21].

dimensions to the P-POD (CubeSat launcher) design, the 3U CSD has key features that provide higher payload mass capability (6 kg), tabbed preload system to guarantee a stiff and modelable load path to the CubeSats, a higher ejection velocity, lower overall volume, and mounting features to allow fastening of the CSD at any of the six faces. Moreover, it has a 15-pin in-flight disconnect allowing battery charging and communication from the outside of the CSD into the payload while in the launch pad or on orbit, and a reusability that allows a separation test to be conducted at will without any consumables hundreds of times to guarantee reliability. The CSD is qualified to levels exceeding Mil-Std-1540 for thermal vacuum, vibration, and shock [18], and has a technology readiness level (TRL) of 9. Further details of 6U CSD are shown in Figures 18.15 and 18.16. CDSs of various sizes can also be stacked on a single mounting plate, Figure 18.17.

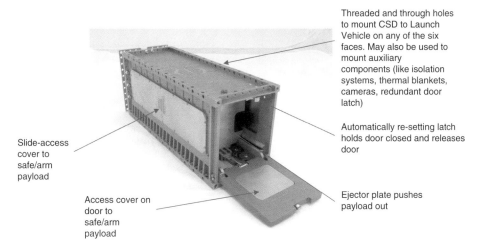

Threaded and through holes to mount CSD to Launch Vehicle on any of the six faces. May also be used to mount auxiliary components (like isolation systems, thermal blankets, cameras, redundant door latch)

Automatically re-setting latch holds door closed and releases door

Slide-access cover to safe/arm payload

Access cover on door to safe/arm payload

Ejector plate pushes payload out

Figure 18.15 3U CSD and associated features [20, 21].

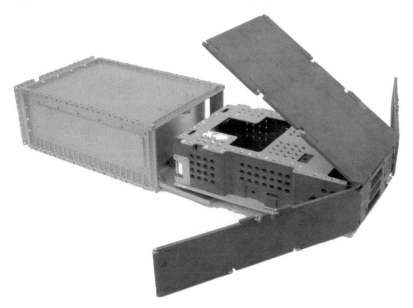

Figure 18.16 6U CSD, the CSD internal walls may constrain deployables until ejection [20, 21].

Figure 18.17 Five CSDs stacked on a single mounting plate [21].

18.4.8 JEM-Small Satellite Orbital Deployer (J-SSOD)

Japanese Experiment Module-Small Satellite Orbital Deployer (J-SSOD) of Japan Aerospace Exploration Agency (JAXA) of Japan is used to deploy CubeSats from the ISS. In October 4, 2012 five CubeSats were successfully deployed from the ISS. The small satellites were transported to the ISS in the HTV-3 (Kounotori 3) cargo vessel that blasted off on an H-IIB rocket from the Tanegashima Space Center of Japan on Saturday, July 21 at 02:06 UT. The cargo vessel arrived at the ISS on July 27 and the ISS Canadarm2 robotic arm was used to install the HTV-3 to its docking port on the Earth-facing side of the Harmony module at 14:34 UT. The CubeSats were then unloaded by the Expedition 32 crew. The CubeSats were mounted in a J-SSOD. TechEdSat, F-1 (NanoRacks), and FITSAT-1 were there in one pod, while WE-WISH and a scientific 2U CubeSat RAIKO were in the second pod, Figure 18.18. Japanese astronaut Akihiko Hoshide (http://iss.jaxa.jp/en/kiboexp/news/small_satellites_ deployment_fr.html) put the J-SSOD into an airlock, which was depressurized and exposed to the vacuum of space via an automatic door. The Kibo robotic arm was then used to grapple the J-SSOD in the airlock and move it out away from the station, so that the satellites could be deployed. Previous deployments of amateur radio satellites have only been possible when astronauts have performed an extravehicular activity (EVA). The Kibo robotic arm and the J-SSOD enabled a larger number of satellite deployments, already. Shown in Figure 18.19 is the UBAKUSAT [22] in JAXA Tsukuba Space Center before it was deployed on 11th May 2018. The NanoRacks company (http://nanoracks.com) also makes use of Kibo module to deploy CubeSats. Once deployed from ISS, the CubeSats could have a life-time of up to a couple of years before they burn up in the Earth's atmosphere. As an example, one of 32 ISS deployed QB50 CubeSat HAVELSAT [22, 23] that was deployed from Kibo module using NanoRacks deployer (shown in Figure 18.20 with BeEagleSAT inside) on May 16, 2017 re-entered earth and vanished on March 7, 2019.

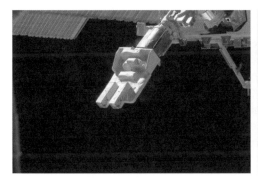

Figure 18.18   The Kibo robotic arm and the J-SSOD. Deployment of TechEdSat, F-1, and FITSAT-1 from ISS by the JAXA ground station. Source: Photo courtesy of JAXA, TSUKUBA.

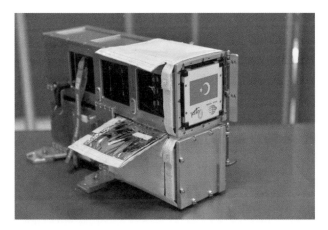

Figure 18.19   J-SSOD with UBAKUSAT inside. Source: Photo courtesy of JAXA, TSUKUBA.

Figure 18.20 BeEagleSAT in NanoRacks deployer.

18.4.9 Tokyo Tech Separation System and AxelShooter

AxelShooter [24] (Figure 18.21) is a lightweight, flexible, easy-to-customize separation mechanism especially suitable for picosatellites, nanosatellites, and microsatellites (1–30 kg class satellites). The design concept of AxelShooter is based on past separation mechanisms developed by Tokyo Institute of Technology (Tokyo Tech). Until now, Tokyo Tech has experienced three nanosatellite launches with their original separation mechanisms (CUTE-I in 2003, Figure 18.22) [25], CUTE-1.7 + APD in 2006, and CUTE-1.7 + APD II in 2008, and one demonstration mission of the separation mechanism itself in 2005. All of their missions were totally successful as far as the separation mechanisms are concerned. This flight-proven know-how is applied to AxelShooter, but several points are revised from these heritages, in order for users (satellite developers) to handle the system more easily and safely [24].

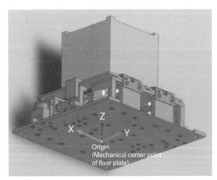
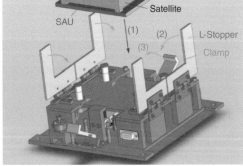

Figure 18.21 1U CubeSat AxelShooter of Axelspace of Japan.

The basic specifications of 1U AxelShooter are summarized in Table 18.4. The coordinate system is depicted in Figure 18.15. The CubeSat separation systems L-Stoppers and Clamps are also shown.

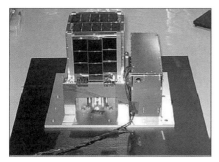

Figure 18.22   Left: CUTE-I separation mechanism; right: CUTE-1.7 + APD II separation mechanism.

Table 18.4   The basic specifications of 1U AxelShooter [24].

	With 1U CubeSat (1.5 kg)			W/o satellite		
Mass (kg)	2.99			1.49		
Dimensions [mm] $x \times y \times z$	$200 \times 200 \times 152$			$200 \times 200 \times 63.5$		
Center of gravity [mm] x, y, z	$(-0.06, -0.07, 63.42)$			$(0.05, -0.13, 24.25)$		
Moment of inertia tensor [kg mm²] (calculated around the center of gravity) $[xx\ xy\ xz$ $yx\ yy\ yz$ $zx\ zy\ zz]$	11 851.81	−310.73	2.66	4725.29	−310.71	1.78
	−310.73	12 319.20	−5.90	−310.71	5250.05	−13.75
	2.66	−5.90	12 176.10	1.78	−13.75	8883.87
Material	—			Aluminum (96%) Stainless steel (4%)		

18.5   Future Prospects

As CubeSats get larger in terms of form factors (greater than 3U), the deployers must adapt. The 6U, 12U, and 27U payloads will feature advanced technologies that streamline integration and ensure mission success. Another interesting field of application of the CubeSats is to create constellations; hence, the future deployers should be customized for the purpose of launching several tens of CubeSats at the same time in an optimal manner. Longer life of the deployers would be required for its application in the upcoming interplanetary missions. The data and power interface between the CubeSats and the main platforms can be improved to perform activities before deployment of more complex missions. Eventually, the concept of deployer should evolve into an autonomous container with solar panels, Attitude and Orbit Control System (AOCS), telemetry, tracking and command (TTC), and propulsion capabilities more in the direction of a stand-alone spacecraft.

Acknowledgments

The information and documentation provided by Professor Shinichi Nakasuka of University of Tokyo, Japan, Ms. Rei Kawashima of UNISEC and Mr. Nojiri of Axelspace of Japan, Freddy Pranajaya of University of Toronto, Canada, Karin Ulrich of Astro- und Feinwerktechnik Adlershof GmbH of Germany, and Mike Whalen of Planetary Systems Corporation for the preparation of the present notes is gratefully acknowledged.

References

1 "CubeSat Design Specification Revision 13," California Polytechnic State University, 2015. http://www.cubesat.org/images/developers/cds_rev13.pdf

2 Puig-Suari, J., Turner, C., and Ahlgren, W. (2001). Development of the standard CubeSat deployer and a CubeSat class Picosatellite. In: *2001 IEEE Aerospace Conference Proceedings (Cat. No. 01TH8542)*, vol. 1, 1/347–1/353.

3 R. Nugent, R. Munakata, A. Chin, R. Coelho and J.Puig-Suari, 2008. "The CubeSat: The Picosatellite Standard for Research and Education", AIAA Space 2008 Conference and Exposition, San Diego, CA, AIAA paper 2008–7734.

4 T-POD 2015. Personal communication with Prof. Shinichi Nakasuka of University of Tokyo, Japan, January 2015.

5 UTIAS SFL: University of Toronto Institute for Aerospace Studies Space Flight Laboratory, "*Webpage:* http://utias-sfl.net"

6 Mohamed R. Ali, 2009. "Design and Implementation of Ground Support Equipment for characterizing the Performance of XPOD and CNAPS & Thermal Analysis of CNAPS Pressure Regulator Valve," MSc Thesis, Graduate Department of Aerospace Engineering, University of Toronto.

7 F. Pranajaya, 2012. "Nanosatellite Launch Services Past, Present and Future Launches", 1st Canadian Nanosatellite Workshop, the CASI Astro 2012 conference, Quebec, Canada.

8 R. Bretfeld, (2014). "Technical Specification, PicoSatellite Launcher (PSL)", Document PSL-SP01 Issue 1–0, 01.10.2014, Astro- und Feinwerktechnik Adlershof GmbH, Germany, www.astrofein.com

9 Thomas, Hellwig, 2018. "Technical Specification, PicoSatellite Launcher Pack (PSL-P-6U)", Document PSL-6U-SP001, 05.03.2018, Astro- und Feinwerktechnik Adlershof GmbH, Germany, www.astrofein.com

10 R. Bretfeld, 2018. "Technical Specification, PicoSatellite Launcher Pack (PSL-P-12U)", Document PSL-P-12U-SP001, 03.05.2018, Astro- und Feinwerktechnik Adlershof GmbH, Germany, www.astrofein.com

11 ISIS: Innovative Solutions In Space B.V., "*Webpage:* http://www.isispace.nl"

12 Launch Services Program, 2011. Program Level Poly-Picosatellite Orbital Deployer (P-POD) and CubeSat Requirements, LSP-REQ-317.01 Revision A, National Aeronautics and Space Administration John F. Kennedy Space Center, Florida Launch Services Program, October 2011.

13 *"General Environmental Verification Standard (GEVS) For GSFC Flight Programs and Projects (GSFC-STD-7000A)",* NASA Goddard Space Flight Center, 2014.

14 Aslan, A. R., 2013. "CubeSat Deployers and Deployment Systems", VKI Lecture Series on "CubeSat Technology and Applications", 29 January–1 February 2013, von Karman Institute, Belgium.

15 *"Small Spacecraft Technology State of the Art",* NASA/TP–2015–216648/REV1, *Mission Design Division Staff, 2015. NASA Ames Research Center, Moffett Field, 94035 California, December.*

16 Poly Picosatellite Orbital Deployer Mk. III Rev. E, User Guide (CP-PPODUG-1.0-1), 2014, www.cubesat.org (accessed on 11.07.2019).

17 Michael Christopher Ligori, 2011. "Next Generation Nanosatellite Systems: Mechanical Analysis and Test", MSc Thesis, Graduate Department of Engineering Science, University of Toronto.

18 A. R. Aslan, et.al., 2013. "Development of a LEO Communication CubeSat", 6th International Conference on Recent Advances in Space Technologies—RAST2013, Istanbul, 12–14 June 2013.

19 M. van Bolhuis and C. Bernal, 2012. "QB50 Deployment System", 3rd QB50 Workshop, von Karman Institute, Sint-Genesius-Rode, Belgium, 2 February 2012.

20 W. Holemans, R. G. Moore, and J. Kang, 2012. "Counting Down to The Launch of POPACS (Polar Orbiting Passive Atmospheric Calibration Spheres)", SSC12-X-3, 26th Annual AIAA/USU, Conference on Small Satellites.

21 Canisterized Satellite Dispenser (CSD) Data Sheet, 2018. 2002337F, 3 August 2018. www.planetarysys.com.

22 A. R. Aslan, (2018). "TÜ-SSDTL Contributions to National and International Space Technology Development and Capacity Building with CubeSat and CanSat", United Nations/Brazil Symposium on Basic Space Technology "Creating Novel Opportunities with Small Satellite Space Missions, 11–14.09.2018, NATAL, BRAZIL.

23 Aslan, A. R., Karabulut, B., Uludağ, M. Ş., et al., 2016. BEEAGLESAT and HAVELSAT: Two 2U CubeSats for QB50, The 7th Nanosatellite symposium, Deorbit Device Competition, the 4th Mission Idea Contest, and the 4th UNISEC-Global Meeting, Kamchia, Varna, Bulgaria, 18–23 October 2016.

24 AxelShooter 2013, personal communication with Axelspace, Japan, January 2013.

25 http://lss.mes.titech.ac.jp/ssp/cute1.7/index_e.html, "Cute-1.7 + APD II FM & Separation System".

19 I-4c

Mission Operations
Chantal Cappelletti

Faculty of Engineering, University of Nottingham, United Kingdom

19.1   Introduction

The term "mission operations" is used to refer to all activities required to operate a space-craft and its payloads. Thus, mission operations are strictly related to mission success, which can be defined as the achievement of mission goals, or in other terms, can be considered as quantity, quality, and availability of delivered mission products and services within a given cost envelope.

To achieve mission success, it is essential to have a clear definition of the mission goals since the beginning. What is necessary to perform, how to operate the satellite, how much data per week should be downloaded, and so on are requirements that should be well defined since the first phases of the mission design. A wrong definition of goals will result in increased mission costs, time, and complexity, and in reduced chances of success of mission.

Designing and launching a satellite mission is a long and complex process that covers the definition, design, production, verification, validation, postlaunch operations, and postop-erational activities, involving both ground segment and space segment elements. To ensure mission success, it is not enough just to complete all these single phases with success. More than that, it is necessary to have a close link between all the mission design phases and between segments. The mission operator is the key person to ensure the connection between the different segment teams.

Considering this viewpoint, ground systems and operations are key elements of a space system and play an essential role in achieving mission success. It is the duty of satellite operators to participate in all the mission phases to ensure that the mission goals, in terms of operations, will not be sacrificed or modified during the mission phases. Only by attending all the process phases, the operator can support the team suggesting the best way to proceed and to ensure that the operations in space will be successfully performed.

In this chapter, an introduction to mission operations, rules, and functions is given, emphasizing the particular aspects that concern nanosatellite operations.

Nanosatellites: Space and Ground Technologies, Operations and Economics, First Edition.
Edited by Rogerio Atem de Carvalho, Jaime Estela, and Martin Langer.

19.2 Organization of Mission Operations

Following the definition of Consultative Committee for Space Data Systems (CCSDS) [1], mission operations include:

- Monitoring and control of the spacecraft subsystems and payloads.
- Spacecraft performance analysis and reporting.
- Planning, scheduling, and execution of mission operations.
- Orbit and attitude determination, prediction and maneuver preparation.
- Management of on-board software (load and dump).
- Delivery of mission data products.

These activities are typical functions of the Mission Control Center (MCC) and are performed by operations teams during the flight phase of the mission.

In the same time, the operations team performs design activities of prelaunch operations, including development of the mission operations concept, policies, data flows, training plans, staffing plans, cost estimates, and capability to archive and distribute mission operations data.

The operations team composition varies considering the mission, needs, and resources available. Sometime during the development of the mission, it is common to see a variation of the team composition due to different factors. For example, for nanosatellite missions, the limited budget or the natural turnover of people working in the project can cause changes in the team. In a university satellite mission, it is normal that students, after finishing their courses, leave the team, and normally the project is not scheduled in accordance to student life. In addition, nanosatellites are still a small business in aerospace and people frequently prefer to quit the project to follow better opportunities. These factors should be considered during the operations team establishment to guarantee the project continuity and success. A well-defined documentation of all activities can help to substitute people without the risk of losing all the knowledge achieved by the team.

The operation team can be composed by:

- One or more operations managers
- Spacecraft operators
- Payload operators
- Mission planners
- Flight dynamics engineers
- Ground system operators
- Mission exploitation personnel
- Ground system maintenance engineers

In nanosatellite missions, where normally budget and personnel are extremely reduced, one person can have different responsibilities. Thus, for example, he or she might be the ground system operator and the ground system maintenance engineer at the same time. In order to reduce the costs, it is important to reduce the number of people employed in the operations phases. To do that, it is important that from the beginning of the mission design, the operation team must be involved in the decision phases, in order to suggest solutions to avoid possible problems during the in-orbit phases.

In nanosatellite missions, the operations manager is normally the more experienced person in the team, able to have an overall picture of all the mission phases and all the possible

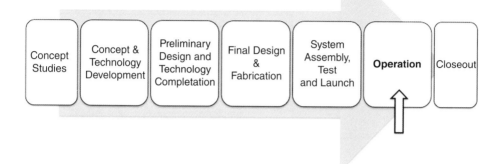

Figure 19.1 Phases of satellite missions.

failures in the missions or in the process phases. The operations manager is the person that is in charge of giving suggestions and supervises the overall project.

The main activities of this team concern the postflight mission operations, but it is extremely important to involve them since the beginning of the project. As shown in Figure 19.1, the operation phase is immediately after the launch, but work in mission operations should not be limited to this phase. Mission operations are not restricted to the activities related to the operations planning, execution, and evaluation of the combined space segment and ground segment during last phases (E and F for European Cooperation for Space Standardization (ECSS) standard) of a space project, but must be included in all the mission design phases, with different rules and experts involved.

During the design phase, mission operators can support the design team suggesting strategies and solutions. It has been studied and proved that by involving the operation team from the beginning of the design, it is possible, depending on the experience of the mission's operations team, to optimize the resources, reducing the costs, and increasing the mission reliability. Experienced mission operators know what will occur after launch and what could be the possible risks for the mission.

The mission operator's approach is substantially different from that of the mission designer. During the design, the engineer is more focused on having the system working. Contrary to that, the operator puts his attention on what can fail during the mission. This different approach changes completely the results of investigations. The cooperation between these two mission experts and the capability to share different points of view result in an increase of mission success rate.

19.3 Goals and Functions of Mission Operations

Ground system engineering could be divided in two major engineering disciplines, operations engineering and ground segment engineering, which are separated in the process chain, but with many links between them starting at the definition/requirements phase. Both engineering disciplines contribute to the definition of requirements and on the design review of the space segment. Operations engineering is related to all the functions that are

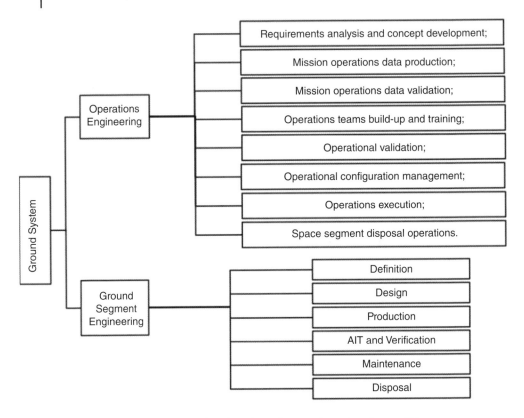

Figure 19.2 Ground system engineering phases in accordance to reference [2].

needed to ensure the preparation and execution of the mission operations, and it is performed from the preliminary mission design phase onward. Ground segment engineering is more related to the ground segment design and maintenance itself as part of the overall project. In Figure 19.2, a flow chart shows the ground system engineering in the two subgroups and their main roles.

In other literature, the mission operations system is divided into the functions that the operators should carry out in the mission design and development. Functions can be shared between the engineering segments since there are many links between them.

In the two following subparagraphs, the operations' functions are described as suggested by reference [3]. This approach is commonly used in the USA and it is different from the engineering separation approach described in the ECSS standard (see reference [2]). Essentially, the two approaches capture the same: space operators should be integrated in all the mission phases, playing a fundamental rule on the mission design, development approach, and final phases, and hardware, software, people, and procedures operate together to complete them [4]. Automating some of these functions can lead to lower operations costs and, in most cases, lower life-cycle costs. Analyzing the Wertz approach in reference [3], mission operations functions can be divided in two groups: mission database operations functions and mission operations support functions.

19.3.1 Mission Database Operations Functions

The operation engineering functions consist of the data share through the mission database and the space elements' avionics. This group is divided in nine different functions:

1. Mission planning
2. Activity planning and development
3. Mission control
4. Data transport and delivery
5. Navigation and orbit control
6. Spacecraft operations
7. Payloads operations
8. Data processing
9. Archiving and maintaining mission database

Sometimes, particularly in nanosatellite missions, two or more functions can be merged considering mission goals and budget.

19.3.2 Mission Operations Support Functions

The role of operations team is not strictly restricted to the postlaunch phases. The operators should support all the mission phases in order to guarantee the success of mission operations.

The reliability of mission operations does not depend on the operations phases itself but on how the operations have been designed and planned and how all the subsystems work together in order to accomplish the mission goals.

A satellite can be considered as an orchestra where all musicians, the subsystems, work together in order to reach the perfect music execution, the mission. Every musician should follow the instructions of the orchestra maestro, as every satellite subsystem should be designed following the suggestions of the mission operator.

In the mission, system engineering can be divided in four main mission operations support functions:

1. System engineering, integration, and test
2. Computer and communications support
3. Developing and maintaining software
4. Managing mission operations

The four functions have a temporal sequence (see Figure 19.3). At the beginning, the operators should work in the system engineering design, integration, and tests. In this phase, the main role of the mission operators is to give suggestions and recommendations to the team to ensure that the mission operations concepts will be maintained. The second function is related to computer and communications, and the operator has a more active role. He should plan the computer and communications infrastructure maintaining the system working. Mission operators are also in charge to develop and maintain the operations software, updating it periodically, and also consider the results of test campaign. In this phase, the operator should ensure the functionality of all the subsystems. Last but not the least is the mission operations management that is performed after the launch. In this

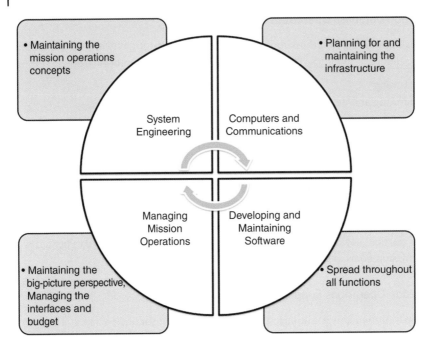

Figure 19.3 Ground segment engineering: mission operations support functions.

phase, considering the knowledge developed during the participation to all the mission phases, the operators should be able to manage the satellite in every kind of situation and be able to prevent failures and invertible damages to the space segment.

19.4 Input and Output of Mission Operations

In space mission design, documentation is one of the key points to ensure continuity and increase the link between different mission segments. Each mission phase should have a clear definition of what is the reference document and what documents should be defined in that step. The same rule is applied to mission operations: operations team should produce a set of data and information that is collected in different dedicated documents.

The document considered essential for the mission operations is the mission operations plan (MOP), described in detail in Section 19.5. The ECSS standards [2] indicate a series of reference documents for MOP and other operations' document output, illustrated in Figure 19.4. The principal ones are the mission analysis report (MAR), the mission operations concept document (MOCD), and space segment user manual (SSUM).

19.4.1 MAR

The objective of the MAR is to provide all the information on the mission characteristics, in particular its orbit and attitude, in sufficient detail for the planning and execution of mission operations. The MAR also provides input to the spacecraft and mission design.

Mission Operations Data Production Process

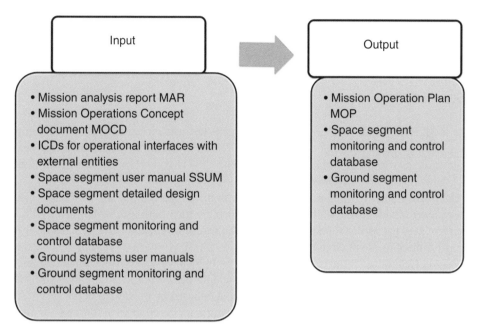

Input	Output
• Mission analysis report MAR • Mission Operations Concept document MOCD • ICDs for operational interfaces with external entities • Space segment user manual SSUM • Space segment detailed design documents • Space segment monitoring and control database • Ground systems user manuals • Ground segment monitoring and control database	• Mission Operation Plan MOP • Space segment monitoring and control database • Ground segment monitoring and control database

Figure 19.4 The Mission Operations Data Production Process.

The MAR shall identify all pertinent requirements and constraints arising from the following:

• The mission definition, including the scientific objectives.
• The spacecraft platform and payload design.
• Data recovery and data circulation.
• The selected launcher.
• Orbit and ground station network.

19.4.2 MOCD

The objective of the MOCD is to define the operations concepts for the various phases of the mission, covering both the space segment and the ground segment. The MOCD shall describe the mission operations concepts for each distinct mission phase, also taking into account that not all these processes are applicable for each mission phase and the corresponding process can therefore be omitted.

In order to keep the mission costs low in this phase, the possibility to reuse existing operational capabilities and services should also be identified, possibly modified or extended for the mission.

19.4.3 SSUM

The objective of the space SSUM is to provide the space segment design, operational information, and data that are used by the ground segment supplier and operations supplier to prepare and implement mission operations. The operation team normally participates in the editing of this document. In nanosatellite missions, since normally the principal user is the developer itself, this document can be skipped, but it is extremely important to remind that sensible contents such as mission objectives and operational information should be written clearly in the MOP.

19.5 MOP

The MOP is the fundamental reference document for mission operations. It acts as trait d'union between users and operators. For this reason, it should be written in a way to be understandable for different kinds of readers. Thus, the jargons of computer scientists and programmers should be avoided, if possible.

The work on the MOP begins significantly prior to the spacecraft launch. Mission objectives trigger the identification of key tasks, which are then built into a MOP. This plan is then reviewed to ensure that it is compatible with spacecraft capabilities and that appropriate safety margins are maintained.

The MOP contains all the rules, procedures, and timelines necessary to implement mission operations during all mission phases. It should be written in accordance with the mission objectives and respecting any constraints imposed by the design of the space and ground segments. It also contains rules and procedures for the conduct of contingency operations.

The MOP's introduction shall list the applicable and reference documents in support of the generation of the document and provides a summary of the mission, the ground segment, and the overall operations concept. It shall also present the overall MOP structure and rationale.

In reference [2], a set of rules to write a MOP is given. In particular, the MOP shall contain the rules and criteria governing the conduct of mission operations, including:

- Launch hold criteria.
- Principles and mechanisms for the routine and foreseen contingency operations of the mission.
- Rules and criteria for emergencies and anomalies not covered by existing contingency procedures.
- Rules and criteria for transferring operations between different authorities (where applicable).
- Mission termination criteria.

The MOP shall present the organization, responsibilities, and location of the parties and teams involved in the mission operations during each mission phase, as for example control centers or control rooms.

To foster the communication between the different teams, a common practice is to provide an organizational chart for each operational team, including the names and positions of everyone.

19.5.1 Suggestions to Write a MOP

Several standards and techniques can be used to write a correct MOP. Following Larson suggestions in reference [3], the MOP divides the development in 13 steps:

1. *Identify the mission concept, supporting architecture, and key performance requirements*: The first step is dedicated to the definitions of mission requirement and performances. This is the most critical phase. A clear definition of the mission requirements and constraints, since the beginning of mission design, results in cost reduction and in increase of mission success. A wrong definition of mission requirements will drastically affect the mission's cost and complexity and time schedule.

   In analogy with the "five Ws," five points are identified that should be analyzed during this first step:

   • Mission objectives: What to do with the satellite?
   • Mission meaning: Why this mission must be achieved?
   • Mission timeline: When should each system be developed?
   • Mission users: Who will be the payload data users?
   • Mission orbit: Where will the mission be exploited?

   In particular, the mission description gives information about the trajectory, launch dates and windows, trajectory profile, maneuver profile needed to meet mission objectives, mission phases, and the activities required during each phase.

2. *Determine scope of functions needed for mission operations*: In this step, the mission is divided into discrete, workable phases, such as launch, early orbit, normal operations, entry, descent, and landing. Each workable phase concurs to a specified function with distinct goals, objectives, and operational requirements. It is extremely important in this phase to take into account the other mission documents such as Interface Control Document's (ICD) or MAR in order to understand characteristics of the spacecraft bus, the payload, and the overall mission, and identify correctly the goals and objectives of each workable phase.

3. *Identify ways to accomplish functions and whether capability exists or must be developed*: Responsibilities are identified and defined for each function, dividing them in three blocks:

   • On-board autonomy
   • Ground-based operators
   • Automated functions

   The decision taken during this phase will affect the costs and operability of the mission (as described in Section 19.6). Exploiting the functions, we should be clearly aware of system capabilities on board and on ground, as well as degree of automation on the ground and degree of autonomy on the spacecraft. It is also important to know if there are possibilities for reuse of flight crew and software.

4. *Do trades for items identified in the previous step*: The options previously identified in step 3 drive either performance or cost. In step 4, a small group of operators, crew, and designers need to do trades and decide how to carry them out, looking for approaches

that truly minimize life-cycle cost without jeopardizing the safety and reliability of the mission and systems.

5. *Develop operational scenarios and flight techniques*: The mission operations are typically divided in three types of scenarios:
   - User scenario: How the user interacts with the system elements and receives data.
   - System scenario: How systems and subsystems within an element work together.
   - Element scenario: How the elements of the space mission architecture work together to accomplish the mission.

   The user scenario is developed with the help of the user itself. During the early study phases of a mission, users show how they want to acquire data and receive products from the payload. In a science mission, the user is the principal investigator or science group or an individual observer.

   The system scenario should be developed in detail after the definition of the operations architecture. The preliminary steps needed to conduct the mission are identified. The overall description of system scenario with more detailed information and subsystems will be given during element design. Once scenarios for the user system and element are created, they must be integrated to understand what happens throughout the mission and include all key activities and eliminated overlaps. The scenarios give us our first look at how our system operates as a unit to produce the mission data.

6. *Develop timeliness for each scenario*: After the definitions of all scenarios, times needed will be added to do each set of steps and determine which steps can run in parallel or have to be in series. Here events and their frequency of occurrence are identified, and which organization is responsible for them.

7. *Determine the resource*: For each step of each scenario, resources must be assigned in terms of machines or people. The main constraints to decide the resource are performance requirements and available technology. The idea is toward automation on the ground and autonomy in space taking into account the safety and reliability of each system. This phase is extremely sensitive because if on one hand the system automation can be increased, on the other hand mission risks must not be increased.

8. *Develop data-flow diagrams*: A data-flow diagram is a tool that helps understanding the origin and flow of data throughout the system to the end user or customer and to the archive. Each diagram shows processes, points toward data storage, and inter-relationships. To reduce overall life-cycle cost is important to have a clear picture of data-flow in order to modify it in a safe manner maintaining the same mission reliability level.

9. *Characterize responsibilities of each team*: Another key point to maintain the mission cost is the team establishment and responsibilities. Different steps have been assigned to different operating teams. The goal of this step is to identify the number of people required for each mission operation phase, their responsibilities, and how costs can be reduced involving the same team in different phases or decreasing the number of people in each team.

10. *Asses mission utility, complexity, and operations cost drivers*: To each operational activity, a level of complexity that can be low, medium, or high is assigned. Considering this

classification, the number of operators for each activity can be defined to generate a model based on previous missions of the same class. During design, this model gives us rules for trade studies to reduce operations costs. However, personnel reduction should be avoided in highly complex operational activities to ensure the mission's reliability.

11. *Identify derived requirements*: The MOP should identify new or derived requirements necessary to reduce cost and complexity and enhance the safety and reliability of operations. Here, the derived requirements are only identified and no information about how to implement them is given.

12. *Generate a technology development plan*: The MOP should also identify all the technology used to support the mission operations concept. This technology may not exist or may be focused and prototyped appropriately for mission approval. Here, it is important to identify needed or risky technology.

13. *Iterate and document*: The last MOP development step is to document the results of the iteration through the MOP. The documentation should at least include requirements and mission objectives, key constraints, scenarios, timelines for each key scenario, ground and flight crew tasks, hardware and software functions, responsibilities and structure of organization and team, data-flow diagrams, and payload and derived requirements.

19.6 Costs and Operations

Mission success is strictly related to estimation and management of mission cost. Available budget drives design and development decisions, leading to a successful mission or failure. This is something common to all satellite missions, but it is more tangible in nanosatellite projects where, due to the limited budget, every single acquisition can have a fundamental importance for the future mission development.

This consideration should be very clear to the mission operators who should help the team and safeguard the satellite operations from preliminary mission design phases onward. The operations manager has a double role to ensure mission success in a cost-effective manner: on one side, he should help the rest of the team during design and mission development to take the right decisions, maintaining expenses under control. On the other side, he should minimize mission operations costs.

Some rules can help the operators to take the right decisions, limiting the mission costs and increasing the mission success, such as:

1. Define mission requirements and constraints in detail.
2. Participate in all mission design phases.
3. Avoid complexity.
4. Design for operability.
5. Provide operation description document.
6. Share knowledge.
7. Use standard solutions for equipment and test procedures.
8. Perform tests.
9. Use existing facilities.

1. *Define requirements and constraints in detail*: A clear definition of mission requirements and constraints results in a strong knowledge of what the mission should perform. This is a first point to ensure that time, and money, is not wasted. In this phase, it is important to identify and fix the goals and what is expected from the satellite. The mission costs stay under control if changes are not applied later, affecting the decision taken in this phase. Operators should participate in the definition phase ensuring that the choices made are feasible from the operations point of view. As understood, it is also important to determine the payload requirements, such as how, and how often, data are required. This is a fundamental point to identify key tasks to operate the payload successfully.

   To satisfy this requirement, the use of mission planning standardization illustrated by several institutions (NASA, ESA, JAXA, etc.) can help the team to not incur in errors and to avoid falling in never-ending decision loops.

2. *Participate in all project design phases*: As already said, the operators should participate in the mission's payloads and spacecraft design phases ensuring that the decisions are taken considering how they will affect the operations phases. Operations often become expensive when these design decisions are taken first, without considering how they will affect operations. Instead, the operations concept should be developed concurrently with other elements, so that the best overall approach can be selected based on affordability and life-cycle cost.

   Simple suggestions in this phase can ensure to have easy solutions and low costs to operate the satellite. One of the common choices that should be taken in this phase is the use of a beacon signal to identify the satellite after the launch [5]. Nanosatellite launches are normally shared with other satellites in piggyback or cluster launches. In this kind of launch, tens of satellites are separated from rocket fairing with a temporal separation of seconds, or even tenths of a second. For example, the Dnepr 2014 cluster launch, performed by Russian/Ukrainian company Kosmotras on the 19th of June 2014, successfully inserted 33 satellites into orbit [6]. Adding four satellites that separated from their mother satellite UniSat-6 [7], a total of 37 satellites were deployed. In cases similar to this one, becoming more frequent every day due to the increase of nanosatellite market, it is extremely difficult to distinguish the satellites considering their orbital positions. The two line elements (TLE) determination is not highly accurate and operators can follow in vain the different objects not identifying their own satellite. A simple beacon signal can avoid this situation, helping the team to understand on the first passage the satellite position and health status.

   In the design of beacon signal, it is important to do a trade-off between the power consumption and the minimum amount of data and transmission time required to get enough information about satellite status.

3. *Avoid complexity*: Operations cost naturally increases in proportion to the complexity of operation activities. In particular, the costs increase considering two aspects: on the one hand, increased mission operations complexity means to enlarge the operations team; on the other hand, a more complex system requires more cautious operations, more expensive equipment, and a greater budget.

   For these reasons, the complexity of spacecraft operations control should be minimized, reducing the number of telecommands of states, of conditional relationships, flight rules, and the frequency and lengths of station contacts as much as possible.

This step is also related to on-board autonomy, since the spacecraft capability to perform certain operations functions on its own without direct support from ground also reduces spacecraft operations tasks.

In first approach, it looks obvious that a way to reduce operations cost is to build a spacecraft that does not require any control from ground. Unfortunately, this is not an easy solution, since an increase in on-board autonomy instead of mission operations complexity often results in an increase of mission vulnerability.

Technology advances in spacecraft computing and improved software capabilities allow to perform functions on orbit that were not possible earlier, such as monitoring alarms or processing payload data.

Of course, migrating these functions from the ground to the spacecraft can reduce operations cost and complexity, but what are the effects on the satellite design? Sometimes, an increase of on-board capabilities results in larger satellite dimensions, more mass and/or power consumption, and a higher vulnerability, since a more complicated system has to be designed. Mission operators must compare cost increases for development and maintenance with operational cost savings.

The reduction of ground operations tasks should take into account that some processes are peculiarities of ground segment, such as satellite reprograming, and cannot be transferred on-board.

On the other hand, on-board activities can be performed without time limitations related to satellite passage over the ground station or signal propagation delays.

The mission operator is the key person to solve the trade-off between spacecraft design simplicity and operations budget reduction, since he should know exactly how each specific function could affect mission costs and reliability.

Each mission has different constraints and requirements that result in different benefits related to the on-board autonomy implementation. In nanosatellite missions, characterized by a short duration, increase in on-board autonomy can be not so effective and results only in jeopardizing of the project.

Nevertheless, new technological solutions are under implementation to increase on-board capabilities in nanosatellite projects. One idea under implementation is to apply Swarm intelligence to satellite constellations. Swarm intelligence is a form of computer control that is patterned, in principle, on the actions of collections of relatively unintelligent insects. The fundamental notion of swarm intelligence is that a group, which practices a set of consistent patterns, in aggregate, will produce a desired behavior as, or more effectively than, if central control was implemented [8]. Applying this technique to a nanosatellite constellation, the single limited processing capability can be increased, sharing the functions between the different satellites of the constellations.

4. *Design for operability*: While designing the spacecraft, operability issues are too often ignored or considered only a duty for the mission operations team. It is also important to remind that if the operations are temporally located after the launch, the design should be done considering the operability of the satellite itself and operator team should be integrated in the design team.

Robust spacecraft design and resource margins, in terms of on-board capabilities and budget, are some of the design techniques that increase mission reliability.

Robust approach guarantees flexibility, versatility, and resilience of the spacecraft design, increasing the system response capability to possible in-orbit failures. The system should be capable to be subjected to minor changes without losing its reliability.

Another way to reduce costs is to simplify the operations by doing less modification in spacecraft subsystems. For example, in several missions, the use of an omnidirectional antenna in uplink and downlink allows to eliminate problems and complexity related to attitude determination and control system and, as a consequence, operations preparation and monitoring.

During critical phases, a clear knowledge of on-board resources helps the operators to take the right operational decisions that sometimes can result in mission recovery. Having margins on resources, such as on-board power and thermal capabilities and data storage or processor memory, gives more opportunity and flexibility in case anomalies arise and may limit the resulting damage.

5. *Provide operation description document*: In nanosatellite operations, teams are susceptible of people turnover. This can jeopardize the mission and increase the costs. A solution to this problem is to provide a set of important documents describing all the operations phases. Some of these documents are described in detail in Sections 19.4 and 19.5. Others can be produced in accordance to the goals of mission operations.

At the same time, it is important to guarantee the continuity in the team, for example, overlapping work periods or including one or more people periodically in the team to watch and learn how to operate the ground station, the satellite, and the payload.

This situation is a classical problem in university missions where students involved in the mission leave at the end of the course. Knowing this from the beginning helps the project manager to arrange solutions to avoid issues during the project development.

6. *Share knowledge and experience*: The capability to create a network of people able to share knowledge and experiences in satellite design is a useful talent that leads the mission to the success. Beginner teams face problems that more experienced people already had solved in the past. Receiving support from more experienced people avoids making mistakes and reduces project time and costs. In nanosatellite projects, it is possible to ask for support from users' community, such as the CubeSat Community. A simple email sometimes can help the team to take the right decision or to select the right component.

Participation in workshops or conferences allows the team to increase knowledge about the state of the art of the problem and at the same time to build a network of contacts.

The operations team should also learn from past experiences and in particular from failures. A correct failure analysis helps the team not to fall in the same mistakes again.

7. *Use standard solutions for equipment and test procedures*: Standardization has a dual role from the point of view of the small-satellite community operations: on one hand, it coordinates mission operations phases avoiding wastage of time, while on the other hand, it increases mission complexity since the actually available standards are not specifically dedicated to nanosatellite missions.

The nanosatellite community is trying to identify rules and standards applicable to small satellite missions.

The new standards should consider different aspects of nanosatellite missions, [8] such as:

- Operator training
- Material composition
- Structural integrity
- Control
- Mitigation of risk to manned and unmanned
- Operating security
- Proximity to other satellite during the launch
- How to avoid disruption of other satellites
- Privacy of individuals and other spacecraft
- Information accuracy
- Warning of proximity
- Warning of danger
- Registration and coordination
- Debris prevention

Unfortunately, not all these points are always taken into account by operations team. This results in an increase of mission-related risks. For example, not all the small satellites provide a debris mitigation analysis or are aware of procedures to register their satellite frequencies. Lack of this information results in loss of time and increases spacecraft design-related issues. Let us consider for example the case of a CubeSat to be launched in low Earth orbit (LEO) with an altitude higher than 700 km. In order to satisfy the Inter-Agency Space Debris Coordination Committee (IADC) 25 Debris Mitigation Guidelines [9], the satellite should be equipped with a deorbiting device, such as a drag sail or a propulsion system. Also, in the case of a completely autonomous deorbiting system, such as the one boarded in EduSat satellite [10], modifications on the satellite are requested. Unfortunately, in not all the cases the needed modifications can be performed at last minute, since not always it is possible to have spare room to install a new mechanism.

Operators should be aware of all the standards and documentations needed since the beginning and ensure that the team performs the needed analysis and procedures to solve all the critical points.

8. *Perform tests*: Testing the spacecraft and the ground segment functionality before launch is helpful for the operations team. Each time a test is performed, the team gains more information about its own system. Two different kinds of tests can be identified:

   - Verification test: to demonstrate the conformity of the system with its design specification and interfaces.
   - Validation test: to provide a preliminary confirmation that it is fit for use.

   This classification can also distinguish between environmental and functional tests. Qualification/acceptance environmental tests shall be performed in accordance with the request of launch provider and the target conditions in orbit. Functional tests should consider not only normal operations but also dangerous situations that may occur during the mission.

In general, to perform ground segment validation, operational simulator and space segment models are used. The more tests are performed, the better it is possible to ensure the compatibility between space segment and ground stations.

In order to provide feedback to the engineering of future space missions, reports addressing mission operations experience, lessons learned, spacecraft in-orbit performance, and end-of-life tests shall be produced.

In case of replacement of staff, the new members of the operations teams shall undergo formal training and operational exercises in a test environment (simulations) before working autonomously.

9. *Use existing facilities*: A strategy that can be implemented for reducing mission operations costs is obviously the use of same facilities and teams for several missions. Many sectors of space industry, such as the telecommunications, are using this strategy allowing an operator to control a whole fleet of satellites. The idea to use existing facilities to operate the satellite offers significant advantages in terms of the efficient utilization of staff and facility resources, but at the same time can also increase the amount of data possible to download in a mission. As an example, the cooperation established between GAUSS Srl (Italy) and Morehead State University (USA) allows both teams to have multiple satellite passages and double the amount of data possible to download during the mission. Different ground station networks for nanosatellite missions are under development in order to give the possibility to the operators to use multiple mission ground stations. Nowadays, the key issue to be solved is the development of standards to be used by these facilities in order to guarantee that every nanosatellite operator can use each ground station on the network.

References

1 The Consultative Committee for Space Data Systems (CCSDS). 2006. Mission Operations Services Concept, Informational Report, CCSDS 520.0-G-2—Green Book, August 2006.

2 European Cooperation for Space Standardization ECSS-E-ST-70C, 2008. Space Engineering: Ground Segment and Operations, 31 July 2008.

3 Wertz, J.R. and Larson, W.J. (2008). *Space Mission Analysis and Design*, 3e. Space Technology Library.

4 Fortescue, P., Swinerd, G., and Stark, J. (2011). *Spacecraft Systems Engineering*, 4e. Wiley.

5 Robert J. Twiggs, James Cutler, Introduction to Space Systems and Spacecraft Design, SmallSat Space Education, Morehead State University Lectures.

6 http://www.gaussteam.com/unisat-6-mission.

7 Brandon C. Wood, 2001. "An Analysis Method for Conceptual Design of Complexity and Autonomy in Complex Space System Architectures", Master of Science in Aeronautics and Astronautics and Master of Science in Technology and Policy, Massachusetts Institute of Technology, June 2001.

8 Jeremy Straub, (2015). "In Search of Standards for the Operation of Small Satellites" AIAA 2015 SciTech Conference.

9 IAA Position Paper on Space Debris Mitigation, 2006. ESA Technical Report SP 1301, February 2006.

10 Cappelletti C., Di Lauro R., 2012. "Edusat Completely Passive Deorbiting System", IAC-12.A6.4.14. 63rd International Astronautical Congress.

Further Reading

Doug Edsey, Josh Berk, Jeremy Straub, David Whalen, and Scott Kerlin (2013). "Mission Operations for an Earth Imaging CubeSat," University of North Dakota, Graduate School Scholarly Forum.

van der Ha, J., Marshall, M.H., and Landshof, J.A. (1996). Cost-effective mission operations. *Acta Astronautica* 39 (1–4): 61–70.

van der Ha, J. (2003). Trends in cost-effective mission operations. *Acta Astronautica* 52 (2–6): 337–342.

Landshof, J.A., Harvey, R. J. and Marshall, M.H., (1994) "Concurrent Engineering: Spacecraft and Mission Operations System Design", NASA Technical Report N95-17614.

Wertz, J.R. and Larson, W.J. (1996). *Reducing Space Mission Costs*. Microcosm, Inc., Kluwer Academic Publishers.

http://www.kosmotras.ru/en/launch15.

20 I-5

Mission Examples

Kelly Antonini, Nicolò Carletti, Kevin Cuevas, Matteo Emanuelli, Per Koch, Laura León Pérez, and Daniel Smith

GomSpace A/S, Aalborg, Denmark

20.1 Introduction

When it comes to details on missions, the best way to get them is from their primary source. Thus, it was decided in this book to provide mission details from a single source so as to avoid any misinterpretation. Therefore, this chapter presents practical examples of existing nanosatellite missions, highlighting the objectives of each of them. All the satellites and missions described are developed by GomSpace, a Danish space company established in 2007. GomSpace is a manufacturer and supplier of nanosatellite solutions for customers in the academic, governmental, and commercial markets [1].

GomSpace develops nanosatellites according to the CubeSat specification. This standard defines the size of the satellite to be of one or more $10 \times 10 \times 10\,cm^3$ size cubes or units with each unit weighing up to 1.33 kg. The CubeSat specification has allowed spacecraft to be more standardized. This standardization along with the general low-mass when compared to old-space spacecraft has made nanosatellites attractive not only to academic but also to professional space agencies and commercial customers [2].

In recent years, new business solutions and technology miniaturization are pushing the boundaries of the payloads that can be hosted in nanosatellites. However, payload power demands are also increasing, which in turn causes an increase in size, mass, and complexity. The mission applications for nanosatellites are varied: micro-g experiments, Internet of Things (IoT), Earth observation, communication, surveillance, and tracking, just to name a few. Nanosatellites are predominantly deployed into low Earth orbit (LEO), however higher orbits and even deep space missions exist with NASA at the forefront and ESA following closely behind.

Nanosatellites: Space and Ground Technologies, Operations and Economics, First Edition.
Edited by Rogerio Atem de Carvalho, Jaime Estela, and Martin Langer.
© 2020 John Wiley & Sons Ltd. Published 2020 by John Wiley & Sons Ltd.

20.2 Mission Types

Most nanosatellite missions can be categorized into four mission types:

- Educational
- Technology demonstration
- Science
- Commercial

In this section, all four mission categories are presented, and their interlinking explained.

20.2.1 Educational Missions

Educational missions are for the most part concentrated around the technical universities or governmental institutions worldwide that want to build up their expertise and skills within the space technology and industry domain. For universities, the educational missions are not only the means to train and educate their students in the space domain but also a relatively low-cost and fast way to enable students to test some of the theory in practice.

University educational missions are an excellent way to give students practical experience with designing, testing, launching, and operating satellites. Educational missions usually make use of relatively small nanosatellites, generally CubeSats, referred to as 1U up to typically a maximum of 3U in size.

20.2.2 Technology Demonstration Missions

The purpose of conducting technology demonstration missions is to gain knowledge, build experience, and mature technologies. These can lead to more advanced and sophisticated missions by proving the feasibility of the technologies they develop.

Technology demonstration missions can be initiated and managed from a wide range of organizations and they are typically fully or partly funded from space agencies, such as ESA in Europe or NASA in the USA. The organizations that conduct technology demonstration projects range from universities to governmental institutions, but also include small privately owned companies and large international corporations. Technology demonstration missions funded by, for example, the EU are normally required to have partners from both universities and companies.

Technology demonstration missions in the nanosatellite domain focus either on the design, development, and testing of platform and payload components for nanosatellites or to demonstrate technologies intended for larger satellites and missions on conceptual basis. This is enabled by the fact that it is relatively fast and less costly to launch nanosatellites to space.

Examples of technology demonstration missions are:

- *Test of miniaturized instruments known from the traditional space industry*: Demonstration of miniaturized hyperspectral cameras and star trackers for improved pointing accuracy, for example.
- *Radio communication between nanosatellites:* To demonstrate the feasibility and usages of this capability in commercial and science missions.

- *Acquisition of air traffic and maritime traffic data for a broad variety of user applications*: To prove the ability to receive Automatic Dependent Surveillance Broadcast (ADS-B) and Automatic Identification System (AIS) data in a reliable manner.
- *Space docking mechanism for larger satellites*: To demonstrate a technology on a small setup to prove that it can be used on missions with larger spacecraft, for example.

The main reason for using nanosatellites for technology demonstration is that it is a relatively fast, flexible, and low-cost method of performing testing in space.

20.2.3 Science Missions

The purpose of using nanosatellites for scientific missions is overall identical to the purpose of using a larger satellite: capturing data of a certain nature for the specific mission, analyzing the data, and gaining new knowledge about Earth, space, etc. Many science missions are organized in a consortium comprised of a space agency, one or more universities, and, in some cases, also contributions from private companies.

Examples of science missions are:

- *Mapping of the magnetic field of the Earth:* The data can be used for mapping the changes in the magnetic field and provide input for researchers trying to understand the core of the Earth.
- *Deep space missions*: Asteroid exploration, interplanetary missions, and lunar exploration, for example.
- *Astronomy missions*: A swarm of nanosatellites can provide solutions such as distributed detectors, which are used in place of large-surface detectors for bigger satellites, for example. The solution with larger satellites is difficult, time-consuming, and very expensive. A swarm of interconnected nanosatellites with gamma-ray detectors can be employed to localize short gamma-ray bursts, which are believed to be the electromagnetic counterparts to the gravitational waves, for example.
- *Research activities in the space environment:* Microgravity experiments, for example.

Given the size and flexibility of nanosatellites, an entire range of new scientific missions can be conducted, which was not possible until now either from a technical or an economical perspective.

20.2.4 Commercial Missions

Commercial missions are built around a business case and generate revenue based on the output of the mission. These missions can also be initiated by governmental organizations with specific demands to obtain observations of geographical areas, for example. Commercial missions are for the most part conducted by privately owned companies either as a complement to their existing business or as the core business for the company.

Examples of commercial missions are:

- *Data and voice communication missions*: Communication service providers that want to build or expand their service to include voice and data communication via nanosatellites.

- *IoT*: A very fast growing market where the nanosatellite is used to capture and feed the data to ground.
- *Earth observation and surveillance missions*: Detection of pollution, monitoring the oceans for oil spill, and detection of forest fires, for example.
- *Air and maritime traffic*: Utilizing nanosatellites to assess air and maritime traffic management.
- *Spectrum monitoring*: Spectrum monitoring of ships, for example.

The growth in the commercial market is mainly driven by the fact that nanosatellite technologies have reached a maturity level where it has become feasible both from a technical and a financial perspective to conduct these missions.

20.3 Mission Examples

This section provides mission examples for the categories listed earlier: educational missions, technology demonstrations, science missions, and commercial missions.

20.3.1 Educational Missions

In the educational mission examples shown in the following text, students are involved for the whole duration of the project. They first receive an introductory course given by aerospace engineers and experts in the various subsystems. The students must assemble, integrate, and test the satellite. Often they are also in charge of installing the ground station at the university facilities, and once launched the students command the nanosatellite. The missions enable students to get hands-on experience in a very large variety of satellite-related disciplines. Students thereby have first-hand experience of what is often only taught in theory. They graduate with real-life experience in the field and are, ultimately, more employable.

20.3.1.1 Delphini-1
Delphini-1 is an educational 1U nanosatellite developed by the University of Aarhus, Denmark. The aim of the project is to ensure the usage of space technology for university research and education [3]. Delphini-1 is a proof of concept. The university will use its first mission to pave the way to potential bigger platforms in the future. As is often the case for educational institutes, the university is granted a free launch opportunity on-board the International Space Station (ISS). The project is sponsored by ESA, the Aarhus University Research Foundation, and the Danish Agency for Higher Education and Science. Delphini-1's payload is an optical camera, and the satellite will communicate to ground using a Ultra High Frequency (UHF) antenna. The nanosatellite will be in Sun-synchronous orbit (SSO) and is capable of downloading 2 MB of data per day using one ground station located in Aarhus. The mission duration will likely be less than 6 months due to the satellite's low orbit altitude (Figure 20.1).

20.3.1.2 FACSAT
The FACSAT mission is a 3U satellite, which was developed for the Colombian Air Force Academy. The mission seeks to educate the young student cadets about the process of

Figure 20.1 The Delphini-1 nanosatellite and team. Source: Reproduced with permission from Aarhus University.

designing a satellite from requirement formulation to in-orbit operations. FACSAT's payload is an optical camera, which will be controlled via an amateur frequency radio in UHF. The CubeSat will be launched to a SSO orbit [4] (Figure 20.2).

Figure 20.2 Flight model of FACSAT.

20.3.2 Technology Demonstration

Nanosatellites are perfect candidates to demonstrate new technologies due to their low cost and quick turnaround time. The successful demonstration of these technologies oftentimes leads to commercialization on larger platforms.

20.3.2.1 GOMX-3

GOMX-3 was the first ESA in-orbit demonstration (IOD) nanosatellite to be launched [5]. It was intended to establish a solid base in the extended use of nanosatellites for experimentation and commercial purposes in Europe. The idea of the project was motivated by the planned mission of the Danish astronaut Andreas Morgensen on board the ISS in September 2015, which offered the opportunity to enable a Danish astronaut to release a Danish nanosatellite [6].

The mission consisted of a 3U nanosatellite with a very compact platform solution and the capability of three-axis pointing. It accommodated three different payloads for X-band transmission and L-band and ADS-B signals reception. The platform used for this satellite included numerous failure mitigation and propagation avoidance techniques, such as latch-up protections, current limitations, and watchdogs, which allowed automatic control of the satellite to recover and to protect the platform from failures of payload or subsystems. The main challenge of GOMX-3 was the schedule, allowing less than 1 year from the project kick-off to launch delivery. Utilizing the tailored European Cooperation for Space Standardization (ECSS) for IOD CubeSat standard enabled GomSpace to reach a good balance between reducing cost and a challenging timeframe, on the one hand, and increasing the quality as well as mitigating risk, on the other (Figures 20.3 and 20.4).

The satellite was deployed from the ISS on 19 August 2015 to a near-circular orbit at 400 km altitude and 51.6° inclination. Its planned lifetime was intended to be 3 months long. Long enough to enable the demonstration of L-band/UHF signals acquisition from satellites at higher altitude, ADS-B data collection in space, and X-band transmission capabilities, all of them coordinated with a new precise attitude and determination control system. At the time, CubeSats were extensively used for educational purposes, but their reliability rates and capabilities for professional applications were limited in Europe. The noticeable success of this IOD mission enabled GomSpace to extend the mission lifetime until its natural deorbit around 12 months after its in-orbit deployment.

GOMX-3 was one of the first CubeSat missions to include a software-defined radio (SDR). The spectrum monitoring was tested by UHF signal tracking and L-band spectrum recording using a common SDR based on advance field-programmable gate array (FPGA) technology. During several experiments, the radio was capable of keeping stable temperature ranges while capturing specific geostationary satellites at the 1592 MHz frequency band. As third party experiment, the satellite accommodated the Syrlinks EWC27 X-band transmitter connected to the FPGA via low-voltage differential signaling (LVDS) for fast data transmission to specific ground stations located in Kourou and Toulouse. This transmission was tested reaching up to 115.5 Mb of data transfer in 5.77 min [7]. All these radio experiments required an accurate three-axis pointing, reaching a pointing error of 1° (1σ) in sunlight. This Attitude Determination and Control System (ADCS) accomplishment enables future nanosatellite missions to accommodate

Figure 20.3 GOMX-3 flight model in its
fully deployed configuration.

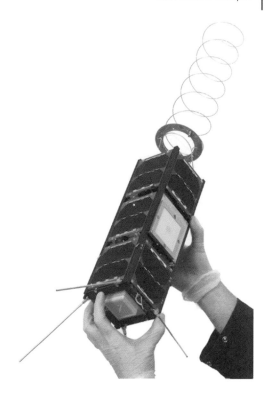

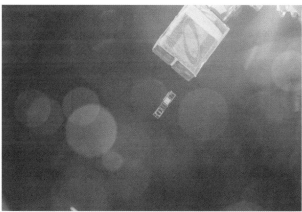

Figure 20.4 Deployment of the GOMX-3 nanosatellite (center) from the ISS on 5th October 2015.
Source: Photo courtesy of Astronaut Scott Kelly.

a large variety of optical and directional payloads that require accurate pointing [8, 9]
(Figure 20.5).

The satellite was collecting ADS-B data from planes crossing the Atlantic Ocean where
there is no terrestrial coverage, and the data were integrated in a well-known air traffic
database showing real-time flight tracking in areas with no terrestrial equipment coverage
[10]. This test shows the ability of nanosatellites to perform real-time signal monitoring

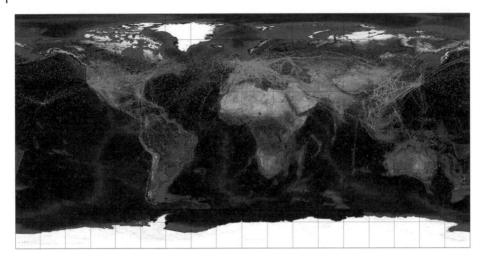

Figure 20.5 Global representation of the ADS-B data collected by GOMX-3 where each dot corresponds to a plane location captured and downloaded by the satellite. (*See color plate section for color representation of this figure*).

with a multitude of applications, such as automatic surveillance [11]. The success of this mission contributed to extend the nanosatellite technology program in the European Space Agency [12]. The mission also drove the continuation of ESA nanosatellite missions lead by GomSpace, and it promotes the maturity of the commercial nanosatellite sector in Europe [13].

20.3.2.2 GOMX-4
The GOMX-4 mission was kicked off just after the completion of the GOMX-3 satellite to apply and extend the lessons learned and functionalities demonstrated. The mission consists of two twin satellites, GOMX-4A and GOMX-4B. GOMX-4A (or Ulloriaq) is developed for maritime and aerial monitoring above the Arctic region for the Danish Acquisition and Logistic Organization (DALO). GOMX-4B is the second ESA nanosatellite launched for demonstration of technologies [14, 15]. The two satellites share the common goal of being precursor of CubeSat constellations, thanks to their intersatellite link (ISL), as well as station-keeping functionality and relative orbit control between them.

The primary objective of the mission is to control the relative drift and separation distance between satellites, while exchanging data at different separation distances up to line of sight visibility. Both satellites have the same SDR capabilities, transmitting, and receiving between them in the S-band. GOMX-4B accommodates an innovative propulsion module, which allows orbit control relative to GOMX-4A. The ISL operations are checked and characterized by a variety of SDR configurations with different separation distances ranging from 200 km up to 4500 km. The results of these experiments are useful for a better understanding of the minimum required number of satellites and the feasible latency for data downlink in future similar constellation missions (Figure 20.6).

Figure 20.6 Flight models of GOMX-4A (left) and GOMX-4B (right).

GOMX-4B also accommodates three secondary payloads: the Chimera payload developed by ESA to evaluate the behavior of different memories under radiative environment [16, 17], the HyperScout from Cosine to capture hyperspectral images in space with a miniaturize camera flying for the first time [18], and a star tracker from Innovative Solutions in Space (ISIS) to evaluate its performance in space [19].

The Chimera payload is continuously operating and the failures or latch-up events in the on-board memories are being recorded to study its hardening in radiative environment. The HyperScout payload is operated activating different functionalities to obtain in-orbit results ranging from image acquisition above specific target regions on Earth, compression algorithms, and data processing. The selection of software and parameters is updated and optimized during this instrument's IOD with the goal of releasing a flight-proven instrument to the market. The star tracker payload is used to obtain dark sky images and attitude determination results. The outcomes from its in-orbit operations allow to determine the instrument accuracy in LEO.

These different technologies are interfaced to a new powerful avionics platform, designed for 6U CubeSats. This platform contains ADCS, electrical power system (EPS), on-board computer (OBC), and telemetry, tracking, and command (TTC) subsystems, which only occupy approximately 2U of the internal volume, and it is designed to be easily extended to 8U, 12U, 16U, and even larger systems.

Both GOMX-4 satellites integrate the same platform components differentiated only by the power distribution configuration, components layout, payloads interfaces, and their specific software functionalities. The innovative on-board EPS is a modular system with independent input and output modules fully adaptable to the size and needs of each spacecraft. The ADCS system is designed to accommodate interfaces to a large number of sensors, ranging from sun sensors and magnetometers, to accurate gyro sensors or star trackers, and powerful actuators, as well as propulsion module connections. The external components of the satellites, such as solar panels, patch and dipole antennas, and radiative surfaces or shields, were designed to be accommodated in future satellites without the need of customization (except for the mechanical interface).

Both satellites have been designed for more than 3 years lifetime where, in addition to the competition of the main mission objectives, GOMX-4A will continuously collect ADS-B and AIS data as well as images from its on-board camera. GOMX-4B will demonstrate the new technologies from the different European companies during the first 6 months of operations and, once the IOD goals are finalized, its functionality shall be extended to ADS-B data collection from the additional SDR included for this commercial purpose. This approach shows the plurality of nanosatellite application, combining experimental and commercial purposes to optimize the return and profit of the mission.

Both satellites were successfully launched by Long March 2D on 2nd February 2018 to a 500 km sun synchronous orbit. Since then, the two satellites were in Launch and Early Orbit Phase (LEOP) during the first 2 months when all the subsystems were being commissioned, including all payloads. The ADCS system was in-orbit calibrated and characterized, and the orbit altitude difference between them was corrected by a propulsive maneuver.

After the successful completion of LEOP, the satellites were nominally operated satisfying the goals of the primary and secondary payloads without major issues beyond few software debugging actions. The extensive variety of objectives and technologies included in both GOMX-4 satellites is enabling GomSpace to improve its business offerings. During this mission, GomSpace implemented several new processes, improving integration, verification, and automatization for larger-scale constellations. The system design of these two satellites ensures direct scalability and flexibility, preceding the new generation of nanosatellites and a step forward when compared to GOMX-3 and predecessor missions [20, 21] (Figure 20.7).

20.3.2.3 CubeL

The CubeL mission, a joint project between Deutsches Zentrum für Luft- und Raumfahrt (DLR), TeSat, and GomSpace, seeks to demonstrate the smallest laser communication terminal on the market called "OSIRIS4CubeSat" [22].

Currently, the CubeSat sector is bottlenecked when it comes to downlink technologies. Various factors influence this bottleneck, but arguably the limitations in mass and power along with regulatory difficulties such as frequency allocation are the main inhibitors. Overcoming the downlink shortcoming is especially important, as CubeSats are proving to be a force to be reckoned with when it comes to generating vast amount of data. CubeL will be able to downlink data up to a rate of 100 Mbps.

CubeL is a 3U nanosatellite with two primary payloads: a visible-spectrum camera to generate the large volumes of data to be later downlinked by the "OSIRIS4CubeSat" laser. CubeL is a stepping stone for future projects that will set out to achieve high-volume data

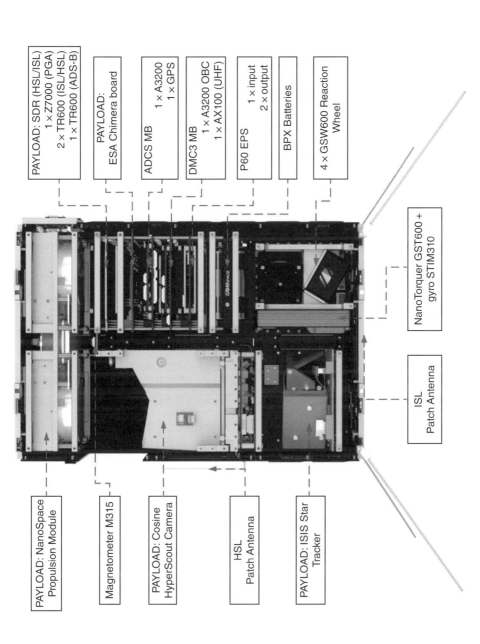

PAYLOAD: SDR (HSL/ISL)
1 × Z7000 (PGA)
2 × TR600 (ISL/HSL)
1 × TR600 (ADS-B)

PAYLOAD:
ESA Chimera board

ADCS MB   1 × A3200
1 × GPS

DMC3 MB
1 × A3200 OBC
1 × AX100 (UHF)

P60 EPS   1 × input
2 × output

BPX Batteries

4 × GSW600 Reaction
Wheel

PAYLOAD: NanoSpace
Propulsion Module

Magnetometer M315

PAYLOAD: Cosine
HyperScout Camera

HSL
Patch Antenna

PAYLOAD: ISIS Star
Tracker

NanoTorquer GST600 +
gyro STIM310

ISL
Patch Antenna

Figure 20.7 GOMX-4 internal layout: CAD system design (top) and flight model satellite (bottom). (*See color plate section for color representation of this figure*).

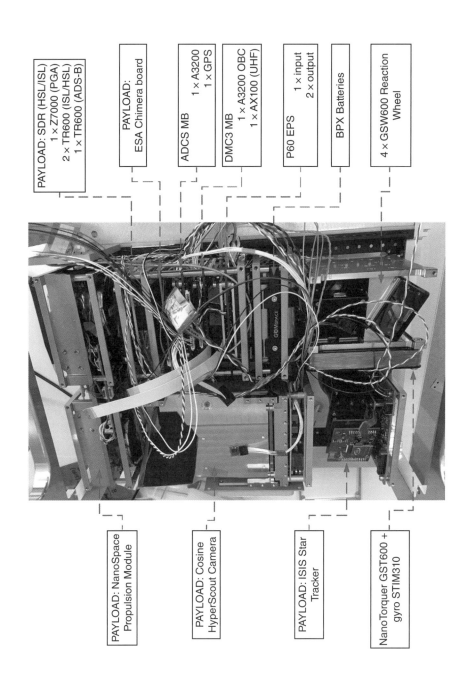

PAYLOAD: SDR (HSL/ISL)
1 × Z7000 (PGA)
2 × TR600 (ISL/HSL)
1 × TR600 (ADS-B)

PAYLOAD:
ESA Chimera board

ADCS MB   1 × A3200
               1 × GPS

DMC3 MB
1 × A3200 OBC
1 × AX100 (UHF)

P60 EPS   1 × input
           2 × output

BPX Batteries

4 × GSW600 Reaction
Wheel

PAYLOAD: NanoSpace
Propulsion Module

PAYLOAD: Cosine
HyperScout Camera

PAYLOAD: ISIS Star
Tracker

NanoTorquer GST600 +
gyro STIM310

Figure 20.7 (Continued)

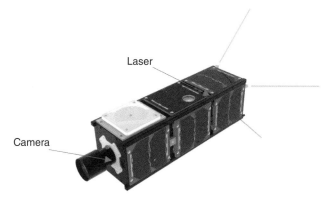

Figure 20.8 Illustration of CubeL with payloads.

rallying via intersatellite laser link. This technology will expedite the downlink of actionable information to locations that are out of line of sight for satellites (Figure 20.8).

20.3.3   Science Missions

20.3.3.1   DISCOVERER
Nanosatellites are often used to perform scientific missions. Considering they generally cost less than larger, more traditional satellites, they can be a test platform for payloads that may one day be used on larger satellites. SOAR (Spacecraft for Orbital and Atmospheric Research), a scientific nanosatellite, is an example of such a mission and is part of the DISCOVERER (Disruptive Technologies for Very Low Earth Orbit Platforms) project, funded by European Union's Horizon 2020 research and innovation program. The objectives of the SOAR mission are to investigate the interaction between different materials and the atmospheric flow environment in very low Earth orbit (VLEO) and to test novel attitude and orbital control mechanisms using aerodynamic forces and torques.

The Earth remote sensing market is expanding day by day, and lower orbits enable higher-resolution imagery with smaller payloads. A key issue with VLEO is that atmospheric density and drag are higher and the lifetime before reentry is therefore typically short, on the order of days to months. Without reducing or compensating for the effect of drag, it is often not economically worth launching a satellite into such an altitude. In addition, the aerodynamic perturbations can negatively affect image quality [23] (Figure 20.9).

SOAR is a 3U nanosatellite with two payloads: a set of four steerable fins, which allow different materials to be exposed to the flow at varying angles and also provide aerostability and the capability to test novel aerodynamic control maneuvers; and an ion and neutral mass spectrometer (INMS), which is used to characterize the oncoming flow composition, density, and velocity [24].

20.3.3.2   TESER
The exponential increase in number of satellites launched into LEO in recent years has driven the need to develop a mechanism capable of addressing the space debris issue. Technology for Self-Removal of Spacecraft (TESER) is an on-going project intended to design

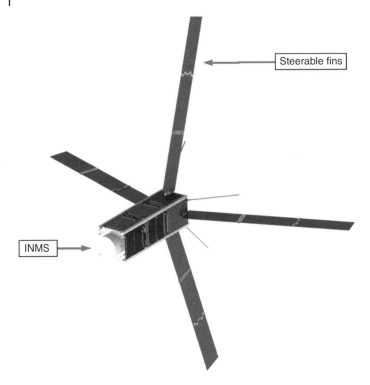

Figure 20.9 Illustration of SOAR with payloads.

a removal module, which can be adapted to any spacecraft to ensure it will deorbit at the end of its lifetime. The mission is led by a consortium of 10 European entities, with experts ranging from nanosatellites to space law [23] (Figure 20.10).

The need to remove space debris has sparked a range of initiatives competing to develop a cost-effective and reliable method for debris removal, with the focus being on reduced complexity. Consequently, TESER's objective is to produce a solution versatile enough to be included on any spacecraft design, while balancing cost and reliability. In this context, the project is based on flight-proven nanosatellite systems.

The first project phase has developed and tested three configurations of removal subsystems using a standard 6U nanosatellite. The different removal mechanisms are:

- A passive solution based on deployable wings that increase the surface area to accelerate deorbit.
- An active mechanism based on a solid rocket motor, which drives the spacecraft to a lower orbit where it is deorbited.
- A hybrid mechanism based on drag sail acting as large surface electrode and an electro-dynamic tether.

The removal subsystem is composed of a common nanosatellite platform, which includes full functionality for OBC, EPS, ADCS, and COM where any of the three removal modules can be integrated under a flexible design philosophy. The solution allows the use of the removal subsystem or deorbiting module attached to the main spacecraft even when

Figure 20.10 Self-removal sequencing. Source: Reproduced with permission from Airbus Defence and Space.

the satellite finishes its lifetime due to an irrecoverable failure [24]. The application of nanosatellite technology enables a cost-effective contribution to one of the biggest space problems of the moment: the protection and cleanliness of space.

20.3.4 Commercial Missions

20.3.4.1 STARLING

Following a successful demonstration of a space-based ADS-B receiver on-board GOMX-3, GomSpace and Investeringsfonden for Udviklingslande (IFU, Danish Investment Fund for developing countries) created Aerial and Maritime (A&M). A&M investigated the commercialization of ADS-B data for areas of the globe that are not fully covered by ground-based radar. The focus was directed especially toward the equatorial zone to provide up-to-date ADS-B data [25].

The aim of the mission is to enable countries in the equatorial area to charge airlines flying over their national airspaces. In addition, building on experience in space-based AIS systems, an AIS payload was also added early in the project. The project was set to reuse as many commercial off-the-shelf (COTS) systems as possible, available and tested in the GOMX-3 mission. Moreover, since it ran in parallel with the GOMX-4 program, it benefitted from the development carried out around the GomSpace SDR products.

Originally designed to be a four-satellite constellation flying at 500 km, it was later upgraded to an eight-satellite constellation. The project was then named STARLING and, as per the birds known for flying in an elaborate formation in the sky, the STARLING spacecraft have been designed to acquire and maintain their constellation positions on the orbital plane using drag management.

To maintain a 100% duty cycle on the payload side, under the STARLING project, GomSpace developed double deployable solar panels which brought the average power generated by the spacecraft to around 13 W. Additional concerns regarding the thermal

Figure 20.11   Illustration of STARLING.

management were also addressed by performing detailed thermal analyses and designing adequate subsystems. Moreover, STARLING benefited from the GomSpace "Hardening Program," which qualifies all subsystems for a 5-year mission in LEO, considering total ionizing radiation and thermal cycles. In general, the satellites' commercialization brings on the project additional requirements connected to continuous operations, automatization, availability, and reliability, which are not always addressed in IOD missions (Figure 20.11).

20.3.4.2   Three Diamonds and Pearls

Thanks to the economic advantages of using CubeSats, it is becoming a common strategy to use an IOD mission to prove a concept and explore the challenges of the following extended mission. An example of this is the mission of the company Sky and Space Global (listed as SAS in the Australian Securities Exchange). On June 23rd 2017, SAS launched their first satellites in space to test intersatellite communication and ground communication, among several other functionalities.

This mission is based on three satellites, individually called Red, Blue, and Green, and together referred as the "Three Diamonds." The Diamonds successfully realized the world's first narrowband intersatellite communication for CubeSats, providing services, such as phone calls, instant messaging, and data/image transfer [26]. The mission includes other functionalities, such as IoT, machine to machine (M2M), and push to talk (PTT). The first financial transaction via nanosatellites was also made possible by using the three satellites and the messaging app BeepTool [27].

The goal of SAS is to connect the regions of the world that today live without this service. The "Three Diamonds" allowed the technical development to plan for a 200 satellites

Figure 20.12 Three Diamonds: Red, Green, and Blue.

mission, named "Pearls" [28], which will grant full coverage of the equatorial region. This second mission will begin in 2019, when the first satellites will be launched, and by 2020 the full constellation will be deployed [29] (Figure 20.12).

20.4 Constellations

20.4.1 STARLING

The eight STARLING spacecraft are considered part of a constellation despite the lack of intersatellite capability. The constellation has three main operational phases (Figure 20.13):

- Acquisition phase: Satellites drift relative to each other to separate into their nominal orbits over time.
- Station-keeping phase: Keeping the satellites at their nominal locations relative to one another given disturbances, such as drag, J_2, etc.
- End-of-life phase: After the mission has reached conclusion, the satellites must deorbit. Thus, they can be either left tumbling or they can be placed in high-drag mode. Both solutions would free the LEO in less than 25 years, as prescribed by the international code of conducts [30]. The high-drag mode would speed up re-entry. A decision between the two solutions will be taken later in the mission, considering the health of the satellites, the possible conjunctions with human outposts or other assets in LEO, and changes of the space law/code of conduct.

20.4.1.1 Constellation's Operational System

The constellation's operational system consists of a fleet of satellites, a ground segment, which is performing offline orbit determination, and a control system, which derives differential drag maneuvers required to ensure ground coverage of the constellation.

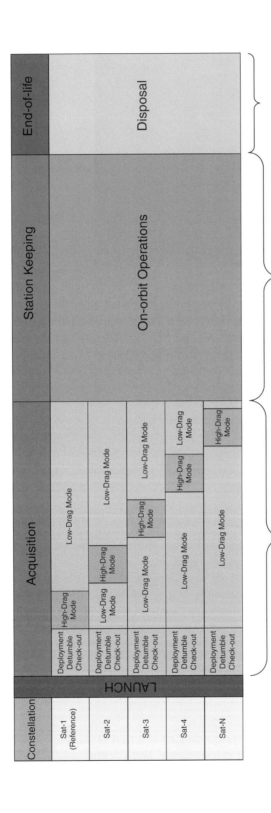

Figure 20.13 Relation between satellites' operation phases and constellation's phases. (*See color plate section for color representation of this figure*).

To compute the required maneuvers, the constellation control system collects the orbital parameters of each satellite in the constellation. To retrieve this data, two methods are used:

- *Two line elements (TLE)*: Generated by North American Aerospace Defense Command (NORAD) and provided by US Air Force (USAF) Joint Space Operations Center (JSpOC), the TLE is a parameter string describing the orbit for any given satellite. The orbital elements in line 2 are mean values, ignoring periodic variations in the satellite's orbit. This causes large uncertainties in position estimates, which may vary from 1 km to several hundred kilometers in error. The large error margin may also complicate the distinction between closely positioned satellites during LEOP.
- *Global positioning system (GPS)*: Position telemetry from the individual satellite in the constellation, which delivers a 1σ position knowledge within 30 m accuracy.

Both methods proposed are available for the STARLING mission, with the most appropriate being chosen based on a compromise between accuracy, power consumption, and complexity.

20.4.1.2 Orbit Determination and Propagators

Depending on the method used for retrieving orbital knowledge for the individual satellites in the constellation, the following determination methods and propagators are used:

- *TLE—SGP4 propagation*: The simplified general perturbation (SGP4) model is used with the TLE as input and predicts the position and velocity of the given satellite. Comparing SGP4 propagation with GPS precision has shown an error accumulation of up to 50 km over a period of 15 days. The model includes perturbations such as the J_2 model for changes in the Earth's shape, drag, and radiation. The SGP4 model uses a fixed ballistic coefficient given by the TLE, which means that it does not account for changes in the drag force acting on the satellite.
- *GPS—orbit determination and propagation*: GPS positions can be combined with UHF ranging to inform a Keplerian orbit model. Furthermore, the following parameters for known perturbations are needed:
  - Atmospheric drag
  - Solar pressure
  - Gravitational variations J_2 and J_3

The output of the orbit determination is the satellite position and velocity.

20.4.1.3 Constellation Control System

The estimated position and velocity for each satellite in the constellation are used to find the optimal control sequence of high-/low-drag maneuvers. The goal being to utilize the satellite constellation by maintaining the desired relative distance between the satellites.

An optimal Singular Bang-Bang (SBB) control scheme is used, which tries to reduce the amount of delta V and separation error acquisition time. The optimal SBB control sequence is a binary control and in an ideal case limits each satellite to only perform one high-drag maneuver during the control horizon.

With SBB control, only one satellite performs high-drag maneuvers at the time, which is a good feature if the payload is direction dependent for a defined duty cycle.

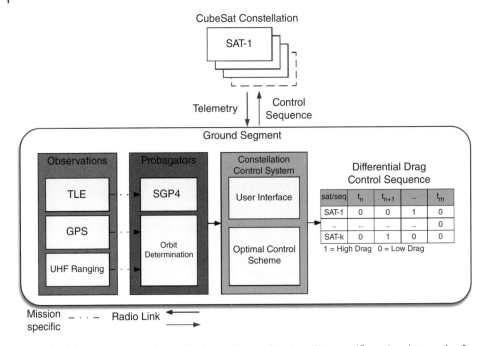

Figure 20.14 Constellation determination and control system diagram. (*See color plate section for color representation of this figure*).

A more general overview of the orbit control steps is listed here:

- Fetch the initial state values from the satellite constellation (orbit determination).
- Find the optimal high-/low-drag schedule by optimizing relative to the utility function (SBB control).
- Update and perform high-/low-drag schedule for each satellite in the constellation.
- Wait for a defined period.

The high-/low-drag schedule consists of a list of binary states for each satellite in the constellation for a given time t_n to a predefined control horizon t_m. It should be noted that at time t_m, each satellite in the constellation should be in the state of low drag (Figure 20.14).

20.4.2 Sky and Space Global

The Sky and Space Global missions are a great example of the challenges to face when "upgrading" from an IOD, such as a small constellation of three satellites (Three Diamonds), to a large constellation of 200 satellites (Pearls). Standardization needs to be a predominant factor in the design: in a mission with few satellites, it is possible to introduce customizations both on hardware and software; a sizeable constellation requires to have all the functionalities needed on all the satellites, with hardly any difference but an identification code. Furthermore, the standardization implies that the design must consider the worst-case scenario of the whole constellation. An example of this is the thermal design that needs to be adequate, at the same time, for all attitudes and orbits that the constellation operates in.

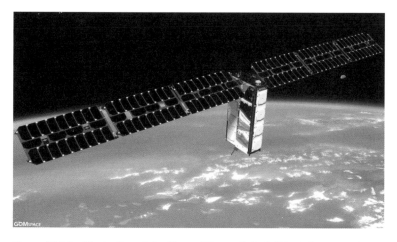

Figure 20.15 Illustration of the Sky and Space Global's "Pearls" spacecraft.

The different requirements between an IOD and a commercial satellite impose a different approach toward lifetime and redundancy. For example, the acceptable depth of discharge of the batteries is lower for a longer lifetime satellite, requiring additional cells compared to an IOD. Even though the functionalities might be the one tested in the IOD mission, the duty cycles of a commercial mission are more demanding, increasing the required power generation and the stored energy to survive the eclipse. One more aspect to consider is that the utilization of the whole constellation necessitates relaying commands through the constellation. This requires an analysis of the constellation network, to efficiently exchange data between satellites and ground.

The production of the satellites is heavily affected by the quantity of units. When one or few satellites are to be integrated, it is common to integrate one satellite completely before moving to the following one. When the number increases, the efficiency of the production line is a key factor: integration stations are created, where small steps of the integration are performed, to allow a parallel and more efficient process. In addition, the procurement of the components can cause heavy delays if not carefully planned and verified.

The differences are not only in the design and manufacturing of the satellites. A major change in effort comes in the LEOP phase: in a mission consisting of a small number of satellites, the tests and early operations can be performed manually, interacting directly with each satellite. In a constellation of 200 satellites, it is unrealistic to do so, and therefore requires a high level of automatization (Figure 20.15).

20.5 Perspective

The chapter presents examples of the variety of nanosatellite missions, from small educational satellites to missions that would have not been possible using conventional satellites due to the high cost and long development time or simply due to the nature of conventional satellites.

Since the early days of nanosatellites, CubeSats have been a tool for university space education. Nowadays, more and more universities around the world are using nanosatellites in their educational programs. Starting from technology demonstration missions and the consequent maturity that these have brought, nanosatellites found their way into the commercial market with a large variety of applications. In addition, research institutions around the world have adopted the nanosatellite capabilities in their missions to explore Earth and space. Outer space is the newest challenge in the nanosatellite world. Considerations need to be made with regard to radiation protection, power generation, and lifetime. Lunar and deep space missions are currently being planned, and it will be interesting to see the discoveries that will be enabled by this technology.

The nanosatellite world is still in its early days and several technical challenges are to be resolved. However, the increasing momentum and investments in nanosatellite technologies will facilitate the utilization of space technologies for a much wider field.

References

1 GomSpace, "About GomSpace," [Online]. Available: https://gomspace.com/UserFiles/about_gomspace_1page.pdf.

2 European Space Agency—Space Engineering & Technology, 2017."Technology CubeSats," 15 November 2017. [Online]. Available: http://www.esa.int/Our_Activities/Space_Engineering_Technology/Technology_CubeSats.

3 Aarhus University, 2017 "The Aarhus University nano-satellite project," Aarhus University, 10 October 2017. [Online]. Available: http://projects.au.dk/ausat/delphini1. [Accessed 14 June 2018].

4 GomSpace A/S, 2017"The Colombian Air Force orders its second advanced Nanosatellite platform from GomSpace," GomSpace, 1 December 2017. [Online]. Available: https://gomspace.com/news/the-colombian-air-force-orders-its-second-adv.aspx. [Accessed 14 June 2018].

5 European Space Agency—Space Engineering & Technology, 2015 "ESA's first technology nanosatellite reporting for duty," 16 October 2015. [Online]. Available: https://www.esa.int/Our_Activities/Space_Engineering_Technology/ESA_s_first_technology_nanosatellite_reporting_for_duty.

6 M. Fernandez, A. Latiri, T. Dehaene, G. Michaud, P. Bataille, C. Dudal, P. Lafabrie, A. Gaboriaud, J.-L. Issler, F. Rousseau, A. Ressouche and J.-L. Salaun, 2016. "X-band transmission evolution towards DVB-S2 for Small Satellites," in *Small Satellites*, Utah.

7 D. Gerhardt, M. Bisgaard, and L. Alminde, 2016 "GOMX-3: Mission Results from the Inaugural ESA In-Orbit Demonstration CubeSat," in *AIAA/USU Conference on Small Satellites*, Logan, Utah.

8 NASA, "International Space Station—NanoRacks-GOMX-3," ISS Program Science Office, [Online]. Available: https://www.nasa.gov/mission_pages/station/research/experiments/2078.html.

9 J. A. Larsen, D. Gerhardt, M. Bisgaard, L. Alminde, R. Walker, M. Fernandez, and I. Jean-Luc, 2016 "Rapid results: The GOMX-3 CubeSat path to orbit," in *The 4S Symposium*, Valletta.

10 Flightradar24, 2016 "Successfully Testing Satellite-based ADS-B Tracking," 01-07-2016. [Online]. Available: https://www.flightradar24.com/blog/tracking-flights-with-satellite-based-ads-b-receivers.

11 European Space Agency—Space Engineering & Technology, 2016 "Tiny CubeSat tracks worldwide air traffic," 07 April 2016. [Online]. Available: https://www.esa.int/Our_Activities/Space_Engineering_Technology/Tiny_CubeSat_tracks_worldwide_air_traffic.

12 European Space Agency—Space Engineering & Technology, "Technology CubeSat hitch-hiker on today's HTV launch," [Online]. Available: http://www.esa.int/Our_Activities/Space_Engineering_Technology/Technology_CubeSat_hitch-hiker_on_today_s_HTV_launch.

13 F. Ongaro, 2017 "The impact of new markets on ESA technology priorities (& management)," in *Swiss Space Industry Days*, Lausanne.

14 European Space Agency—Space Engineering & Technology, 2018 "ESA's GOMX-4B CubeSat relaying data across space from Danish twin," 18 April 2018. [Online]. Available: http://www.esa.int/Our_Activities/Space_Engineering_Technology/ESA_s_GomX-4B_CubeSat_relaying_data_across_space_from_Danish_twin.

15 European Space Agency—Space Engineering & Technology, 2018 "The size of a cereal box: ESA's first satellite of 2018," 02 February 2018. [Online]. Available: https://www.esa.int/Our_Activities/Space_Engineering_Technology/The_size_of_a_cereal_box_ESA_s_first_satellite_of_2018.

16 European Space Agency—On-board computer and data handling, 2017 "Chimera board," 24 May 2017. [Online]. Available: https://www.esa.int/Our_Activities/Space_Engineering_Technology/Onboard_Computer_and_Data_Handling/CHIMERA_Board.

17 European Space Agency—Space Engineering & Technology, 2018 "Putting everyday computer parts to space radiation test," 29 January 2018. [Online]. Available: http://www.esa.int/Our_Activities/Space_Engineering_Technology/Putting_everyday_computer_parts_to_space_radiation_test.

18 M. Esposito, S. Conticello, C. Van Dijk, and N. Vercryssen, 2018 "Initial Operations and First Light from a Miniaturized and Intelligent HyperSpectral Imager for Nanosatellites," in *The 4S Symposium*, Sorrento.

19 S. Writers, 2008 "ISIS to develop star tracker for nanosatellites," SpaceDaily, 07 October 2008. [Online]. Available: http://www.spacedaily.com/reports/ISIS_To_Develop_Star_Tracker_For_Nanosatellites_999.html.

20 L. León Pérez and P. Koch, 2018 "GOMX-4—the twin European mission for IOD purposes," in *AIAA/USU Conference on Small Satellites*, Logan, Utah.

21 L. León Pérez, P. Koch, D. Smith and R. Walker, 2018 "GOMX-4, the most advance nanosatellite mission for IOD purposes," in *The 4S Symposium*, Sorrento.

22 Tesat-Spacecom, 2018 "World's smallest Laser Communication Terminal from Tesat on track—CDR for CubeL successfully held," 12 April 2018. [Online]. Available: https://www.tesat.de/en/media-center/press/news/699-pi1288-world-s-smallest-laser-communication-terminal-from-tesat-on-track-cdr-for-cubel-successfully-held. [Accessed 14 June 2018].

23 J. Virgili-Llop et al., 2014 "Very Low Earth Orbit mission concepts for Earth Observation. Benefits and challenges," in *Reinventing Space Conference*, London, UK.

24 P. C. E. Roberts et al., 2017 "DISCOVERER—Radical Redesign of Earth Observation Satellites for Sustained Operation at Significant Lower Altitudes," in *68th International Astronautical Congress (IAC)*, Adelaide, Australia.

25 "Teser Project," [Online]. Available: http://teserproject.eu.

26 P. Voigt, C. Vogt, L. León Pérez, L. Ghizoni, R. Förstner, K. Konstantinidis, A. Wander, M. Valli, S. Brilli, A. Kristensen, C. Underwood, H. Stokes, T. Lips et al., 2018 "Teser—Technology for self-removal—status of a horizon 2020 project ensure the post-mission-disposal of any future spacecraft," in *International Astronautical Congress*, Bremen.

27 GomSpace A/S, 2016 "GomSpace and IFU will Collaborate on a Big Satellite Project in Africa," GomSpace, 6 June 2016. [Online]. Available: https://gomspace.com/news/gomspace-and-ifu-will-collaborate-on-a-big-sa.aspx. [Accessed 14 June 2018].

28 Sky and Space Global Ltd., 2017 "Partnership Discussions Underway with Strategic Industry Players," 25 September 2017. [Online]. Available: www.investi.com.au/api/announcements/sas/54f82796-f2e.pdf. [Accessed 14 June 2018].

29 C. Mbaka, 2018 "Sky and Space Global is using nano-satellites to provide affordable connectivity in Africa," 20 February 2018. [Online]. Available: https://techmoran.com/sky-space-global-using-nano-satellites-provide-affordable-connectivity-africa. [Accessed 14 June 2018].

30 D-Orbit, 2017 "D-Orbit to collaborate with Sky and Space Global on Launch and Deployment of the SAS Nanosatellite Constellation," D-Orbit, 18 October 2017. [Online]. Available: http://spaceref.com/news/viewpr.html?pid=51723. [Accessed 14 June 2018].

21 II-1

Ground Segment

Fernando Aguado Agelet and Alberto González Muíño

Signal Theory and Communications Department, University of Vigo, EE Telecomunicación, Spain

21.1   Introduction

The ground segment is a key element of any space system [1, 2]. It provides the essential link with satellites to perform operations or provide system services. In this chapter, an introduction to ground segment in nanosatellite systems is presented.

First of all, ground segment functionalities are identified, and a typical architecture is shown. Within this configuration, hardware and software elements of a ground segment are introduced to provide the reader with a complete view of its design for nanosatellite missions.

To conclude, new trends in the ground segment for nanosatellites, such as software-defined radio (SDR) solutions, are introduced.

21.2   Ground Segment Functionalities

The ground segment provides the communication link with the satellite or satellites to perform operations, receive housekeeping telemetry and scientific data, and send telecommands to execute platform and payload operations on board.

Figure 21.1 shows the main functionalities of the ground segment:

- Control and communicate with the space segment, this involves:
  - Generating the operation plans.
  - Controlling the spacecraft according to the operation plans.
  - Obtaining the requested data by the user.
- Communicate with the user segment, this involves:
  - Receiving the requests from the users.
  - Distributing the scientific data among users.

Nanosatellites: Space and Ground Technologies, Operations and Economics, First Edition.
Edited by Rogerio Atem de Carvalho, Jaime Estela, and Martin Langer.
© 2020 John Wiley & Sons Ltd. Published 2020 by John Wiley & Sons Ltd.

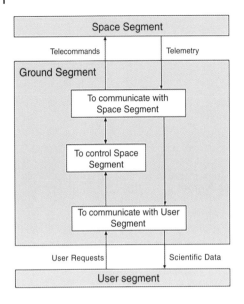

Figure 21.1 Ground segment functional architecture.

21.3 Ground Segment Architecture

Figure 21.2 presents typical architecture of a standard ground segment for small satellite missions. The name of the facilities and their interfaces may vary from mission to mission, since the requirements of the mission may impose a different architecture. These services are often located in the same center in this type of missions.

- The ground station (GS) (or a network of ground stations) is in charge of providing the link between the ground segment and the satellite or satellites. It manages the physical layer and, sometimes, also the link layer. The ground station performs the satellite tracking during the communication accesses with the spacecraft.
- The ground control facilities (GCFs), in case of a distributed ground segment topology, communicate the different ground stations with the mission operation facilities through the communication management facility (CMF).
  - The CMF is in charge of the data communication function. It communicates with a directly connected ground station or a ground station network coordinating with the ground station in use at each moment. It maintains and supervises the communication link between the ground station that is tracking the spacecraft and the spacecraft operation center.
  - The flight dynamics facility (FDF) performs flight dynamics tasks propagating the spacecraft orbit in order to know its predicted position. It is a shared facility between the ground control and the mission operation facilities, since its products are used by both of them. In the case of the ground control facilities, the FDF provides accurate tracking data to point the antennas and adjusting the ground station equipment.
- The mission operation facilities (MOF) are in charge of the satellite and payloads operation, mission planning, and mission data archiving and distribution.

- The spacecraft operations facility (SOF) oversees the monitoring and control of the spacecraft. It analyzes the telemetry received from the satellite and executes the operation plans generated in the mission planning facilities.
- The mission planning facility (MPF) is in charge of the generation of operation plans for the mission. These plans are based on the users' requests and the predicted position of the spacecraft; thus, the predicted positions of the spacecraft are retrieved from the FDF. The MPF provides an interface to the users to collect their requests.
- The data archiving and distribution facility (DADF) archives and distributes the mission data. These data include the housekeeping data (engineering data) used by the SOF and the user data (scientific). The DADF provides an interface to the users to access this data.

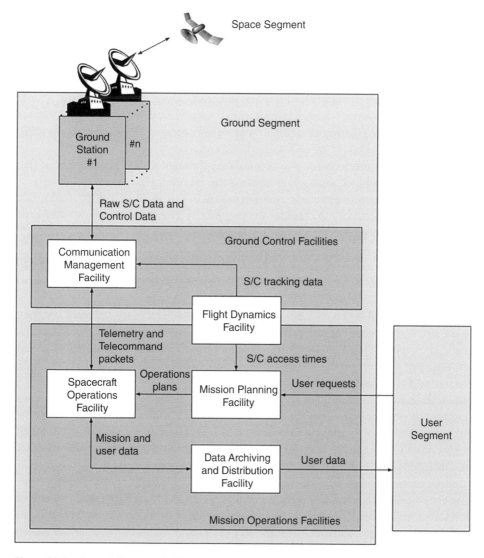

Figure 21.2 Ground Segment Architecture and key facilities.

21.4 Ground Station Elements

The main elements of a ground station are illustrated in Figure 21.3, including the radio frequency (RF) equipment, the structural elements, and the ground segment software. Nanosatellites usually operate in radio amateur bands, on VHF, UHF, and L-bands for bidirectional communications and S-band for one-way downlink communication.

21.4.1 Radio Frequency Equipment

From the communication point of view, RF equipment aims to provide a transparent path between the baseband signals and the satellite and vice versa. RF equipment includes all the physical layer communications elements along with the transceiver, the hardware control software interface, and the telemetry, tracking, and command (TTC) modems.

Cross Yagi antennas are widely used for VHF (144–146 MHz) and UHF (432–438 MHz) bands. Antenna gains typically range from 12 to 18 dBi for VHF and from 16 to 18 dBi for UHF. The antenna gain depends on the number of elements of the antenna array, thus, higher gains require larger and heavier antennas, a more robust shaft, and a more powerful rotor. Cross Yagi antennas give either vertical and horizontal polarizations or left-hand circular polarization (LHCP) and right-hand circular polarization (RHCP) simultaneously.

The antennas are connected to the preamplifier through coaxial pigtails. The pigtail lengths shall be computed taking into account the maximum antenna rotation both in the horizontal and vertical planes. In addition to this, in case of a LHCP/RHCP cross Yagi antenna, it is also necessary to consider the coaxial characteristics in order to select the proper lengths to preserve a correct wave phasing between both polarizations.

Depending on the received power by each polarization, the GS operator shall select it through a switch. In advanced nanosatellite ground stations, UHF and/or VHF chains may be duplicated to receive both polarizations simultaneously and transmit the uplink signal using the best one.

For L-band (1–2 GHZ) and S-band (2–4 GHZ), helical antennas are the most common choice, although parabolic antennas may be used in case of a desired higher reception gain. S-band is currently used for downloading telemetry and scientific data in radio amateur GSs. Figure 21.4 shows a radio amateur ground station including a dual polarization UHF scheme and VHF, L and S antennas.

From the communication point of view, low-noise high-gain preamplifiers (LNA) must be located as close as possible to the antenna system to minimize the overall noise factor of the ground station. For S-band frequency, downconverters are commonly used, which in addition to amplifying the received signal also convert it into the UHF or VHF standard band included in all the radio amateur transceivers.

Coaxial cables are used to interconnect the different RF elements present in a GS. Typically, low-loss coaxial cables with a diameter between 10 and 15 mm are used. These thick cables are not only used to minimize the communication chain losses but also to reduce the GS system noise factor. Table 21.1 summarizes the typical values for the communication chain in a radio amateur ground.

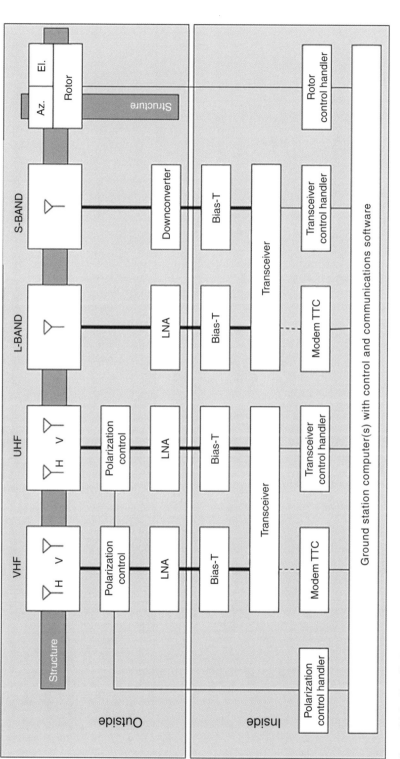

Figure 21.3 Detailed ground station block diagram.

Table 21.1 Typical values for the communication chain in a radio amateur ground station.

Band	TX/RX	Antenna	Pigtail (5 m)	Lightning protector	Preamplifier	Coaxial cable	Lightning protector	Transceiver	Mod.
VHF	RX	G = 12.2 dBi SWR <1.6	L = 0.2 dB	L = 0.2 dB SWR <1.3	G = 20 dB NF = 0.9 dB	L = 0.9 dB	L = 0.2 dB SWR <1.3	a)	b)
	TX				L = 0.2 dB			From 5 to 100 W	
UHF	RX	G = 16.2 dBi SWR <1.6	L = 0.5 dB	L = 0.2 dB SWR <1.3	G = 20 dB NF = 0.9 dB	L = 1.5 dB	L = 0 0.2 dB SWR <1.3		b)
	TX				L = 0.2 dB			From 5 to 75 W	
L-band	RX	G = 13.2 dBi SWR <1.5	L = 0.8 dB	L = 0.2 dB SWR <1.3	G = 20 dB NF = 0.9 dB	L = 2.9 dB	L = 0.2 dB SWR <1.3		b)
	TX				L = 0.2 dB			From 1 to 10 W	
S-band	RX	G = 18.2 dBi SWR <1.5	L = 1.2 dB	L = 0.2 dB SWR <1.3	G = 32 dB[c] NF = 0.7 dB	L = 0.9 dB[c]	L = 0.2 dB SWR <1.3	a)	b)

a) Transceiver parameters for VHF, UHF, and L-band (S-band is downconverted to VHF):
- FM:
  - Sensitivity: 12 dB, SINAD, 0.18 μV
  - Selectivity: 15 KHz/−6 dB
- Narrow FM:
  - Sensitivity: 12 dB, SINAD, 0.18 μV
  - Selectivity: 6 KHz/−6 dB
- SSB, CW:
  - Sensitivity: 10 dB, SINAD, 0.11 μV

b) Typical modulation: AFSK 1200 bps, FSK 9600 bps, MSK 1200 bps, BPSK 1200 bps.

c) Frequency downconversion to VHF band, in case of an S-band preamplifier: G = 20 dB, NF = 1.2 dB, Frequency band: 2300–2400 MHz. Coaxial cable losses in S-band: 4 dB.

The power module, not drawn in Figure 21.3 for simplicity, includes a power supply and provides power to all the active components. The power is distributed through dedicated cables, or, in the case of the preamplifiers, through the coaxial cable. In this case, the coaxial cable is connected to the power module through a particular interconnection device called "bias-T."

The transceiver is the radio equipment responsible for modulating, amplifying, and upconverting the baseband signal received from the TTC modems to VHF, UHF, or L-bands for the uplink. For the downlink, it makes the complementary tasks: it amplifies, downconverts, and demodulates the RF signal received from the satellite into a baseband signal, which is sent to the TTC modems. Commercial radio amateur radios have three RF interfaces: VHF, UHF, and L-band (S-band uses the VHF or UHF interface, since it is previously downconverted to one of these bands). Also, there are two external connections, one to the modem, and another to the computer.

The TTC modem is in charge of the communication part: its central element is the terminal node controller (TNC) or modem, which provides the communication interface to the transceiver. It receives telecommand data frames from the computer, modulates them in baseband, and sends them to the transceiver. On reception, it receives the baseband signal from the transceiver, demodulates it, and sends the decoded frames to the computer in order to be processed.

Hardware control handlers automatically adjust both the rotor and the transceiver during a pass, based on the orbit prediction provided by the FDF. This module retunes the radio during the overall pass compensating the frequency Doppler shift on the ground station because the spacecraft transmits and receives using its nominal frequency. Furthermore, the control software updates the satellite elevation and azimuth to the rotor control hardware, allowing a proper antenna pointing to the satellite during the overall pass.

21.4.2 Structural Elements and Rotor

The structure includes the shaft, which supports the rotor with the antenna array, the supporting antenna masts, and the applicable ladders and service footbridges (Figure 21.4).

The rotor is a software-controlled engine, which points the antennas to the satellite during each pass above the ground station. The azimuth and the elevation are typically updated every degree or half of a degree. Most of the commercial rotors allow up to 360° + 180° turns in azimuth, and up to 180° turns in elevation.

The rotor selection should take into account the physical properties of the antennas and the type of mission orbit. The higher the orbit is, the less angular speed the rotor needs. Common available commercial rotors are entirely suitable for typical low Earth orbits (LEO) higher than the International Space Station. On the other hand, missions using lower orbits than the International Space Station should carefully select a proper rotor fast enough to track the satellite.

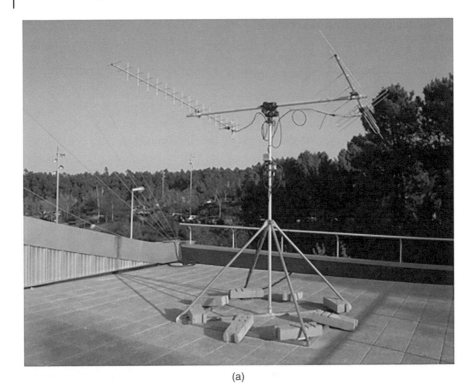

(a)

(b)

Figure 21.4 (a) A small satellite ground station including cross Yagi VHF and UHF antennas and L and S helical antennas. (b) Ground station control room.

21.5 Ground Segment Software

A significant effort of the design of a ground segment should be in the software selection and development. This software manages all the hardware elements and provides tools to the operators to execute their work.

The software present in a ground segment (Figure 21.5) is used for different stages of the operation. During the preparation of the operation, software to predict the orbital position of the spacecraft is needed as well as software to prepare the telecommand sequences to be sent to the satellite. During the communication accesses with the satellite, the software shall be able to manage the pointing of the antennas, the adjustment of the radio equipment, and the maintenance of the communication link with the satellite. It is also necessary for a software tool to send telecommand sequences and to monitor the telemetry received from the satellite. After the execution of a communication access, software tools analyze the housekeeping and scientific data retrieved from the satellite and, in some cases, a software tool to reconstruct the orbit and attitude of the satellite shall be available.

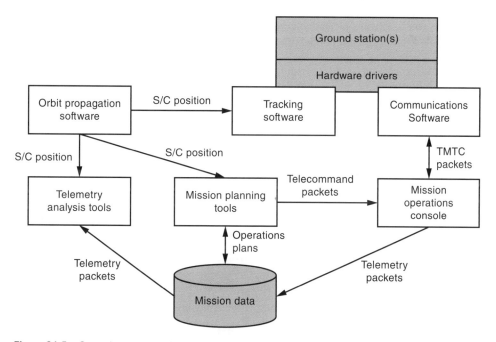

Figure 21.5 Ground segment software architecture.

21.5.1 Orbit Propagation Software

One of the essential software tools in a ground segment is a software package that allows the operators to calculate the estimated position of the spacecraft at a precise moment. This information can be used both during the planning of the operations, the execution of a communication access, or during telemetry analysis.

The software package retrieves the orbital parameters of the satellite from the Internet in the form of two-line elements (TLEs). The main TLE information is composed by the

North American Aerospace Defense Command (NORAD), and the data can be accessed through different websites, such as CelesTrak [3] or Space-Track [4]. NORAD generates the TLE from radar measurements of the object in orbit. Each time that an object modifies its orbit above a defined limit, the NORAD publishes an updated TLE.

Several software packages are available for free, and they are suitable for the operation of a nanosatellite.

21.5.2 Tracking Software

The antenna pointing and the radio equipment adjustment (Doppler shift correction, output power, etc.) are managed by the tracking software. This software uses data from the orbital propagator data to know the satellite position relative to the ground station antennas. Using this information, the antenna is pointed toward the satellite and the Doppler shift is compensated in the reception and transmission frequency of the radio equipment.

The tracking software is connected to the station hardware with proper drivers specially developed for the equipment present in the station.

This kind of software is available with integrated orbital propagation capabilities and compatible drivers for commercial off-the-shelf (COTS) station components [5, 6].

21.5.3 Communications Software

As mentioned before, the maintenance of the communication link between the spacecraft and the operators is one of the main functionalities of the ground segment. A software component shall be available to monitor and control the communication link.

This component can manage the retransmissions, error control and correction, and routing of packets from/to different ground stations.

21.5.4 Mission Planning Tools

One of the primary activities of the operation is the mission planning. The ground segment shall provide software tools to the operators in order to plan the experiments, observations, or housekeeping operations of the satellite. These software tools use as inputs the information provided by the orbit propagation software.

The output of the mission planning tools is the telecommand sequences that are used to program the on-board computer.

21.5.5 Mission Operations Console

During the execution of the communication pass, operators need a console to monitor the health of the satellite, to send telecommand sequences, and to verify the correct reception of these telecommand sequences by the on-board computer.

Since the logical interfaces of each mission and the functionalities implemented by the on-board software vary from mission to mission, standard software valid for every mission is not available. This software tool shall be developed and validated specifically for each mission/satellite.

21.5.6 Telemetry Analysis Tools

Each mission shall develop its own tools to analyze the housekeeping and scientific data retrieved from the satellite. These tools may use the reconstruction of the orbital position provided by the orbit propagation software in order to correlate the experimental data (observations) with the position of the satellite in every moment.

21.6 Ground Segment Operation

21.6.1 Usage Planning

The ground segment represents a limited resource. Its usage shall be carefully planned in order to optimize it and to cover the needs of the mission.

In a typical ground station, only one satellite can be simultaneously attended. The antennas can only point to one region of the sky and, therefore, tracking another satellite in another sky area at the same time is not possible. The access time of the satellites operated from a ground station shall be predicted in advance using orbital propagation software. Conflicts between different satellites (access times overlap) shall be solved in advance.

It is important to take care of the temporal constraints of the ground station when planning access time. The operators will only be available during canonical working hours in some cases, or perhaps there will be other time constraints that shall be taken into account.

There are some commercial software tools that facilitate the planning of the ground station. The operator has to introduce all the constraints of the system, the satellites to be tracked, and some data about the ground station. Once the software has this information, it can generate a preliminary planning showing all the conflicts that the operator shall solve.

21.6.2 Communication Access Execution

No improvisation is allowed in a space mission. All the activities to be performed with the satellite shall be planned in advance, and these include the communication access with the ground station.

For each communication opportunity, the procedures to be executed shall be decided and prepared beforehand. This preparation includes the generation of the telecommand sequences to be sent to the satellite and the planning of what and how much telemetry will be retrieved during the communication with the spacecraft.

Once the communication procedure is completely prepared, it is time to set the ground station. With enough time in advance, all the equipment shall be configured for the communication and the antennas shall be pointed to the predicted acquisition of signal (AOS) azimuth.

The operators shall be ready at the ground station to perform all of these tasks. Typically, one operator manages the ground station activities (tracking, communication link monitoring, etc.) and another operator is in charge of the operation of the satellite (telecommand sequences transmission, spacecraft health monitoring, etc.). For complex operations that require a fine coordination, there may be an operations coordinator during the communication. In bigger missions, there may be several operators controlling and commanding the spacecraft.

Once the satellite rises above the horizon, the planned procedures shall be executed. If everything goes well, all the planned activities must be executed, and the desired telemetry be retrieved from the satellite. Sometimes, an anomaly or a contingency situation appears during the operations. It is important to have suitable procedures to respond to these situations and to train the operator to deal with them.

After the completion of the communication access, once the satellite signal is lost, it is time to park the antennas in a safe position, switch the equipment to the standby mode or off, and to archive the data received from the satellite. These data are analyzed later by the operations engineers.

The operators shall log all the activities in order to have future references about the operations performed by the satellite. These logs allow future anomaly investigations and provide a view of what happened in the mission.

21.7 Future Prospects

In this section, some of the future prospects in the ground segment engineering will be introduced. The technology involved in a ground segment, both software and hardware, is continuously improving in order to optimize the use of the resources and, definitely, to reduce the cost of implementation and operation of the missions [7].

In particular, there are a lot of advances in the field of ground station implementation for nanosatellites in the amateur bands since the explosion of CubeSat-based missions. Several universities and particulars are working in order to bring improvements to the ground stations. Two of them are SDR technologies and ground station automation.

21.7.1 SDR

The main idea behind SDR is to replace some of the hardware components of the ground station by software components. This can be achieved placing analog to digital conversion as close as possible to the antenna and implementing signal processing techniques. This software-based architecture provides the flexibility to change easily the definition of the modulation, type of signal, or bandwidths. With SDR, the ground stations increase their flexibility and compatibility and reduce their cost, since the hardware is not mission-specific and it is possible to use it across different missions.

In Figure 21.6, a typical ground station with SDR is shown:

The main change in the hardware is the replacement of the radio transceivers and modems by a RF frontend in charge of:

- Band selection (filter).
- Upconversion/downconversion from/to baseband to/from the desired operation band.
- Power amplification for transmission.
- Low noise amplification for reception.
- Analog to digital conversion and vice versa.

Figure 21.6 Common ground station versus SDR ground station.

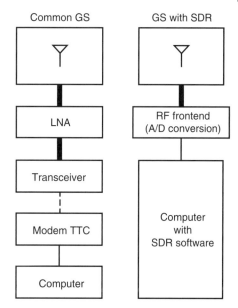

Modulation/demodulation, encoding/decoding, encryption/decryption is done in software, which implements digital signal processing techniques. In order to adapt ground stations for a new mission with a new modulation, frame format, or encoding, it is only necessary to add the specific processing blocks to the ground station software.

A ground station with SDR can also allow receiving several spacecraft signals simultaneously using the same RF frontend if the baseband bandwidth is enough. Each of the signals can be processed by different software blocks, allowing them to use different modulations and encodings at the same time. This is useful, for example, for Launch and Early Orbit Phase (LEOP) operations when a constellation of nanosatellites is carried with the same launcher.

Nowadays, there are commercial RF frontends [8, 9] suitable for a nanosatellite ground station. Some development is needed regarding RF electronics in some cases but most of the issues can be solved with COTS components.

Regarding the software for SDR, in the last years, several ready-to-use tools for the reception of satellite signals appeared. In the case of transmission, the development of specific processing software for missions is needed. GNU Radio [10] is a development framework that makes the development of complete SDR systems easier, using already developed building blocks, and thus significantly helps while setting up an SDR ground station.

21.7.2 Ground Station Automation

With the evolution of the hardware and software components of a ground station, and the integration among them, it is now possible to automatize most of the activities of the station. The automation of the ground station will allow the reduction of costs in terms of required staff present in the station for repetitive tasks, optimization of the use of limited resources, and reduction of human errors.

Some of the activities that can be automated are:

- *Station scheduling:* As stated in the software section, it is possible to use software tools that generate a master schedule. Based on the station constraints and the spacecraft to be tracked, the software can predict the access times between the ground station and the spacecraft and schedule tracking times.

- *Station preparation for access time:* All the preparations for a satellite pass (antenna pointing, equipment configuration, etc.) can be automated if the station is adapted to do it. All the equipment shall be "software configurable" and proper software tools or script shall be developed for this purpose. These preparations are triggered by the master scheduler of the station.

- *Signal acquisition and communication link setup:* Once the tracked satellite enters into the field of view of the antenna, the ground station shall acquire the spacecraft signal and set the communication link between the satellite and the station. The communication software shall be able to manage the AOS and the upcoming actions to set the communication link (manage transmission power, polarizations, protocols synchronization, etc.). It is important to note that for this to be automated, it is necessary to design properly the communication protocols between space and ground and this affects the spacecraft design too.

- *Operations automation:* With the communication link working, the planned operations with the satellite can be executed. It is possible to automate the ground station activities (tracking of the spacecraft, register of GS events and data, maintaining the communication link, etc.) or to automate also the operations with the satellite. The automation of the satellite operations introduces high challenges regarding the monitoring and control of the satellite without human intervention and requires a complex software development in close cooperation with the on-board software developers. It can be as simple as a batch of telecommands to be sent to the spacecraft if certain conditions are met, or as difficult as autonomous responses generated in base of the satellite status and planned operations.

- *Communication access finalization:* Once the operations are finalized, the ground station shall go to a standby configuration. This includes parking the antennas in a safe position, shutting down or setting the equipment in standby mode, and archiving the logs and received data. All of these tasks can be automated by software scripts.

- *Fault detection and handling:* During any of the automated activities, a failure can appear in the hardware of the station or in the execution of the software. It is important to take this into account when designing the ground station automation, since a failure can cause the loss of the communication access or, even worse, the damage of equipment. Proper failure detection rules shall be implemented by the automation software in order to detect and handle these failures. In some cases, the failure is easily corrected by software but in other cases the human intervention is required; for these cases, the automation system shall implement some kind of notification to the station operators.

As can be observed, the automation of a ground station is a matter of software development and proper hardware/software integration. Some of the automated tasks can be solved with existing software tools and others require specific software developments. In any case, it is necessary to work on the automation from the beginning of the mission, since it imposes requirements over the ground station and over the spacecraft design. Once implemented, the complete ground station shall be tested in different scenarios, including possible failures, in order to be sure that the automation does not introduce any risk for the mission, and it work as expected.

References

1 65 Authors from the Astronautics Community (2011). *Space Mission Engineering: The New SMAD* (eds. J.R. Wertz, D.F. Everett and J.J. Puschell) (Space Technology Library, Vol. 28).

2 *Spacecraft Systems Engineering*, 2011, Peter Fortescue, Graham Swinerd, John Stark. United Kingdom, Wiley.

3 Celestrak: https://celestrak.com

4 SpaceTrak: https://www.space-track.org

5 Orbitron: http://www.stoff.pl

6 GPredict: http://gpredict.oz9aec.net

7 *International Study on Cost-effective Earth Observation Missions*, 2005, Rainer Sandau. Germany, DLR.

8 USRP: http://www.ettus.com

9 HackRF project: https://greatscottgadgets.com/hackrf

10 GNURadio: http://gnuradio.org

22 II-2

Ground Station Networks

Lucas Rodrigues Amaduro and Rogerio Atem de Carvalho

Reference Center for Embedded and Aerospace Systems (CRSEA), Polo de Inovação Campos dos Goytacazes (PICG), Instituto Federal Fluminense (IFF), Campos dos Goytacazes, Brazil

22.1   Introduction

A ground station, earth station, or earth terminal is a terrestrial radio station designed for extraplanetary telecommunication with spacecraft (constituting part of the ground segment of the spacecraft system), or reception of radio waves from astronomical radio sources. Ground stations may be located either on the surface of the Earth or in its atmosphere. Earth stations communicate with spacecraft by transmitting and receiving radio waves in the super high-frequency or extremely high-frequency bands (microwaves). When a ground station successfully transmits radio waves to a spacecraft (or vice versa), it establishes a telecommunications link. Ground stations may have either a fixed or itinerant position, specialized satellite earth stations are used to telecommunicate with satellites. Other ground stations communicate with manned space stations or unmanned space probes. A ground station that primarily receives telemetry data, or that follows a satellite not in geostationary orbit, is called a tracking station. When a satellite is within a ground station's line of sight, the station is said to have a view of the satellite. It is possible for a satellite to communicate with more than one ground station at a time [1].

22.2   Technological Challenges

In recent years, the tremendous progress in computational power has brought a number of innovations to the ground segment. This also implies technological challenges for ground stations and offers new approaches due to recent advances in digital signal processing or network control. Hence, this section points out the specific technological challenges and implications for the ground segment to operate multisatellite systems.

In this context, the terms multisatellite systems, distributed satellite systems, and satellite networks are used interchangeably, and describe in general a space segment with more than one spacecraft, used to achieve a common goal. The definitions of a formation, constellation, or cluster are used to distinguish between different topologies or satellite control strategies. Besides the space segment, the ground segment is an integral part of each space mission, in fact, a pair formed by one satellite and one ground station is a distributed system consisting of two nodes communicating with each other during contact. However, considering a more general case, having a multisatellite system connected to a ground station network is more suitable. Therefore, two different distributed systems can be identified, one in space and another on ground, in close relationship. The topology of the ground segment plays an important role for telemetry and telecommand due to the movement of the satellites on their orbits, the communication links between ground stations and satellites change frequently, adding new challenges to the operation of a satellite network, and demanding careful planning and scheduling [2].

22.3 Visibility Clash Problems of Stations and Satellites

The most common constraint in ground station scheduling is the clash of visibility windows caused by multiple spacecraft in relation to a single ground station. Visibility clashes are situations when two or more spacecraft passing over a ground station have overlapping visibilities and ground station resources must be apportioned equitably among the spacecraft so as to generate maximum value in the mission. This paper describes metaheuristic methods to optimally resolve visibility clashes. The methods apply both to telemetry, tracking, and command (TTC) and payload (PL) operations support. Initially we use a simple, linear objective function to set the stage for the procedure. Subsequently, we describe a method to derive the applicable nonlinear optimization function to help resolve visibility clash with realistic constraints and objectives. Such a payoff function can capture preferences, penalties, soft constraints, mission priorities, and other factors as articulated by space mission scientists. This leads to a general solution to the clash resolution.

Broadly defined, a "visibility clash" is said to occur when low Earth orbit (LEO) spacecraft have either overlapping visibilities at a ground station or their respective loss of signal (LOS) and acquisition of signal (AOS) difference is less than the station reconfiguration time. Nonclashing visibilities may be assigned tasks by priority-dispatch rules. However, "clash resolution" must find the optimum apportioned support schedule that maximizes the total value or utility generated from supports provided to each clashing spacecraft. In this process, some spacecraft may receive no support at all, yet the total schedule is optimal [3] (Figure 22.1).

Different types of objectives can be formulated, namely, maximizing matching of visibility windows of spacecraft to communicate with ground stations, minimizing the clashes of different spacecraft to one ground station, maximizing the communication time of the spacecraft with the ground station, and maximizing the usage of ground stations. A communication clash represents the event when the start of one communication task happens before the end of another one on the same ground station. The objective is to minimize the clashes of different spacecraft to one ground station. To compute the number of clashes, spacecraft are sorted by their start time [4].

Figure 22.1 Clash problem of multiple satellites and stations. (*See color plate section for color representation of this figure*).

22.4 The Distributed Ground Station Network

The distributed ground station network (DGSN) is a relatively novel network concept of small ground stations and connected via the Internet for performing automatic scans for satellites and other beacon signals. By correlating the received signal with the precise global navigation satellite system (GNSS) synchronized reception times of at least five ground stations, it enables the positioning of the signal's origin. Thus, a global tracking of small satellites becomes possible in this "reverse GPS" mode. It allows mission operators to position and track their small satellites faster after piggyback commissioning, when the final orbit is yet undefined and could differ from the specified orbit. Furthermore, it allows permanent communication in "data-dump" mode. In this mode, DGSN ground stations relay the received data to the servers and thus to the operator.

22.5 Infrastructure

From the very beginning of the space era, different facilities for satellite operation were grouped together in order to form the ground segment. The ground station network is here often associated with the receiving stations. Nevertheless, there is not always a clear distinction made between the ground network and appending facilities for mission control. These facilities are the mission control centers (MCCs), which are responsible for telemetry and telecommand of a spacecraft, and the science control centers (SCCs), which are dedicated to the science payload of a mission. These centers are not only logically separated; often they are also geographically distributed in different locations. All facilities are typically connected through a data network responsible of collecting, exchanging, and archiving data, and to grant access to end users.

In the scope of many space missions, a broad spectrum of ground infrastructure was established. The available resources were steadily upgraded and emerged over time to support different satellite missions. Therefore, those networks and appended facilities have grown to a complex system containing a variety of network architectures and technologies. Hence, it is hard to derive a generic architecture of a traditional ground station network. In the context of low-cost ground stations, a dedicated mission and SCC on a distant location are mostly not set up. Each individual ground station originates from a small satellite project, and the ground station itself performs all necessary tasks for satellite operation and can be seen as a stand-alone system. When these stations are combined to form a network, the tasks of the operation centers are performed from software applications, using the Internet for data exchange.

22.6 Planning and Scheduling

One of the most limited resources of a satellite in LEO is for sure the communication time. The contact window between a satellite and ground station can be determined from the orbit elements and the location of the ground station. For a LEO, satellites are typically five or six contact windows between 5 and 15 min available each day. This implies two major drawbacks: First, the satellite is only visible for a few minutes each day, which limits the amount of transferable data dramatically. Moreover, the ground station is not utilized for a large fraction of each day. Second, the contact windows have fixed start and end times due to orbit geometry, and a ground station can serve only a single satellite at a time. Therefore, overlapping contact windows of individual satellites leads to the question which of these satellites should be operated first. To handle conflicts respectively overlapping contact windows, scheduling comes into play.

22.7 Generic Software Architecture

It is possible to define a generic software architecture that can be used as a guide to develop a GSN. The functions of such an architecture are defined according to the tasks that must be performed by a GSN, which are planning the communications, orchestrating the network, and controlling the nodes of the network. The type of the network will be defined by the way the decisions regarding synchronization of the nodes are made in the network; if decisions are centralized, it will be a client-server network, if they are decentralized, it will be a peer to peer network. Figure 22.2 presents a generic, client-server architecture, thus based on centralized decision-making. This type of architecture is here introduced because

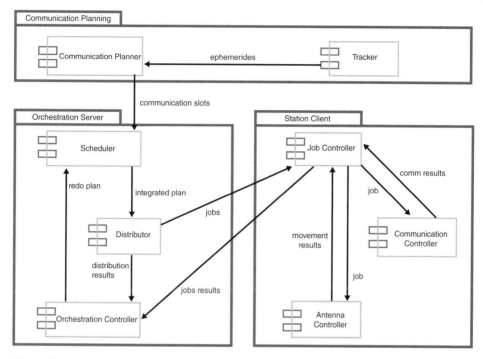

Figure 22.2 Generic software architecture.

it is closer to the traditional control center concept, and therefore easier to understand. It is divided into three main blocks: communication planning, orchestration server, and station client, described in the following text. Communication planning is composed by the communication planner and tracker. The tracker gets the satellite ephemeris and sends it to the planner, who does the important job of crossing the satellites' passing data on the stations with the energy they will have available, and the demands of data and telecommands to transmit. In this way, the planner defines for each orbit of each satellite how much data it will have to transmit and receive, depending on its power—who must take into account the tasks that the satellite has to perform in space, the level of its batteries, and whether they will be recharged during the run over the network.

The orchestration side is responsible by coordination of the network itself. With the planning carried out, it is necessary to define the network programing, which is done by a scheduler, who must define in an optimal way how much data and remotes each station will assume from each satellite. The scheduler receives the communication demand information already adjusted by the available powers and matches them again with the ephemeris, optimally determining the jobs for each station. With jobs determining what each station should do for each satellite, the distributor takes care of distributing these jobs over the network to clients, returning the result of that distribution to the orchestration controller, who in turn keeps a log of how the distribution and, later, the orchestration of the network itself are performing, that is, how the execution of what was planned in terms of communication between the space and the ground segments is really happening. The orchestration controller must have the ability, with the results of the distribution and execution tasks, to issue alerts so that in case of failures in those activities that compromise the optimization, a new scheduling is provided.

Finally, on the client side, at each station, the job controller receives the jobs and translates them to two physical systems that must operate in synchronized physical conditions, namely, the antenna controller, who controls the antenna and its movements, and the communication controller, who manages the operations of a digital radio, including frequency adjustments, Doppler compensation, and other parameterizations, as well as its interaction with the computer connected to the network, in order to obtain the data and commands to be transmitted. The job controller reports to the orchestration controller any anomaly in the physical environment that could lead to noncompliance in the jobs determined.

Although other architectures can be designed, the functions should be approximately the same as those presented here. The main point to consider is that, in the case of decentralized decision-making models, the server package modules would be replaced by client-hosted algorithms based on negotiation methods. In the case of the communication planner, it could be replaced by a global planning data pool, where the constraints and demands of all satellites and stations would be posted, so that conflicts and opportunities for collaboration could be collectively identified.

22.8 Example Networks

In order to bring knowledge from different experiences, three example GSNs are introduced in this chapter, one considered to use a traditional approach, with basic procedures for exchanging of information, and the two others based on the more recent advances of informatics, being one considered heterogeneous, or made by different stations, and the other built on top of the same model of station. The next topics describe these networks.

22.9 Traditional Ground Station Approach

From the first space missions, the ground segment consisted of several entities; exchanging information between these parts can be considered already as networking. The aggregation of different stations was already performed in the beginning of the space era to achieve better coverage or to increase redundancy. A good example for this "traditional" or "classic" approach is the ESA tracking stations (ESTRACK) system, consisting of nine ground stations combined to a network, and it was initiated already in the early 1970s. Also NASA started quite early with the combination of different stations to networks, the Deep Space Network (DSN) was established in 1958 [5].

Before addressing the traditional ground station approach, an important aspect related to the taxonomy of the term ground station needs to be clarified: In literature, the ground segment is typically divided into mission control and ground station network. The mission control is aggregated in control center (mission control center [MCC], spacecraft operations control center [SOCC]) and is responsible for mission planning and mission operations, for example, monitoring and commanding the spacecraft. The ground station network, composed of different ground stations, is on the other side dealing with signal reception and transmission, orbit tracking, etc. The mission control and the ground station network are not only logically divided but also often geographically separated from each other. A typical ground station contains in this context several receiving stations including different types of antennas as well as corresponding hardware equipment. For example, the ground station in Weilheim (part of the ESTRACK system) contains six different antennas ranging from 6 to 30 m to support deep space missions as well as near Earth missions. So, a ground station describes in a traditional sense a sophisticated system containing a larger number of facilities for satellite communication, acting as the interface between satellite and mission control center. The traditional ground station network contains only a few, but therefore highly specialized stations, to support a broad spectrum of space missions. On the contrary, in the context of small satellite projects, a ground station is considered rather as a single entity used to access one single satellite, and consists here typically only of one antenna system and corresponding equipment for communication with a single satellite in LEO [2].

The ESTRACK system is used to explain the overall organization of a large ground station network (see Figure 22.3). More specifically, the architecture of a typical ground station, that is, a satellite receiving station will be explained. The ESTRACK core ground network consists currently of nine ground stations located in Australia, Africa, Europe, and South America. Since 1968, ESTRACK supported more than 60 missions. The ground stations of the ESTRACK system contain many highly specialized stations. Thus, dedicated stations for deep space missions and geostationary satellites exist and can in general not be replaced with each other. The ESTRACK system has for example dedicated 35 m diameter antennas for deep space missions, and for near Earth missions 15 m diameter antennas are

Figure 22.3 Map showing locations of ESTRACK stations. Source: Reproduced with permission from ESA.

provided. The frequency bands range from S-band to k_a-band [6], thus all kinds of missions can be supported by ESTRACK, but dedicated stations for specific type of missions need to be requested.

An exemplary ground station for communication with a LEO satellite consists typically of the following units: antenna system, transmitting and receiving equipment, as well as TTC equipment: The antenna system contains different-sized apertures, mainly large parabolic dish antennas for signal reception (the size depends on several factors, important is to provide the segment, hence not all satellites need to be operated at the same time). A different strategy is to use a network of highly distributed ground stations to efficiently operate a distributed space segment. A new concept of low-cost ground station networks has evolved in the recent years, mirroring the distributed satellite approaches in the ground segment, too. It is especially well suited for the operation of huge number of satellites. The advantage of these low-cost receiving stations is that they resemble in architecture and are therefore compatible with most small satellite platforms. Also, larger satellites promise highly distributed ground systems, a new step in satellite operations. The antenna systems of low-cost stations were designed for specific frequency bands, the signal processing can be performed with commercial off-the-shelf (COTS) components. The main challenge is an intelligent network concept for a distributed space segment in combination with loosely coupled ground stations.

By contrast, the classic ground station approach contains typically a broad spectrum of specialized hardware, which limits the utilization to a specific kind of satellite mission or type. These dedicated ground stations cannot be grouped together arbitrarily for the operations phase of a distributed system. Concepts from the field of distributed intelligence and control for distributed space missions are of special interest, unfortunately only some of them are transferable to the traditional ground station concept [2].

22.10 Heterogeneous Ground Station Approach

SatNOGS Network is a global management interface to facilitate multiple ground station operations remotely. An observer is able to take advantage of the full network of SatNOGS ground stations around the world. It is part of the SatNOGS project [7].

SatNOGS project is a complete platform of an open source networked ground station. The scope of the project is to create a full stack of open technologies based on open standards, and the construction of a full ground station as a showcase of the stack.

SatNOGS provides the basis for:

- Bulk manufacturing and deployment of affordable satellite ground stations.
- Modular design for integration with existing and future technologies.
- A platform for a variety of instrumentation around satellite ground station operations.
- A firm platform for a ground station collaborative network (one to one, one to many, many to many).
- A community-based approach on ground station development.
- A solution for massive automation of operator-less ground stations based on open standards.

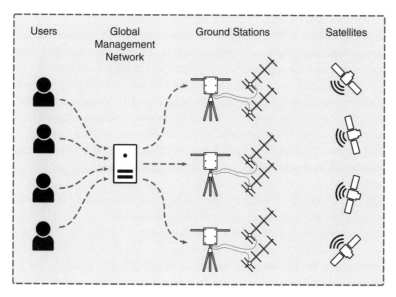

Figure 22.4 SatNOGS explanation. Source: Reproduced with permission from SatNOGS 2015.

There are a number of elements to the project that integrate hardware and software in a way that allows multiple observers to be connected to multiple ground stations so that tracking and monitoring satellites from multiple locations is possible. The data collected are publically available through the production environment. Figure 22.4 demonstrates how SatNOGS works.

At the center is the global management network consisting of:

Users:
   A user is any validated member of the SatNOGS system. All you need is to sign up and with a bit of familiarity you will be able to use the system.
Global management network:
   SatNOGS Network: Our observations, scheduling, and discovery server. SatNOGS Network is a web application for scheduling observations across the network of ground stations. It facilitates the coordination of satellite signal observations, and schedules such observations among the satellite ground stations connected on the network.
   SatNOGS Client: An embedded system that receives the scheduled operation from Network, records an observation, and sends it back. SatNOGS Client is the software to run on ground stations, usually on embedded systems, which receives the scheduled observations from the Network, receives the satellite transmission, and sends it back to the Network web app.
   SatNOGS Ground Station: The actual ground station instrumentation is with tracker, antennas, Low Noise Amplifiers (LNAs), and connected to Client. SatNOGS Ground Station is an open source hardware ground station instrumentation with a rotator, antennas, electronics, and connected to the Client. It is based on 3D printed components and readily available materials.

Ground stations:

A ground station is considered to be exactly as the name suggests. It is a receiver and antenna combination connected to the SatNOGS Global Management Network with some form of computer. There are two main types of ground stations, these are:

Fixed or nonrotator: Their antennas wait for the satellites to pass over them (these are simpler types and commonly have a Raspberry Pi and RTL-SDR combination coupled to an antenna, such as a turnstile, eggbeater, or other similar fixed antenna type).

Steerable or rotator: They point a Yagi antenna at the satellite and track it as it passes overhead. These stations can use the simpler Raspberry Pi and RTL-SDR combination but need a rotator such as the SatNOGS rotator or a commercial alternative.

Satellites:

Simply any satellite in the SatNOGS Database. This is a growing record of educational, amateur, and commercial satellites commonly referred to as the DB.

SatNOGS DB: Our crowd-sourced suggestions transponder info website. SatNOGS Database is a crowd-sourced application allowing its users to suggest satellite transmitter information for currently active satellites. Its data are available via an application program interface (API) or via a web application interface, allowing other project to use its satellite transmitter information [7].

The network is open to anyone. Any observer is able to utilize all available ground stations and communicate with satellites. All observation results are public and all data are distributed freely under the Creative Commons Attribution Share-Alike license. Whether you own satellite ground station equipment or you want to build one, you can head to SatNOGS Project site to get up to date documentation and info on how to build a SatNOGS ground station.

22.11 Homogeneous Ground Station Approach

The Rede Integrada Brasileira de Rastreamento de Satelites (RIBRAS) Brazilian Integrated Satellite Tracking Network started in 2014 as an official initiative of the Brazilian Space Agency, aimed at supporting current and future national and international space missions. Each RIBRAS ground station is equipped with two independent towers, one for S-band and other for UHF and VHF communications [8]. These antennas are driven by robust industrial motors, connected to drivers that were specifically programed to provide smooth movements of $10°-2°$ precision. Both antennas are connected to a single analog-digital ICOM-9100 radio, a multiband equipment with two independent receivers, making it possible of receiving signals in two bands simultaneously. This radio is controlled by software running on an industrial PC, which in turn makes the logical connection of the station to the network. The automation of the network is done through the RIBRAS system, which is a set of software aimed at synchronizing the ground stations through a previously generated work plan and a set of control routines, to make sure that the maximum amount of data is collected during the satellites overflight [9] (Figure 22.5).

22.11.1 Automation and Optimization

Based on the study of the networks that preceded it, the RIBRAS developers decided to create it on top of two paradigms, namely, the automation and optimization of its

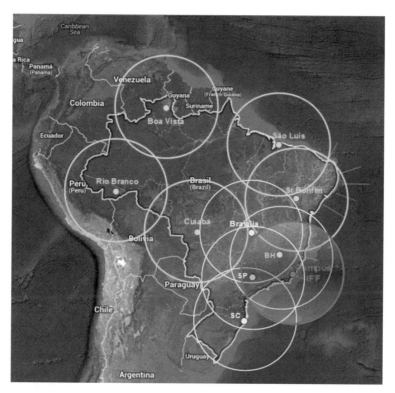

Figure 22.5 RIBRAS ground stations coverage in Brazil, representing footprints for satellites flying at 200 km terminal orbits.

operations. Based on the conceptual architecture presented in Figure 22.2, the respective software packages were created, using an incremental approach that initially aimed at the automation of the towers and the operation of the radios. Automation would allow to overcome the limitation of multiple times of passage of several satellites and also to guarantee the accuracy of the tasks as a whole, allowing a safer data input for an optimization model. Having achieved the goal of automating the systems, the next step was to develop this model, according to a certain philosophy of network operations management.

The RIBRAS approach to network operations management occurs at three levels: planning, programing, and control. For the first two levels, an innovative, integrated optimization model was developed that aims at providing orchestration solutions for optimizing the total volume of downloaded data by a cluster of ground stations, considering a constellation of satellites. The main result of the integrated model is to determine the optimal sequence of antennas' movements, with respective starting and ending azimuths and elevations, for a given overflight of the constellation, independent of the geometry of the flight. This model supplies objective functions that maximizes the download and minimizes the movements, as well as it takes into account the communication schedule constraints of each satellite, as well as the physical constraints of the antennas' motors and wiring. This model's implementation was tested against many simulation scenarios, and for instance, in an ordinary personal computer, for 25 satellites and six stations, the solution times are in order of tens of seconds, while on doubling the number of stations it grows to hundreds of seconds. Taking into account that for most missions there are a few stations and satellites, an optimal solution can be found in a few seconds [10].

The model supplies an objective function that considers the quality of transmission, parameterized by the angle of view, and takes into account the following constraints:

- Overlapping of ground stations range.
- Antenna movement delays and limitations.
- Data available to download.

Additionally, it is possible to prioritize the download from a given satellite to a given ground station by setting up the appropriate coefficients in the objective function. The scheduler uses the data provided by the tracker (Figure 22.2), which provides satellite's two-line elements (TLEs), with the precision of 1° and 1 s. It generates a schedule that supplies detailed aiming and azimuthal movements for each tower of each station. The next step is to transfer this program to the job distributor that functions according to the following steps:

- Transforms the schedule generated by the scheduler into a set of jobs for each station involved in the plan.
- Sends the jobs to the job controller of each station.
- Reports distribution errors to the orchestration controller.

A job is composed by the station ID, the satellite ID, satellite TLE, and passes initial and final times and angles. With these data, the receiving station is able to get into the right position at the right time for each pass, as well as to calculate its tracking velocity.

The distributor has a configuration file containing every single station's network address and port, making it easier to add or remove a station from the network, simply by editing this file. Therefore, for each batch of the jobs, the distributor searches for its respective ID on the configuration file and generates a message directed to the correct address.

The orchestration controller is the module responsible for controlling the execution of jobs distributed by distributor, receiving feedback from stations, and sending runtime errors to the scheduler, who uses sensitivity analysis to judge whether a new plan should be generated by restarting the whole cycle. The feedback received consists in distribution and physical errors (network, communications, and mechanical) generated by the stations. RIBRAS system was developed on top of a flexible distributed core, which deals with the exchange of messages between all modules [11].

In 2015, the first ground station at the Reference Center for Aerospace and Embedded Systems (CRSEA/IFF, Brazil) was set up, and the Station Client package and parts of the Communication Planning and Orchestration Server packages, in a very similar fashion of what was presented in Figure 22.2, were deployed. Since then, National Oceanic and Atmospheric Administration (NOAA) and Sistema Espacial para Realização de Pesquisas e Experimentos com Nanossatélites (SERPENS – Space System for Conducting Research and Experimentation with Nanosatellites), as well as the International Space Station (ISS), were tracked successfully, in both automatic and semi-automatic ways. After this long period of tests, where a series of adjustments were made in the automation of the stations, which allows a high point precision, as well as a complete automation of the towers and radios, the next steps are to physically implant the other nine stations, distributed in Brazil, as shown in Figure 22.5.

22.12 Conclusions

The number of ground station networks is growing every year, and besides the traditional government and academic consortia, commercial solutions offering renting of time slots of specific set of stations are already available. Even so, there is still room for creating new networks, due to political strategic issues, legislation, and even geographic positioning—for instance, the Southern Hemisphere is still little served by this kind of communication infrastructure. This chapter sought to briefly introduce the main concepts and challenges encountered in the development of a GSN, offering as example three different approaches, among the several possible, and serving as a basis for the implementation of methods and technologies for the management of operations of this type of networks.

References

1 Wertz, J.R. and Larson, W.J. (2005). *Space Mission Analysis and Design*, 3e. Library of Congress Cataloging-In-Publication Data.

2 D'Errico, M. (2013). *Distributed Space Missions for Earth System Monitoring*. Space Technology Library. Published jointly by Microcosm Press and Springer.

3 Sanjay Kumar, Tapan P. Bagchi 2016. "Modeling and results of Satellite Ground Support Optimization." University of Texas in Dallas, Dallas, Texas, USA, Indian Institute of Technology Bombay, Mumbai, India.

4 Xhafa, F., Sun, J., Barolli, A. et al. (2012). Genetic algorithms for satellite scheduling problems." Hindawi Publishing Corporation. *Mobile Information Systems* 8: 351–377.

5 Fisher, F., Mutz, D., Estlin, T. et al. (1999). The past, present, and future of ground station automation within the DSN. In: *Proceedings of the 1999 I.E. Aerospace Conference* (ed. R.P. Wright), 315–324. Aspen, CO: IEEE.

6 Maldari, P. and Bobrinskiy, N. (2008). Cost efficient evolution of the ESA network in the space era. *Space Operations Communicator* 5: 10–18.

7 Daniel J. White, 2015. "SatNOGS: Satellite Networked Ground Stations." Digital Communications Conference.

8 Rogerio A. Carvalho, Weslleymberg Lisboa, Lucas R. Hissa, Luiz G. Moura, Lucas R. Amaduro, Cedric S. Cordeiro, 2014. "The Software Architecture for the RIBRAS Ground Station Network." In: Proceedings of the 1st IAA Latin America CubeSat Workshop, Brasilia, Brazil.

9 Lucas R. Amaduro, Rogerio A. Carvalho, Luiz G. Moura, 2016. "A Nation-Wide Ground Station Network: The RIBRAS Project." In: Proceedings of the 1st IAA Latin American Symposium on Small Satellites, Buenos Aires, Argentina.

10 Rogerio A. Carvalho, Cedric S. Cordeiro, Luiz G. Moura, 2013. "Optimizing Data Download using Clusters of Networked Ground Stations: A Model for the QB50 Project." In: 5th European Cubesat Symposium, Brussels, Belgium.

11 Lucas R. Hissa, Lucas R. Amaduro, Rogerio A. Carvalho, Luiz G. Moura, 2015. "Orchestration and Controlling of an Automated Ground Station Network." In: Proceedings of the 2nd IAA Latin America CubeSat Workshop, Florianópolis, Brazil.

23 II-3

Ground-based Satellite Tracking

Enrico Stoll[1], Jürgen Letschnik[2,3], and Christopher Kebschull[1]

[1] *Institute of Space Systems, Technical University of Braunschweig, Germany*
[2] *Institute of Astronautics, Technical University of Munich, Germany*
[3] *Information Management System Architect, Airbus Defence and Space, Taufkirchen, Germany*

23.1   Introduction

Sixty years of human spaceflight has left its traces in Earth orbit. The number of objects has significantly increased since the era of Sputnik, the Apollo program, and the Space Shuttle. Figure 23.1 shows the evolution divided into low Earth orbit (LEO), medium Earth orbit (MEO), geostationary orbit (GEO), highly eccentric Earth orbit (HEO), LEO–MEO crossing orbits (LMO), MEO–GEO crossing orbits (MGO), GEO transfer orbit (GTO), navigation satellites orbit (NSO), extended geostationary orbit (EGO), and other orbits. It is evident that the population in LEO dominates the number of objects. This trend of the steady increase of objects will be amplified in the future by the launch of mega constellations [2] that intend to provide a new global knowledge infrastructure by supporting Internet coverage everywhere on the planet.

If an object in Earth orbit is detectable from ground (see Section 23.3), it will be tracked and logged by the Joint Space Operations Center (JSpOC) via a network of space surveillance sensors. More than 40 000 objects were cataloged that way since the days of Sputnik, many of which have since re-entered. The catalog contains the position and velocities of the objects in a two-line element (TLE) format that was originally designed to work with punch card systems. The unclassified part of the TLE catalog is publicly available on the Internet and is updated on a daily basis. It is a common misconception that the first spacecraft was the Russian Sputnik satellite in 1957. Actually, the object with TLE number 00001 was the Sputnik rocket that brought the satellite with the same name into orbit and decayed on 1st December 1957.

Usually, the information embedded in TLE is the basis for tracking satellites and maintaining communication between the ground and the space system. The tracking information is also often the basis for Space Situational Awareness (SSA) analyses [3] and subsequent collision avoidance maneuvers [4]. This chapter gives an overview of ground-based satellite tracking. Therefore, the following section gives an introduction to TLE and state vectors, in which the different orbital elements are explained. Section 23.3 depicts the TLE/tracklet generation from ground measurement optical telescopes and

Nanosatellites: Space and Ground Technologies, Operations and Economics, First Edition.
Edited by Rogerio Atem de Carvalho, Jaime Estela, and Martin Langer.
© 2020 John Wiley & Sons Ltd. Published 2020 by John Wiley & Sons Ltd.

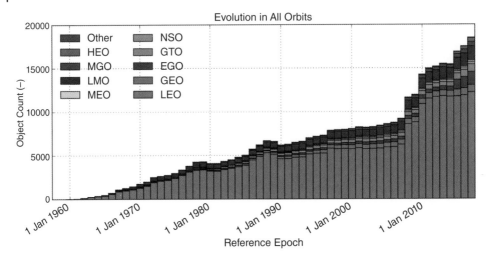

Figure 23.1 The evolution of number of objects in geocentric orbit-by-orbit class [1]. (*See color plate section for color representation of this figure*).

radars. Section 23.4 shows how the TLE information can be utilized to track CubeSats with ground stations (GS), whereas Section 23.5 gives an overview on propagation techniques. The chapter concludes with Section 23.6, which depicts the principles of operations of ground stations and Section 23.7 that provides a summary.

23.2 Orbital Element Sets

The position and velocity of a spacecraft on orbit can be characterized by different sets of orbital elements. Three different representations are shortly discussed in this section—state vectors, two-line elements, and Keplerian elements, since they are the element sets that are most commonly used.

23.2.1 State Vectors

State vectors, consisting of the radius vector $\underline{r}$ and the velocity vector $\underline{v} = \underline{\dot{r}}$, are one possible set (see Table 23.1) and can be determined in an arbitrary coordinate system. They can be derived from on-board Global Navigation Satellite System (GNSS) devices, which additionally allow for information in a second coordinate frame, that is position in latitude, longitude, and altitude and velocity in east, north, and up representation. The time stamp is provided by Global Positioning System (GPS) week (since the GPS epoch on 6 January 1980) and GPS seconds since the beginning of the week. Note that the GPS seconds are not adjusted for leap seconds, such that GPS is currently ahead of Coordinated Universal Time (UTC) by 18 s.

The World Geodetic System (WGS), with its latest version WGS84, usually is the standard reference system for GPS applications. The coordinate origin of WGS84 coincides with the Earth's center of mass. The according error is believed to be less than 2 cm. In WGS84, the

Table 23.1 Different orbital element sets.

State vectors	Two-line elements	Keplerian elements
r_x	M: mean anomaly	θ: true anomaly
r_y	n: mean motion	a: semi-major axis
r_z	e: eccentricity	e: eccentricity
v_x	i: inclination	i: inclination
v_y	Ω: RAAN	Ω: RAAN
v_z	ω: argument of perigee	ω: argument of perigee

meridian of zero longitude X-axis is the International Earth Rotation Service (IERS) reference meridian, which lies 5.31 arc sec east of the Greenwich Prime Meridian. The Z-axis is the direction of the Conventional Terrestrial Pole[1] (CTP) for polar motion, as defined by Bureau International de l'Heure (BIH). The Y-axis completes a right-handed, orthogonal Earth-centered, Earth-fixed (ECF) coordinate system. This is measured in the plane of the CTP equator, 90° east of the X-axis.

23.2.2 Two-line Elements

The disadvantage of using state vectors is that all of the six elements change quickly over time, and therefore, the orbital element set is not intuitively manageable. Thus, for general orbit data presentation and orbit data storage, the system of orbital TLEs has become accepted, which uses an intuitively understandable characterization of the orbit shape, attitude, and dimension. The TLE set (see Figure 23.2) of general space objects

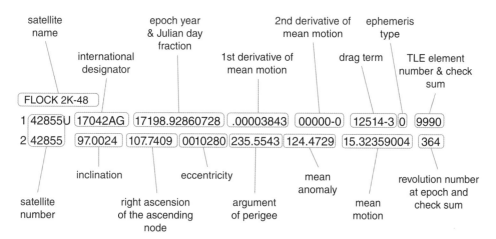

Figure 23.2 Nomenclature of two-line element set in compliance with NORAD.

---

1 The conventional terrestrial pole is the mean position of the Earth's spin axis between 1900.0 and 1905.0. All instantaneous positions of the pole are referenced to this position. The IERS maintains a record of the instantaneous pole in relation to the CTP in units of seconds.

(spacecraft, space debris, etc.) is generated by the North American Aerospace Defense Command (NORAD). Besides the TLE orbital elements (M, n, ε, i, Ω, ω), an epoch time t_0 at which the set is valid, and means for the propagation of the elements are provided. Latter will be discussed in Section 23.5. Contrarily to the state vector representation, only one element, the mean anomaly,[2] is constantly changing over time, whereas the remaining five elements are only subject to small changes and can be considered as constant for a first approximation.

23.2.3 Keplerian Elements

The third system of Keplerian elements also consists of six elements (a, e, θ, i, Ω, ω) where the first two (the eccentricity e and the semi-major axis a) characterize the orbit shape according to Figure 23.3. The true anomaly θ, as the angular value between the perigee and the radius vector (measured in flight direction), defines the position of the spacecraft on the orbit and changes within one orbit period between 0 and 2π.

The remaining three of the six orbital parameters describe the orientation or attitude of the orbit with respect to the Earth-centered inertial frame (IJK) (see Figure 23.4). This coordinate frame has its origin at the center of the Earth with its first I axis pointing toward the vernal equinox.[3] Being fixed in space (inertial), its third axis K coincides with the Earth's rotational axis and the second axis J completes a right-handed Cartesian coordinate system.

The fourth element, the inclination i (of the orbit plane), determines the angle between the angular momentum vector, which is perpendicular to the orbital plane, and the K axis of the reference frame. As Figure 23.4 shows, the intersection between the equatorial plane and the orbital plane is called the line of nodes. A special point on this line is the ascending node, which marks the location where the satellite passes through the equatorial plane moving from south to north.

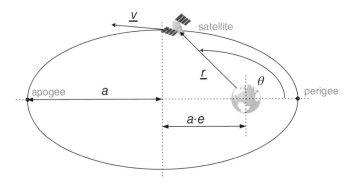

Figure 23.3 Geometry of an elliptical orbit.

---

2 If the spacecraft's orbit ellipse would be circumscribed with an auxiliary circle and a second fictitious spacecraft would orbit that circle, the true anomaly of the auxiliary spacecraft is the mean anomaly of the original spacecraft.

3 The vernal equinox is a fixed point in space. It is characterized by the intersections between ecliptic and equatorial planes. The Sun passes the vernal equinox from south to north.

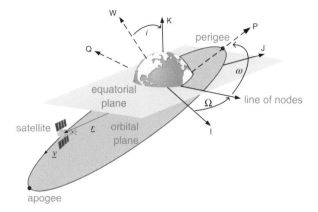

Figure 23.4 Earth-centered inertial (*IJK*) and perifocal system (*PQW*). (*See color plate section for color representation of this figure*).

The right ascension of the ascending node Ω is the angular value between the vernal equinox (which has the same orientation as the *I* axis of the reference frame) and the line between the Earth's center and the ascending node. It is measured according to a right-hand rotation about the *K* axis.

The argument of perigee ω is the angle from the line of nodes to the perigee measured in the direction of the motion of the satellite. It is the sixth Keplerian element.

Note that the sets given in Table 23.1 are not exhaustive. A variety of other representations of orbital elements exist. All of these sets are interchangeable, as will be shown in Section 23.4, where the TLE set of a spacecraft will be converted into the state vectors in commonly used coordinate systems.

23.3 Tracklet Generation from Ground Measurements

The so-called state $\underline{X}(t)$ of a satellite in space requires six quantities as depicted in Table 23.1. It is commonly written in vector form either using the Keplerian elements:

$$\underline{X}(t) = (a(t), \varepsilon(t), i(t), \Omega(t), \omega(t), \theta(t))^T \tag{23.1}$$

or in the Cartesian coordinate system:

$$\underline{X}(t) = (\underline{r}(t), \underline{v}(t)) = (x(t), y(t), z(t), \dot{x}(t), \dot{y}(t), \dot{z}(t))^T \tag{23.2}$$

Given the knowledge of an initial state $\underline{X}_0$, extrapolation techniques are needed to derive a state $\underline{X}_1$ in future instances in time, for example, during a pass of the satellite over a ground station, so that the antenna can be pointed accurately using azimuth (Az) and elevation (El).

23.3.1 Perturbations

Based on the use case, different techniques can be applied. For high-precision applications, such as supporting rendezvous and docking maneuvers or conjunction analysis to identify close approaches, numerical theories should be used. Analytical theories usually do

Table 23.2 Satellite perturbations.

Gravitational perturbations	Nongravitational perturbations
• The deviation of the gravitational potential from spherical shape • Third-body gravitation potential (mainly through the Sun and the Moon) • Solid earth and ocean tides	• Atmospheric drag • Solar radiation pressure • Earth's albedo • Thruster forces

not reach the accuracy of their numerical counterparts but they yield sufficient accuracy for most applications that involve the short-term extrapolation for finding a passing object in the sky. The extrapolation of a satellite's state vector is also referred to as propagation. Several propagation approaches exist for example to simulate the space debris environment [5]. Orbit propagation algorithms take the two-body problem into account. Thus, the motion of the satellite is primarily defined through the gravitational potential of the central mass. However, accelerations through perturbing forces need to be regarded for increased accuracy of the extrapolated state. Perturbing forces can be divided into gravitational and nongravitational perturbations (or accelerations), as Table 23.2 shows.

Based on the orbit of the satellite, different perturbing forces may be dominant. In LEOs, accelerations from nonspherical part of gravitational potential and the atmospheric drag should be regarded, while on higher altitudes, such as GEO orbits, forces from third-body gravitational potential and the solar radiation pressure are more important. For satellites with higher eccentricities, such as GTO orbits, a full set of perturbations should be considered in the propagation process. As any orbit propagation is based on the abstraction of the real world, through the use of models to describe the two-body problem, or the listed perturbing forces in the earlier text, the extrapolated state will be error-prone. Advancing further into the future from the initial state $\underline{X}_0$ using any not perfect extrapolation technique will lead to a decrease in the accuracy of the state and thus the loss of the satellite's position. In order to avoid the degradation of the state to a point where the satellite cannot be reacquired, updates through real world position measurements need to be performed.

23.3.2 Sensor Types

Position and velocity measurements can be performed using different types of sensors. Both, in situ and on-ground means can be applied. As the miniaturization advanced, GNSS modules have become a robust and affordable option to retrieve in situ state measurements directly from the satellite. There are multiple upsides to using GNSS hardware, one of which is the availability of and accuracy of measurements, especially for smaller satellites. The downside is the need of extra hardware on-board the satellite, which has an impact on the power and mass budget. Additionally, the data are available only to the operator of the (active) spacecraft.

Alternative ways to track objects are given through sensors that are placed on the surface of the Earth. Telescope, radar, and satellite laser ranging (SLR) stations can perform independent tracking of objects. The stated sensor types provide different kinds of observational data and accuracies, as listed in Table 23.3. Optical telescopes are able to provide angular measurements as azimuth and elevation or right ascension and declination values. The accuracy of the measurements is in the order of arc seconds. The operation of a telescope is time and weather dependent. Noisy factors such as Sun, Moon, and ambient light will interfere with the sensitive electronic light sensor of the camera system. Furthermore, the satellite must reflect the sunlight so that it is brought to the focal point of the telescope. This requires a good alignment of the Sun, Earth, and satellite toward the observer. Due to the small field of view (FOV) of telescopes, they cannot be used in a scanning capacity in LEO. Usually, they are used for tracking objects or scanning for objects in MEO to GEO regions.

In addition to angular measurements, radars are able to provide range (R) and range rate (RR) measurements. They also bring the advantage of working independent of time and weather. However, due to the need of power for sending out a signal so that it is reflected of the satellite and returned to the receiver, a range constraint comes with the application of radars. They are primarily used in the LEO region. Different kinds of radars can serve different purposes. Phased array radars, which use multiple electronically steered beams, can cover larger volumes of space scanning for objects. Single beams from phased arrays or mechanically steered antennas have a smaller FOV. They can be used to track objects and generate a higher number of observations. The accuracy in the range measurement can be in the order of meters to centimeters, depending on the frequency and signal processing. For a more precise measurement of the range, component lasers can be used. They reach accuracies in the order of millimeters. They can supplement telescopes, as they can supply an additional range measurement.

SLR systems transmit laser pulses to a satellite and facilitate the turnaround light time of the laser to distance measurements of high accuracy. They are mainly used for scientific and geodetic missions that require such a high precision [6]. Being an active system, energy is required. Further, the weather conditions are important since an optical medium is used. SLR systems provide ranges only, however, they can be used together with a telescope and data fusion techniques have to be applied.

Table 23.3 shows that three out of four sensor types produce less than the required six quantities to define a satellite's position and velocity in space, which means that additional processing is needed to derive a full state vector of six Keplerian elements or Cartesian coordinates.

Table 23.3 List of sensor types that can be used to retrieve measurement data of a satellite. Different sensor types are able to produce different fidelities of measurements. The given accuracy values consider the order of the magnitude that the systems are able to achieve.

Sensor type	Observational data Quantities	Coordinate system	Order of position accuracy
GNSS module	$x, y, z, \dot{x}, \dot{y}, \dot{z}$	Earth-centered, Earth-fixed (ECF)	Meters
Radar	Az, El, R, RR	Topocentric horizon	Arc seconds and meters
Telescope	Az, El	Topocentric horizon	Arc seconds
SLR	R	Topocentric horizon	Millimeters

23.3.3 Orbit Determination

The process of deriving a state vector from measurements is called orbit determination. Based on the observational data that are available, different techniques can be used. For cases where no a priori knowledge is available, initial orbit determination (IOD) techniques have to be applied. For updates of states, statistical orbit determination techniques should be used. The methods are discussed in detail in literature [7, 8]. The commonly used and robust statistical method of the nonlinear least squares is briefly shown in the following text.

Statistical orbit determination, based on a nonlinear least squares approach, is an iterative process in which the initial state $\underline{X}_0$ is updated until a cost function is minimized. The differences between the predicted state $\underline{Y}_p(\underline{X}_0) = [Az_p \ El_p]^T$ and a measured state $\underline{Y}_m = [Az_m \ El_m]^T$, using telescope observation data as an example, lead to a residual r:

$$r = \underline{Y}_p - \underline{Y}_m = \begin{bmatrix} Az_p - Az_m \\ El_p - El_m \end{bmatrix}. \tag{23.3}$$

The residuals for multiple measurements $i = 1, \ldots, n$ are accumulated in a vector $\underline{b}$:

$$\underline{b} = \sum_{i=1}^{n} \underline{r}_i \tag{23.4}$$

Finding the minimum error requires the computation of the matrix $\underline{\underline{A}}$, which is the Jacobian of $\underline{Y}_m$ and establishes the connection between the observation space (i.e. the data from the telescope $\underline{Y}_{m_i}$) and the state space for the initial state $\underline{X}_0$ (in Earth-centered inertial (ECI) coordinates):

$$\underline{\underline{A}} = \frac{\partial \underline{Y}_{mi}}{\partial \underline{X}_0} = \begin{bmatrix} \dfrac{\partial Az_{m_i}}{\partial r_{I_0}} & \dfrac{\partial Az_{m_i}}{\partial r_{J_0}} & \dfrac{\partial Az_{m_i}}{\partial r_{K_0}} & \dfrac{\partial Az_{mi}}{\partial v_{I_0}} & \dfrac{\partial Az_{m_i}}{\partial v_{J_0}} & \dfrac{\partial Az_{mi}}{\partial v_{K_0}} \\ \dfrac{\partial El_{m_i}}{\partial r_{I_0}} & \dfrac{\partial El_{m_i}}{\partial r_{J_0}} & \dfrac{\partial El_{m_i}}{\partial r_{K_0}} & \dfrac{\partial El_{m_i}}{\partial v_{I_0}} & \dfrac{\partial El_{mi}}{\partial v_{J_0}} & \dfrac{\partial El_{m_i}}{\partial v_{K_0}} \end{bmatrix} \tag{23.5}$$

Please note that based on the mathematical models of the extrapolation technique finding the partial-derivatives matrix can be a challenging task. To reduce the complexity, the matrix $\underline{\underline{A}}$ can be expressed as an observation partial-derivatives matrix $\underline{\underline{H}}$ and the error state transition matrix $\underline{\underline{\Phi}}$:

$$\underline{\underline{A}} = \frac{\partial \underline{Y}_{mi}}{\partial \underline{X}_0} = \frac{\partial \underline{Y}_{mi}}{\partial \underline{X}_i} \frac{\partial \underline{X}_i}{\partial \underline{X}_0} = \underline{\underline{H}} \cdot \underline{\underline{\Phi}} \tag{23.6}$$

The observational partial derivatives matrix $\underline{\underline{H}}$ carries information how changes in the state space effect the elements of the observational space at the time t_i of the measurement:

$$\underline{\underline{H}}(t_i) = \begin{bmatrix} \dfrac{\partial Az_{m_i}}{\partial r_{I_i}} & \dfrac{\partial Az_{m_i}}{\partial r_{J_i}} & \dfrac{\partial Az_{m_i}}{\partial r_{K_i}} & \dfrac{\partial Az_{mi}}{\partial v_{I_i}} & \dfrac{\partial Az_{m_i}}{\partial v_{J_i}} & \dfrac{\partial Az_{mi}}{\partial v_{K_i}} \\ \dfrac{\partial El_{m_i}}{\partial r_{I_i}} & \dfrac{\partial El_{m_i}}{\partial r_{J_i}} & \dfrac{\partial El_{m_i}}{\partial r_{K_i}} & \dfrac{\partial El_{m_i}}{\partial v_{I_i}} & \dfrac{\partial El_{mi}}{\partial v_{J_i}} & \dfrac{\partial El_{m_i}}{\partial v_{K_i}} \end{bmatrix} \tag{23.7}$$

In order to determine the partials in $\underline{\underline{H}}$, multiple coordinate transformation matrices have to be derived and chained. For the case that ECI coordinates are available, the $\underline{\underline{H}}$ matrix becomes an identity matrix. If GPS coordinates are available, only the transformation matrix from the ECF to the ECI systems is needed. The more complex case is when observational data in the topocentric horizon azimuth elevation (AzEl) system are given:

$$\underline{\underline{H}} = \frac{\partial \underline{Y}_m}{\partial \underline{r}_{ECI}} = \frac{\partial \underline{Y}_m}{\partial \underline{r}_{AzEl}} \frac{\partial \underline{r}_{AzEl}}{\partial \underline{r}_{ECF}} \frac{\partial \underline{r}_{ECF}}{\partial \underline{r}_{ECI}} \tag{23.8}$$

A connection between the initial state $\underline{X}_0$ to the nominal state $\underline{X}_i$ is created by the error state transition matrix $\underline{\underline{\Phi}}$. It can be accomplished by imprinting small changes to the initial state $\underline{X}_0 + \delta$ and propagating them to the time of the measurement t_i, creating a modified state vector $\underline{\hat{X}}_i$ at the epoch. Each component of the initial state is modified individually and thus six $\delta_{I...K}$ values are needed for the whole error state transition matrix. This technique is known as finite differencing [7]:

$$\underline{\underline{\Phi}} = \begin{bmatrix} \frac{\hat{X}_I - X_I}{\delta_I} & \frac{\hat{X}_I - X_I}{\delta_J} & \frac{\hat{X}_I - X_I}{\delta_K} & \frac{\hat{X}_I - X_I}{\delta_{\dot{I}}} & \frac{\hat{X}_I - X_I}{\delta_{\dot{J}}} & \frac{\hat{X}_I - X_I}{\delta_{\dot{K}}} \\[2ex] \frac{\hat{X}_J - X_J}{\delta_I} & \frac{\hat{X}_J - X_J}{\delta_J} & \frac{\hat{X}_J - X_J}{\delta_K} & \frac{\hat{X}_J - X_J}{\delta_{\dot{I}}} & \frac{\hat{X}_J - X_J}{\delta_{\dot{J}}} & \frac{\hat{X}_J - X_J}{\delta_{\dot{K}}} \\[2ex] \frac{\hat{X}_K - X_K}{\delta_I} & \frac{\hat{X}_K - X_K}{\delta_J} & \frac{\hat{X}_K - X_K}{\delta_K} & \frac{\hat{X}_K - X_K}{\delta_{\dot{I}}} & \frac{\hat{X}_K - X_K}{\delta_{\dot{J}}} & \frac{\hat{X}_K - X_K}{\delta_{\dot{K}}} \\[2ex] \frac{\hat{\dot{X}}_I - \dot{X}_I}{\delta_I} & \frac{\hat{\dot{X}}_I - \dot{X}_I}{\delta_J} & \frac{\hat{\dot{X}}_I - \dot{X}_I}{\delta_K} & \frac{\hat{\dot{X}}_I - \dot{X}_I}{\delta_{\dot{I}}} & \frac{\hat{\dot{X}}_I - \dot{X}_I}{\delta_{\dot{J}}} & \frac{\hat{\dot{X}}_I - \dot{X}_I}{\delta_{\dot{K}}} \\[2ex] \frac{\hat{\dot{X}}_J - \dot{X}_J}{\delta_I} & \frac{\hat{\dot{X}}_J - \dot{X}_J}{\delta_J} & \frac{\hat{\dot{X}}_J - \dot{X}_J}{\delta_K} & \frac{\hat{\dot{X}}_J - \dot{X}_J}{\delta_{\dot{I}}} & \frac{\hat{\dot{X}}_J - \dot{X}_J}{\delta_{\dot{J}}} & \frac{\hat{\dot{X}}_J - \dot{X}_J}{\delta_{\dot{K}}} \\[2ex] \frac{\hat{\dot{X}}_K - \dot{X}_K}{\delta_I} & \frac{\hat{\dot{X}}_K - \dot{X}_K}{\delta_J} & \frac{\hat{\dot{X}}_K - \dot{X}_K}{\delta_K} & \frac{\hat{\dot{X}}_K - \dot{X}_K}{\delta_{\dot{I}}} & \frac{\hat{\dot{X}}_K - \dot{X}_K}{\delta_{\dot{J}}} & \frac{\hat{\dot{X}}_K - \dot{X}_K}{\delta_{\dot{K}}} \end{bmatrix} \tag{23.9}$$

δ can be based on a relative value, for example, $\delta_I = \underline{X}_I \cdot 3\text{‰}$.

Using the residuals and the partial-derivatives matrix yields the state error:

$$\Delta \underline{x} = (\underline{\underline{A}}^T \underline{\underline{A}})^{-1} \underline{\underline{A}}^T \underline{b} \tag{23.10}$$

The state error $\Delta \underline{x}$ is subsequently used to update the initial state $\underline{X}_0$ and iterate the process until the abort criterion based on the root mean square of the residuals (RMS):

$$\text{RMS} = \sqrt{\frac{\underline{b}^T \underline{b}}{n}} \tag{23.11}$$

is reached. n is the number of measurements used in the process times the number of components available in the observation vector. For optical observation vectors, there are two components in a measurement $\underline{Y}_m = \begin{bmatrix} Az_m & El_m \end{bmatrix}^T$. The iteration can end when the RMS converges:

$$\left| \frac{\text{RMS}_i - \text{RMS}_{i+1}}{\text{RMS}_i} \right| \leq \varepsilon \tag{23.12}$$

The RMS tolerance ε can be defined as desired. Values between 10^{-4} and 10^{-3} are recommended. An improvement of this method can be achieved when knowledge of the quality of the measurement is introduced. Through calibration and testing a sensor, an understanding on the magnitude of uncertainties in the measurement space can be obtained. These can be expressed in the measurement space as 1-sigma standard deviation values. In the scope of the least squares approach, a so-called weighting matrix is introduced, carrying this information in the diagonal:

$$\underline{\underline{W}} = \begin{bmatrix} \dfrac{1}{\sigma_{Az}^2} & 0 \\ 0 & \dfrac{1}{\sigma_{El}^2} \end{bmatrix} \tag{23.13}$$

In this case, where only azimuth and elevation angles are given, $\underline{\underline{W}}$ is a 2×2 matrix. The matrix shape can vary based on the type of observations that are available. Adding range and range rate information would result in values on the diagonal of a 4×4 matrix.

The formulation of the state error is then updated to incorporate the knowledge of the error in the measurements:

$$\Delta \underline{x} = \left(\underline{\underline{A}}^T \underline{\underline{W}} \cdot \underline{\underline{A}} \right)^{-1} \underline{\underline{A}}^T \underline{\underline{W}} \cdot \underline{b} \tag{23.14}$$

In consequence, the RMS formulation is extended correspondingly:

$$\text{RMS} = \sqrt{\dfrac{\underline{b}^T \underline{\underline{W}} \cdot \underline{b}}{n}} \tag{23.15}$$

As a by-product of the state error determination, the resulting uncertainty in the state due to measurement errors is found, which is expressed as a covariance matrix $\underline{\underline{P}}$:

$$\underline{\underline{P}} = \left(\underline{\underline{A}}^T \underline{\underline{W}}\, \underline{\underline{A}} \right)^{-1} = \begin{bmatrix} \sigma_x^2 & \sigma_y\sigma_x & \sigma_z\sigma_x & \sigma_{\dot{x}}\sigma_x & \sigma_{\dot{y}}\sigma_x & \sigma_{\dot{z}}\sigma_x \\ \sigma_x\sigma_y & \sigma_y^2 & \sigma_z\sigma_y & \sigma_{\dot{x}}\sigma_y & \sigma_{\dot{y}}\sigma_y & \sigma_{\dot{z}}\sigma_y \\ \sigma_x\sigma_z & \sigma_y\sigma_z & \sigma_z^2 & \sigma_{\dot{x}}\sigma_z & \sigma_{\dot{y}}\sigma_z & \sigma_{\dot{z}}\sigma_z \\ \sigma_x\sigma_{\dot{x}} & \sigma_y\sigma_{\dot{x}} & \sigma_z\sigma_{\dot{x}} & \sigma_{\dot{x}}^2 & \sigma_{\dot{y}}\sigma_{\dot{x}} & \sigma_{\dot{z}}\sigma_{\dot{x}} \\ \sigma_x\sigma_{\dot{y}} & \sigma_y\sigma_{\dot{y}} & \sigma_z\sigma_{\dot{y}} & \sigma_{\dot{x}}\sigma_{\dot{y}} & \sigma_{\dot{y}}^2 & \sigma_{\dot{z}}\sigma_{\dot{y}} \\ \sigma_x\sigma_{\dot{z}} & \sigma_y\sigma_{\dot{z}} & \sigma_z\sigma_{\dot{z}} & \sigma_{\dot{x}}\sigma_{\dot{z}} & \sigma_{\dot{y}}\sigma_{\dot{z}} & \sigma_{\dot{z}}^2 \end{bmatrix} \tag{23.16}$$

Along-track, cross-track, and radial errors can be extracted from the covariance matrix, looking at the diagonal. Each element represents a 1-sigma deviation in state space. The updated state vector is of osculating nature. It represents the true estimate of the orbital parameters, either in Cartesian or converted into Keplerian elements. However, osculating elements show high frequency variations, which makes it difficult to characterize the orbit. Using averaging techniques, the osculating elements can be converted into mean elements, which reflect secular and long-periodic variations. One method to create mean elements is to employ a differential correction approach once more, by removing the short-periodic variations in an iterative process using a simple (e.g. analytical) orbit propagation theory. Over multiple orbits, osculating elements are created by a more complex theory (e.g. semi-analytical or numerical simulation) to create a reference trajectory from an initial state ($\underline{X}_0$). The residuals between the reference trajectory and the results from the simple theory are used to derive a state error ($\Delta\underline{x}$) to correct the initial state. After the convergence of the least squares process, the new (mean) elements ($\widetilde{\underline{X}}_0$) represent the initial state for a trajectory best fitting the trajectory of the more complex orbital theory. This step also brings the benefit that a simplified propagation approach can be applied, as the equations of motion are integrated. The secular trends are integrated numerically, while the high-frequency and periodic terms are derived analytically. This is called semi-analytical propagation. The publicly available TLE sets of the space catalog contain the mean elements. The TLE sets are commonly used and should always be used in conjunction with respective propagator that is compatible with the prediction method that has been used to derive them [9, 10].

23.4 Tracking CubeSats with Ground Stations

The basis for tracking CubeSats with antenna systems from ground and maintaining communications are usually the TLEs. This section shows how to convert TLE data into azimuth and elevation values that can be used by the rotator control of a ground station antenna.

23.4.1 Vector Rotations

In order to convert the different coordinate systems into each other, rotational matrices are utilized. $\underline{\underline{R}}_j(\alpha)$ denotes a rotation of a vector about the system main axis $j \in \{1, 2, 3\}$ by an angle α. Subsequently, rotations about the main coordinate axes are defined by:

$$\underline{\underline{R}}_1(\alpha) = \begin{bmatrix} 1 & 0 & 0 \\ 0 & \cos\alpha & -\sin\alpha \\ 0 & \sin\alpha & \cos\alpha \end{bmatrix}; \underline{\underline{R}}_2(\alpha) = \begin{bmatrix} \cos\alpha & 0 & \sin\alpha \\ 0 & 1 & 0 \\ -\sin\alpha & 0 & \cos\alpha \end{bmatrix}; \underline{\underline{R}}_3(\alpha) = \begin{bmatrix} \cos\alpha & -\sin\alpha & 0 \\ \sin\alpha & \cos\alpha & 0 \\ 0 & 0 & 1 \end{bmatrix}$$

$$(23.17)$$

Note that the matrix definition is chosen in a way that an application of such a matrix to a vector will result in an active rotation of that vector within the coordinate system.[4] Rotational matrices are elements of the special orthogonal group in three-dimensional vector space $\underline{\underline{R}}_j \in SO(3)$. Therefore, a backward transformation can be simply accomplished by utilizing the properties of the group:

$$\underline{\underline{R}}_j^{-1}(\alpha) = \underline{\underline{R}}_j^T(\alpha) = \underline{\underline{R}}_j(-\alpha) \tag{23.18}$$

---

4 The y-axis $\underline{y}$ of an xyz-system is obtained, for example, by turning the x-axis $\underline{x}$ by 90° about the z-axis:

$$\underline{y} = \underline{\underline{R}}_z(\pi/2) \cdot \underline{x} = \begin{bmatrix} 0 & -1 & 0 \\ 1 & 0 & 0 \\ 0 & 0 & 1 \end{bmatrix} \cdot \begin{bmatrix} 1 \\ 0 \\ 0 \end{bmatrix} = \begin{bmatrix} 0 \\ 1 \\ 0 \end{bmatrix}.$$

23.4.2 TLE to Keplerian Elements

The elements e, i, Ω, and ω can be directly extracted from TLE (Table 23.1). The semi-major axis a, which is not part of a TLE set, can be calculated by means of the Earth standard gravitational parameter μ_E and the mean motion n [11], which is a measure for the velocity of the spacecraft on its orbit:

$$a = \sqrt[3]{\frac{\mu_E}{n^2}} \tag{23.19}$$

The true anomaly θ can be derived [7] by using the eccentric anomaly[5] E:

$$\theta = 2 \cdot \arctan\left(\sqrt{\frac{1+e}{1-e}} \cdot \tan\left(\frac{E}{2}\right)\right) \tag{23.20}$$

which is a function of the mean anomaly[6] M:

$$M = E - e \cdot \sin(E) \tag{23.21}$$

that given by the TLE set. Since there is no explicit formulation for E available, a Newton iteration has to be used to determine the eccentric anomaly:

$$f(E) = E - e \cdot \sin(E) - M \tag{23.22}$$

$$E_{j+1} = E_j - \frac{f(E_j)}{\frac{d}{dE}f(E_j)} = E_j - \frac{E_j - e \cdot \sin(E_j) - M}{1 - \cos(E_j)} \tag{23.23}$$

A fast convergence is guaranteed by using the following iteration start [12]:

$$E_0 = M + 0.85e \cdot sgn(\sin(M)) \tag{23.24}$$

23.4.3 Keplerian Elements to Perifocal Coordinates

Now that the Keplerian elements are derived, the transformation to state vectors has to follow. State vectors can be expressed in different coordinate systems. If tracking and CubeSat communication are considered, the coordinate system of interest is a local antenna-centered system. In order to derive the state vector in this system, four intermediate steps have to be taken:

- Keplerian elements to perifocal coordinates.
- Perifocal coordinates to ECI.
- ECI to ECF.
- ECF to local AzEl coordinates, which will be discussed in the following text.

The perifocal coordinate system, denoted by PQW (see Figure 23.4) has its origin at the center of the Earth. This system lies in the orbital plane with its first axis pointing toward P the perigee and the third axis W being perpendicular to the orbital plane. The second axis Q

---

5 When projecting the spacecraft position on its orbit onto the ellipse's circumscribing circle (auxiliary circle), perpendicularly to the major axis, then the eccentric anomaly is the angle between the direction of perigee and the radius vector of the projection, measured from the central body.

6 The mean anomaly is the true anomaly of an auxiliary spacecraft, orbiting on the orbit's auxiliary circle.

completes the right-handed Cartesian system. The Keplerian elements can be directly used to derive position and velocity vectors [13] in the perifocal (*PQW*) system:

$$\underline{r}_{PQW} = \frac{a(1 - e^2)}{1 + e \cos \theta} \begin{pmatrix} \cos \theta \\ \sin \theta \\ 0 \end{pmatrix} \tag{23.25}$$

$$\underline{v}_{PQW} = \sqrt{\frac{\mu}{a(1 - e^2)}} \begin{pmatrix} -\sin \theta \\ e + \cos \theta \\ 0 \end{pmatrix} \tag{23.26}$$

23.4.4 Perifocal to ECI Coordinates

The perifocal state vectors of a spacecraft can be transferred into the Earth-centered inertial coordinate system, denoted by *IJK* (see Figure 23.4), by a series of transformations [14, 15] using the rotation matrices defined in Section 23.4.1:

$$\underline{r}_{IJK} = \underline{\underline{R}}_3 (\Omega) \cdot \underline{\underline{R}}_1 (i) \cdot \underline{\underline{R}}_3 (\omega) \cdot \underline{r}_{PQW} =: \underline{\underline{A}}_{PQW \rightarrow IJK} (\Omega, i, \omega) \cdot \underline{r}_{PQW} \tag{23.27}$$

$$\underline{v}_{IJK} = \underline{\underline{A}}_{PQW \rightarrow IJK} (\Omega, i, \omega) \cdot \underline{v}_{PQW} \tag{23.28}$$

23.4.5 ECI to ECF coordinates

Since the state vectors are required in a ground-based coordinate system, the vectors have to be transferred to the ECF. The origin is at the center of the Earth and the third axis Z coincides with the rotational axis of the Earth (and thus with K). X and Y lie in the equatorial plane, whereas X points toward the zero meridian (Greenwich). Subsequently, the ECF coordinate frame and the ECI coordinate frame differ only in one rotation angle. This is the angle between the vernal equinox and the Greenwich meridian, considered from the Earth center and is the so-called Greenwich sidereal time θ_{GST}. It can be calculated using the number of Julian centuries[7] elapsed from epoch J2000 and thus, is a function of epoch time η and the elapsed time since:

$$\theta_{GST} = f(\eta + \Delta t) \tag{23.29}$$

After obtaining Greenwich sidereal time θ_{GST}, the state vectors in ECF coordinates can be found as follows [16]:

$$\underline{r}_{XYZ} = \underline{\underline{R}}_3 (\theta_{GST}) \cdot \underline{\underline{A}}_{PQW \rightarrow IJK} \cdot \underline{r}_{PQW} =: \underline{\underline{A}}_{PQW \rightarrow XYZ} (\Omega, i, \omega, \theta_{GST}) \cdot \underline{r}_{PQW} \tag{23.30}$$

$$\underline{v}_{XYZ} = \frac{d}{dt} \underline{r}_{XYZ} + \underline{\omega}_E \times \underline{r}_{XYZ}$$

$$\approx \frac{d}{dt} \underline{\underline{R}}_3 (\theta_{GST}) \underline{\underline{A}}_{PQW \rightarrow IJK} \cdot \underline{r}_{PQW} + \underline{\underline{A}}_{PQW \rightarrow XYZ} (\Omega, i, \Omega, \theta_{GST}) \cdot \underline{v}_{PQW} + \underline{\omega}_E \times \underline{r}_{XYZ} \tag{23.31}$$

---

7 For further details on the Julian date and the Greenwich sidereal time, refer to literature [13].

This takes into consideration the relative movement of the systems with the angular velocity of the Earth $\underline{\omega}_E = (0\ 0\ 7.2921 \times 10^{-5})$ rad s^{-1} without considering precession, nutation, or polar motion of the Earth. For the time, derivate of the rotational matrix $\frac{d}{dt}\theta_{GST} = |\underline{\omega}_E|$ holds. Furthermore, it is assumed that the right ascension of the ascending node Ω, the argument of perigee ω, and the inclination i of the orbit are constant for the time (satellite pass over ground station) considered.

If highly precise coordinate transformation is required, the movement of Earth's rotational axis across its surface (polar motion), the change in the orientation of the rotational axis (precession), and small periodical oscillation of the rotational axis (nutation) have to be considered with their respective rotational matrices [17].

23.4.6 ECF to Ground Station AzEl Coordinates

Ground-based antenna usually uses AzEl-based coordinate system from tracking purposes. The elevation is the angle between the local horizon and pointing direction. The azimuth is the angle between the pointing direction and local north, measured eastward. The transformation of the ECF system (XYZ) into the antenna system (uvw) depends on antenna position latitude (μ) and longitude (λ). Both the satellite vector $\underline{r}_{sat}$ (due to Eq. (23.26)) and the ground station (GS) vector $\underline{r}_{GS}$ (due to its position) are known in ECF coordinates. Accordingly, the GS to satellite vector is known:

$$\underline{r}_{GS \to sat} = \underline{r}_{sat} - \underline{r}_{GS} \tag{23.32}$$

The communication direction vector can be transferred into a GS intrinsic fixed coordinate system uvw in which the position of the satellite is known. This coordinate system is fixed at the surface of the Earth with its third axis w being the extension of the radius vector (zenith). The first u axis points toward north and the second axis v completes the right-handed Cartesian coordinate frame and points toward west. As Figure 23.5 shows, this

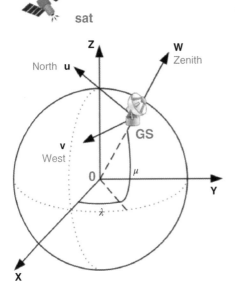

Figure 23.5 Relationship between the ECF frame (XYZ) and the GS frame (uvw).

transformation can be fulfilled using three rotations containing latitude μ and longitude λ of the GS ($s(\alpha) := \sin(\alpha)$, $c(\alpha) := \cos(\alpha)$):

$$\underline{r}_{GS \to sat,uvw} = \left\{ \underline{R}_3(\pi) \cdot \underline{R}_2\left(\frac{\pi}{2} - \mu\right) \cdot \underline{R}_3(\lambda) \right\}^{-1} \cdot \underline{r}_{GS \to sat,XYZ}$$

$$= \begin{bmatrix} -c\lambda \cdot s\mu & s\lambda & -c\lambda \cdot c\mu \\ s\lambda \cdot s\mu & -c\lambda & s\lambda \cdot c\mu \\ -c\mu & 0 & s\mu \end{bmatrix} \cdot \underline{r}_{GS \to sat,XYZ} = \begin{bmatrix} r_u \\ r_v \\ r_w \end{bmatrix} \quad (23.33)$$

The azimuth and elevation of the antenna system can eventually be derived from $\underline{r}_{GS \to sat,uvw}$ as follows:

$$Az = 2\pi - \arctan 2(r_v, r_u); \; El = \arcsin r_w. \quad (23.34)$$

23.5 Orbit Propagation

Using TLE, the state vector $\underline{X}(t_0) = (a(t_0), \varepsilon(t_0), i(t_0), \Omega(t_0), n(t_0), M(t_0))^T$ of the spacecraft is known at the epoch or time stamp of the TLE set. This is usually calculated to be coinciding with the ascending node. However, for establishing communication with the satellite at t_1, it is necessary to propagate the vector to $\underline{X}(t_1)$ and derive antenna degrees of freedom Az(Δt) to Az($t_1 + \Delta t$) and El(t_1) to El(t_1) accordingly. Here, t_1 is the start of a satellite pass over the ground station and Δt the time in which communication can be maintained.

Orbit propagation algorithms take into account the two-body problem. Thus, the motion of the satellite is primarily defined through the gravitational potential of the central mass. However, accelerations through perturbing forces need to be regarded for increased accuracy of the extrapolated state. Perturbing forces that are summarized in Table 23.2 have to be taken into considerations when propagating orbits. In general, propagation methods can be divided into two main categories (and mixed approaches thereof):

- Numerical methods, also known as special perturbations, use numerical integration approaches. The precision of the method is usually bought by an increase of computational time.
- Analytical methods, also known as general perturbations, use analytical integration approaches. They are usually a very fast option, however lack the precision of numerical approaches.

23.5.1 Numerical Orbit Propagation

Several numerical integration methods, such as Encke [18], Gauss–Jackson [19], Runge–Kutta [6], Cowell [20], and so on exist. Generally, an increasing number of perturbing forces leads to an increased accuracy of the propagated state, when looking at a multiple-day propagation (Figure 23.6). In the shown example, the greatest error from the reference is created when reducing the degree and order of the geopotential to 2° × 2°, instead of 70° × 70°. When staying at 70° × 70° and order geopotential modeling, but disregarding the drag, the error based on the orbit of the satellite, different perturbing forces may be dominant. In LEOs, accelerations from nonspherical part of gravitational

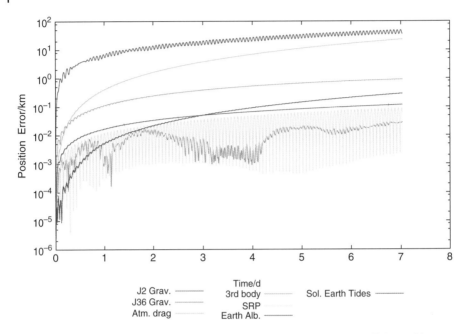

Figure 23.6 Comparison of the position error from a propagated state over 7 days with a numerical propagator for a sun-synchronous object on a $640 \times 619\,km^2$ orbit. The J_2 and J_{36} gravitational models are compared against a J_{70} model. The remaining lines are created by disregarding the respective force models in the propagation of the object. (*See color plate section for color representation of this figure*).

potential and the atmospheric drag should be regarded, while on higher altitudes, such as GEO orbits, forces from third-body gravitational potential and the solar radiation pressure are more important. For satellites with higher eccentricities, such as GTO orbits, a full set of perturbations should be considered in the propagation process.

23.5.2 Analytical Orbit Propagation

An analytical orbit theory is available through the Simplified General Perturbations-4 (SGP4) propagator, which is the standard propagator that should be used when processing TLE sets with an orbital period of $T < 225\,min$. For orbital periods $T \geq 225\,min$, the Simplified Deep-space Perturbations-4 (SDP4) propagator should be used. SGP4 is based on an analytical solution of the two-body problem and assumes a power density function for its atmospheric modeling [9]. The program can be used to propagate the state of an orbit forward and backward, converting a mean to an osculating state in either Keplerian elements or Cartesian coordinates. The latter is important when using a TLE set with a different propagator than SGP4, like a more precise numerical orbit theory. First, the osculating orbital elements must be rebuilt from the mean elements of the TLE set. As SGP4 is a propagator compatible with the orbit prediction method used to remove the periodic variations, it is also capable of restoring these variations. With a rebuilt osculating orbital element set, the data can be used with other propagators. SGP4 uses a true equator and mean equinox (TEME) coordinate system, which needs to be considered when processing the retrieved state vector. Several different propagators are available with OREKIT 8 [21]. The software framework offers analytical, semi-analytical, and numerical propagators.

23.6 Principle of Operations of Ground Stations

In general, tracking of ground stations has been required in situations where the communications link specification dictates that the received and transmitted signal levels over the satellite link be maintained within defined limits. This usually becomes necessary when the antenna beam width is lower than the visibility angle (orbit section) of a satellite pass. The required pointing accuracy is primarily determined by the beam width of the antenna (which is a function of antenna diameter and operating frequency). To understand the beam width of an antenna, it is necessary to know some fundamentals about an antenna diagram in general.

23.6.1 Fundamentals of Antenna Technology

If a simple dipole is used, the transmitted power of this dipole will be spread spherically in all directions. By using special antenna types (helix, Yagi, or parabolic reflector antennas), the power will be focused and transmitted in a certain direction only. Figures 23.7 and 23.8 illustrate this effect. The effect of focusing in reference to the transmission of the dipole is called antenna directivity or antenna gain. The maximum antenna gain G_{max} [22] can be calculated by

$$G_{\text{max}} = \eta \cdot \left(\frac{\pi \cdot D}{\lambda} \right)^2 \tag{23.35}$$

Figure 23.7 Antenna directivity in comparison to an isotropic dipole radiation pattern (0 dB sphere).

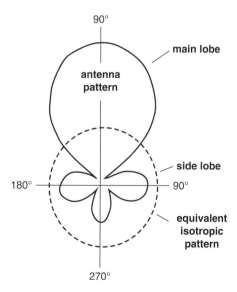

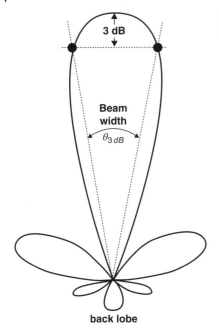

Figure 23.8 Definition of the 3 dB beam width of an antenna pattern ($\Theta = 3$dB).

where η is the antenna efficiency (typically 0.4–0.6), D is the antenna diameter, and λ is the wavelength.

By increasing the diameter of the antenna or the used frequency, the antenna gain will increase. That also implies that the focusing effect also increases. To get a better definition of the focusing effect, the 3 dB (half power) beam width is used (Figure 23.7). The half power beam width is the angle between the half power (−3 dB) points of the main lobe, when referenced to the peak effective radiated power of the main lobe.

In other words, the angular range of the antenna pattern in which at least half of the maximum power is still emitted is described as a "3 dB beam width." Bordering points of this main lobe are therefore the points at which the field strength has fallen by 3 dB regarding the maximum field strength. This angle is then described as beam width or aperture angle or half power (−3 dB) angle—with notation Θ (also φ) [22]. The beam width Θ is exactly the angle between the marked directions (dashed lines) in Figure 23.7. The angle Θ can be determined in the horizontal plane (with notation Θ_{Az}) as well as in the vertical plane (with notation Θ_{El}). Table 23.4 gives characteristic values for typical antennas, which will be used for CubeSat operations.

It can be seen that the 3 dB beam width is diverting dramatically, depending on the frequency. The mentioned frequencies in the table are typical frequencies that are used for CubeSat operations (VHF: 133 MHz, UHF: ~433 MHz, S-band: 2300 MHz). Depending on which antenna and frequency will be used for the ground station and for communication, it implies a direct impact on the tracking system of the ground station. For an antenna with a beam width of the same order as the satellite motion angle, it is possible that no tracking will be required and the antenna can be fixed. If an antenna is used with a much narrower beam width, then tracking again becomes necessary. Four classes of

Table 23.4 Characteristics of different antennas and by using different frequencies.

Antenna type	Frequency (MHz)	Antenna gain (dBi)	3 dB beam width (degree)
Helix antenna (length 0.8 m)	430	2	135
	2300	16.5	25
Yagi antenna (length 3 m)	144	6.9	77
	433	16.5	26
Reflector antenna (3 m)	2300	35	3

tracking systems have evolved to meet the various needs of satellite communications and operations:

- Program tracking
- Monopulse or simultaneous sensing
- Sequential amplitude sensing
- Electronic beam squinting

The latter three techniques can be classified under the general heading of automatic tracking or autotracking and are normally not used for CubeSat operations. The most common tracking technique in CubeSat operations is program tracking.

Program tracking uses prepared data describing the path of the satellite, seen from the Earth station as a function of time. This information is fed to the antenna positioning system that points the antenna in the appropriate direction (azimuth, elevation) at a specific time. The calculation of azimuth and elevation (Section 23.4.6) can be done as a precalculation, before the satellite pass, or directly during the pass. The time to perform calculations and provide steering commands can be neglected with modern Central Processing Unit (CPU) power. It is important to note that on high elevation passes that necessitate high angular velocities of the ground station antenna, the update rate of sending steering commands should be high enough to keep the pointing loss as small as possible. Sometimes, sophisticated interpolation techniques are used on subsequent transits of the satellite pass.

23.6.2 Tracking Software Examples and Features

Off-the-shelf software uses most of the functionalities that were described. There are a lot of software solutions, commercial and available free of charge, for example, Ham Radio Deluxe, Orbitron, Gpredict, etc. These tools typically use program tracking as a standard tracking solution and are in that case a real-time satellite tracking and orbit prediction application. Most of the software tools are multiplatform implementations and available for modern computer desktop environments, such as Linux, BSD, Windows, and macOS. These software tools provide a different number of features/software elements to support the CubeSat operations, some of them described shortly in the following text.

Satellite tracking is a software element that calculates the satellite pass (azimuth, elevation, time). Typically, the NORAD SGP4/SDP4 algorithms are used for fast and accurate real-time tracking calculations. Prediction parameters and conditions can typically be fine-tuned by the user to allow both general and very specialized predictions. The update of the Keplerian elements or TLEs from the web via Hypertext Transfer Protocol (HTTP), File Transfer Protocol (FTP), or from local files is usually performed automatically. Some softwares also predict the time of future passes for a satellite, and provide with detailed information about each pass. Satellites and their positions are displayed and also other data are visualized lists, tables, maps, and polar plots (radar view). Typically, it is possible to add more than one satellite or ground station and organize it individually for our operations concept. Context sensitive pop-up menus allow quickly predicting the future passes by clicking on any satellite.

Rotor control is a software element that supports most popular rotator controllers. In some programs, the radio control is also included, which allows and is needed for autonomous tracking.

Logbook is a software element for logging purposes. All station activities can be logged starting for the station's preparation over the real tracking up to the all nominal and contingency station activities. Figure 23.9 gives an example of the main page of the software Orbitron. Central element is the world map with the animated satellites and their foot prints. Upcoming satellite passes are visualized below this map. On the right side, a polar diagram of the current satellite pass is shown together with important parameters below this diagram.

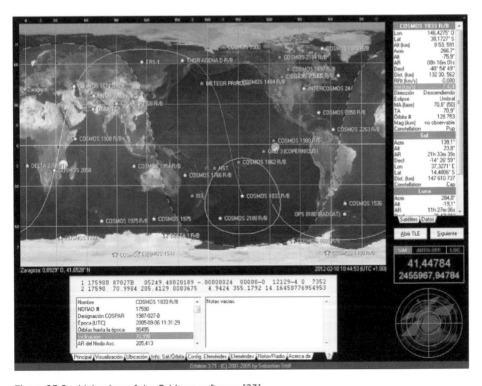

Figure 23.9 Main view of the Orbitron software [23].

23.6.3 Challenges in CubeSat Tracking

Although a large number of freely available software solutions are available, proprietary developments are constantly being implemented. One reason is that most of the software tools are not able to handle a real multisatellite tracking option. This plays a role if it is required to track and operate with different satellites over time. It also implies that the software is able to configure the ground equipment, such as converters, modems, etc. Additionally, it is necessary to handle conflicts of different satellite passes, which can become rather complex.

As briefly explained in Section 23.6.2, the handling of satellite passes with high elevations can be a challenging exercise. The azimuth angular rate of an antenna turntable (elevation over azimuth) increases with higher elevations. At an elevation of 90°, the azimuth angular rate can converge to infinity if a satellite passes directly over the ground station zenith. The typical angular rates of commercial turn tables (rotor systems) for CubeSat applications are in the range of 3–9° s^{-1}. As depicted in Figure 23.10, this leads to a maximum elevation of 80°–85° that can be supported. Depending on the used antenna type and the resulting beam width, link losses have to be taken into account on high elevations.

The high azimuth angular velocities at high elevations are opposed to the low angular velocities at low elevations. The focus is often placed on the high elevations in operation, usually driven by the better link balance. However, the low elevations are of great interest when the duration of a satellite pass is concerned. The contact time increases dramatically if low elevation passes can be supported, as Figure 23.10 (right) shows. The disadvantage of the low elevations is that disturbances due to reflections on the ground and multipath propagation can take place, especially if the antenna beam width is very large. A compromise between contact time and link stability must therefore be found.

The combination of using antennas with a small beam width and using program track at the same time can lead to depointing effects. The reasons behind such effects are various. Causes can be found in the mechanical installation of the antenna system. A misalignment of the orthogonality between azimuth and elevation axis or a little installation of the central tower out of the plumb are typical mechanical causes. Figure 23.11 shows the case of such a misaligned antenna. The blue line illustrates the elevation before, during, and

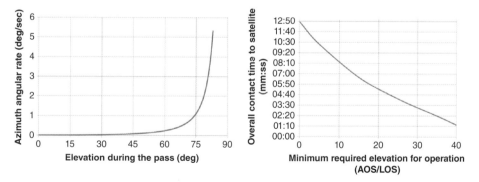

Figure 23.10 (Left) Azimuth angular rate of the antenna turn table over elevation (ground station toward satellite) during a satellite pass. (Right) Contact time to satellite over minimum required elevation for possible satellite operations.

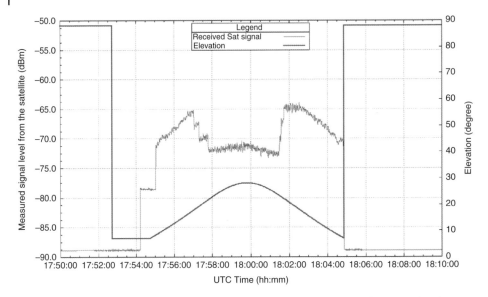

Figure 23.11   Measurement data from a satellite pass: received signal level measured at the ground station (red) and elevation of the ground station during the pass (blue). (*See color plate section for color representation of this figure*).

after the pass. It can be seen that the elevation starts and ends at an elevation of 90°, which is typical for small ground station operation. Driving the antenna elevation to 90° keeps the antenna safe in case of strong winds and other environmental effects. The red line shows the received signal from the satellite, measured at the ground station. The signal jumps just before the pass to a stable level, which indicated that the low-noise amplifier (LNA) was switched on. During the first part of the pass, the received signal increases as expected. The signal increases because of the decreasing distance between the satellite and the ground station. At a certain level of elevation, the signal drops for a period of time and raises back again at about the same elevation of the initial drop. It is a typical indicator that the pointing of the antenna toward the satellite is not perfect. During the low-signal period, the signal level is high enough that the receiver module on ground will be in a lock state. In that case, it could be that the operator would never recognize this effect by only monitoring the lock state of the receiver.

Problems can also occur, if there are unplanned thruster activities on the satellite, which lead to a change of the orbit and would require an update of the ephemeris for the ground station tracking. This effect can happen if again an antenna with a small beam width in combination with program tracking is used. However, since typical CubeSat missions do not feature thrusters, the antenna beam is most likely very wide, and such a scenario is very unlikely.

23.7 Summary

With technological advances, satellites in general became more versatile and the number of launches seems to increase on a yearly basis. Long-term projections of the space debris environment are commonly used to assess the launch trends [24]. Especially, CubeSats enjoy a great popularity, since they are more efficient than ever and thus can be used in a wide area of space missions. This advancement comes with new challenges, both on-ground and in-orbit. On the one hand, the size and weight of this satellite class limit the power that can be provided to its subsystems. This has a direct impact on the communication capabilities. Limits in the size reduce the gain of the satellite's antenna, while the power budget limits the output of the system. To compensate, ground stations have to be pointed more accurately to acquire a signal and track the satellite.

On the other hand, due to the small form factor the CubeSats are at the detection limit of current surveillance systems and pose a challenge to frequently update object catalogs with their current position and velocity information. Good knowledge of an orbit is needed to track small-sized objects and perform orbit determination such that information stays fresh in the catalog. Currently, CubeSats are part of collision warnings, but most satellites of this size are not able to maneuver and thus cannot react with avoidance maneuvers. This task is often left to a potential collision partner.

In order to approach these challenges and reduce the risks, steps can be taken in the design phase of CubeSats:

1. Add GNSS capabilities and perform proper orbit determination where possible.
2. Proactively share the orbit information to reduce the risks toward other parties in space.
3. Additionally, add measures that increase detectability, such as retroreflectors.

Another advantage of the latter is the possibility to better discern CubeSats from one another. Especially, after the bulk deployment, the risk of confusing a satellite with another one is high, due to the closely spaced release of many. In this early mission phase, the objects are very hard to be identified and correlated for catalog maintainers. Using a retroreflector with a unique reflection pattern increases the chance of correctly identifying the own satellite. This in turn also increases the chance of finding the own satellite and establishing the first contact.

References

1 ESA Space Debris Office, 2017. "ESA's Annual Space Environment Report," GEN-DB-LOG-00208-OPS-GR, Issue 1, 27 April 2017, https://www.sdo.esoc.esa.int/ environment_report/Reference 2.

2 Radtke, J., Kebschull, C., and Stoll, E. (2017). Interactions of the space debris environment with mega constellations—using the example of the OneWeb constellation. *Acta Astronautica* 131 (2000): 55–68.

3 E. Stoll, K. Merz, H. Krag, B. D'Souza, and B. B. Virgili, 2013. "Collision Probability Assessment for the RapidEye Satellite Constellation," in Space Debris Conference, Darmstadt, Germany, April.2013.

4 E. Stoll, R. Schulze, B. D'Souza, and M. Oxfort, 2011. "The impact of collision avoidance maneuvers on satellite constellation management," in Space Surveillance Conference, Madrid, Spain, June 2011.

5 Kebschull, C. et al. (2017). Simulation of the space debris environment in LEO using a simplified approach. *Advances in Space Research* 59 (1): 166–180.

6 Montenbruck, O. and Gill, E. (2000). _Satellite Orbits. Models Methods Applications._ Germany: Springer.

7 Vallado, D.A. (1997). _Fundamentals of Astrodynamics and Applications._ McGraw Hill.

8 Escobal, P.R. (1997). _Methods of Orbit Determination._ Krieger Pub Co.

9 F. R. Hoots, R. L. Roehrich, 1980. "SPACETRACK REPORT NO. 3—Models for Propagation of NORAD Element Sets."

10 D. A. Vallado, P. Crawford, R. Hujsak, 2016. "Revisiting Spacetrack Report #3: Rev 1," AIAA/AAS Astrodynamics Specialist Conference and Exhibit, 21–24 August 2016, Keystone, Colorado.

11 Wertz, J.R. and Larson, W.J. (1999). _Space Mission Analysis and Design._ Boston, USA: Kluwer.

12 Walter, U. (2007). _Astronautics._ Darmstadt, Germany: Wiley.

13 Fasoulas, S. (2003). _Bahnmechanik für Raumfahrzeuge._ Institut für Luft- und Raumfahrttechnik, Technische Universität Dresden.

14 Kaula, W.M. (1966). _Theory of Satellite Geodesy: Applications of Satellites to Geodesy._ Dover Publications Inc.

15 Curtis, H.D. (2005). _Orbital Mechanics for Engineering Students._ Amsterdam, The Netherlands: Elsevier.

16 E. Stoll, 2008. "Ground Verification of Telepresence for On-Orbit Servicing," PhD thesis, Institute of Astronautics, Technical University of Munich.

17 Gérard Petit and Brian Luzum (eds.) (2010). (IERS Technical Note: 36) Frankfurt am Main: Verlag des Bundesamtes für Kartographie und Geodäsie, 179 pp., IERS Conventions. ISBN 3-89888-989-6.

18 Fukushima, T. (1996). Generalization of Encke's method and its application to the orbital and rotational motions of celestial bodies. _The Astronomical Journal_ 112: 1263–1277.

19 Berry, M.M. and Healy, L.M. (2004). Implementation of Gauss–Jackson integration for orbit propagation. _The Journal of the Astronautical Sciences_ 52: 331–357.

20 Lambert, J.D. and Watson, A. (1976). Symmetric multistep methods for periodic IVPs. _Journal of the Institute of Mathematics and its Applications_ 18: 189–202.

21 Anon, 2017. "Orekit Features," https://www.orekit.org/features.html, visited September 2017.

22 Maral, G., Bousquet, M., and Sun, Z. (2009). _Satellite Communication—Systems, Techniques and Technologies,_ 5e. New York: Wiley.

23 Orbitron—Satellite Tracking System, [Online] http://www.stoff.pl, [Accessed: 15 March 2019].

24 Radtke, J. and Stoll, E. (2016). Comparing long-term projections of the space debris environment to real world data—looking back to 1990. _Acta Astronautica_ 127: 482–490.

24 II-4a

AMSAT[1]

Andrew Barron (ZL3DW)

Christchurch, New Zealand

24.1 Introduction

AMSAT was founded in 1969 by radio amateurs working at NASA's Goddard Space Flight Center and other "radio hams" from the Baltimore–Washington DC region. Their aim was to continue the efforts begun by Project OSCAR (Orbital Satellite Carrying Amateur Radio), which had already built four successful amateur radio satellites. The first AMSAT project was to coordinate the launch of Australis-OSCAR 5, constructed by students at the University of Melbourne, Australia.

AMSAT has now become a worldwide group of amateur radio operators who share an active interest in building, launching, and then communicating with each other through amateur radio satellites. By any measure, AMSAT's track record has been impressive. Since its founding 50 years ago, AMSAT has used predominantly volunteer labor and has donated resources to design, construct, and, with the assistance of international government and commercial agencies, successfully launch, more than 60 amateur radio satellites into Earth orbit. There are now AMSAT groups in 19 countries. Each AMSAT group uses the AMSAT name followed by a suffix indicating the group's country or region. For example, AMSAT-NA is the North American group, AMSAT-DL is the German group, AMSAT-UK is self-explanatory, and AMSAT-ZL is the New Zealand AMSAT group [1].

It is important to remember that AMSAT is a volunteer organization with an amazing track record. Without AMSAT and the relationships that the group has formed with NASA, US government agencies, the European Space Agency, and commercial space industry companies, the hundreds of CubeSats that are built by schools, universities, and amateur radio groups would not have been possible.

---

1 Parts of this chapter have been previously published under the title AMSATs & HAMSATs: Amateur Radio and Other Small Satellites. Used with permission of the author, and the publisher, Radio Society of Great Britain.

Nanosatellites: Space and Ground Technologies, Operations and Economics, First Edition.
Edited by Rogerio Atem de Carvalho, Jaime Estela, and Martin Langer.
© 2020 John Wiley & Sons Ltd. Published 2020 by John Wiley & Sons Ltd.

24.2   Project OSCAR

Project OSCAR Inc. was started in 1960 by members of Foothill College and radio amateurs from the Thompson Ramo Wooldridge (TRW) Radio Club of Redondo Beach, California, many of whom worked at TRW.[2] Their goal was to investigate the possibility of putting an amateur satellite into orbit. Project OSCAR was responsible for the construction of the first four amateur radio satellites, OSCAR, OSCAR II, OSCAR III, and OSCAR IV.[3]

Since the beginning, the group has focused on supporting and promoting amateur radio satellite related projects. Today, more than 50 years later, Project OSCAR's mission is "To initiate and support activities that promote the satellite amateur radio hobby." Project OSCAR's primary goal is to reach out and provide logistical support, training, and in some cases equipment to amateur radio associations, schools, and the public at large [1].

24.2.1   OSCAR 1 Satellite (1961)

Just 4 years after the USSR started the Space Age by launching the Sputnik satellite, OSCAR was launched from Vandenberg Air Force Base on 12 December 1961 as ballast on a Thor-Agena rocket. OSCAR was the first satellite to be ejected into space as a secondary payload. The primary payload was a military satellite called Discoverer 36 (Corona 9029).

It is very impressive that amateur radio enthusiasts were able to build a successful satellite only 4 years after the launch of Sputnik. OSCAR was the first amateur radio satellite and also the first non-government-built satellite. It was a curved rectangular box, about $30 \times 25 \times 12 \, cm^3$. The two approximately square surfaces were curved but concentric with each other. The satellite was shaped to fit the available space on the Thor-Agena booster and its weight was critical because it replaced one of the weights necessary for balancing the payload in the rocket stage. A monopole-transmitting antenna about 60 cm long extended from the center of the convex surface. Reflective stripes were applied to the outside surfaces to reduce the temperature inside the satellite (Figure 24.1) .

The OSCAR satellite was built by the volunteer efforts of a group of amateur radio operators. It was battery-powered, with no solar panels or attitude control system. Its launch positioned OSCAR in an elliptical orbit ranging from 244 to 475 km above Earth's surface. Known today as OSCAR 1, it was the very first ham radio "HAMSAT."

OSCAR was not capable of two-way communications. Its radio transmitted the letters "HI" in International Morse code with 140 mW of power on a frequency of 144.983 MHz. While this does not seem like a lot of transmitter power, it was 14 times the power transmitted by the 10 mW radio in Explorer 1, America's first satellite.

There was some scientific value in OSCAR's "BEEP-BEEP-BEEP-BEEP," "BEEP-BEEP" greeting. The speed of the Morse code transmission was relative to the temperature inside the satellite. At that time, this was important information, critical to the design of subsequent satellites. OSCAR had no solar panels to recharge the batteries, which only had

---

2 TRW built many spacecraft between 1958 and 2002, including the Pioneer 1 and Pioneer 10 satellites, and the Atlas and Titan rockets. The company was purchased by Northrop Grumman in 2002.

3 The Roman numeral numbering scheme was dropped after OSCAR IV, and these days the first four OSCAR satellites are normally called OSCAR 1, OSCAR 2, OSCAR 3, and OSCAR 4.

Figure 24.1 Lance Ginner, K6GSJ, with the flight model of OSCAR 1. Source: Reproduced with permission from AMSAT.

enough capacity to power the transmitter for 22 days. During that time, 570 amateurs in 28 nations around the globe picked up OSCAR's call from space and mailed in reception reports. The satellite's low orbital altitude meant that OSCAR could only stay in orbit above Earth for 50 days before it slipped down into the atmosphere and burned up on January 31, 1962 [1].

24.2.2 OSCAR 2 Satellite (1962)

OSCAR 2, earlier known as OSCAR II, was the second amateur radio satellite launched by Project OSCAR into low Earth orbit. It was launched on June 2, 1962, using a Thor-DM21 Agena-B rocket from Vandenberg Air Force Base. Like OSCAR 1, OSCAR 2 was launched as a secondary (ballast) payload. The primary payload was another Corona reconnaissance satellite, Corona 43. OSCAR 2 looked the same as OSCAR 1, but it had a different surface coating intended to reduce the temperature inside the satellite and its transmitter was set to a lower output power of 100 mW in order to increase the battery life. OSCAR 1 had stopped transmitting a long time before the satellite eventually re-entered the atmosphere and it was hoped that OSCAR 2 would transmit for a longer time. But unfortunately, the orbit was lower, only $207 \times 394\,\mathrm{km}^2$, so the satellite re-entered early. OSCAR 2 transmitted on a frequency of 144.983 MHz for 18 days ceasing operation on June 20, 1962. It re-entered the atmosphere the following day on June 21, 1962.

A third satellite was built to the same design, but it was never flown because the OSCAR team decided to concentrate on building a satellite that included a transponder [1].

24.2.3 OSCAR 3 Satellite (1965)

OSCAR 3, earlier known as OSCAR III, was launched on 9 March 1965 into an $884 \times 917\,\mathrm{km}^2$ orbit, much higher than the first two OSCAR satellites.

Figure 24.2 The OSCAR 3 satellite. Source: Reproduced with permission from AMSAT.

It was the first amateur radio satellite to carry a linear transponder. More than a thousand radio hams reported having QSOs (ham radio conversations) via OSCAR 3. The satellite was used for the first trans-Atlantic amateur radio contacts via satellite contacts across Europe, and between New York and Alaska. The transponder remained active for 18 days and the two beacon transmitters remained working for several months. The satellite had solar panels to help power the beacon transmitters, making it the first amateur-built satellite to use solar power. It had a $20 \times 30 \, cm^2$ rectangular shape with four monopole antennas. The uplink was 145.975–146.025 MHz and the downlink was 144.325–144.375 MHz. OSCAR 3 is no longer active, but it is still in orbit [1] (Figure 24.2).

24.2.4 OSCAR 4 Satellite (1965)

OSCAR 4, earlier known as OSCAR IV, was the fourth satellite to be built by the California-based Project OSCAR team. It was launched on the 21 December 1965 on a Titan rocket. It was the first amateur radio satellite to be launched into a high orbit and the first to be fully powered by solar panels. It was supposed to go into a geostationary orbit, but unfortunately, the rocket's fourth stage failed, and the satellite was left in an unintended highly elliptical "transfer" orbit. OSCAR 4 was a regular tetrahedron with edges 48 cm long. It had four independent monopole antennas and contained a tracking beacon transmitter and a communications repeater. It was powered by a solar cell array and batteries. The unplanned $161 \times 33\,000 \, km^2$ orbit prevented nominal use. Only about 12 two-way communications were established through the transponder, but one on December 22, 1965 was the first direct satellite communication between the USA and the USSR. The transponder operated for 85 days, but the satellite remained in orbit for 11 years. OSCAR 4 was the first HAMSAT to use two amateur bands. It had a 3 W VU mode transponder. The uplink was in the V band (144.300–144.310 MHz), the downlink was in the U band (432.145–432.155 MHz) [1] (Figure 24.3).

Figure 24.3 An internal view of OSCAR 4. Source: Reproduced with permission from AMSAT.

24.3 AMSAT Satellite Designations

Since the early days, it has been traditional for amateur radio satellites and satellites carrying transmitters on amateur radio frequencies to be awarded an OSCAR number. OSCAR stands for "Orbital Satellite Carrying Amateur Radio." OSCAR numbers or "AMSAT designators" are allocated by AMSAT-NA. To be awarded an OSCAR number, the satellite must carry an amateur radio payload and its operating frequencies must have been coordinated by the International Amateur Radio Union (IARU), usually with help from AMSAT. The satellite must have successfully achieved orbit and at least one transmitter must be operating. Finally, the satellite's owner must request an OSCAR number. AMSAT numbers are not reallocated when a satellite fails or re-enters the atmosphere.

Most amateur radio satellites have multiple names. They have a project name that reflects the project's scientific goals, a name relating to the amateur radio project they were created under, and after a successful launch they are awarded an AMSAT "OSCAR" designation. For example, the RadFxSat, "Radiation Effects Satellite," carries the Fox-1B amateur radio package and has an OSCAR designation of AO-91 [2].

Many recent CubeSats do not have AMSAT designations, especially if they only use amateur radio frequencies for telemetry and control. But almost all CubeSats that carry amateur radio transponders, digipeaters, or Frequency modulation (FM) repeaters do receive an AMSAT OSCAR designation.

The first letter indicates the origin of the satellite. The second letter is always O for OSCAR. For example:

AO: AMSAT OSCAR is used for satellites constructed by AMSAT.
DO: Used for Digital Orbiting Voice Encoder (Dove) OSCAR 17 from Brazil.
EO: European OSCAR or Emirates OSCAR, includes the FUNcube satellites.
FO: Fuji OSCAR. The FO designation is used for many satellites built in Japan.
IO: Used for ITAMSAT OSCAR 26, the first HAMSAT from Italy.

LO: Used for LUSAT OSCAR 19, from AMSAT-LU in Argentina.

NO: Navy OSCAR. The NO designation is used for satellites constructed by the US Naval Academy.

SO: Used for three Saudi satellites, SO-41, SO-42, and SO-50.

UO: University OSCAR. Used to designate satellites constructed by University of Surrey, UK.

WO: Used for Webersat OSCAR 18 from Weber State University in the USA.

GO-32: Gerswin OSCAR from Israel.

MO-30: Constructed by the National University of Mexico.

MO-46: Malaysian OSCAR 46.

SO-33: Used for SEDS SAT from Alabama, USA.

An RS designation indicates Radio Sputnik. Most of the Russian HAMSATs were and are not issued with AMSAT designations. They are known by their "RS" designations. An exception is RS-14, which was a joint Russian and German project, known as both AO-21 and RS-14.

24.4 Other Notable AMSAT and OSCAR Satellites

24.4.1 OSCAR 7 Satellite, AO-7 (1974)

OSCAR 7, commonly known as AO-7 (AMSAT OSCAR 7), is arguably the most successful OSCAR satellite ever. It is still working after an amazing 44+ years in space. It was launched on a Delta rocket, from Vandenberg Air Force Base on 15 November 1974. The AMSAT project name for AO-7 was "Phase II-B."

After a stunning 6½ years of service, the satellite died due to battery failure in 1981. But then in 2002, after more than 27 years of silence, the satellite miraculously sprang back to life. A radio ham, Pat Gowen, G3IOR, came across a beacon sending slow 8–10 wpm Morse code on 145.973.8 MHz. It sounded like old OSCAR satellite telemetry, with the familiar HI followed by a string of numbers in groups of three. But he was unsure which satellite it could be. Amateur radio satellite enthusiasts were stunned to find that it was coming from long-dead OSCAR 7. It is believed that the shorted batteries must have corroded to the point where they went open circuit, allowing the solar panels to power the satellite while it is in sunlight. Very luckily, AO-7 was the first amateur satellite to include a voltage regulator to charge the batteries from the solar panels. Without this, the solar panels would probably have failed when the batteries became shorted, or the transponders would have failed due to overvoltage when the batteries became open circuit 20 years later (Figure 24.4).

24.4.2 UoSAT-1 Satellite (UO-9) (1981)

UoSAT was built by the University of Surrey, Guildford, England. It was a scientific and educational low Earth orbit (LEO) satellite containing experiments and beacons but no amateur transponders, making it the first AMSAT rather than being a HAMSAT.

UoSAT was launched into a 560 km LEO on October 6, 1981, by a Thor-Delta launcher from Vandenberg Air Force Base. It was a secondary payload alongside the "Solar

Figure 24.4 AO-7 satellite and the project team. From left are Dick Daniels, W4PUJ (SK); Jan King, W3GEY; "hired hand" Marie Marr; and AMSAT Founding President Perry Klein, W3PK. Source: Reproduced with permission from AMSAT.

Mesosphere Explorer" satellite. It remained fully operational until it re-entered the atmosphere on October 13, 1989, from a decaying orbit after 9 years of service. UoSAT was the first amateur satellite to carry an S-band beacon and the first with an on-board computer, the "integrated housekeeping unit" (IHU) that could be reprogramed from the ground. The satellite featured a camera, a digitalker speech synthesizer, battery management, attitude management, remote control, and on-board experiments designed to study radio propagation from high frequency (HF) through to the microwave bands. As well as the main telemetry beacon on 145.826 MHz, 1200 baud audio frequency-shift keying (AFSK), there were phase-related beacons on the 7, 14, 21, and 28 MHz HF bands and other beacons on 146 MHz, 435 MHz, 2.5 GHz, and 10.5 GHz. The satellite had one of the earliest two-dimensional charge-coupled device (CCD) arrays forming the first low-cost CCD television camera in orbit. The resulting images transmitted from space were spectacular considering the freshness of its technology. UO-9 was not a stabilized Earth-pointing satellite, so the areas covered by its photos were random.

24.4.3 ISS (ARISS) (1998–Present)

AMSAT has also been involved in supporting amateur radio activities on manned spacecraft including the Space Shuttles and the International Space Station (ISS), which has two amateur radio stations aboard.

SAREX (1983 to 1999) was the Shuttle Amateur Radio Experiment. It was a program set up to enable amateur radio operators, schools, guides and scouts, and other youth groups with the ability to talk to Space Shuttle astronauts while they were in orbit.

ISS has been active since November 20, 1998. Since then it has been manned continuously. Building on the success of SAREX, it includes provision for amateur radio. The amateur radio program on the ISS is known as Amateur Radio aboard the International Space Station (ARISS) [3]. The primary role of the amateur radio equipment is educational outreach. It is used for contacts between astronauts or cosmonauts on the station and school or youth groups on Earth:

> ARISS lets students worldwide experience the excitement of talking directly with crewmembers of the International Space Station, inspiring them to pursue interests in careers in science, technology, engineering, and math, and engaging them with radio science technology through amateur radio. [4]

24.4.4 OSCAR 40 (AO-40) (2000)

AO-40 is the most complex amateur radio HAMSAT built to date, with transponders on frequencies ranging from 21 MHz up to 24 GHz. It was launched from Kourou, in French Guiana on November 16, 2000 at a reported cost of US$4.5 million. At the time, it was the heaviest Ariane rocket ever launched into space and the first to carry four satellites. On 13 December 2000, the main motor on AO-40 was fired for testing prior to a planned burn to move the satellite to its intended highly elliptical orbit (HEO), but an explosion occurred, and the satellite ceased transmitting data to the control station. It was later inferred that a mistake in failing to remove a protective cap before the launch had caused control valves to malfunction. The satellite was left in an unintended equatorial orbit and it was not until 25 December, after many attempts, that command team member Ian Ashley, ZL1AOX, successfully sent a "reset" signal to the satellite and regained control. On-board cameras were used to establish the attitude of the satellite and the "magnetorquer" system was used to spin-stabilize the satellite. It must have been a great Christmas present to get control back after 12 days of wondering if they had lost a US$4.5 million satellite.

AO-40 carried five receivers in the HF, VHF, UHF, L-, S-, and C-bands and seven transmitters in the HF, K, VHF, UHF, S-, and X-bands. It also had several on-board experiments including two cosmic ray monitors and a passive ionospheric sounder to scan the 0.5–30 MHz band. There were two cameras, which would send images on request, in response to a command on the amateur radio uplink. It was the first amateur-built satellite and possibly the first satellite of any kind to use a GPS receiver to determine its location.

On 25 January 2004, downlink telemetry indicated that the main battery was at an extremely low voltage and to save the battery bank, the IHU computer cut power to the transponders. The S-band beacon continued to transmit for a day or two. Unfortunately, the computer switched the auxiliary battery into the circuit in parallel with the main battery effectively shorting it out, and due to the low voltage, the control station was unable to operate the relays to isolate the main battery bank. If it had been possible, the satellite might have been saved. On 9 March 2004, Colin Hurst, VK5HI, reported the last known reception of a signal from the AO-40 beacon (Figure 24.5).

24.4.5 SuitSat (AO-54) (2006)

Probably the most unusual AMSAT so far is SuitSat, a redundant Russian "Orlan" space suit, which was fitted with a radio transmitter, some instrumentation, and padded out with unwanted laundry.

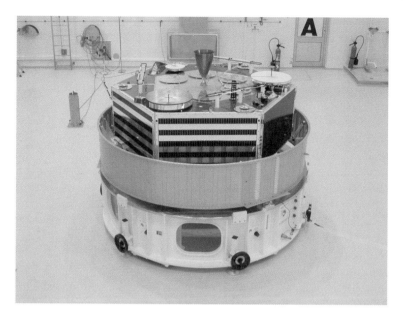

Figure 24.5 OSCAR 40 in its specific bearing structure (SBS) prior to launch. Source: Reproduced with permission from AMSAT.

The "ISS Expedition 12" crew launched SuitSat from the International Space Station on February 3, 2006 and AMSAT designated it as AMSAT OSCAR 54 (AO-54). SuitSat transmitted a digital recording of the voices of school children on 145.990 MHz, plus telemetry data, and an image using the Robot 36 Slow Scan Television (SSTV) format. Unfortunately, the signal from SuitSat was much weaker than expected, probably due to a broken antenna. You needed a fairly good setup to hear anything at all and image reception was poor. The suit-satellite was last heard on 18 February 2006. It has now burned up in the Earth's atmosphere.

24.5 The Development of CubeSats

Amateur-built satellites have been constructed in many different sizes and shapes. Some of the amateur early radio satellites were quite large and heavy. For example, AO-40 weighed 400 kg. The very early AMSAT OSCAR satellites were designed and constructed so that they could be fitted into the launch vehicle as weight-balancing ballast. Since about 1990, the number of amateur-built satellites being constructed and launched has risen dramatically. It is now common for 10 or more amateur-built satellites to be launched at a time. Some launches have deployed a hundred CubeSats.

Dozens and then hundreds of universities, clubs, and amateur radio groups began to build satellites. The variation in designs was becoming a big problem for the organizations that were responsible for launching them into orbit. Each satellite had a different set of requirements. There were different shapes, sizes, weights, cradles, and deployment methods. Some sort of standard was required. In 1999, California Polytechnic State University (Cal Poly) and Stanford University jointly developed a set of standard specifications for CubeSats. They are available at http://www.cubesat.org [5].

These days most satellites launched into low Earth orbit are built following the CubeSat guidelines. They carry one or two scientific instruments as their primary mission payload and some carry amateur radio transponders. CubeSats are based around 10 cm × 10 cm × 10 cm units weighing no more than 1.33 kg. They can be constructed in configurations from 1 to 12 units. Inside, the CubeSats have several circuit boards. The outside surfaces are usually covered with solar panels and antennas for the uplink and downlink radios. Sometimes they carry items such as bigger antennas, fold-out solar panels, or sails that can deploy once the satellite is in space.

The launch cost for a CubeSat is generally directly proportional to its weight. It is quite expensive, so many institutions now build even smaller "pico" satellites typically based on collections of $10 \times 10 \times 10 \, mm^3$ modules. It is a big challenge to fit the experimental or educational payload, radios, a power supply, solar panels, and antennas into such a small space. DC power is a big challenge for these very small satellites, as there simply is not enough room for big solar panel arrays or for a large battery.

Building a CubeSat has never been easier. You can buy "off-the-shelf" CubeSat frames and a range of commonly used technology cards and panels. Unfortunately, launching a CubeSat remains much more difficult than building one. If you are a well-funded University or organization, you can pay full price and get your satellite into space relatively quickly. Many countries offer commercial satellite launching services, including USA, Russia, China, India, and even New Zealand. The alternative is to persuade NASA to launch your satellite from the International Space Station for free or at a discounted rate. There is a huge demand for these cheaper opportunities and a limited NASA education budget. To qualify, your satellite must have a valid educational and/or scientific purpose, such as an experiment that meets NASA's educational objectives. Even if your project is approved, you will have to wait for about 2 years for your project to reach the top of the heap.

24.6 FUNcube Satellites

In October 2009, a team of volunteer experts from AMSAT-UK, in collaboration with their colleagues in the Netherlands at AMSAT-NL, started work on the FUNcube satellite concept. The initial plan was to design, build, and launch a single spacecraft but further flight opportunities became available and the project now comprises five satellite missions [6].

The FUNcube satellites have a dual role, functioning as amateur radio transponders during night time and weekend passes over the UK and as educational outreach satellites for school programs during daylight passes. The educational telemetry transponder runs higher power during the daytime when the amateur radio transponder is turned off.

FUNcube-1 (AO-73) is a complete educational 1U CubeSat with the goal of enthusing and educating young people about radio, space, physics, and electronics. It was launched from Russia on a Dnepr rocket on November 21, 2013.

FUNcube-2 on UKube-1 was a follow-on project comprising of a set of FUNcube boards mounted on the UKube-1 triple CubeSat. It was launched on a Soyuz rocket from Baikonur on July 8, 2014.

FUNcube-3 on QB50p1 (EO-79) is a 2U CubeSat. It was launched on June 19, 2014 into a $621 \times 604 \, km^2$ sun-synchronous orbit, inclined at $97.98°$.

FUNcube-4 is to be a payload on the European Student Earth Orbiter (ESEO) satellite mission. ESEO is a 20 kg class microsat, which incorporates experimental payloads from a number of universities around Europe. The FUNcube payload will provide similar telemetry to its predecessors but will have a more powerful transmitter. The amateur radio payload will be an FM voice L/V mode repeater.

FUNcube-5 (EO-88) on Nayif-1 is a 1U CubeSat project developed by the Emirates Institution for Advanced Science and Technology (EIAST) in partnership with students at the American University of Sharjah (AUS). It carries an enhanced FUNcube communications package to provide educational outreach telemetry and an amateur transponder. Nayif-1 was successfully launched from India on February 15, 2017 together with 103 other spacecraft.

24.7 Fox Satellites

The Fox satellites are the most recent series of nanosatellites built by AMSAT-NA. They each include an analog FM repeater that will allow simple ground stations using a handheld transceiver and a simple handheld dual band Yagi antenna to make contacts using the satellite [7].

Fox-1A (AO-85) is a U/V mode 1U CubeSat with an FM voice repeater for amateur radio use. It was launched on 8 October 2015.

The RadFxSat launched on 17 November 2017, it carries the Fox-1B amateur radio package. It has an OSCAR designation of AO-91.

Fox-1C was the flight spare for Fox-1A. It has not yet been launched into orbit.

Fox-1D (AO-92) is a 1U CubeSat with a UV[4] and LV mode FM voice repeater. It was launched on 12 January 2018.

Fox-1E will be launched as a package attached to RadFxSat-2, the second RadFxSat. It will be equipped with a 30 kHz wide VU mode linear transponder for amateur radio, and radiation experiments for Vanderbilt University's Institute for Space.

24.8 GOLF Satellites

In October 2017, AMSAT-NA announced a new AMSAT program called "Greater Orbit–Larger Footprint (GOLF)." The project is dedicated to achieving the deployment of amateur radio CubeSats into higher orbits so that they cover more of the Earth's surface and allow longer distance and more transcontinental ham radio contacts. The aim is to develop a series of three-unit CubeSats that can be held ready for rapid deployment into medium Earth orbit (MEO), HEO, or geosynchronous orbits, as launch opportunities become available.

The initial "GOLF-T" satellites will be test satellites launched into low Earth Orbits. They will be designed to test the technologies that will be required for future GOLF satellites

---

4 Note: UV mode indicates a UHF 435 MHz uplink and VHF 146 MHz downlink. VU mode indicates a VHF uplink and UHF downlink. LV mode indicates a UHF L band 1.2 GHz uplink and VHF downlink.

destined for higher orbits. The GOLF-T satellites will be used to investigate the use of commercial off-the-shelf components to build satellites capable of withstanding the higher level of radiation that satellites are exposed to in MEO or HEO orbits. They will also test new IHU management computers, new attitude determination and control systems, CX mode 5 GHz uplink to 10 GHz downlink software-defined radio (SDR) transponders, and deployable solar panels.

24.9 The IARU and ITU Resolution 659

The IARU is a global federation of national amateur radio associations from more than 150 countries. For many years, the IARU has provided a free frequency coordination service for amateur radio and amateur-built satellites that include radios that operate on amateur radio frequencies. Frequency coordination is essential to limit interference between satellites or ground stations and to ensure that satellite uplinks and downlinks are placed on sensible frequency allocations within the amateur radio "satellite sub-bands."

Traditionally, nearly all amateur-built low Earth orbit satellites, whether carrying amateur radio transponders or not, have used frequencies in the 2 m and 70 cm amateur radio bands for uplink control and downlink telemetry. However, since 2014, the large number of amateur-built satellites being launched has made it impossible for IARU to coordinate interference-free frequencies in the fairly busy 144–146 MHz 2 m amateur radio band.

The International Telecommunications Union (ITU) is the international group that coordinates radio spectrum usage worldwide. It makes recommendations about what frequencies should be allocated to various radio services. National regulators usually adopt the ITU recommendations. In this way, television, broadcast radio, amateur radio, satellite, maritime, emergency services, land mobile, aeronautical, and many other bands are standardized worldwide. Nearly every year, the ITU holds a World Radio Conference (WRC), which votes on any changes to the ITU recommendations.

In 2012 (WRC-12), the ITU noted the sudden proliferation of "nanosatellites and picosatellites" and suggested that this should be discussed at a future conference. At WRC-15 (2015), the ITU adopted "resolution 659," which stated that the use of 144–146 and 435–438 MHz by nonamateur radio satellites is not in accordance with the international radio regulations. Previously, a blind eye had been turned to this. The resolution goes on to say that nonamateur radio satellites should operate in the licensed part of the radio spectrum that is allocated to "space operation." There are satellite bands above and below the 2 m and 70 cm amateur radio bands at 137–138, 148–149.9, 174–184, 267–273, 400.15–402, and 1427–1429 MHz. There are also a few other "space operation" bands that are restricted as to their use or only available in certain countries.

Forcing future amateur-built, nonamateur radio satellites to use these licensed bands instead of the amateur radio bands could cause problems, such as satellites not being licensed to operate while they are over particular countries. Control stations wishing to transmit on the satellite uplink frequencies may be required to pay a license fee. For amateur radio operators and other hobbyists, it means that you might need different antennas and possibly different radios to receive signals from the amateur-built "AMSATs" than they do for the amateur radio "HAMSATs," which will continue to use amateur radio frequencies.

On 30 June 2017, the IARU issued a news release stating that it will be issuing revised guidelines for satellite frequency coordination.

The IARU believes that the definitions in article 25 of the Radio Regulations, which governs the "amateur services," are broad enough to encompass nearly all educational satellite projects that include hands-on experience with radio communication and are conducted under an amateur radio license. Their strong preference is for all satellites using amateur radio frequencies to be operated under amateur licenses. This means that the operation of a satellite control station would have to be under the direct control of a licensed amateur radio operator and satellite constructors who propose to use amateur radio frequencies for their satellites would be required to use the IARU frequency coordination process.

The IARU will not coordinate frequencies for satellites operating in the "space operation" bands. It is unclear what organization will take over that responsibility. However, it will continue to coordinate satellites with combined amateur radio and nonamateur radio missions.

The change proposed by the ITU could create a division between amateur radio satellites and experimental satellites. At present, most amateur-built nanosatellites in low Earth orbit only have one transmitter. It is used to downlink satellite control and information telemetry, send data from on-board cameras or experiments, and possibly for amateur radio. If the telemetry and the educational or experimental downlink have to be on "space operation" frequencies, it is unlikely that the satellite will be able to carry an amateur radio repeater or transponder as well.

On the other side of the coin, satellites built by amateur radio groups that carry amateur radio transponders or repeaters often carry an educational experiment as well, as this is a good way to get funding toward the launch cost. I believe that there will be no problem operating this type of satellite on amateur radio frequencies provided any uplink traffic is managed by a licensed amateur radio operator. It is possible that in the future more nanosatellites will be designed to carry an amateur radio payload so that the constructors can justify using amateur radio frequencies.

References

1 AMSAT, A brief history of AMSAT by Keith Baker, KB1SF/VA3KSF, and Dick Jansson, KD1K, https://www.amsat.org/amsat-history.

2 Project OSCAR, https://projectoscar.wordpress.com.

3 ARISS, Amateur radio aboard the International Space Station, http://www.ariss.org.

4 ARISS, http://www.ariss.org.

5 CubeSat, Standard specifications for CubeSats are available at http://www.cubesat.org.

6 FUNcube satellites, http://funcube.org.uk.

7 Fox satellites, https://www.amsat.org/meet-the-fox-project.

24 II-4b

New Radio Technologies[1]
Andrew Barron (ZL3DW)

Christchurch, New Zealand

24.10   Introduction

University and amateur radio groups have been building nanosatellites and more recently CubeSats since the early 1960s. They have always been at the forefront of innovation and have achieved many firsts over the years. OSCAR 1 was the first nongovernment satellite ever launched and it also was the very first satellite to be ejected as a secondary payload from a primary launch vehicle and then enter a separate orbit. OSCAR 3 was the first amateur radio satellite to carry a linear transponder. The satellite was used for the first trans-Atlantic amateur radio contacts via satellite. And UoSAT-1 (UO-9) was the first amateur-built but nonamateur radio satellite. It was also the first amateur-built satellite to carry an S-band (2.4 GHz) beacon and the first with an on-board computer, the IHU "integrated housekeeping unit," that could be reprogramed from the ground [1].

The high-cost to nonprofit, noncommercial construction of launching satellites has continued to drive innovation to ensure that amateur nanosatellites are as small and light as possible while the scientific, educational, and occasionally amateur radio project goals are achieved. The relentless trend to miniaturize electronic components so that ever more complex circuits can be incorporated into integrated circuits means that the computers and digital signal processing power of modern nanosatellites are many times more powerful than the primitive computers flown on early OSCAR satellites such as UO-9. The new radios are smaller and lighter than the early ones, they are more frequency agile, and they are reconfigurable from the ground while the satellite is in orbit. The availability of these new components means that CubeSats and other nanosatellites can now be constructed entirely from commercial off-the-shelf (COTS) circuit boards and components. In at least one case, a satellite was constructed using a cellphone. After all, a cellphone has a reasonably powerful computer, transmitters and receivers on several radio bands, a GPS receiver, a color camera, WiFi and Bluetooth for communicating with a control computer, and rechargeable batteries.

1 Parts of this chapter have been previously published under the title AMSATs & HAMSATs: Amateur Radio and Other Small Satellites. Used with permission of the author, and the publisher, Radio Society of Great Britain.

Nanosatellites: Space and Ground Technologies, Operations and Economics, First Edition.
Edited by Rogerio Atem de Carvalho, Jaime Estela, and Martin Langer.
© 2020 John Wiley & Sons Ltd. Published 2020 by John Wiley & Sons Ltd.

Some recent satellites use software-defined radios (SDRs) for control and telemetry and as repeaters or transponders in the case of amateur radio satellites. There have been experiments with more efficient transmitters, new digital modes of transmission, electrical propulsion, solar sails, bright light-emitting diodes (LEDs) visible from Earth, and the use of supercapacitors rather than batteries.

24.11 SDR Space Segment

There are several advantages to using SDR technology on nanosatellites. Because the received radio signals are converted to digital data early in the signal chain, they have better dynamic range and noise performance than traditional superheterodyne receivers. The lack of tuned radio frequency (RF) stages, hardware mixers, and crystal oscillators makes them more resistant to the detuning effects of temperature changes within the satellite. Most importantly, a SDR can be reconfigured by uploading a new firmware file while the satellite is in orbit. For example, a satellite could be changed from a role as an FM voice repeater to a digital mode transponder. Or the mode could be changed from the AX.25 protocol at 1200–9600 bauds, or even to binary phase-shift keying (BPSK). This ability to reconfigure a satellite while it is in orbit is a major advancement over the old fixed radio systems. In satellites with multiple transponders, it can help with resilience. For example, if the telemetry and control downlink transmitter failed, the satellite could be reconfigured to downlink housekeeping telemetry on a different frequency or band.

The radios in older nanosatellites were pretuned to frequencies allocated by the International Amateur Radio Union (IARU). But after the satellite has been placed into orbit, they typically end up a few kilohertz off frequency. This is usually due to the temperature inside the satellite being different to what was expected, or changing as the satellite orbits, or enters and exits periods of eclipse. SDRs can be reconfigured to the correct frequencies using commands or uploaded firmware sent from the control station.

SDR receivers are inherently wideband devices, so depending on their design they may be able to be shifted to operate on a completely different band if necessary. There are limitations if the radio has a fixed tuned preselect filter before the analog-to-digital conversion stage. Transmitters are less frequency agile due to the requirement for low-pass filters following the RF power amplifier. A software-defined transmitter can be used on several bands provided low-pass filters are provided for each band. SDR circuit boards are also smaller and lighter than traditional superheterodyne receivers and transmitters. Weight and the available space inside a nanosatellite are always a prime consideration for any satellite designer.

One possible disadvantage of using SDRs rather than conventional architecture radios is that SDRs are essentially computer devices. They use a microprocessor or more commonly a field-programmable gate array (FPGA) to perform digital signal processing. These devices are sensitive to the particle radiation in space. Faults can include temporary disruptions and hard faults caused by cosmic radiation, and solar flares, X-rays, gamma rays, electrons, and protons ejected from the Sun. Satellites in medium Earth orbit or highly elliptical orbits are also exposed to high-energy particles (mostly electrons and protons), as they traverse the Van Allen belts.

Computer technology is advancing all the time. Shrinking the size of components within the computer devices to less than 0.1 μm and at the same time increasing the device complexity make them more vulnerable to impact damage from solar particles and radiation. The voltages inside the semiconductors are being reduced to reduce the amount of heat that the devices generate internally, and this reduces the margin between digital one and zero, making the data more susceptible to being altered by radiation noise. The new processors are faster so that they can process more data, allowing higher levels of parallel processing and the ability to manage a larger radio bandwidth, but also making them more susceptible to "single event" or transient disruption.

Old-fashioned conventional radios are virtually immune to damage from these types of radiation, although the satellite IHU computer controlling the satellite is at risk. Satellites using conventional radios with digital signal processing are at the same risk as satellites using SDRs, since digital signal processor (DSP) chips are very similar to the computer chips employed for SDRs.

Some of the frequencies used for nanosatellite communications are becoming difficult to use due to congestion and interference. Often the bands are designated as "multiple use" so there is no regulatory protection from signals that cause interference to the satellite when it is over highly industrialized countries. The 2 m (144–148 MHz) amateur band is heavily used and the unregulated 2.5 GHz "industrial scientific and medical" (ISM) band suffers considerable spectrum pollution from high-power devices, such as RF heating equipment and medical diathermy machines, as well as millions of Wi-Fi routers and devices. SDRs can help by providing extremely sharp bandpass filters around the wanted bandwidth and band-stop filters to reduce interference signals.

24.12 SDR Ground Segment

The proliferation of SDR receivers and wideband low-noise amplifiers has made the reception of microwave band signals cheaper and easier. Cheap SDR receivers and micropower transceivers are available for frequencies up to 6 GHz.

Due to size constraints, most nanosatellites in low Earth orbits use linear polarized antennas. As the satellite orbits, the plane of the polarized radio signal changes with respect to the polarization of the ground station antennas. Also, Faraday rotation skews the polarization of the radio wave as it passes through the atmosphere. Use of an SDR with two synchronous receivers allows the radio software to combine signals from both polarizations to get the best overall signal. Phase synchronous SDR receivers have two receivers clocked from the same oscillator. Two analog-to-digital converters sample at the same time and the FPGA firmware runs in parallel so that the resulting digital data streams are synchronized. One receiver is connected to the horizontally polarized antenna and the other receiver is connected to the vertically polarized antenna. The phase of the received signal from one antenna can be manipulated in software so that it combines with the signal from the other antenna to achieve a signal enhancement. This process eliminates losses due to cross-polarization fading between the satellite and the ground station. It can also be used for noise or interference reduction. Rather than being used to improve the received signal level, changing the phase relationship between the signals received two antennas can be used to cancel noise or an interfering signal. "Nulling" the unwanted signals will usually have a negligible effect on the reception of the wanted signal from the satellite, improving the overall signal-to-noise ratio.

The National Oceanic and Atmospheric Administration (NOAA) satellites transmit weather map images on 137 MHz, but the signal is quite wideband. The filters in most conventional FM receivers are too narrow for good image reception. Special wideband receivers preset for the six NOAA channels are available, but using an SDR is recommended. SDR receivers are cheap and they have adjustable bandwidth, so they are ideal for NOAA satellite reception and they can be used for receiving other bands as well. The NOAA satellites also transmit high-resolution pictures using the high-resolution picture transmission (HRPT) format on channels in the 1.7 GHz band. HRPT stands for high-resolution picture transmission. The mode uses L-band downlink frequencies around 1.7 GHz. An SDR receiver can be used that can reach 1.7 GHz, such as the FUNcube dongle or a HackRF SDR. Reports online suggest that RTL dongles are not good enough for this job. HRPT pictures have a resolution of 1.1 km, much better than the automatic picture transmission (APT) images.

Inmarsat satellites transmit in the L-band at around 1.5 GHz. With an SDR dongle receiver and a patch, dish, or helix antenna, one can receive data from an Inmarsat and decode the STD-C NCS channel. This channel is mainly used by vessels at sea. It contains enhanced group call (EGC) messages, which are like short text messages containing information, such as search and rescue and coast guard messages as well as news, weather, and incident reports.

A future possibility is the establishment of distributed satellite systems (DSS), which will utilize the wideband capability of SDR receivers at the ground station to simultaneously receive and decode signals from clusters of small picosatellites as they work together. Clustered satellites can each transmit or repeat a low bandwidth of data that can be combined at the ground station. Using interferometric synthetic aperture techniques, a cluster of satellites can capture radar or radio images that would otherwise require a large satellite dish.

APT = Automatic Picture Transmission system is an analog image transmission system developed for use on weather satellites in the 1960s.
RTL dongle receiver = an SDR receiver designed for digital television reception.
FUNcube dongle = an SDR receiver specifically designed for satellite reception.
HackRF = an SDR transceiver designed for hackers. Very wide frequency coverage.
STD-C NCS channel = STD-C is an Inmarsat data or message-based system used mostly by maritime operators.

24.13   Modern Transmitter Design

In nanosatellites, power is often the most limited resource available. There is limited space for solar panels and batteries. The strong desire to achieve a high-power transmission which will make it easier for ground stations to receive signals from the satellite must be balanced against the power capability of the satellite and its ability to dissipate the heat generated by the transmitter and the RF power amplifier. The goal is to have a very efficient RF power amplifier without compromising the transmitted bandwidth and quality requirements. Good linearity is important if nonconstant envelope modulation is to be used. BPSK and Quadrature Phase-Shift Keying (QPSK) digital modulation and FM voice do not need a linear power amplifier. But, single sideband (SSB) and high-order quadrature amplitude digital modulation modes such as 16QAM or 64QAM do need a linear amplifier.

Digital signal processing and SDR techniques have allowed great improvements in transmitter and power amplifier design and performance. Software modification of the modulating signal data can dynamically correct for nonlinearity in the power amplifier, significantly improving the level of transmitter nonlinearity. This reduces the chances of splatter or intermodulation products affecting receivers on-board the satellite and maximizes the RF power carrying the wanted signal.

Highly efficient nonlinear switching amplifiers can be manipulated to produce RF power with a linear response. The techniques are still experimental and considerable effort is going into the development of high-power linear amplifiers using nonlinear amplification modes.

Envelope control of the modulating signal can allow the use of efficient nonlinear class E power amplifiers for transmissions that require linear amplification. The technique uses control of the supply voltage to affect the signal amplitude and direct phase modulation of the power amplifier final stage. Switching off the amplifier metal-oxide-semiconductor field-effect transistors (MOSFETs) is performed at zero crossing points of the RF waveform, minimizing power dissipation inside the devices.

Class G or H amplifiers are a more efficient variation of linear class AB amplification. They work by dynamically changing the power supply voltage supplied to the power amplifier transistors so that they are not operating at full voltage when the input level is low. This reduces the amount of energy wasted as heat when the input level is low. When the input level is high and full power is required, the DC rail is increased so that the amplifier can reach full output without clipping. Significant efficiency gains can be achieved at the expense of a more complicated power supply control system and slightly reduced total harmonic distortion (THD) performance.

The class S power amplifier is a nonlinear switching mode amplifier similar in operation to the class D (pulse width modulated) amplifier. A class S amplifier converts analog input signals into square wave pulses using a delta-sigma modulator and uses the resultant pulse stream to switch the power transistors. The output of the amplifier stage is the same square wave digital stream as the input signal with higher voltage levels, in other words, an amplified digital signal. A low-pass filter recovers the analog waveform. As the amplifier MOSFET transistors are either fully on or off, there is very low heat dissipation within the transistors. Efficiencies approaching 100% are possible.

Delta-sigma modulation is a process where an analog-to-digital converter outputs a digital pulse stream representing the change in the level of the input signal compared to the previous sample, rather than sampling the actual input level at each sample time like a conventional analog-to-digital converter (ADC). The process can be referred to as pulse frequency modulation. The larger is the difference in input level (delta voltage), the higher is the frequency of the output pulses.

Reference

1 AMSAT, A brief history of AMSAT by Keith Baker, KB1SF/VA3KSF, and Dick Jansson, KD1K, https://www.amsat.org/amsat-history

25 III-1a

Cost Breakdown for the Development of Nanosatellites

Katharine Brumbaugh Gamble

Washington, DC, USA

25.1 Introduction

Satellites have been built and launched for over 60 years. Initially, satellites were small with simple payloads. Over time these payloads became bigger, necessitating larger, more complex, and expensive spacecraft. In the past decade, small satellites have re-emerged as a lower-cost alternative space platform. There are four classifications, summarized in Table 25.1, of small satellites, which are gaining in popularity. Those spacecraft having a mass between 0.1 and 1 kg are considered "picosatellites," while those with masses between 1 and 10 kg are called "nanosatellites." Larger classes of spacecraft include "microsatellites" and "minisatellites," having masses between 10 and 100 kg and larger than 100 kg, respectively.

California Polytechnic State University (Cal Poly) has established a standard launching mechanism for nanosatellites called the Poly-Picosatellite Orbital Deployer (P-POD). The P-POD is frequently flown as a secondary payload on unmanned launch vehicles, making it easier for small satellites, which use the system to obtain launches. In order to use the P-POD, the spacecraft must be in the shape of 10 cm cubes—called CubeSats. One Cube-Sat is called a 1-unit (1U) cube. Multiple CubeSats may be combined to form various size configurations of units, such as 1U, 2U, and 3U. The P-POD and CubeSat standard was first demonstrated in June 2003 with the launch of two P-POD devices and a total of six 1U CubeSats [1].

Cost of a spacecraft mission can be broken down into many different elements, but primarily consists of: hardware; integration and testing; personnel and management; launch, operations, and facilities costs. Each of these categories will be explored further throughout this chapter. It is the identification and estimation of these cost elements that is crucial to cost analysis of a spacecraft mission. The mission managers strive to estimate all the costs associated with a given mission, so the budget is not exceeded. Because of the nature of unforeseen difficulties, many missions face the risk of exceeding the budget. A few risk reduction strategies are offered in Section 25.5.

The cost analysis of spacecraft missions and the associated systems engineering processes has always been an important parameter for mission planners. Nanosatellite

Nanosatellites: Space and Ground Technologies, Operations and Economics, First Edition.
Edited by Rogerio Atem de Carvalho, Jaime Estela, and Martin Langer.

Table 25.1 Satellite classification by typical mass ranges [30].

Satellite classification	Typical mass range (kg)
Picosatellites	0.1–1
Nanosatellites	1–10
Microsatellites	10–100
Minisatellites	>100

projects, however, are a new development in the domain of traditional spacecraft programs and many engineering processes and practices which have been successfully implemented on larger-scale satellite projects have never been applied, and may not be appropriate on the smaller scale. Lacking a suitable body of evidence that can be applied to small satellites, many small satellite projects are operated with purely ad hoc or even without established engineering practices.

This chapter will first detail the differences between recurring and nonrecurring costs associated with small satellite missions along examples of each type, as available. Then, industry-used small satellite cost-estimating methodologies—including a discussion on parametric and grassroots models—will be described to provide readers concrete steps with which to begin their cost-estimating process.

Before beginning the description of cost breakdown for nanosatellites, a few key definitions must first be made clear. "Life-cycle cost" is the total cost of the system over its intended life cycle [2]. Blanchard and Fabrycky go on to outline 12 steps of a typical life-cycle cost analysis process; though not specific to satellites or even small satellites, the list is nevertheless useful for the discussions of this chapter:

1. Define system requirements and technical performance measures.
2. Specify the system life cycle and identify activities by phase.
3. Develop a cost breakdown structure (CBS).
4. Identify input data requirements.
5. Establish costs for each category in the CBS.
6. Select a cost model for analysis and evaluation.
7. Develop a cost profile and summary.
8. Identify high-cost contributors and establish cause-and-effect relationships.
9. Conduct a sensitivity analysis.
10. Identify priorities for problem resolution.
11. Identify additional alternatives.
12. Evaluate feasible alternatives and select a preferred approach.

"Cost model" is often used to describe the methodology employed to determine the costs of a spacecraft mission. These costs include personnel, hardware, facilities, operations, and management costs. Some specific cost models used in the satellite industry are detailed in Section 25.4. "Recurring cost" refers to the project cost elements that do not carry over into additional projects or add-on missions; these are typically costs associated with a specific mission and only that mission. Examples of recurring costs are spacecraft components, integration, and testing hardware, which may be specific to a particular mission, or test, and personnel costs associated with the specific mission. Conversely, "nonrecurring cost" refers to the cost elements, which can be shared between missions, or can be applied to subsequent projects. Sometimes these nonrecurring costs are also called "one-time costs." Examples of nonrecurring costs include laboratory facilities, ground station equipment, and generic testing hardware.

25.2   Recurring Costs

As explained by Larson and Wertz [3], recurring hardware costs are costs associated with flight hardware manufacturing, integration, and testing. Nonrecurring hardware costs are then associated with the design, drafting, and engineering of the protoflight Engineering Design Unit (EDU). Similar distinctions may be made across other aspects of the mission cost, which can be divided into repeating (recurring), and one-time (nonrecurring) costs. This section will detail four categories of recurring costs: hardware, integration and testing, launch and operations, and personnel. Because of the sensitive nature of project costs and the limited number of small satellite missions, little information is available for case study analysis. However, what little is known will be provided for context as appropriate.

25.2.1   Spacecraft Hardware

The hardware included in a spacecraft varies depending on the mission objective, the space-craft size, and how the mission is to be accomplished. For example, a spinning spacecraft may not need to have the same level of pointing control as a three-axis stabilized spacecraft. Furthermore, the type of components aboard a CubeSat differs from those aboard larger nanosatellites. These differences dictate which components are necessary to accomplish the mission objective. Given the limited space within small satellites, component redundancy is often not a choice, as only the strictly necessary components are able to fit within the spacecraft. Because of this difference in necessary hardware, the spacecraft hardware cost varies largely across the spectrum of nanosatellite missions. This section will highlight a few options, along with the price, for typical components aboard nanosatellites. For more information on many of these components, it is recommended to visit the manufacturers' websites.

25.2.1.1   Attitude Determination and Control System (ADACS)
The purpose of Attitude Determination and Control System (ADACS) is to determine where the spacecraft is in space, and make sure the satellite maintains a particular attitude or spin per mission requirements. This is typically accomplished via an actuator and sensor suite. Some companies, such as Maryland Aerospace, sell an ADACS suite for a variety of mission needs.

Actuators are those devices aboard a spacecraft that allow the spacecraft to move, either in a rotational or translational way. These components typically include magnetorquers, reaction wheels, and thrusters. Magnetorquers, also known as magnetic torque rods, interface with the magnetic fields of the planetary body and use the forces to provide torque. Often

Table 25.2 Examples of ADACS components and associated information. Note that many of the manufactures listed here also sell other spacecraft components.

	Manufacturer	Approx. size	Approx. mass (g)	Typical cost
Actuators				
Reaction wheel	Sinclair Interplanetary	$50 \times 50 \times 40$ mm^3	185	US$25 000
	Astrofein RW1	$21 \times 21 \times 12$ mm^3	24	Not available
	Maryland Aerospace MAI-400	$33 \times 33 \times 38$ mm^3	90	US$7100
Magnetic torque rod	Andrews Space	228 mm length	400	US$11 000
	SSBV Space & Ground Systems	10 mm diameter $\times$ 70 mm length	30	Not available
Thruster	IFM nano thruster	$94 \times 90 \times 78$ mm	870 grams	30,000 Euros
	Vacco Propulsion Unit for CubeSats	Size varies: 0.25–1U	Varies based on size	Not available
Sensors				
Magnetometer	Honeywell HMR2300	$10.67 \times 2.54 \times 2.23$ cm^3	25	US$1000
Sun sensor	Sinclair Interplanetary	$34 \times 32 \times 21$ mm^3	34	US$12 000
	Solar MEMS SSoC-A60	$30 \times 30 \times 12$ mm^3	25	€7200
	Space Micro Coarse	12.7 mm diameter $\times$ 9 mm height	10	Not available
	NewSpace Systems NCSS-SA05	$33 \times 11 \times 6$ mm^3	5	Not available
Star tracker	Berlin Space Technologies ST200	$30 \times 30 \times 38$ mm^3	40	Not available
	Blue Canyon Technologies Standard NST	$100 \times 55 \times 50$ mm^3	350	Not available
	Sinclair Interplanetary	$62 \times 56 \times 38$ mm^3	158	US$120 000

this torque is used for detumbling maneuvers, or for desaturizing reaction wheels. Reaction wheels spin one direction to enable the spacecraft to spin the opposite direction, due to the conservation of angular momentum. Thrusters allow translational movement through the emission of gas or particles, depending on the particular design. There are many different types of each of these components, some of which have been listed in Table 25.2 along with their approximate size, mass, and cost. Note that in many cases, the manufacturer does not list the price of the component on the datasheet, and instead the cost is listed as "not available" in Table 25.2.

ADACS sensors are the components that determine where the spacecraft is in orbit. This can be done in a number of ways. Many nanosatellites employ sensors such as magnetometers to detect the magnetic fields of the Earth, sun sensors to determine the location of the sun, and star trackers to calculate the satellite's attitude with respect to the stars visible throughout the orbit. Examples of these components are provided in Table 25.2 along with their approximate size, mass, and cost.

25.2.1.2 Avionics

The spacecraft avionics typically comprises the flight computer, communications, and power system, including the solar panels. These are the components necessary to gather and process data on-orbit and send this information down to waiting ground stations. As with the ADACS system, there are a variety of components for each of the avionics sub-categories. Table 25.3 provides some examples of manufacturers along with typical size, mass, and cost for each component, if these values are known.

Table 25.3 Avionics components and associated information.

	Manufacturer	Approx. size	Approx. mass (g)	Typical cost
Flight computer				
	phyCORE LPC3250	$58 \times 70 \, mm^2$	50	Not available
	Bluetechnix TCM-BF527	$31.5 \times 36.5 \, mm^2$	5	Not available
	ISIS on-board computer	$96 \times 90 \times 12.4 \, mm^3$	94	€4400
Communications				
UHF/VHF radio	ISIS TRXUV UHF/VHF transceiver	$96 \times 90 \times 15 \, mm^3$	85	€8500
	AstroDev Helium	$96 \times 90 \times 16 \, mm^3$	78	US$4900
	AstroDev Lithium	$65 \times 33 \times 10 \, mm^3$	52	US$5000
S-band radio	Microhard n2420	$57 \times 98 \times 38 \, mm^3$	210	Not available
	ISIS TXS	$90 \times 96 \times 33 \, mm^3$	300	Not available
UHF/VHF antenna	ISIS deployable	$98 \times 98 \times 7 \, mm^2$ (stowed)	100	€5000
Power system				
EPS	ClydeSpace 1U	$95 \times 90 \times 15.4 \, mm^3$	86	Not available
	GomSpace	$89 \times 93 \times 26 \, mm^3$	200	Not available
Solar panels	ISIS (1U)	Customizable	50	>€2500
	GomSpace (1U)	$60.36 \, cm^2$ effective cell area	>26	Not available
	ClydeSpace (1U)	Customizable	Not available	Not available

Choice of flight computers, communication, power system, and solar panels differ based on the mission requirements. Depending on the necessary processing power, a mission may choose one flight computer over a competitor. The frequency band needed for communication dictates which radios and antennas the spacecraft has on board. The power system chosen depends upon the amount of power needed to accomplish mission objectives. Note that in Table 25.3, some of the data are provided as a minimum value because alternative versions increase, which are larger and more powerful. The solar panels come in a variety of shapes and sizes, from body-fixed to deployable. The power system and solar panel components listed in Table 25.3 represent the base model. The choice between sets of components is left to the mission designer to make after completing thorough trade studies.

25.2.1.3 Structure and Payload

The previous sections detailed options for components aboard the spacecraft, of which there are a finite quantity. However, there are greatly more combinations of structural design and manufacturing. Some organizations buy ready-made structures, such as the CubeSat structures sold by Pumpkin, Inc. Many more organizations have their own in-house structural designers and machinists, while others send drawings and metal to an outside shop.

The payload of a spacecraft is often specially built and falls under its own cost-estimating process (see Section 25.4.4 for payload cost models). Many universities receive these payloads as in-kind donations, because companies or other organizations wish to flight test their payload. Many organizations specifically design a payload and then a spacecraft to fly it.

Additional considerations for nanosatellite missions include thermal protection and propulsion. Neither of these elements is necessary for many missions, however some spacecraft have strict thermal requirements, or perhaps have mission objectives that require the use of translational or rotational movement from a thruster. There are many ways to employ thermal control and/or thrusters, and it is left to the reader to determine whether these components are necessary for their mission and the most efficient method of incorporating them onto the spacecraft.

Despite the difficulty to provide cost estimates for the structure and payload of a nanosatellite, the cost of the spacecraft structure and payload should be included in the budget early on, especially with contingency.

25.2.2 Integration and Testing

Before delivering to the launch provider, a nanosatellite must be thoroughly tested via functional as well as environmental testing. While organizations have their own methods of progressing through these tests, it is worth noting that recurring costs exist with the integration and testing phase and it is prudent to include cost estimates in a program's budget as early as possible. Primarily, these costs are associated with additional components, perhaps to replace broken ones, as well as ground support equipment (GSE). Furthermore, the mission designer should not neglect the materials necessary for integration in the budget assessments. These materials are components such as nuts, bolts, screwdrivers, coatings, adhesives, etc. There are various elements of GSE that can split into categories: shipping containers and testing mechanisms.

The size of the nanosatellite determines the size of the shipping containers. It is also advised to include G-sensors to determine if the satellite experiences any sudden accelerations or decelerations, such as potholes or being dropped, which could harm the components. CubeSats, for example, easily fit in Pelican cases, which also have sufficient padding to keep the spacecraft stable during shipping.

Many mission designers often neglect to include testing mechanism design and cost estimation in their early budget. Unless the location of testing provides the test rigs, it is often up to the mission to provide their own rigs on which the spacecraft will be attached for testing. Costs and designs can vary greatly and are highly dependent upon mission requirements.

25.2.3 Launch, Operations, and Personnel

The ability for nanosatellites to obtain no-cost launches into space has drastically improved over the past decade. It may be easier for satellites following the CubeSat standard to find free launches because of the CubeSat Launch Initiative and Educational Launch of Nanosatellites (ELaNa) in addition to dedicated deployment mechanisms, such as the Cal Poly P-POD, Innovative Solutions In Space (ISIS) ISIPOD, NanoRacks, and Planetary Systems deployers. More about these systems was provided in Chapter 18 I-4b. However, many small satellite launch mechanisms are in development, and some have already launched, such as the Vega small satellite launch vehicle [4]. For now, the majority of non-CubeSat nanosatellites are launched as secondary payloads aboard military or commercial resupply launches. In many cases, the developer need not pay any launch costs, as long as they deliver the launch-ready spacecraft on schedule. It is advised, though, to check with the launch provider early in the mission life cycle to determine launch costs as well as environmental requirements the spacecraft will need to satisfy prior to launch vehicle integration.

Recurring operations costs are primarily associated with facility or personnel costs. These costs greatly differ based on the environment, that is, whether the organization is an academic institution, government entity, or corporation. The cost also varies depending on the amount of training required for personnel to operate any of the facilities, including the ground station. Much of the operations costs are in the nonrecurring category, assuming that the hardware bought is applicable, or easily reconfigurable, to many different missions.

The cost associated with personnel is always a large recurring and nonrecurring cost. Personnel cost is recurring in the sense that the mission needs to pay the employees for the recurring tasks of integrating and testing spacecraft. The personnel cost greatly differs based on the organization. However, in most circumstances, the personnel cost grossly outweighs the cost of the spacecraft itself. It is vital to include the personnel cost, with contingency, in the initial budgets.

25.3 Nonrecurring Costs

Nonrecurring costs are those costs associated with items that are one-time purchases. This section will describe some typical examples of nonrecurring costs: spacecraft testing, integration and testing facilities, ground station equipment, as well as nonrecurring personnel costs. Once again, because of the sensitive nature of project costs and the limited number of small satellite missions, little information is available for case study analysis. However, what little is known will be provided for context as appropriate.

25.3.1 Spacecraft Testing

Because Larson and Wertz [3] describe nonrecurring hardware costs as those associated with the design, drafting, and engineering of a protoflight unit, much of the spacecraft testing cost belongs in the "nonrecurring" category. Protoflight components or satellites are often confused with the EDU or flight spacecraft components. EDU components are primarily used for the interfacing between components in order to design and test the systems similar to those, which will be flown on the flight spacecraft. Many times, these EDU components are placed onto a "FlatSat"—arranged with all components on a flat board for functional testing. Protoflight components are components that are interfaced with on a regular basis but are ultimately part of the EDU spacecraft. These components may, however, be flight-qualified to fly on the flight unit and are treated with the same care as flight components.

Typically, the protoflight and/or EDU spacecraft cost more than the flight unit did. This is primarily due to the inherent nature of interfacing with a component for the first time. That is, a component may break, and replacements may be necessary. Thus, the higher cost is due to purchasing a protoflight unit and a few replacements or alternative components. See Tables 25.2 and 25.3 for examples of ADACS and avionics components and their prices.

25.3.2 Integration and Testing Facilities

The integration phase may require materials to simplify, accelerate, and make the integration process easier while at the same time following all cleanliness requirements for spacecraft integration. For example, if the organization does not yet have a clean room or clean bench, it must procure a clean environment prior to integrating the flight spacecraft. However, once it is procured, it may be used for future missions, making this a nonrecurring cost. Another nonrecurring integration cost example would be the purchase of other cleanliness items, such as electrostatic discharge (ESD) jackets, wrist straps and grounding cords, workstation mats, and gloves. While some of these supplies may need to be replaced from time to time, they are generally usable for many years. One example of making the integration process easier would be an integration support jig useful for keeping the spacecraft in a certain orientation without someone needing to hold it. Additionally, the use of liquid nitrogen tanks, such as a Dewar, to freeze conformal coating and staking compounds in batches is useful rather than requiring a fresh batch to be made each time it is needed. General organizational materials are also useful to keep the integration area clean and organized. Finally, many organizations may purchase supplies in order to set up their own wiring and soldering stations.

The purchase of testing equipment is dependent upon the organization and their capabilities. For example, universities may not have the funds to purchase much of the equipment industry or government entities are able to secure. However, many universities have agreements with industry or government organizations for low-cost use of their facilities. Some examples of nonrecurring testing equipment include:

- Thermal vacuum chamber
- Shake table for vibration testing
- Air-bearing table
- Center of gravity/moment of inertia testing rig

Many low-cost missions may find it more cost-efficient to work with Department of Defense (DoD) or NASA partners to obtain access to these testing facilities.

25.3.3   Ground Station

Once the spacecraft is launched, the mission team communicates from at least one ground station. Many times, missions rely on partner ground stations to recover as much data as possible. However, many missions decide to build their own ground station. While the equipment can be quite expensive, the ground station is a nonrecurring cost, since it can theoretically be reconfigured for additional missions. A simple ground station might consist of an antenna; receiver and transmitter, or transceiver; computers; and cabling to connect all the components. The choice of the specific brand and component for each of these elements depends on the mission requirements. That is, the frequency band on which the ground and spacecraft will communicate, the amount of data needed to downlink, and other orbit parameters. A simple full ground station kit is made available by typical CubeSat vendors for approximately €45 000. Some missions, however, may wish to purchase specific components. Many ground stations have begun using software-defined radios, which allow reconfiguration to different frequency bands. The ground station is a large expense for many universities, so some universities have taken to relying on industry or government partners for spacecraft communications.

25.3.4   Personnel

As with the recurring cost, the personnel cost is a large portion of the total mission cost. The one-time set of labor hours spent developing interfaces, documentation, and the one-time aspects of spacecraft integration and testing may be considered the nonrecurring personnel costs, since they are not transferable from mission to mission. If necessary, reference [5] provides a method to separate the recurring and nonrecurring personnel costs by calculating the reusability value of the mission. However, in most cases, it is difficult to distinguish between recurring and nonrecurring personnel costs, unless team members directly charge their time to distinct codes.

25.4   Satellite Cost-estimating Models

According to Larson and Wertz [3, 6], when estimating cost, there are four major engineering and program requirements that influence the total mission cost—size, complexity, technology availability, and schedule. These four mission parameters determine the class of the mission, and drive the procurement approach and the level of government oversight. Industry cost models tend to rely upon a top-down approach called parametric cost estimating in which these four characteristics help establish cost-estimating relationships (CERs) for each subsystem of the satellite. In addition to parametric models, there are two other cost-estimating approaches, which will be described in this section: bottom-up estimating, sometimes called grassroots, and analogy-based estimating. Because of their overwhelming popularity in industry, most of this section will focus on parametric models.

In parametric estimating models, the CERs are a mathematical equation with input variables, such as mass, volume, power output, etc., and yield an output of the total cost for that subsystem. The CERs are different for each model and sometimes vary within the specific model depending upon the input variable. For example, some of the models have different CERs if the total spacecraft mass is between 0 and 100 kg versus a total mass between 100 and 1000 kg, such as the small satellite cost model (SSCM) to be discussed in Section 25.4.2.

The CER equations themselves are based upon historical data from previous missions. Thus, as more and more space missions take place, these CERs are updated to more accurately reflect growing trends in space mission costs. Because these CERs are based upon actual data, the relationships reflect the impact of schedule and engineering changes as well as other programmatic issues that typically arise during mission planning and executing.

The industry parametric models may be broken down into subcategories of cost models and instrument models. Cost models are used to estimate the cost of the spacecraft, payload excluded, and encompass the entire mission—from design to fabrication to operations. The cost models mentioned in this chapter have been selected because they encompass unmanned spacecraft missions. Should this technique be used for human-rated spacecraft, these models would be irrelevant. The instrument models are typically only used in the designing, building, and operating of the instrument itself as opposed to the entire spacecraft.

Each model has its own set of assumptions and rules. These rules and assumptions typically deal with what each component or subsystem of the satellite encompasses. Naturally, each cost model is only as good as the database of satellites used to determine the CERs. Additionally, models tend to differ on their definitions of "recurring," "nonrecurring," and other cost definitions. For these reasons, it is suggested that a user read and understand the model's user's guide before using the cost model for analytical purposes.

In each of the following sections, the most widely used cost models will be explained. For more information on how to use the cost model, refer to the individual model user's guides. Also, for an example of applying the SSCM and the NASA/Air Force Cost Model (NAFCOM) to a 3U CubeSat mission, see reference [5]. In some cases, the CERs are not available due to the proprietary nature of the models. Instead, the cost model is a type of computer program where the input variables are entered, and the program calculates and displays the total cost for that subsystem of the satellite.

25.4.1 Nonparametric Cost-estimating Methods

A bottom-up, or grassroots, estimating approach starts with determining costs at the component level and working up to the spacecraft and ultimately the mission level. This method tends to result in the highest fidelity cost estimate. An innovative method of estimating the cost associated with a student-built 3U CubeSat was completed in reference [5] and uses a grassroots, or "bottom-up," approach to determining the total cost of a mission. The grassroots approach accounts for all hardware, personnel, and integration and testing costs as they apply to student projects. Many of these costs are minimal to nonexistent when compared to the associated costs in industry. Chapter 6 of reference [5] provides a summary of the student grassroots cost approach and compares the results with typical industry models to be explained in the following sections.

The grassroots approach taken in reference [5] included listing out each piece of hardware and its associated price for both the Development/EDU satellite as well as the flight unit. Following the modularity philosophy of the missions, the components were placed into their respective modules in order to obtain costs per module. Flight system integration costs are calculated based upon previous flight experience, whereas the integration costs of the development satellite were based upon purchased items needed to fabricate the EDU. Because most of the mission-level tests will be completed at the expense of government partners, the majority of integration and testing (I&T) costs need not be accounted for in the grassroots approach. The facilities cost estimate provided in Chapter 4 of reference [5] gives an idea of the costs accumulated at the university level as of February 2012 when buying supplies necessary for integration and testing.

Analogy-based estimating uses the cost of another, similar mission as the basis. The cost of the new mission is then calculated by making adjustments for differences in size, complexity, and many other factors. Inherent in this cost-estimating methodology are, of course, a number of assumptions and questions. How does one determine which missions are similar? How does one adjust for the size and complexity? As Larson and Wertz summarize, "This method also presumes that a sufficiently similar item exists and that we have detailed cost and technical data on which to base our estimate" [3].

25.4.2 Small Satellite Cost Model

The SSCM was created by The Aerospace Corporation to parametrically estimate the costs associated with small satellites (less than 1000 kg) using the CERs mentioned earlier [7]. The SSCM includes two models. The first is for spacecraft that have a mass less than 100 kg, the typical size range of nanosatellites, and the second model is for those small satellites with masses between 100 and 1000 kg. However, as detailed in reference [5], the cost model for satellites less than 100 kg may not accurately portray costs of the picosatellites such as CubeSats, typically having masses less than 30 kg. Note that access to the SSCM is currently restricted to US citizens, but the methodology and assumptions are listed in the following text for comparison to the other models mentioned in this section. For more information on any of these points, see reference [7].

1. Estimates assume the cost of developing and producing one spacecraft, corresponding to NASA Phases C/D and DoD Phases B/C. Concept development and operations are not included. Also, the emphasis is on spacecraft bus costs; payload, launch vehicles, upper stages, and associated GSE are not included.
2. All costs estimated by CERs are contractor costs.
3. CER estimates are in FY05 $K.
4. Nonrecurring and recurring costs can be estimated separately, using the provided factors. Nonrecurring costs include all efforts associated with design; drafting; engineering unit integration, assembly, and testing (IA&T); GSE; and program management/systems

engineering costs that can be identified as nonrecurring. This includes all costs associated with design verification and interface requirements. Recurring costs cover all efforts associated with flight hardware manufacturing, IA&T, program management, and systems engineering that can be identified as recurring.

5. Spacecraft system-level cost estimates for program management, systems engineering, integration, assembly and checkout, GSE, and system test operations are separate from the subsystem level.

6. CERs are statistical fits to data derived from actual costs of recent small satellite programs. Use of CERs to estimate costs of future programs relies on the assumption that historical trends will accurately reflect future costs.

7. Most costs in the small satellite database are actual program costs at completion, gathered from the spacecraft operators. In a few cases, however, cost data provided was for satellites that were nearly complete but had not yet been launched. In those cases, contractor estimate at complete (EAC) costs were used.

8. CERs estimate burdened costs including direct labor, material, overhead, and general administrative costs.

9. Most programs in the database rely on some degree of hardware commonality to previous units, but limited quantitative data were available. CERs therefore yield costs that represent a mathematical average amount of heritage, level of technology complexity, and amount of schedule delays and engineering changes. Cost estimates derived from the CERs should therefore be accompanied by a comprehensive cost risk assessment to estimate potential effects of a level of complexity below or beyond average.

25.4.3 NASA Air Force Cost Model (NAFCOM)

NASA and the United States Air Force created the NAFCOM cost model and released the newest version in 2011. The cost model is a computer program, which must be installed with a short list of other programs in order to be fully operational. Additionally, government regulations require the user to be a US citizen and contain the program on their personal computer rather than install it on public-access computers such as in a laboratory setting. In order to gain access, one must contact the NAFCOM representative with valid paperwork and a government sponsor willing to vouch for the appropriate usage of the NAFCOM program [8].

The NAFCOM cost model is based upon historical data from previous missions and may be divided into several different categories:

- D&D represents the design and development cost.
- STH is the cost of the system test hardware.
- Flight unit corresponds to the costs associated with the flight unit effort including the period beginning with the start of production initiated by long lead procurements and ending with the delivery of the first unit. Flight unit costs represent only the cost of the first unit to fly.
- Design, development, test, and evaluation (DDT&E) encompasses the design and development through the factory checkout of the first flight article. This value is determined by adding the D&D and STH values.

- Production refers to the cost of a flight unit multiplied by the quantity.
- Total reflects the total development and production cost of all systems required for the program. This value is obtained by adding the DDT&E and production costs.

The NAFCOM model refers to the technology maturity index (TMI), which is based upon the technology readiness level (TRL). The TMI value ranges from 1 to 12 and is based upon experience with the technology, flight experience, test experience, and the application of the technology. The TMI scale, applied in the calculations for each subsystem or work break-down structure (WBS) category, is defined by the NAFCOM model as:

1. Technology research has begun to be translated into applied research and development.
2. Technology is in the conceptual or application formulation phase.
3. Technology has been subjected to extensive analysis, experimentation, and/or a characteristic proof of concept, but has no flight experience.
4. Technology has been validated in a lab/test environment but has no flight experience.
5. Technology has experience, but not in a space environment.
6. Technology has flight experience, but not recent flight experience.
7. Technology has recent flight experience (<5 years), and the application of technology is at the edge of experience.
8. Technology has recent flight experience (<5 years), and the application of technology within realm of experience.
9. Technology is approaching maturity (5–10 years) of flight experience encompassing at least three missions, and the application of technology is at the edge of experience.
10. Technology is approaching maturity (5–10 years) of flight experience encompassing at least three missions, and the application of technology within realm of experience.
11. Technology is mature (>10 years) of flight experience encompassing at least five missions, and the application of technology is at the edge of experience.
12. Technology is mature (>10 years) of flight experience encompassing at least five missions, and the application of technology within realm of experience.

The NAFCOM model has very specific questions, which the user must answer. The model, like many parametric cost models, bases the CERs on previous mission data. Furthermore, NAFCOM lets the user select which previous missions to use in the CER calculations. Additionally, the database provides an explanation of each mission, so the user may select missions, which are similar to their own. For example, one may select all of the unmanned, Earth-orbiting, scientific explorer-class missions.

25.4.4 Other Models

The SSCM and NAFCOM models discussed in the previous sections represent the most common government cost models that are used for initial cost estimates of many spacecraft missions. There are, however, multitudes of other cost models in existence. These models can be classified into two categories: mission cost models and instrument cost models. Each spacecraft model is listed in the following text, with a reference to find more information and reasons why one may or may not consider using it.

1. Mission cost models:
   a. Unmanned Space Vehicle Cost Model, 8th edition (USCM8) is only applicable if the input parameters are within the database ranges. The USCM8 model is managed by the Air Force Space Command and access may be requested by visiting their website [9].
   b. Planetary Data Systems Archiving Cost Analysis (PDS) may be found online but is mainly meant for planetary missions [10].
   c. Space Operations Cost Model (SOCM) may be found online but seems to be out of date and no contact was listed to whom a request could be made for additional information and/or the newest version [11].
   d. Systems Evaluation and Estimation of Resources (SEER) is an industry-used cost model, which is not based upon flown space systems. Moreover, there is a fee required to use the software [12].
   e. Parametric Review of Information for Costing and Evaluation (PRICE) is widely used in industry, but not primarily based upon space missions and requires a fee to use [13].
2. Instrument cost models do not accurately reflect the complexity of designing, building, testing, and operating a small satellite. Rather, these models exist purely for modeling the costs associated with developing the instruments placed on the spacecraft. Instrument models are widely used in industry. The most common models are:
   a. NASA Instrument Cost Model (NICM) [14].
   b. Scientific Instrument Cost Model (SICM) [15].
   c. Multivariable Instrument Cost Model (MICM) [16].
   d. Passive Sensor Cost Model (PSCM) [17].

25.5 Risk Estimation and Reduction

According to the NASA Risk Management Procedural Requirements, "risk is the potential for performance shortfalls, which may be realized in the future with respect to achieving explicitly established and stated performance requirements." [18] For this chapter, cost risk is the shortfall of interest, but the risk management process detailed here could easily be expanded for other mission risks, as detailed in reference [19–21]. Risk management is the process of risk identification, analysis, mitigation planning, and tracking of the root cause of problems and their ultimate consequences. Risk management plans improve the likelihood of mission success by identifying potential failures early and planning methods to circumvent any issues. However, in the aerospace industry to date, risk management plans have typically only been used for larger and more expensive satellites and have rarely been applied to smaller satellites with a mass less than 10 kg, called nanosatellites. For this class of smaller satellites, which is becoming of greater interest to the aerospace industry, these larger-scale risk management plans need to be adapted to provide a suitable risk management methodology for nanosatellites.

Fortunately, initial research has been completed leading to a set of small satellite software tools useful for identifying mission risks and methods, to reduce the likelihood and consequence of the occurring mission risk [19–21]. According to reference [19], a risk management plan entails three major steps each consisting of sub-steps, as detailed in Table 25.4. The three major steps are to identify the mission risks, determine the appropriate mitigation techniques, and to closely monitor the progress of the risks. By identifying, mitigating, and tracking the risks, it is believed that the mission will have a higher chance of success. There exist many examples of large-scale missions using the risk management process [22–28]. Many of these missions, however, use high-fidelity models, including quantitative assessments, such as probabilistic risk assessment (PRA). Because of the limited resources and short program life cycle of small satellite missions, it is desirable to avoid the more expensive and detailed methods of risk analysis such as PRA by employing analytical methods of identifying and tracking mission risks using common low-cost software tools.

According to reference [20], the generic cost risk is identified as the possibility of a combination of four different root causes: running out of money, a delay or not receiving promised funding, incomplete understanding of projected total mission cost, or the increase in component prices beyond expectation. More root causes may be possible for a given mission configuration, but these four suffice for the sake of generality. Once the root causes of risk have been fully identified according to the sub-steps of Table 25.4, the appropriate mitigation techniques may be determined to help reduce the likelihood or consequence of the occurring of root cause. For example, with the four general root causes described earlier, mitigation techniques may include: adding contingency to the budget, working with customers to obtain emergency funds, creating component alternative lists, or de-scoping components. It is advised that missions select mitigation techniques from each of the four categories listed in Table 25.4. By having a diverse set of mitigation techniques to choose from, the mission managers can decide the best course of action based on their knowledge of the mission, and/or by using a tool such as the Decision Advisor [21].

Table 25.4 Steps of a risk management plan.

Main step	Sub-steps
A. Risk identification	1. Review the mission concept of operations 2. Identify root causes 3. Classify priority of risk 4. Name responsible person 5. Rank likelihood (L) and consequence (C) of root cause 6. Describe rationale for ranking 7. Compute mission risk likelihood and consequence values 8. Plot mission risks on L–C chart
B. Determine mitigation techniques	Choices consist of: 1. Avoid the risk by eliminating root cause and/or consequence 2. Control the cause or consequence 3. Transfer the risk to a different person or project 4. Assume the risk and continue in development
C. Track progress	Plot the mission risk values on an L–C chart at key life cycle or design milestones to see progress.

To monitor the progress of the mission risks via the mitigation strategies described, re-evaluate the likelihood and consequence (L–C) values at key life cycle or design milestones, such as design reviews. The program manager and systems engineer should consult with subsystem or task leads identified as the "responsible person" to obtain the most recent status of each root cause when completing the re-evaluation. Ideally, both of the L–C values will decrease with each successive re-evaluation. However, if the mission risk increases in either likelihood or consequence, this re-evaluation will capture the change.

25.6 Conclusions

The interest in nanosatellites has boomed over the past decade, but the systems engineering processes are yet to catch up to the demand in these small satellite platforms. When it comes to cost estimation, no good model exists. While this chapter has outlined typical costs associated with a nanosatellite mission in terms of recurring and nonrecurring elements, the chapter by no means provides a template cost estimation because it would be impossible to represent all the various missions possible with nanosatellites. From the diverse range of spacecraft components to the wide variety of spacecraft shapes and sizes to the differences in ground stations and testing facilities, each mission must determine the best method of estimating and accounting for its costs. Many resources exist for cost estimating [6, 29], and it is left to the reader to find the best resource for their mission. In the author's humble opinion, it is best to use a combination of cost models to provide a rough estimate of expected costs and keep track of all costs in a detailed grassroots approach.

References

1 Nugent, R., Munakata, R., Chin, A., Coelho, R., Puig-Suari, J., (2008). "The CubeSat: The Picosatellite Standard for Research and Education." AIAA paper 2008-7734. AIAA Space 2008 Conference. September 9–11, 2008. San Diego, California.

2 Blanchard, B.S. and Fabrycky, W.J. (2006). *Systems Engineering and Analysis*, 4e. New Jersey: Prentice Hall International Series in Industrial and Systems Engineering.

3 Larson, W.J. and Wertz, J.R. (1999). *Space Mission Analysis and Design*, 3e. California: Microcosm Press.

4 "Vega," European Space Agency. (2013). http://www.esa.int/Our_Activities/Launchers/Launch_vehicles/Vega. Last updated: 10 May 2013. Last accessed: 12 January 2015.

5 Brumbaugh, K. M., (2012). The Metrics of Spacecraft Design Reusability and Cost Analysis as Applied to CubeSats. Master's thesis. The University of Texas at Austin, January 2012. Print.

6 Wertz, J.R. and Larson, W.J. (1996). *Reducing Space Mission Cost*. Microcosm Press and Kluwer Academic Publishers Print.

7 "Small Satellite Cost Model." (2012). The Aerospace Corporation. Retrieved from http://www.aero.org/capabilities/sscm/index.html. Date accessed: 20 April 2012.

8 Winn, S. NASA/Air Force Cost Model (NAFCOM). (NASA Technical Document 20020048607). Marshall Space Flight Center.

9 "Unmanned Space Vehicle Cost Model." United States Air Force Space Command. Retrieved from www.uscm8.com. Date accessed: 13 March 2012.

10 "PDS: Planetary Data System." NASA. Retrieved from http://pds.nasa.gov/tools/cost-analysis-tool.shtml. Date Accessed: 13 March 2012.

11 "Space Operations Cost Model (SOCM)." NASA. Retrieved from http://cost.jsc.nasa.gov/SOCM/SOCM.html. Date accessed: 13 March 2012.

12 "SEER by Galorath." Galorath. Retrieved from http://www.galorath.com. Date accessed: 13 March 2012.

13 "PRICE." PRICE Systems. Retrieved from www.pricesystems.com. Date accessed: 13 March 2012.

14 Habib-Agahi, H.; Mrozinski, J.; Fox, G. (2010). "NASA Instrument Cost/Schedule Model." Jet Propulsion Laboratory. IEEEAC paper #1158. 29 October 2010.

15 PRC System Services, (1990). "Scientific Instrument Cost Model (SICM)," Technical Brief 125, PRC D-2327-H, February 1990.

16 Dixon, B. and Villone, P., (1990). "Goddard Multi-Variable Instrument Cost Model (MICM)," RAO Note #90-1, May 1990.

17 Parametric Cost Estimating Handbook. NASA. (1995). 14 September 1995. Retrieved from http://cost.jsc.nasa.gov/pcehhtml/pceh.htm. Date accessed: 13 March 2012.

18 Agency Risk Management Procedural Requirements (2008). NASA procedural requirements. *Natural Product Reports* 4A: 8000.

19 Brumbaugh, K.M. and Lightsey, E.G. (2013). Application of risk management to university CubeSat missions. *Journal of Small Satellites* 2 (1): 147–160.

20 Gamble, K.B. and Lightsey, E.G. (2014). CubeSat mission design software tool for risk estimating relationships. *Acta Astronautica* 102: 226–240.

21 Gamble, K. B., Lightsey, E. G., (2015). "Decision Analysis Applied to Small Satellite Risk Management," SciTech 2015 Conference, Orlando, FL, 5–9 January 2015.

22 Frank, M.V. (1995). Choosing among safety improvement strategies: a discussion with example of risk assessment and multi-criteria decision approaches for NASA. *Reliability Engineering and System Safety* 49: 311–324.

23 Smith, C., Knudsen, J., Kvarfordt, K., and Wood, T. (2008). Key attributes of the SAPHIRE risk and reliability analysis software for risk-informed probabilistic applications. *Reliability Engineering and System Safety* 93: 1151–1164.

24 Perera, J.S. (2002). Risk management for the International Space Station. In: *Joint ESA–NASA Space-Flight Safety Conference* (eds. B. Battrick and C. Preyssi), 339. European Space Agency, ESA SP-486. ISBN: 92-9092-785-2.

25 "Orion Crew Exploration Vehicle Project Integrated Risk Management Plan," NASA, (2008). CxP 72091, Rev. B, Released: November 18, 2008.

26 (2009). *International Space Station Risk Management Plan*. NASA Johnson Space Center, SSP 50175, Revision C, September.

27 (2006). *Space Shuttle Risk Management Plan*, vol. XIX. NASA Johnson Space Center, NSTS 07700.

28 NASA, (2010). "NASA Risk-Informed Decision Making Handbook," NASA/SP-2010-576, Version 1.0, April 2010, Office of Safety and Mission Assurance, NASA HQ.

29 Fox, B., Brancato, K., Alkire, B., (2008). "Guidelines and Metrics for Assessing Space System Cost Estimates," Technical Report for the US Air Force, RAND Corporation. Print.

30 NASA. (2012). "Edison Small Satellite Flight Demonstration Missions." Office of the Chief Technologist. 2 February 2012. OMB Number 2700-0085.

26 III-1b

Launch Costs

Merlin F. Barschke

Institute of Aeronautics and Astronautics, Technische Universität Berlin, Germany

26.1 Introduction

The launch costs for a nanosatellite mission play an important role in the overall cost estimate, as they account for a significant fraction of the total mission costs. Apart from the mass of the spacecraft to be launched, there are several factors that influence the final cost of the entire launch campaign. These include volume of the spacecraft's envelope on the launcher, its form factor, the target orbit, as well as the nationality of satellite developer and launcher. This chapter gives an overview over the current launch market for small satellites, as well as over emerging trends such as the upcoming generation of microlaunchers that might significantly influence the launch market in the future. Furthermore, it covers topics such as launch sites, the different milestones from launch reservation to the checkout campaign, and the costs associated to the launch of a nanosatellite.

26.2 Launching Nanosatellites

Sputnik 1, the first satellite that was ever launched into orbit had a launch mass of just over 80 kg and can therefore be considered a small satellite. However, although some of the spacecraft that followed after Sputnik also had launch masses below 100 kg, the focus soon shifted toward larger spacecraft and thus to larger launchers. It was only in the 1970s that universities started to investigate the capabilities of small satellites. Here, the aim was exploiting financial advantages that result from launching lower masses and reduced complexity of the spacecraft. Since then, the number of small satellites launched to orbit increased continuously, with dramatic growth predicted in the coming years.

Until recently, the only option for satellites with a launch mass far below 100 kg to be delivered to orbit was by "hitchhiking" a larger spacecraft's launch. However, today new options start to emerge, backed by the ever-growing number of small satellites in need for launch. As the terminology used for small satellite launch options differs widely in literature, the definitions used in this chapter are summarized in the following text (cf. reference [1]).

Nanosatellites: Space and Ground Technologies, Operations and Economics, First Edition.
Edited by Rogerio Atem de Carvalho, Jaime Estela, and Martin Langer.
© 2020 John Wiley & Sons Ltd. Published 2020 by John Wiley & Sons Ltd.

26.2.1 Dedicated Launch

The common launch concept for larger spacecraft is the dedicated launch, where one or two spacecraft are brought to their target orbit by the launcher. Although such spacecraft are already developed with the available launch mass and volume of a certain launcher in mind, significant resources are often left unused in this case. As of 2010, the smallest satellite that was commercially launched in a dedicated launch had a mass of 110 kg and was launched aboard the Pegasus launch vehicle in 1998 [2]. However, with the upcoming microlaunchers (i.e. launchers with a launch mass below 500 kg [3]), this situation is expected to change (cf. Section 26.2.4.2).

26.2.2 Piggyback Launch

The excess resources of many dedicated launches described earlier lead to the introduction of piggyback launches, where one or more small satellites are carried along to orbit with the main passenger (e.g. by Arianespace [4] or United Launch Alliance (ULA) [5]). Here, larger piggyback payloads are called secondary payloads, while smaller ones such as CubeSats are referred to as tertiary payloads. While this concept made the launch of small satellites affordable, it carries the drawback of the piggyback satellites being tied to the main passenger in terms of target orbit and launch date.

As a current record, India's Polar Satellite Launch Vehicle (PSLV) successfully launched 103 piggyback passengers along with the main payload Cartosat-2 within one single launch in early 2017 [6]. Figure 26.1 shows a Russian Fregat upper stage carrying 72 piggyback satellites shortly before encapsulation.

Figure 26.1 Soyuz Fregat piggyback launch with 72 satellites in addition to the main passenger. Here, the majority of the piggyback satellites were 3U CubeSats, four of which were launched out of one 12U container. Source: reproduced with permission from Roscosmos, Russia. (*See color plate section for color representation of this figure*).

26.2.3 Rideshare or Cluster Launch

With the growing number of small satellites in need for a launch opportunity, the concept of rideshare or cluster launches emerged. Here, similar to the piggyback launch, multiple spacecraft share one launch vehicle in order to aggregate the required mass and volume. However, no single satellite qualifies as main passenger, due to each ones' individual mass being far below the rocket's launch capability. One example of such a rideshare mission is the SSO-A launch of Spaceflight in 2018 [7].

26.2.4 ISS Deployment

A particular form of launching a nanosatellite that emerged in recent years is the deployment from the International Space Station (ISS), a service that is for example provided by NanoRacks LLC [8]. While satellites that are to be launched from the ISS naturally need to be brought to orbit by a rocket like any other spacecraft, the mechanical loads the satellite is subject to are considerably lower due to being transported in a mechanically isolated container. However, the low orbital altitude as well as the given inclination makes launching from ISS only suitable for a limited range of missions. Figure 26.2 shows two CubeSats right after being released from the NanoRacks launcher attached to the Japanese robotic arm on the ISS. As of 2019, the number of satellites that have launched by NanoRacks is 231 [9].

26.2.4.1 Present Launches

There are a number of launch vehicles from several countries that have provided launch capabilities for small satellites to date. Figure 26.3 shows an overview of the number of satellites below 10 kg and CubeSats up to 12U launched until March 2019, as well as the number of launches that carried such spacecraft, grouped by the respective launch vehicle [10]. Here, all vehicles that launched less than 20 satellites of the considered mass range are combined in the category "others." Furthermore, ISS deployments and satellites that

Figure 26.2 Two three-unit CubeSats being deployed by the NanoRacks launcher attached to the end of the Japanese robotic arm on the ISS. Source: Reproduced with permission from ESA.

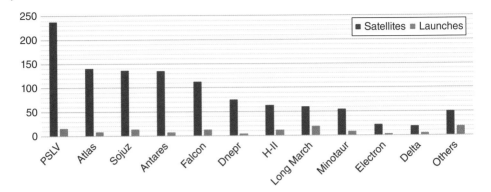

Figure 26.3 Overview of satellites below 10 kg and CubeSats up to 12U launched until March 2019 grouped by launch vehicle. Blue bars indicate the number of satellites that fall in the given mass category brought to orbit by the respective launch vehicle, while red bars illustrate the number of launches that carried such satellites performed by the launcher in question. Source: Reproduced with permission from ESA. (*See color plate section for color representation of this figure*).

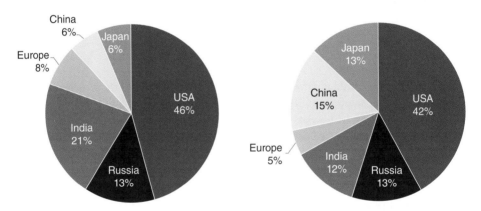

Figure 26.4 Launches of satellites below 10 kg and CubeSats up to 12U until March 2019 sorted by country. The fraction of the total number of satellites launched by each country is shown in panel (a), while panel (b) shows the countries' fraction of the total number of launches. Source: Reproduced with permission from ESA. (*See color plate section for color representation of this figure*).

were separated from other spacecraft on orbit are attributed to the rocket that brought the satellite to orbit.

Figure 26.4 groups all spacecraft below 10 kg launch mass and up to 12U in case of Cube-Sats that were launched until March 2019, as well as the launches that carried satellites of the given mass range in percentage by the launching nations [10].

A selection of currently active launch vehicles relevant to small satellites is shown in Table 26.1. The launchers with a launch mass below 500 kg are listed here to emphasis that such microlaunchers are already in service. However, until the end of 2018 of these micro-launchers, only Electron brought nanosatellites to orbit, while the listed small launchers have repeatedly delivered nanosatellites to orbit (cf. Figure 26.3).

Table 26.1 Selection of active small launchers and microlaunchers with payload capacity and price per kilogram payload mass [3, 4, 11]. Here, the microlaunchers are listed to emphasis that such vehicles are already active; however, as of the end of 2016, no nanosatellites have been brought to orbit by these launchers.

Launcher	Country	Payload (kg)	Cost (k$ per kg)
Electron	USA	225 (LEO) to 150 (SSO)	49
Pegasus	USA	310 (LEO), 210 (SSO)	181–268
Kuaizhou-1A	China	360 (LEO)	25
Long March 11	China	435 (SSO)	23
Epsilon	Japan	700 (LEO), 450 (SSO)	67–104
Vega	Europe	1330 (SSO)	24–34
Soyuz	Russia	1400–1700 (LEO)	12–15
PSLV	India	1750 (SSO)	14–20

The currently available launch options are not always ideally suited for small satellites, especially when it comes to large constellations. Therefore, much activity can be observed in the development of microlaunchers with a launch mass of less than 500 kg to low Earth orbit (LEO). Here, a large number of such vehicles providing launch masses down to 30 kg are currently developed or tested.

26.2.4.2 Future Developments

The small satellite market is expected to grow tremendously in future years. However, mainly due to their comparatively low flexibility regarding schedule and orbit, traditional launch options are not ideally suited for many nanosatellite projects. This is especially true for upcoming mega-constellations with rigid requirements regarding orbit and launch date [11]. Therefore, different approaches are currently evaluated and implemented that aim at providing a suitable level of flexibility for serving various different mission scenarios. The following sections give an overview over future developments in the small satellite launch business.

Microlaunchers Launch systems in the past were nearly exclusively developed for large satellites of many hundreds kilograms or even several tons. Such vehicles would only provide piggyback or cluster launch opportunities to nanosatellites. However, today a large number of microlaunchers are being developed. Here, the payload capacity ranges from 30 to 500 kg. While market analysis reveals that the mass majority of future small satellite launches will be aggregated by spacecraft with a mass above 100 kg, nanosatellites will also be launched in larger numbers [11].

Table 26.2 shows a selection of microlaunchers that are currently under development, listing the payload capacity to LEO, as well as the cost per kilogram payload. One can see that the prices vary considerably and are mostly above the cost per kilogram of larger launchers (cf. Table 26.1). However, as mentioned before, microlaunchers provide more flexibility in launch date and orbit, which is highly desirable for many nanosatellite missions. A more detailed analysis of the microlauncher market can be found in references [3, 11, 13, 14].

Table 26.2 Selection of microlaunchers that are currently under development with LEO payload capacity and price per kilogram payload mass [3, 11–13].

Company	Launcher	Country	Payload (kg)	Cost (k$ per kg)
Generation Orbit	GOLauncher 2	USA	30	83
Vector Space Systems	Vector-R	USA	50	50
Rocket Lab	Electron	USA	150	49
Vector Space Systems	Vector-H	USA	125	28
Virgin Orbit	LauncherOne	USA	300	40
LandSpace	LandSpace-1	China	400	20
Firefly	Firefly	USA	400	20

Figure 26.5 Electron launcher that is developed by Rocket Lab to deliver 100 kg to LEO. CubeSat launches with Electron can be ordered online at a fixed price. Source: Reproduced with permission from ESA.

Figure 26.5 shows the Electron launcher of Rocket Lab that had its first test-flight in 2017 and launched 23 CubeSats till the end of 2018 [10] as an example for an upcoming micro-launcher.

Increasing Flexibility Additionally to the development of small launch vehicles, there are alternative strategies to increase the flexibility of piggyback and rideshare launches. Various types of standardized adapters and accommodation strategies are being developed to host small satellites on established and emerging launchers [15, 16]. Furthermore, adapters that provide propulsion capabilities shall provide more flexibility regarding the parameters of the target orbit [17, 18]. A similar example for an approach to increase the flexibility of a piggyback launch is the InOrbit Now (ION) service offered by the company D-Orbit [19]. Here, the ION CubeSat carrier, a complete satellite in itself, delivers up to 48U CubeSats ranging from 1 to 12U to their specific target orbits, promising more flexibility regarding orbit parameters and faster constellation deployment.

Beyond LEO   With the ever-growing performance of the nanosatellite, more and more missions based on such spacecraft that go beyond LEO are considered today [20]. Hence, launch opportunities for such missions are also of interest for the nanosatellite community. Here, NASA, for example, is offering rideshares for 6U CubeSats on the Orion missions [21].

26.3   Launch Sites

Small launch vehicles that are used for today's piggyback launches use well-established launch sites operated by the USA, Russia (including Baikonur Cosmodrome), India, Japan, China, and Europe (French Guiana). However, the emerging new generation of microlaunchers does also involve the buildup of new launch sites. Rocket Lab, which develops the Electron microlauncher (cf. Table 26.2), for example, offers one launch site in New Zealand and two sites in the USA, namely, in Florida and Alaska. Other launch sites are considered to secure a country's access to space from its own territories (e.g. from the United Kingdom [22, 23]). To this end, even the import of a suitable launcher is considered to avoid long development times [24]. Other countries, such as Sweden or Norway, already possess suitable infrastructure for suborbital launches and seek to expand their launch sites to additionally support orbital launches [25, 26].

Contrary to vertical launched rockets, balloon or air launched systems, such as Orbital ATK's Pegasus [27] or Virgin Orbit's LauncherOne [28], have the advantage to allow operations from a variety of locations, which can significantly reduce the logistics effort associated with the launch and expands the choice of candidate orbits.

26.4   Launch Milestones

Regarding the launch, there are several milestones of importance for a nanosatellite project. The first one here is the launch contract signature, as it usually also determines a launch date. As a next step, the payload interface control documents (ICD) is prepared, which contains all technical agreements between satellite developers and the launcher or broker, a middleman between the two parties. The last step is the hardware delivery and, if applicable, the checkout campaign at the launch site. For the launch of the spacecraft, the team is usually already back home to prepare for the first contract.

26.4.1   Launch Contract

While some providers offer direct contracts with secondary or tertiary payloads, most launches today are contracted through launch brokers, such as Innovative Solutions in Space (ISIS) [29] or Spaceflight Industries [30]. Having extensive experience with export regulations, contracts, and checkout activities on site, such brokers can considerably simplify the launch process.

For piggyback launches, the launch contract is typically concluded around 18 to 6 months before launch [31]. However, often there is still great uncertainty in the development schedule of the satellite as well as in the launch date at this point, as such launches might experience delays between 3 months and 2 years. Therefore, it is recommended to design and plan for multiple launchers, as it might prove advantageous to have the option to change the launcher even shortly before launch [32].

With the service offered by the new generation of microlaunchers, the time frame between contract and launch is expected to decrease considerably, which would diminish the impact of delays in satellite development on the one hand and the launch on the other. Especially, for nanosatellite constellations, the time between contract signature and launch often is a critical element for the deployment. For Planet's Flock constellation, for example, a time frame of 3 months between contract signature and launch is aspired [11]. Therefore, upcoming launch vehicles such as Rocket Lab's Electron also seek for new approaches for contracting. Here, a launch for a satellite that is build according to the CubeSat specification (1–12U) can be booked online at a fixed rate with specifying the year's quarter of the launch [12].

26.4.2 Payload ICD

The payload ICD defines all technical details of launch, hardware delivery, and the checkout campaign at the launch site or the premises of the launch broker. This includes the definition of the interface to the dispenser or, if the satellite is separated by a custom system, the mechanical and electrical interface to the upper stage. The ICD also details the timeline of the checkout campaign and the launch itself. Here, it needs to be specified, for example, at which point in time the satellite may be charged for the last time before launch or when the remove-before-flight items are to be removed. Performing checkout at the launch site means working with the satellite in an environment that is most likely not previously known. Therefore, it is important to specify all requirements for the campaign as precise as possible to ensure a seamless campaign.

26.4.3 Hardware Delivery and Launch Campaign

The launch campaign differs significantly depending on whether the satellite is a CubeSat that is launched from a standardized container or if it is using a custom separation system. CubeSats can nowadays be delivered to the launch broker where they are integrated into the launch container. All further steps are then executed by the broker or the launch provider while the customer can prepare for operations. Thus, the satellite developer does not need to travel and deliver equipment to a potentially remote launch site to perform checkout, which significantly simplifies the launch preparations as the logistics of delivering satellite and equipment to the launch site is not to be underestimated.

26.5 Launch Cost

The costs of launching a nanosatellite are dependent on many different boundary conditions and are subject to fluctuations. However, some launchers or brokers list prices on

Table 26.3 Selection of openly available launch prices stated by Rocket Lab [12], NanoRacks [33] and Spaceflight Industries [30].

Size	Rocket Lab (k$ per kg)	NanoRacks (k$ per kg)	Spaceflight (k$ per kg)
1U CubeSat	77	85	-
3U CubeSat	240	-	295
6U CubeSat	480	-	545
12U CubeSat	960	-	995
50 kg satellite	-	-	1750
100 kg satellite	-	-	3950

the Internet. Table 26.3 gives an overview over launch prices stated by Rocket Lab [12], NanoRacks [33], and Spaceflight Industries [30] that can be openly accessed via the Internet. These prices are considerably higher as the per kilogram prices stated in Tables 26.1 and 26.2 for the individual launchers, as they include a number of services that are not considered for the overall per-kilo-price of the launcher and furthermore usually account for mass and cost of the separation system. All things considered, the launch cost of a specific mission might vary considerably according to the given requirements of the spacecraft and the available launches. Furthermore, price reductions might apply for customers that book into a launch that is not fully booked shortly before the launch is closed.

Payment for a launch is usually subdivided in several installments. A first rate will usually be charged upon reservation of a certain flight opportunity, while the second one is due upon launch service agreement signature. As soon as the ICD is finalized, a third rate applies and the final part is charged when the launch has been performed [31].

References

1 Crisp, N., Smith, K., and Hollingsworth, P. (2014). Small satellite launch to LEO: a review of current and future launch systems. *Transactions of the Japan Society for Aeronautical and Space Sciences, Aerospace Technology Japan* 12: 39–47.

2 Dominic DePasquale, A. C. Charania, Hideki Kanayama, and Seiji Matsuda, 2010. Analysis of the Earth-to-orbit launch market for nano- and microsatellites. In Proceedings of the AIAA SPACE Conference, Anaheim, USA.

3 Wekerle, T., Filho, J.B.P., da Costa, L.E.V.L., and Trabasso, L.G. (2017). Status and trends of smallsats and their launch vehicles—an up-to-date review. *Journal of Aerospace Technology and Management* 7: 269–286.

4 Clayton Mowry and Serge Chartoire, 2013. Experience launching smallsats with Soyuz & Vega from the Guiana Space Center. In Proceedings of the 27th Annual AIAA/USU Conference on Small Satellites, Logan, USA.

5 Jake Szatkowski and David R. Czajkowski, 2013. ULA SmallSat/hosted rideshare mission accommodations. In Proceedings of the AIAA SPACE Conference, San Diego, USA.

6 Durairaj Radhakrishnan, 2017. A century of satellites in a single rocket—PSLVs record breaking mission. In Proceedings of the 68th International Astronautical Conference, Adelaide, Australia.

7 Scott Schoneman, Jeff Roberts, Adam Hadaller, Tony Frego, Kristen Smithson, and Eric Lund, 2018. SSO-A: The First Large Commercial Dedicated Rideshare Mission. In Proceedings of the 32nd Annual AIAA/USU Conference on Small Satellites, Logan, USA.

8 NanoRacks. Satellite deployment. http://nanoracks.com/products/satellite-deployment. [Accessed: 2017-10-06].

9 NanoRacks. NanoRacks Completes Sixth CubeSat Deployment from Cygnus Spacecraft, Continues Historic Program. http://nanoracks.com/nanoracks-completes-sixth-cygnus-deployment-mission. [Accessed: 2019-03-18].

10 Erik Kulu. Nanosats Database. www.nanosats.eu. [Accessed: 2019-03-18].

11 Alan Perera-Webb, Alex da Silva Curiel, and Vadim Zakirov, 2017. The changing launcher landscape—A review of the launch market for small satellites. In Proceedings of the 68th International Astronautical Conference, Adelaide, Australia.

12 Rocket Lab. Book my launch. https://www.rocketlabusa.com/book-my-launch. [Accessed: 2017-10-06].

13 Carlos Niederstrasser and Warren Frick, 2016. Small launch vehicles—A 2016 state of the industry survey. In Proceedings of the 67th International Astronautical Conference, Guadalajara, Mexico.

14 Bill Doncaster, Jordan Shulamn, John Bradford, and John Olds, 2016. SpaceWorks' 2016 nano/microsatellite market forecast. In Proceedings of the 30th Annual AIAA/USU Conference on Small Satellites, Logan, USA.

15 Kaitlyn Kelley, Mitch Elson, and Jason Andrews, 2015. Deploying 87 satellites in one launch: Design trades completed for the 2015 SHERPA flight hardware. In Proceedings of the 27th Annual AIAA/USU Conference on Small Satellites, Logan, USA.

16 Qian Xu, Ling Su, Jie Wu, Cao Li, and Yue Leng, 2017. Structure of payload fairing and adapter for launching multiple satellites in one mission. In Proceedings of the 68th International Astronautical Conference, Adelaide, Australia.

17 Jason Andrews, 2012. Spaceflight Secondary Payload System (SSPS) and SHERPA tug—a new business model for secondary and hosted payloads. In Proceedings of the 26th Annual AIAA/USU Conference on Small Satellites, Logan, USA.

18 Chrishma Singh-Derewa and Ronald Fevig, 2017. Finding NewSpace utilizing LOTUS: Lander/orbiter trans-upper stage. In Proceedings of the 68th International Astronautical Conference, Adelaide, Australia.

19 D-Orbit. InOrbit NOW. http://www.deorbitaldevices.com/products/?slide=4. [Accessed: 2017-10-06].

20 Staehle, R.L., Blaney, D., Hemmati, H. et al. (2013). Interplanetary CubeSats: opening the solar system to a broad community at lower cost. *Journal of Small Satellites* 2: 161–186.

21 Kimberly F. Robinson, 2017. NASAs space launch system: SmallSat deployment to deep space. In Proceedings of the 68th International Astronautical Conference, Adelaide, Australia.

22 Philip Davies, David Riley, Oliver Turnbull, Andy Bradford, and Alex da Silva Curiel, 2016. Vertical launch of small satellites from the UK. In Proceedings of the 67th International Astronautical Conference, Guadalajara, Mexico.

23 Philip Davies, David Riley, Oliver Turnbull, Andy Bradford, Alex da Silva Curiel, Jan Skolmi, Andy Quinn, and Roy Kirk, 2017. Vertical launch of small satellites from the UK—2017 update. In Proceedings of the 68th International Astronautical Conference, Adelaide, Australia.

24 Vadim Zakirov, Alan Perera-Webb, Richard Osborne, and Konstantin Milyayev, 2017. Importing a small Chinese launcher to operate from the UK. In Proceedings of the 68th International Astronautical Conference, Adelaide, Australia.

25 Bertil Oving, Arnaud Van Kleef, Bastien Haemmerli, Adrien Boiron, Markus Kuhn, Ilja Müller, and Marina Petrozzi, 2017. Small innovative launcher for Europe: Achievement of the H2020 project SMILE. In Proceedings of the 68th International Astronautical Conference, Adelaide, Australia.

26 Anne Ytterskog and Anna Rathsman, 2017. SmallSat express—a future launch service for small satellites. In Proceedings of the 68th International Astronautical Conference, Adelaide, Australia.

27 Orbital ATK. Pegasus. https://www.orbitalatk.com/flight-systems/space-launch-vehicles/pegasus. [Accessed: 2017-10-10].

28 Sirisha Bandla, Richard DalBello, Monica Jan, William Pomerantz, and Mandy Vaughn, 2017. LauncherOne: Responsive launch for small satellites. In Proceedings of the 68th International Astronautical Conference, Adelaide, Australia.

29 Innovative Solutions in Space (ISIS). Launch services. https://www.isispace.nl/launch-services. [Accessed: 2017-10-06].

30 Spaceflight. Launch services. http://spaceflight.com/services/launch-services. [Accessed: 2017-10-06].

31 Jason Andrews and Abe Bonnema, 2011. Ticket to space—how to get your small satellite from the cleanroom to orbit. In Proceedings of the 25th Annual AIAA/USU Conference on Small Satellites, Logan, USA.

32 Philip Brzytwa, 2017. Mission assurance through launch flexibility preparing a SmallSat mission on multiple launch vehicles. In Proceedings of the 68th International Astronautical Conference, Adelaide, Australia.

33 NanoRacks. FAQ. http://nanoracks.com/resources/faq. [Accessed: 2017-10-09].

27 III-2a

Policies and Regulations in Europe

Neta Palkovitz

ISIS—Innovative Solutions In Space B.V., Delft, The Netherlands
International Institute of Air and Space Law (IIASL), Leiden University, Leiden, The Netherlands

27.1 Introduction

Small satellites have become increasingly popular over the years. They also took an important role in democratizing space, enabling nontraditional actors to join the space club. The upcoming small satellites' large constellations are currently shaping NewSpace, and the anticipated deep space exploration missions push them to the front line of innovation.

Like any other space activity, small satellite operations are subject to international space law. European satellite operators also have to adhere to domestic space laws and policies.

This chapter aims at explaining the main provisions of space law that apply to small satellite operations, in a "down to Earth" manner. It first focuses on international space law and the UN space treaties. It continues by elaborating on national space law in general, mentioning third party insurance obligations as a specific example that shows the practicalities in applying these laws to small satellites. The chapter ends with recent subjects that raise the need to further regulate small satellite operations.

27.2 International Space Law

27.2.1 General—What Is International Space Law?

International space law is comprised of a set of treaties, UN General Assembly resolutions, and practices that states have accepted. These form a regime that governs the exploration and use of outer space and the celestial bodies, which are not subject to the sovereign power of any state. Furthermore, outer space, the Moon, and other celestial bodies cannot be lawfully acquired by any state or a private entity [1]. For this reason, national space laws can be applied on state nationals or national space objects in the context of activities in outer space, however, not directly to outer space as such [2]. Therefore, international

Nanosatellites: Space and Ground Technologies, Operations and Economics, First Edition.
Edited by Rogerio Atem de Carvalho, Jaime Estela, and Martin Langer.

space law has a very important role in regulating space activities of states and of private entities.

Although many are not familiar with international space law, the efforts to create an international legal framework for space activities began already in the first years of the Space Age. The United Nations Committee on the Peaceful Uses of Outer Space (COPUOS) has been active since 1961, examining matters relating to the use of outer space [3]. The work of COPUOS is generally organized in two subcommittees, the Scientific and Technical Subcommittee and the Legal Subcommittee.

Within COPUOS' scope of work during the 1960s and 1970s, five main space treaties were drafted and accepted by UN member states, these are:

1) The "Treaty on Principles Governing the Activities of States in the Exploration and Use of Outer Space, Including the Moon and Other Celestial Bodies" of 1967 (Outer Space Treaty).
2) The "Agreement on the Rescue of Astronauts, the Return of Astronauts, and the Return of Objects Launched into Outer Space" of 1968 (Rescue Agreement).
3) The "Convention on International Liability for Damage Caused by Space Objects" of 1972 (Liability Convention).
4) The "Convention on Registration of Objects Launched into Outer Space" of 1976 (Registration Convention).
5) The "Agreement Governing the Activities of States on the Moon and Other Celestial Bodies" of 1979 (Moon Agreement).

These five space treaties form the core of international space law. However, it is important to understand that not all the states of the world are parties to these treaties. For instance, while the Outer Space Treaty has more than 100 state parties, and is considered to be the magna carta of space lawyers, the Moon Agreement failed to reach even 20 ratifications by states making it the least popular among the five treaties.

When focusing on satellite operations in general and nanosatellite operations particularly, the most relevant treaties are the Outer Space Treaty, the Liability Convention, and the Registration Convention. The following section will outline some key provisions of these treaties, which are important for every operator to be familiar with.

27.2.2 Key Treaty Provisions

27.2.2.1 Freedom of Exploration and Use of Outer Space and Possible Restrictions

One of the key principles that govern any activity in outer space is that all states are free to access, explore, and use outer space on an equal basis. This basic freedom stands in opposite to the regime that governs states' sovereign airspace, and therefore the drafters of the Outer Space Treaty included this principle in its first Article:

> Outer space, including the Moon and other celestial bodies, shall be free for exploration and use by all states without discrimination of any kind, on a basis of equality and in accordance with international law, and there shall be free access to all areas of celestial bodies. [4]

As almost no freedom is of an absolute nature, the treaty includes different provisions that may restrict it. Main examples are brought as follows:

Article II clarifies that regardless of the activities a certain entity pursues in outer space, this activity will not allow for legitimate appropriation or render the explorer or user with sovereign rights over outer space, the Moon, and other celestial bodies. [...]

Article III repeats Article I in regard to the duty to carry on activities in space in accordance with international law. The former article adds the Charter of the United Nations explicitly as part of general international law, as well as introduces the context of international interests of peace and security while promoting cooperation and understanding in this respect.

Article IV includes specific restrictions that are related to military uses in space. The emphasis is on banning nuclear weapons and other weapons of mass destruction. This explicit mention of such weapons was an important diplomatic achievement, since this provision originated in the Cold War Era. [...]

Article IX contains guiding principles that may restrict states from pursuing certain activities in space as follows:

First, states shall be guided by the principles of "cooperation and mutual assistance and shall conduct all their activities in outer space, including the Moon and other celestial bodies, with due regard to the corresponding interests of all other states parties to the treaty."

Second, there is a need to restrict space activities because of environmental concerns, avoiding harmful contamination or adverse change to Earth's environment.

Third, states are guided to consult other states in case their space activities may cause harmful interference with the activities of the latter. There is no definition in the treaty that would clarify what is a consultation, when it is recommended to perform one (in terms of agreed timing), and what can be regarded as harmful interference. [5]

27.2.2.2 State Responsibility

An important principle that is vital for private entities looking to carry out space activities is the notion that their national state will be internationally responsible for such activities. Article VI of the Outer Space Treaty provides for the following relations between a state party and its nationals:

States parties to the treaty shall bear international responsibility for national activities in outer space, including the Moon and other celestial bodies, whether such activities are carried on by governmental agencies or by nongovernmental entities, and for assuring that national activities are carried out in conformity with the provisions set forth in the present treaty. The activities of nongovernmental entities in outer space, including the Moon and other celestial bodies, shall require authorization and continuing supervision by the appropriate state party to the treaty. [...]

This means that a state will be internationally responsible for activities of its space agency, private commercial companies, universities, nonprofit organizations, and any other governmental and nongovernmental entities that carry the state's nationality. Further, since only states can become full parties to the treaty, the states are responsible to ensure that their nationals act in accordance to the provisions of the treaty. The latter should be achieved by creating regulatory means to allow for state authorization and continuing supervision. In many cases, states have implemented these obligations by creating a licensing regime for private national entities, which allows the state to authorize the space activity in question and to supervise it to some extent. This subject will be elaborated in the section on national space legislation.

27.2.2.3 International Liability

The space treaties provide for the following regime relating to liability for damage caused by space objects:

Article VII of the Outer Space Treaty provides:

> Each state party to the treaty that launches or procures the launching of an object into outer space, including the Moon and other celestial bodies, and each state party from whose territory or facility an object is launched, is internationally liable for damage to another state party to the treaty or to its natural or juridical persons by such object or its component parts on the Earth, in air, or in outer space, including the Moon and other celestial bodies.

In short, since this instrument deals with international law, the state is the liable party toward another state, even if the damage was caused by a space object that is operated by a private entity, and even when the victim is a nongovernmental entity. Therefore, some states made liability arrangements in national space legislation that aim for the licensed operator to indemnify the state, should the latter will be required to pay compensations for damage caused by the operator's activity. In practice, some states require operators to take up a third-party liability insurance policy, to ensure financial means of payment in case compensations are due according to the treaties.

The Liability Convention, which followed the Outer Space Treaty, expanded and developed the provisions in Article VII creating a more detailed legal regime. For instance, while Article VII provides for "general" liability, Article II of the Liability Convention clarifies that the standard of liability for damage on the Earth and in the airspace is absolute liability, while Article III of the latter clarifies that the standard for damage in outer space is based on fault liability.

The provisions of the Liability Convention deal with international liability for damage caused by space objects, which may mean that other theories of liability may be relevant in cases the convention does not apply. A simple example is when damage caused in a collision between two satellites of two different operators, where both operators share the same nationality. In such a case, the convention will not be applicable, since the dispute is not of an international nature.

27.2.2.4 Registration of Space Objects

According to the treaties, space objects, including satellites, have to be registered in a registry of space objects. Article VIII of the Outer Space Treaty provides for the basic arrangements relating to registration, where legally the registration allows for the relevant state to exercise its jurisdiction and control over the space object:

> A state party to the treaty on whose registry an object launched into outer space is carried shall retain jurisdiction and control over such object, and over any personnel thereof, while in outer space or on a celestial body. Ownership of objects launched into outer space, including objects landed or constructed on a celestial body, and of their component parts, is not affected by their presence in outer space or on a celestial body or by their return to the Earth. Such objects or component parts found beyond the limits of the state party to the treaty on whose registry they are carried shall be returned to that state party, which shall, upon request, furnish identifying data prior to their return.

The Registration Convention includes further details on the obligation to register the space object and on the information, which need to be provided by the registering state.

According to Article III of the latter, the register will be maintained by the Secretary General of the United Nations and the information provided shall be available on a full and open access basis. Accordingly, the public register is available online and includes a search form via which it is possible to locate the information by different search categories.

The register includes information about space objects that were not officially registered together with information about objects that were properly registered. The first group is listed within square brackets and highlighted in green, while the second group appears in normal black writing. It is possible to find information about space objects launched as of 1957 and until current time, however, the register does not include information about space debris [6].

According to Article IV of the Registration Convention, the following information should be provided to the Secretary General of the United Nations, "as soon as practicable":

(a) Name of launching state or states.
(b) An appropriate designator of the space object or its registration number.
(c) Date and territory or location of launch.
(d) Basic orbital parameters, including:
    (i) Nodal period
    (ii) Inclination
    (iii) Apogee
    (iv) Perigee
(e) General function of the space object. As part of the efforts to facilitate the registration process and to encourage states to conduct proper registration, the UN has created a standard form that includes instructions to the registering state; the form is available online [7].

After outlining the international obligations mentioned earlier in text, states took upon themselves in order to regulate the exploration and use of outer space. The next section will explain and provide examples for the implementation of these obligations on the national level—in the relation between the state and the satellite operators, focusing on European nanosatellite missions.

27.3 National Laws and Practices in EU Member States

27.3.1 General—What Are National Space Laws?

As shown in Section 27.2.2.2, Article VI creates obligations relating to the conduct of private entities, which are not directly subject to the Outer Space Treaty. Article VI provides the nexus to the fulfillment of such obligations, namely, the state that accepted the obligation has to authorize and continually supervise private activities, ensuring that such activities are in line with the provisions of the treaty.

Europe offers an interesting range of national laws relating to space activities. Different private entities in European states are involved in various space activities, inter alia, telecommunications satellites operation, small satellites operation, dissemination of satellite-generated data, and providing launch services, hence the diversity of the regulatory needs. This diversity resulted with "custom-made" implementation of the obligations under Article VI; however, most domestic laws include a similar regulative core, as elaborated in this section.

An additional variable is the time of legislation. The first space law was enacted in 1969 by Norway, and currently there are several states that have drafted space laws and anticipate completing the legislation in the near future.

While certain states chose to protect their private space industries by creating exemptions and lower standards relating to the obligations on the part of the operator, other states chose to implement Article VI in a stricter manner, with less consideration for the interests of the private space sector. In that sense, national space laws do not implement treaty obligations alone; they implement national space policies as well.

It is common for a national space law to provide clarifications or definitions to key terms as "space activity" and "operator" or "licensee." Furthermore, the duty to obtain a license for certain activities is usually the standing point of such legal instrument. The law then includes terms and conditions relating to the license, licensing criteria, distribution of liabilities between the state and the private operator, satellite registration procedure, and instructions relating to the national satellite register (when it exists). In some cases, there is a reference to insurance for the space activity, space debris mitigation standards, safety matters, and more.

In principle, national space legislation does not distinguish the launch or operation of a traditional satellite from a nanosatellite. This generally makes sense, since the space treaties do not set different provisions for different types of space objects either. Some examples of specific national legal and policy related arrangements relating to small satellites will be brought in the next sections of this chapter.

It is important to clarify that since only some European states have enacted national space legislation, some operators will not be subject to a specific domestic space law or licensing mechanism, while others will, all depending on the applicable national legislation. Further, the licensing requirements, criteria, and standards vary from one state to the other, and therefore, operators of different nationalities will be subject to different requirements even when considering an existing space law. In practice, this may mean that some operators will have to procure mandatory third-party liability insurance for their nanosatellite activity, while others will not (unless they wish to, on a voluntary basis). The last example affects the financial planning of a nanosatellite mission. Another example is related to compliance with debris mitigation standards (see further on this subject in the following text). While some national laws made the guidelines mandatory for licensees to follow, others did not. This may affect the design of the spacecraft or the selected launch when considering the 25-year rule.

Even when a domestic space law is absent, a state party to the Outer Space Treaty is still subject to Article VI, and therefore has the power and duty to regulate national space activities. In such cases, the state can request from the operator to follow an ad hoc authorization procedure. The shortcomings of this situation are the time it may take for the state to consider which documents, guarantees, and attestations it may require from the operator in order to authorize the activity, and the lack of the operator's advance knowledge or prediction of which actions it will have to take for that matter.

27.3.2 Regulations, Official Forms, and Interpreting Guidelines

Since the space industry is rapidly developing, introducing new space activities and applications by an increasing number of players, there is a need to keep the legal framework regulating such activities as flexible as possible. For this reason, typically national space laws are worded in general terms, leaving room for interpretation.

Therefore, in some cases, the legislator furnished practical and specific information for operators via secondary legislative instruments, such as regulations, official application forms for operators, and guidelines. These instruments often specify certain information that is not suitable to be included in a general law, such as sums to be taken as caps-limitation of liability, mandatory insurance coverage, and so on. For example, the UK space agency made official forms of license application for operators available online, which clarify the requirements listed in the Outer Space Act [8].

27.3.3 Additional International Legal Instruments and Their Relevance to National Space Laws

In addition to the binding instruments mentioned earlier in text, there are nonbinding international legal instruments relevant to space activities, such as: international guidelines, UN General Assembly Resolutions, draft treaties, and codes of conduct.

The abovementioned instruments allow for the development of the legal corpus without being formally binding on states as a stand-alone instrument. Considering the ongoing, some states chose to give the provisions of certain nonbinding instruments a binding affect, by incorporating them in their national laws and secondary legislation.

A relevant example is the Inter-Agency Space Debris Mitigation Guidelines [9], which do not enjoy a full international legal binding status, however, were adopted by certain states in the form of a legal binding instrument, and therefore, became binding pursuant and subject to the applicability of the said adopting instrument (for example, the French Space Operation Act, 2008).

27.3.4 Applicability

The next step is clarifying the applicability of national laws. Attention has to be paid to the satellite's national state. Some states have enacted national space legal instrument, which is applicable to "national activities in outer space" or to state nationals.

Assuming there is a national space law that in general is applicable by virtue of the nationality of the satellite or the persons owning it, there is a need to verify that the space activity in question is included under the scope of that law.

In other cases, there might be a need to verify that these activities are indeed included, by referring to following decrees, regulations, forms, and official guidelines aiding the law's interpretation.

An additional element to consider with respect to applicability is simply time. For example, as this regulatory filed is relatively new, the current national law may be amended from time to time to include new provisions or update existing ones that may have an effect on a foreseen space activity. Another case is where there are no space laws at the time of commencement of the satellite's design stage, however, these laws enter into force before the launch of the same satellite, rendering the law applicable nonetheless.

For the reasons mentioned earlier, it is always recommended to inform the relevant administration about the intention to carry out space activities, as soon as feasible.

27.3.5 Examples of European States that Made Specific Consideration for Small Satellite Missions in Their National Space Laws and Policies, with Respect to Third-party Liability Insurance

27.3.5.1 Third-party Liability Insurance and Other Insurance Policies

Third-party liability insurance is generally very common on Earth, and the intention behind this concept of liability is the protection of innocent third parties (meaning parties that are not directly involved in the damaging event or activity) and allowing the compensation of such parties in case damage is caused to them by a certain activity.

In international space law, Article VII of the Outer Space Treaty and the Liability Convention includes provisions on third-party liability of states in case a space object that is attributed to them causes damage to another space object, property, or individuals of a different state (see also Section 27.2.2.3).

In way of practical examples, this may mean that a state A will owe compensations to state B, in case a space object of space A re-entered Earth and injured an individual that holds the nationality of state B. Since that individual is an innocent victim, state B will have the right to claim compensations from state A on his or her behalf. The treaties apply as well to damage that is caused in outer space, for example, where space objects collide in orbit, in this case however, there will be a need to prove which party holds the fault for the collision before there will be a right for compensations.

Since states are internationally liable in case a nanosatellite that is attributed to them causes damage in outer space, the air space, or on Earth, states usually include provisions on the distribution of liability between the state and the private operator who is a state national.

Some states require that the operator will obtain and maintain a third-party liability insurance policy, on specific terms and amounts, as a condition for licensing the space activity. In this manner, states potentially limit their exposure to liability for the acts of their nationals, allowing the first to use payable insurance funds to cover the damage caused by the latter. In any case, according to treaty law, the state is the entity that will have to compensate the victim, and not the operator directly.

The procurement of such an insurance policy means that the operator would have to pay a certain premium in order to secure the coverage of the damage, which is usually limited in a specific amount. A high amount of coverage will usually mean a high premium as well, and vice versa. For this reason, there were cases where states chose to lower the amount of coverage for operators, and specifically to potentially lower or waive the coverage amount for small satellite operators. Examples will be brought in the following section.

In addition to the third-party insurance coverage, which may be mandatory by domestic law, there are other risks that are insurable. These insurances are not mandatory by law, as they do not concern damage to third parties as foreseen by the space treaties. These are some examples:

Prelaunch insurance: This type of insurance coverage may commence by covering the shipment of the satellite from the owner's or manufacturer's facilities to the launch services broker, and/or launch site. It may include coverage against possible damage that can be caused to the satellite during shipment, integration on the launch vehicle, and other processes and activities, up to the launch event.

Launch failure insurance: Lunch activities are still considered to be risky, in the sense that the launch vehicle may be unsuccessfully ignited or launched, causing the loss of the payload(s) on board. Insuring a small satellite against such an unfortunate occurrence will allow the owner to recover its value in case of a failure. Therefore, subject to the policy terms, the owner may be reimbursed with sums that will allow her or him to rebuild a new satellite, without the need to resort to a new financing source for the project.

Insuring the mission: It is possible to insure a satellite in terms of capability to carry out its mission. This type of insurance is very relevant to commercial missions, however, may not be suitable for technology demonstrations or educational missions utilizing small satellites.

27.3.5.2 Examples for National Space Laws and Policies in Europe, Focusing on Small Satellites and Insurance Requirements

As mentioned earlier, national space laws in Europe are diverse and were drafted in different times. Naturally, this means that the drafters of the newer laws could foresee national small satellite activities and make specific provisions accordingly.

A good example for an act that was drafted with specific nanosatellite missions in mind is the Austrian Outer Space Act of 2011 [10]. BRITE-Austria and UniBRITE were the first

Austrian satellites to be launched into outer space in 2013, as a part of the BRITE nanosatellite constellation project [11]. Since this was a pioneering project in Austria, nanosatellite activities were especially considered during the drafting process of the domestic act.

The act requires that operators seeking a license for their planned space activity will insure said activity against third-party liability covering a minimal sum of €60 million. This amount was considered appropriate, since the French Space Act required a similar amount as well (see: Marboe, supra note 11 at 535).

Recognizing that this requirement may be excessively costly to small satellite operators, the Austrian Act includes a special provision that aims to ease the financial burden on the operator. Accordingly, missions that are carried out for the interest of the public, including scientific and educational missions, may be required to take insurance in a lower amount, or be exempted from insurance altogether (Article 4 of the Austrian Act, Marboe, supra note 11 at 536). In this way, the act aims to relief small satellite operators with limited finds, such as universities, without including restrictive definitions on what is a "small satellite," in a way that allows the government to consider different satellite classes without the need to update the act as new classes may emerge.

Another example for specific insurance obligations can be found in The Netherlands. The Dutch Space Activities Act of 2007 sets mandatory third-party insurance for Dutch space activities. However, it does not set a fixed amount that the operator must insure against, as it stipulates that operators must insure for the maximal coverage that can be found in the insurance market [12]. This means that all nongovernmental Dutch small satellites must be insured without exemption.

In 2013, the first insurance policies were negotiated for three Dutch small satellites, launched the same year. The coverage amount was €20 million (per satellite), and since then, all other Dutch small satellites were insured at the same amount. The flexible approach that allows the Ministry of Economic Affairs in The Netherlands to set insurance obligations on a case-by-case basis was beneficial to small satellite operators. On the other hand, there is no guarantee that this amount will not change in the future, which means less legal certainty [13]. The Netherlands also had to issue a special decree in 2015 to clarify that its national space law applies to "unguided satellites," which include small satellites by definition [14].

27.4 Future Regulation and Prospects

One of the questions regarding to the future of national space laws concerns harmonization. Since some states are members to only some of the mentioned treaties, some already enacted a national space law while others lack such specific legal instrument; and considering that each national space law is different in terms of applicability and its substantial provisions, the conclusion is that different legal standards may apply to similar space activities and space objects, depending on the operator's nationality.

Currently, there is no international or supranational entity that has mandate to harmonize laws relating to space activities. In the European context, it can be argued that the EU should have the competence to harmonize domestic space laws of member states. It remains to be seen if such competence will be established in the future.

On a global level, UN COPUOS is currently discussing the applicability of international law to small satellite operations. At this stage, there is only general exchange of information between states on the topic, aiming to understand if there are special regulatory needs for small satellites.

Another subject that regulators will have to consider in the future is small satellites' large constellations and the long-term sustainability of outer space. With the anticipated "mega-constellations" in low Earth orbit (LEO), there is a growing concern regarding congestion and the presence of space debris. UN COPUOS is currently discussing guidelines relating to long-term sustainability in the "Working Group on the Long-term Sustainability of Outer Space Activities" [15].

The use of small satellites in deep space exploration missions is recent and did not trigger the need for special regulatory attention yet. It may be that such a need will be raised in the future.

References

1 Article II of the Treaty on Principles Governing the Activities of States in the Exploration and Use of Outer Space, including the Moon and Other Celestial Bodies, done 27 January 1967, entered into force 10 October 1967, 610 UNTS 205, 6 ILM 386 (1967), hereafter: "Outer Space Treaty".

2 Article VIII of the Outer Space Treaty.

3 UN COPUOS website: http://www.unoosa.org/oosa/en/COPUOS/copuos.html

4 Article I(2) of the Outer space Treaty.

5 N. Palkovitz, 2015. Exploring the Boundaries of Free Exploration and Use of Outer Space—Article IX and the Principle of Due Regard, Some Contemporary Considerations, IISL Proceedings of the 57th Colloquium on the Law of Outer Space (Eleven International Publishing, 2015).

6 UN Outer Space Objects Index: http://www.unoosa.org/oosa/en/osoindex.html

7 The registration form can be found on this UN OOSA webpage: http://www.unoosa.org/oosa/SORegister/resources.html.

8 "License to operate a space object: How to apply:" https://www.gov.uk/apply-for-a-license-under-the-outer-space-act-1986

9 Available at the Inter-Agency Space Debris Coordination Committee website: https://www.iadc-online.org/index.cgi?item=docs_pub

10 Austrian Federal Law on the Authorisation of Space Activities and the Establishment of a National Space Registry, as adopted by the Parliament on 6 December 2011 (hereinafter: "Austrian Act"), an English translation is available at: http://www.oosa.unvienna.org/pdf/spacelaw/national/austria/austrian-outer-space-actE.pdf; I. Marboe, *Austrian Federal Law on the Authorisation of Space Activities and the Establishment of a National Registry (Austrian Outer Space Act),* 2011 Proceedings of the International Institute of Space Law, 530 (2012).

11 BRITE Project's website: www.univie.ac.at/brite-constellation/html/news.html.

12 Law Incorporating Rules Concerning Space Activities and the Establishment of a Registry of Space Objects, 24 January 2007, 80 Staatsblad (2007), Space Activities Act.

An English translation is available at: http://www.oosa.unvienna.org/oosa/en/SpaceLaw/national/state-index.html hereafter: "Dutch Space Act".

13 Palkovitz, N. and Masson-Zwaan, T. (2016). Small but on the Radar: The Regulatory Evolution of Small Satellites in The Netherlands. In: *Proceedings of the International Institute of Space Law* (eds. R. Moro-Aguilar, P.J. Blount and T. Masson-Zwaan), 601–612. The Hague: Eleven International Publishing.

14 See the 'Besluit ongeleide satellieten' 2015. In the Dutch Staatsblad of 28 January 2015, https://zoek.officielebekendmakingen.nl/stb-2015-18.html; for further reading, see: N. Palkovitz, "Small Satellites: Innovative Activities, Traditional Laws, and the Industry Perspective" in Small Satellites: Regulatory Challenges and Chances I. Marboe (Ed.) 47 (2016).

15 UN OOSA website: http://www.unoosa.org/oosa/en/ourwork/topics/long-term-sustainability-of-outer-space-activities.html

28 III-2b

Policies and Regulations in North America

Mike Miller[1] and Kirk Woellert[2]

[1] *Sterk Solutions Corporation, Philipsburg, USA*
[2] *ManTech International, Washington, D.C, USA*

28.1 Introduction

The eventual scope of this chapter will provide guidance for all countries of North America. Such a treatment is beyond the scope of this inaugural edition, and the focus will be on the USA. Future editions of this chapter will be expanded to describe the regulatory environment and licensing regimes for Canada and Mexico. The regulatory discussion will be limited to nonfederal (i.e. commercial, experimental, and amateur) US satellite developers only. NonUS payload developers are expected to comply with the regulations of their country of residence and should refer to the appropriate chapters in this book.

The policies and regulations in the USA governing the operation of nanosatellites are largely the same as those governing operation of satellites in general. These derive from a combination of US obligations under international treaties, agreements, and practices that further US interests in space, and that are encouraging responsible use of the resources involved. The proliferation of nanosatellites to address a variety of opportunities makes this a dynamic topic, as regulations, and regulators, evolve in response to the greatly increasing number and range of functions of this class of satellite.

Nanosatellite technology has commonly been referred to as a disruptive phenomenon [1], lowering barriers and democratizing satellite development [2]. In North America, the phenomenon of nanosatellites (and a sub-category called CubeSats) is enabling a growing user base access to a platform technology with ever increasing utility and capability. Some capabilities such as use of the radio frequency spectrum and imaging of the Earth require nanosatellite developers to understand and comply with legal and regulatory obligations. Concurrently, agencies charged with oversight for activities in space are under increasing pressure from the nanosatellite user community to reform legal and regulatory regimes. The myriad of technology drivers that make nanosatellites disruptive are beyond the scope of this chapter, but two nontechnology drivers are combining to affect the oversight process for these platforms.

The first are launch rates. From the 1960s and well into the twenty-first century, over 1500 small satellites have been launched, at least 958 of which were classified as nanosatellites

Nanosatellites: Space and Ground Technologies, Operations and Economics, First Edition.
Edited by Rogerio Atem de Carvalho, Jaime Estela, and Martin Langer.

[3]. In 2017, over 300 nano/microsatellites were launched, an increase of 200% from 2016 [4]. Contributing to the international increase of nanosatellite launch rates are increased space access opportunities for US nanosatellite developers as secondary rideshare missions and recently, several emerging dedicated small launch service providers. Several US government programs offer launch opportunities, such as the NASA CubeSat Launch Initiative and Venture Class Launch Service [5], the US Department of Defense (DoD) University Nanosat Program, and recently the National Science Foundation [6, 7]. In the private sector, several US companies offer payload integration and CubeSat launch services from either expendable launch vehicles or recently from the International Space Station.

The second is the diverse spectrum of nanosatellite payload developers. On one end, US government sponsored nanosatellites are increasingly sophisticated, and in the private sector there are commercial CubeSat constellation operators—both having resources to fulfill legal and regulatory requirements. On the opposite end, commercial CubeSats have been funded and launched via crowd sourcing services such as Kickstarter [8] and even US primary and secondary schools are pursuing CubeSats [9]—but these groups typically lack dedicated resources for legal and regulatory compliance.

A consequence of the nanosatellite "race to space" is that many relatively inexperienced groups are operating in the global commons of space, alongside more experienced actors. This chapter focuses on current regulatory requirements and best practices for nanosatellite developers, which may be particularly helpful to newcomers. It also examines how agencies are addressing gaps, stress points on the regulatory framework, and some expectations about the directions that will be taken by regulators.

A first step would be to agree on a definition of nanosatellite. Currently, there is no legal standard for classifying different sizes of satellites. NASA has generally classified "small satellites" as having a dry mass of 180 kg or less [10]. These classifications are further divided into:

- *Minisatellites*: 180–100 kg.
- *Microsatellites*: 100–10 kg.
- *Nanosatellites (including CubeSats)*: 10–1 kg.
- *Femto/picosatellites*: less than 1 kg.

International classifications follow a similar structure but are not legally binding. Current international and US space law generally does not differentiate between classes of satellites, although there are clear economic and technical differences between large and small satellites [11].

28.2 Governing Treaties and Laws

28.2.1 The Space Treaties and International Conventions

The space treaties, combined with customary practice since the dawn of the Space Age, serve as a basis for all international space law. As demonstrated in 1957 by the first satellite launched into space, the Russian Sputnik, the launching and operation of satellites by definition has international ramifications. The world reaction to Sputnik, along with the USA and Soviet Union increasing their space capabilities in the context of the Cold War, leads to consensus for international dialogue on space. The USA and Soviet Union, as the principal early spacefaring powers, acted through the United Nations (UN) to codify a set of legal principles of how humanity should utilize outer space, in a sense a Magna Carta for space. In 1963, the UN passed a general resolution on such principles, which was extended to the first international treaty on space, Treaty on Principles Governing the Activities of States in the Exploration and Use of Outer Space (aka Outer Space Treaty or simply OST). The OST is an international, legally binding agreement, to which the USA is party.

The OST, although visionary, was essentially an international first response, and did not treat all issues confronted by the early spacefaring community. Acknowledgment of these gaps led to declaration of the other principal space treaties; The Liability Convention (LC) and the Convention on Registration of Objects in Outer Space. Ratified in 1972, the LC extends beyond OST, the issue of liability for mishaps involving objects launched to space, operated in space, and returned (controlled or uncontrolled) from space [12]. Under Article VI of the OST, the US Government is responsible for activities of private US satellite operators, and hence subject to the terms of the LC. Claims made under the LC may only be made by states against other states. Hence, the LC does not preclude nongovernment entities from pursuit of existing legal avenues of recompense for damages. The LC is a fault-based regime (unique in international law), under which, if damage is caused on the surface of the Earth or to an aircraft in flight (Article II), absolute or strict liability is the only requirement. This means that fault liability requires only proof of responsibility for causation of damage. By contrast, if damage is caused in space (Article III), fault liability additionally requires proof of negligence or intention.

To date the only claim under the LC was made by Canada in 1978 against the former Soviet Union, for its nuclear-powered satellite Cosmos 954 that re-entered and crashed in Canadian territory [13]. The Canadian government issued a claim for damages related to clean up of radioactive material and other hazardous materials scattered over the re-entry corridor. Canada received recompense for damages from the USSR [14]. The US government, as a party to the LC, does not require US nanosatellite operators to obtain on-orbit liability insurance, but does require US launch providers to obtain Maximum Probable Loss liability insurance in order to obtain a launch license [15]. Rather, US government concerns of liability for on-orbit activities by US satellites are mitigated by the operators' compliance with guidelines for space debris and burgeoning traffic management. This policy does not preclude nanosatellite operators seeking liability insurance to satisfy their own program or business requirements.

The Registration Convention (RC) entered into force in 1976. It is a logical consequence of the need to document the orbital parameters and state of jurisdiction of objects placed in space for safety purposes and in the case of mishap, for assignment of liability. Parties to the RC agree to maintain a national database of objects they have placed in space, and provide that data to the UN. The US State Department Office of Space and Advanced Technology (OES/SAT) is responsible for maintaining the official US space object registry [16]. US satellites launched on US launch providers must comply with the RC. The data are provided by the satellite operator to the launch provider, who in turn provides the same to the Federal Aviation Administration (FAA) as part of the launch application. The FAA then forwards this information to the State Department. Assignment of an object to a specific country can become complicated when one state launches a payload on behalf of another. As stated in the RC Article I: "(i) A State which launches or procures the launching of a space object"; or "(ii) A State from whose territory or facility a space object is launched." In practice, registration is typically done by the country to which the satellite owner claims jurisdiction, not the launching country.

Data required by the RC as follows:

(1) The international designator of the space object(s).
(2) Date and location of launch.
(3) General function of the space object.
(4) Final orbital parameters, including: Nodal period; Inclination; Apogee; and Perigee.

A country generally will be either a signatory or a party to a treaty. A signatory is not legally bound by specific treaty provisions and obligations, while a party is legally bound. Another concept is status as a signatory versus ratification. Signing a treaty, aka becoming a signatory, signals a state's intent to comply with a treaty. Ratification is an internal process, once completed whereby a state is obligated to the treaty provisions. For example, in the USA, the President has legal authority to sign a treaty but ratification requires consent by the US Senate. Completion of ratification makes the USA a party. Table 28.1 summarizes the US domestic law implementation of the space treaties.

Table 28.1 Summary of US domestic law implementation of the space treaties.

Space treaties	US status	US policy/regulatory implementation
Outer space treaty	Party	Direct impact on nanosatellite developer/operators
		FCC and NOAA regulatory requirements
Liability convention	Party	
Convention on registration of objects	Party	Required for US-owned satellites, US CFR (§417.19)

28.2.2 International Telecommunications Union/International Organization

The International Telecommunications Union (ITU) is the implementing organization for the international registration requirements of the RC. The ITU receives and publishes the data on behalf of the UN. The ITU currently has 193 member states that also represent over 700 sector members and associates, who are bound by the organization's constitution. The organization was created in 1865 and its primary mission today is to "ensure rational, equitable, efficient, and economical use of the radio frequency spectrum by all radio communication services—including those using the geostationary satellite orbit or other satellite orbits—and to carry out studies on radio communication matters" (ITU CS 78 & 196). It is essentially an international regime for use of the electromagnetic spectrum. Primary responsibilities of the ITU include principles of use of orbit and spectrum, allocation of frequency bands and services, and procedures. The ITU is a United Nations affiliated organization, and in some respects a facilitator of the binding agreements of the OST.

28.2.3 Domestic Policy Within the USA

The United States national space policy is primarily made up of two directives: the US National Security Space Strategy and the National Space Policy of the United States. Both are policy directives at the executive/presidential level that serve as the guiding practices the USA will follow. United States policies are more fluid as presidents set the tone and direction for the US space program. These policies are implemented through presidential executive orders, the most recent of which are to be discussed later in this chapter.

28.3 Orbital Debris Mitigation

Human activity in space has resulted in the accumulation of a large amount of debris in orbit. The general definition of space debris is any artificial, nonoperational object whether created unintentionally or intentionally [17]. Most space debris is the result of practices during the formative years of the Space Age, such as jettisoned hardware, breakups or explosions of improperly saved rocket stages or satellites, rare collisions between satellites or defunct objects, and on two occasions intentional use of antisatellite capabilities by select spacefaring nations. As a result of over 50 years of human activity in space, there is a staggering amount of debris in orbit. As of this writing, more than 21 000 orbital debris objects larger than 10 cm are known to exist. The population of objects 1–10 cm in diameter is estimated at 500 000. Estimate for the number of particles smaller than 1 cm exceeds 100 million [18]. Because of the kinetic energies involved, collisions with millimeter size objects can result in serious damage, catastrophic loss, or life-threating risk to human spaceflight missions.

The US government's position on space debris has evolved considerably since the launch of the first satellite. In the USA, awareness of the hazards of space debris matured throughout the 1970s and early 1980s. By the late 1980s, several dynamics converged, raising the issue to US presidential awareness [19]. In 1988, the USA formally added space debris mitigation as a priority in the US National Space Policy [20]. During the 1990s, a US interagency coordination process culminated in the release of the first US Government Orbital Debris Mitigation Standard Practices in 2001 [21]. Space debris mitigation remains a high priority in US National Space Policy. A coalition of the agencies of spacefaring countries formed the Inter-Agency Space Debris Coordination Committee (IADC) in 1993 [22]. The United Nations, through its Committee on the Peaceful Uses of Outer Space (COPUOS) released a technical report on space debris in 1999. COPUOS mandated in 2001 the IADC to develop a baseline for space debris mitigation techniques, which resulted in publication of the IADC Space Debris Mitigation Guidelines in October 2002 [23].

The economics of satellites of lower nano and pico mass range enables access to space by the average citizen, local group, or school. The characteristics and number of these spacecraft potentially raise new challenges for space sustainability and space situational awareness (SSA). One characteristic is very small cross-sectional area. The typical cross-sectional area of nanosatellites and picosatellites coincides with the 10 cm size limit for detection by active radar sites of the US DoD Space Surveillance Network (SSN) (The overall US SSN-cited resolution limit is 10 cm diameter. Individual SSN sites such as the Massachusetts Institute of Technology's Lincoln Laboratory X-band Haystack Long-Range Imaging Radar (LRIR) can detect objects down to 1 cm diameter at 1000 km) [24, 25]. This makes consistent and accurate tracking and subsequent publishing of orbital ephemeris problematic. The other characteristics such as the severe mass and volumetric and power limits are typical of these satellites. Because of these limits, propulsion systems or other means of orbit maneuvering capability are not widely used by the nanosatellite community (several examples of CubeSat projects mission profile include significant on-orbit maneuver capability using on-board propulsion) [26]. If a possible collision scenario ("conjunction") with another object in space is identified, the only mitigation strategy available is timely notification to the operator of the other object (if that object is an operating satellite with the means to change orbit).

Another characteristic satellites in this mass range share is the space access limitations imposed by the secondary or rideshare concept, where launch providers consider such satellites as secondary payloads and thus developers must defer to the primary payload mission profile. Secondary payloads often must accept longer-lived orbits than their mission would require, and without deorbiting capability, they add to the low Earth orbit (LEO) population of uncontrolled objects. Favorable economics and reduced barriers to entry thus encourage both proliferation and reduced control. The relatively low up-front development and launch costs of a typical CubeSat and other nano and pico form factors [27] offer the potential for an order of magnitude increase in satellite operators. However, the very factors that offer great promise also raise questions on whether LEO space can support potentially thousands of artificial satellites.

Studies by space debris experts show that even if all space launch activity is permanently ceased, the LEO space environment may become unstable through a series of cascade collisions [28, 29]. The Obama administration space policy declared space debris mitigation as a national priority, and more recently the Trump administration's latest space policy directive requires updates to orbital debris mitigation standards, along with new guidelines for satellite design and operation [30, 31].

The procedural implementation of US policy on space debris is formalized in the document "US Government Orbital Debris Mitigation Standard Practices." Under this, the two documents, NASA Procedural Requirements for Limiting Orbital Debris, NPR 8715.6A [32] ("Requirements"), together with NASA Technical Standard 8719.14A, Process for Limiting Orbital Debris [33] ("the Standard"), address the critical subject of orbital debris mitigation for launches over which NASA has purview. Requirements describe in detail the scope of this purview, and responsibilities.

Requirements state that all US nonfederal sponsored satellite missions are required to submit an Orbital Debris Assessment Report, or ODAR, to accompany the Federal Communications Commission (FCC) transmitter license applications, and if applicable, also accompany the National Oceanic and Atmospheric Administration (NOAA) Remote Sensing License application. The standard describes, among other things, how to compose and format an ODAR.

The ODAR will show that the mission:

- Will not release debris during normal operations.
- Will not generate debris by explosions and intentional breakups. This will involve an assessment and failure analysis of all stored energy components, including batteries.
- Will not generate debris by on-orbit collisions during mission operations.
- Will re-enter and burn up in the Earth's atmosphere within 25 years after launch, or maneuver to a storage orbit between LEO and geostationary orbit (GEO) or be retrieved. All known nanosats will re-enter and burn up within 25 years as the disposal strategy, either by natural orbit decay or, less commonly, through the use of deorbiting devices.
- Will limit the number and residual energy of debris fragments that survive uncontrolled re-entry, with calculated limits on human casualty expectation.

The software application "Debris Assessment Software—DAS Version 2.1.1," aka "DAS" (current version as of publication), is available to download from the NASA Orbital Debris Program Office website, and will typically be used to calculate, based on initial orbit, mass and cross-sectional area, the altitude versus time of the spacecraft, and estimate the date when re-entry will occur. This software also calculates the number and size of debris fragments that may survive re-entry, based on user entry of the mass, material, and dimensional properties of each component of the satellite. In addition to supporting the ODAR, for missions involving deployment from the International Space Station, the DAS input data on component properties must be submitted to NASA by the launch integrator.

Practically, nanosatellite mission designers need to be aware of a couple of issues related to submission of the ODAR. The first is the relationship between selection of orbital altitude and inclination and regulatory approval. Operating a nanosatellite at altitudes or inclinations that have conjunction potential with human spaceflight missions such as the International Space Station (ISS), military space assets or high-value commercial assets will likely experience increased regulatory scrutiny. The same applies for selection of orbits within highly congested volumes of space in LEO, such as the sun-synchronous orbits. The second topic is potential for debris during the mission, and passivation, or release of stored energy, at end of mission. Applicants should be prepared to substantiate that satellite failure modes will not lead to high energy release, or if such release occurs, debris will not be generated, or at least will be contained. Case in point, design of batteries and associated power systems are without exception considered in this context. It is incumbent upon the satellite mission sponsor and/or developer to provide the analysis that substantiates conclusions in the ODAR. Therefore, it is important to consider ODAR requirements early in the design phase of a nanosatellite project. Some ODARs are available online by web search; the reader is encouraged to review them. In conclusion, space debris mitigation is the responsibility of all space users.

28.4 Space Traffic Management

Related but distinct from space debris mitigation is the concept of space traffic management (STM). As per the US Space Policy Directive-3 [34], STM refers to the planning, coordination, and on-orbit synchronization of activities to enhance the safety, stability, and sustainability of operations in the space environment.

STM in the USA has been an evolving practice and for many decades has been predominately a military mission. The launch of Sputnik in 1957 and subsequently the imperative to track Intercontinental Ballistic Missiles (ICBMs) developed by the former Soviet Union were drivers for the USA to establish a space object tracking capability. That capability, consisting of various optical and radar sites, has a storied history, experiencing cycles of growth and dilapidation, but today is referred to as the SSN mentioned earlier in this section [35, 36].

The burgeoning US civil space program, early commercial telecommunication satellite operators, and to some extent foreign satellite operators needed space tracking data to deal with the ever-growing problem of potential conjunctions, that is, avoiding space debris (a simple definition of conjunction is two or more objects in space with trajectories exceeding threshold for potential collision). With the SSN as the only effective space surveillance capability, the DoD provided sanitized unclassified data to NASA, who was designated as the responsible agency for coordination with commercial and foreign entities. Under this arrangement, the DoD provided data to NASA on a not-to-interfere basis with the primary national security mission. By 2000, concerns were raised that the arrangement between DoD and NASA was not adequate to meet the ever-growing demand for space-tracking data by commercial and foreign entities [37]. DoD submitted a legislative proposal to Congress for a 3-year pilot program called the Commercial and Foreign Entity (CFE) SSA program, which became law as part of the Defense Authorization Act of 2004 [38]. The US Air Force Space Command was designated as executive agent to manage the CFE, and by 2005 space-tracking data were being provided via the website (www.space-track.org). Also in 2005, the Air Force assigned operational control to the Joint Space Operation Center (JSpOC) for tracking and cataloging all artificial objects in space. One of the JSpOC functions was notifying satellite operators of potential conjunctions. However, effectiveness of the JSpOC conjunction reporting for CFE users was limited by national security concerns, and they lacked adequate information architecture able to automatically process precision ephemeris provided by CFE satellite operators [39]. Hence, JSpOC necessarily prioritized tracking and reporting for military and national security assets, human spaceflight, for example, Space Shuttle missions and the ISS, and select high-value commercial satellites. Conjunction reporting for nonpriority satellites was on a best effort basis [40, 41].

This state of affairs persisted until 2009 when an accidental collision occurred between the US Iridium 33, an operational commercial satellite, and Kosmos 2251, a defunct Russian Strela military communications satellite [42]. The incident was unprecedented and made international headlines as the first collision between two artificial satellites, challenging assumptions about the space debris environment and adequacy for safe operation of satellites.

An outcome of the Iridium–Kosmos collision was that Congress authorized the US Strategic Command (USSTRATCOM) in 2010 to assume responsibility for SSA. The CFE was transitioned to the Space Situational Awareness Sharing Program operated by the JSpOC. Another change was expansion of the JSpOC mission to include conjunction assessment for all active satellites in space and not just US satellites, increasing the number of analysis overhead by a factor of three or four [43].

USSTRATCOM through SSA Sharing Program provides three levels of service: basic, emergency, and advanced. Basic services are free of charge and available to anyone with an account on Space-track.org. Anyone can request an account, subject to approval.

Emergency services are provided in situations where human spaceflight may be endangered or a satellite is at risk due to conjunction. If pre-coordinated, time-sensitive conjunction summary messages are provided to satellite owner/operators, so that they may conduct avoidance maneuvers. Emergency services are non-fee-based and available to anyone with an account. Advanced services require an SSA Sharing Agreement between the parties and include conjunction assessment, launch support, deorbit and re-entry support, disposal and end-of-life support, anomaly resolution, and satellite communication interference support. Advanced services can be requested through Space-track.org.

USSTRATCOM has made significant improvements and expanded services it provides via Space-Track.org. Resources available to members include publishing of daily ephemeris of objects in space (i.e. two-line element (TLE) sets), conjunction notification and technical assessment data (for example, radar cross-section (RCS) data are available on Space-Track .org), various guidance and regulatory documentation, an application programing interface (API) enabling automated delivery of various JSpOC data products, and a laser clearing house for space missions subject to DoD Instruction 3100.11.

Participation in Space-Track.org is voluntary, but highly encouraged. Nanosatellite owner/operators may question the value of registering. A couple of examples provide motivation. Availability of low-cost/footprint Global Positioning System (GPS) receivers provides nanosatellite owner/operators high-fidelity ephemeris. However, if a spacecraft is not flying GPS, or if communications between the ground and the nanosatellite fail or have not yet been established at the beginning of a mission, locational data may be obtained from Space-Track.org, helping efforts to point the ground station antenna when establishing or re-establishing communications. Another example is that propulsion, and hence, maneuver capabilities of nanosatellites are ever increasing. There are gaps in the US SSN and maneuvers by satellites may not be tracked in a timely manner, possibly increasing risk of inadvertent conjunction with other satellites. This scenario can be avoided by prenotification of maneuvers through Space-Track.org.

In 2017, USSTRATCOM directed transition of the JSpOC to the Combined Space Operations Center (CSpOC). The transition completed in July 2018, and was made to increase the scope of operations to include US allies along with commercial and civil partners [44].

Due to concerns about space object tracking and conjunction analysis, nanosatellites less than 10 cm in all dimensions, the aforementioned nominal lower size limit that can be reliably tracked, can expect additional diligence by US agencies. A widely publicized example was the FCC denial of a radio license to SWARM Technologies for their first launch of four SpaceBee satellites, each of which were smaller than a 10 cm cube. The FCC dismissal letter stated that this was based on an insufficient substantiation of trackability in their license application [45].

Responding to tracking concerns, the marketplace is beginning to offer transponders that will attach to the satellite and via satellite communications systems, continuously update the satellite operator with TLE information based on GPS. Like aircraft transponders, these will be powered independently of the satellite, and will be capable of operating independently of the satellite. It remains to be seen how this technology will augment tracking with Earth-based radar, as a method of anticipating potential conjunctions.

Alluded to earlier in this section, 2018 saw the advent of a formal US STM policy. The Trump administration issued Policy Directive-3 (SPD-3) in mid-2018 [46]. The directive observes that the limited STM management and infrastructure provided by USSTRATCOM is inadequate to deal with the increased volume and diversity of commercial space activity while, at the same time, the DoD needs to focus on the ever-increasing contested nature of space. Quoting from a WhiteHouse.gov fact sheet, "The Department of Commerce will make space safety data and services available to the public, while the DoD maintains the authoritative catalog of space objects." SPD-3 sends a clear signal that STM is a civil function focused on orbital safety, preservation of the space environment, and will be the primary interface to the private sector and the public. Complementing that, the DoD will be able to focus on SSA, an inherently military mission. For the better part of the last decade, the DoD was working to transition responsibilities to the FAA, but SPD-3 intent clearly states this to be a commerce function [47].

It remains to be seen how the US will incorporate new policy on orbital debris and STM into the current regulatory framework, and if any new requirements may flow down to satellite owner/operators. Whatever changes may come, they will likely be inclusive of nanosatellites.

28.5 Licensing of Radio Transmission from Space

To date, nanosatellites typically have not returned to Earth intact. Their value lies in performing a task in space and transmitting the results of the task by radio. In the USA, radio transmitters ("transmitters") are typically operated under a license, and transmitters in space are no exception.

28.5.1 Licensing Authorities

FCC and National Telecommunications and Information Administration (NTIA) assure that operation of space-based transmitters conforms to relevant international agreements, as described in Section 28.2 "Governing Treaties and Laws" in this chapter.

28.5.2 NTIA Origins and Range of Authority

Transmitters operated by federal agencies are managed by the NTIA, a branch of the US Department of Commerce. NTIA regulates all federal government spectrum use, and it functions as the executive branch of the telecommunications policy maker, and the President's principal advisor on telecommunications policy. NTIA issues a radio frequency authorization (RFA) to the operator, which licenses the transmission for a specified period. DD1494 is the submittal form used by the applying agency, to request that the NTIA provide the RFA. Details of the necessary procedures are available to agencies of the federal government.

28.5.3 FCC Origins and License Types

Transmitters operated by entities other than federal agencies are licensed by the FCC, an independent agency of the US federal government. FCC regulation extends to all US states and territories. The FCC regulates all spectra other than that allocated to the federal government, and it functions as the legislative branch of the telecommunications policy maker, and it is the Congress's principal advisor on telecommunications policy.

For these missions, FCC is also the national spectrum authority for filing mission data with the (ITU. As per agreement, when a license is granted, the relevant data regarding the satellite and associated transmitters are forwarded to ITU for review and filing, to make the data available internationally. These data are packaged in a SpaceCap database by the applicant, using software downloaded from the ITU website. Attachments to the Space-Cap may include antenna radiation pattern diagrams and an application for an operating symbol, if the applicant does not have such a symbol already assigned. When these data are integrated into the ITU international registry maintained by the ITU, it becomes the international notice for the transmitter.

The license applicant interfaces with the FCC, and the FCC interfaces with ITU. The satellite operator may be invoiced by ITU for reimbursement of ITU costs associated with processing the filing.

FCC satellite transmitter licensing authority originates from FCC's section of the Code of Federal Regulations (CFR), with a separate part for each of the three use categories. License types, corresponding to these three categories, are amateur, experimental, and commercial described as follows:

- Part 97 for amateur radio service: The stated purpose of Part 97 is to ensure the "recognition and enhancement of the value of the amateur service to the public as a voluntary noncommercial service, particularly with respect to providing emergency communications" [48]. The primary mission goal for amateur satellites is "exchanging messages with other amateur stations."
- Title 47, Part 5 for experimental radio service:. Experimental use can include "the purposes of experimentation, product development, and market trials" [49].
- Part 25 for commercial use of satellite communications: Commercial use scope derives from the Communications Satellite Act of 1962, the International Maritime Satellite Telecommunications Act, and the Communications Act of 1934 [50].

FCC indicates that at least 90 days should be allowed between the submittal of a complete amateur or experimental use application and the determination of license. During this time, queries may be sent to the applicant, which must be answered within 30 days, or the license application is subject to cancellation.

28.5.4 Choosing a Frequency

The choice of frequency must be made with regard to applicable regulations, spectrum allocation, and the nature of the intended use, in order to successfully obtain a license. This is documented by the "FCC Online Table of Frequency Allocations," 47 CFR Section 2.106 [51]. It is regularly updated, and the latest version must be used as a reference for determining acceptable frequencies.

Frequency choice will also be influenced by factors such as necessary data bandwidth, available power, equipment availability and cost, required antenna on spacecraft, and selection of earth station equipment. It is necessary to show that interference with existing operations is unlikely, by analysis of databases of existing users, and by coordination with identified users in the frequency range of interest. Frequency choice should be made early in the design process as practical.

28.5.5 FCC License Fee Exemption – Government Entities

FCC fees for the various license types are discussed in the following sections. However, governmental entities are exempt from these fees, as per the following:

§1.1116 General exemptions to charges.

No fee established in §§1.1102 through 1.1109 of this subpart, unless otherwise qualified herein, shall be required for: (f) Applicants, permittees, or licensees who qualify as governmental entities. For purposes of this exemption, a governmental entity is defined as any state, possession, city, county, town, village, municipal corporation, or similar political organization or subpart thereof controlled by publicly elected or duly appointed public officials exercising sovereign direction and control over their respective communities or programs.

Other special categories of applicants or use may also be exempt from fees. The reader is referred to the current version of §1.1116 for further information.

28.5.6 Coordination of Use of Amateur Frequencies

Amateur frequencies are often used by nanosatellite transmitters. Reasons for this include availability of equipment for these frequencies of high quality and reasonable price, and a global community of amateur operators who are able to receive transmissions on these frequencies. Amateur frequencies are used for amateur activities, which include communications among amateurs, and educational activities. Amateur frequencies are not available for commercial use.

Use of amateur frequencies from space transmitters is coordinated with the International Amateur Radio Union (IARU) satellite advisor; see http://www.iaru.org/satellite.html. This coordination serves to minimize the potential for interfering use of a frequency. Coordination is initiated by the operator filing a satellite frequency coordination request with the satellite advisor, who is appointed as a volunteer by IARU.

Coordination is completed when the satellite advisor provides a coordination letter to the operator. The coordination letter is then submitted by the applicant to the FCC. There is no fee for frequency coordination by IARU.

28.5.7 Amateur Licensing for Satellite Transmitters

As per FCC, "The licensed amateur operator should submit a prelaunch notification not later than 30 days after the date of launch vehicle determination, but no later than 90 days before integration of the space station into the launch vehicle. These notifications should be submitted via mail to International Bureau, FCC, Washington, D.C. 20554. In order to facilitate processing, applicants may also provide this material via email to amsat.spacecap@fcc.gov." The notification should include the SpaceCap database with attachments, and the IARU coordination letter, both discussed earlier.

28.5.8 Experimental Licensing for Satellite Transmitters

The experimental operator files an online license application with the FCC Office of Engineering and Technology (OET). The application may be for Special Temporary Authority to operate, if the mission duration is 6 months or less, or for an experimental license, for a duration of up to 2 years. Note that applicants may apply for a license renewal, if operation beyond the original duration granted is desired. The fee for either of these filings is currently US$70. If confidentiality is required, an additional US$70 fee is added [52].

There are no particular frequency ranges identified for experimental operations. In general, the use is licensed on a "noninterference" basis, and operators may be asked to terminate transmission if interference occurs. In addition, applicants may be asked to develop an electromagnetic compatibility (EMC) analysis, showing that interference with existing operators is not likely to occur.

This application is done through the OET website. It includes data describing the applicant, the satellite and the mission, orbital parameters, radio transmitter characteristics, method for positive control of the transmitter, associated ground stations, an ODAR (see the previous section of this chapter), and a SpaceCap database as mentioned previously. The process is extensive, and OET recommends that applicants without experience obtain experienced support.

OET reviews the application, and if there are no issues, they forward the application to the FCC International Bureau (IB). IB reviews the application for conformance with international agreements, and forwards to the ITU for filing. With IB concurrence, OET grants a license. Note that the time is allocated in the process for public notice. With experienced support, the applicant should be allowed a minimum of 3 months from submission of a complete filing, for the licensing process to complete.

28.5.9 Part 25 Licensing for Satellite Transmitters

Applications for uses other than amateur or experimental are filed using Form 312, under Part 25 of the code. Part 25 and Form 312 are for any satellite service that does not fall within the experimental or amateur services. Most are commercial ventures, but that is not a requirement. This process is much more complex than the process for amateur or experimental licenses and may take on the order of a year to complete. Professional support is considered necessary for Part 25 filings.

The applicant must submit Form 312, "Federal Communications Commission Application for Satellite Space and Earth Station Authorizations," along with accompanying exhibits. This is submitted to the FCC online, through the FCC International Bureau Filing System (IBFS), along with the necessary fees. As the title indicates, this form is used both for space transmitters, and for earth stations transmitting to satellites. Required exhibits include an ODAR, Radiation Hazard Analysis, a current Frequency Coordination and Interference Analysis Report, and/or antenna radiation patterns, as well as other documents as required by FCC [53].

Planet Labs became the first nanosatellite operator to obtain a Part 25 commercial satellite license in 2013, and successfully launched a commercial nanosat constellation in 2014 from the ISS [54].

The transition from experimental to commercial licensing for nanosatellites requires a step change in license fees, as well as in the complexity of the licensing process. The license fee for experimental use currently is US$70, as mentioned earlier. The current Part 25 filing fee for a Non-Geostationary-Satellite Orbit (NGSO) constellation (regardless of the number of satellites) is US$471 575. This does not include the filing fee for any associated earth stations. The license will cover a system of nanosatellites up to a specific maximum number. Satellites that are lost from the system can be replenished under the same license [55].

Annual regulatory fees must also be paid to maintain a Part 25 license. The current annual regulatory fee for a NGSO constellation is US$122 775. This does not include the regulatory fee for any associated earth stations, mentioned in Section 1.1156 of the FCC Rules (47 CFR 1.1156) [56].

From the FCC website, a fee payment is required "upon the commencement of operation of a system's first satellite as reported annually pursuant to Sections 25.142(c), 25.143(e), 25.145(g), or upon certification of operation of a single satellite pursuant to Section 25.121(d)." Also, a bond of US$5 000 000 must be posted, which FCC may reduce as certain milestones are met [57].

As for experimental licenses, FCC files the notices produced with the ITU, and the ITU may levy cost recovery fees on the applicant.

Currently nanosats are not handled differently from other satellites in terms of licensing, but at this time there is discussion that the FCC makes special provisions to allow nanosats to be licensed more quickly and at less cost, than is currently the case. Reducing the amount of fees and financial bonds, allowing more flexibility in the definition of schedule milestones, and scheduling deployment rather than launch are some of the changes being suggested to facilitate more efficient handling of nanosat commercial license requests [58].

28.6 Licensing for Remote Sensing Activities from Space

The US Department of Commerce, National Oceanic and Atmospheric Administration, Office of Commercial Remote Sensing Regulatory Affairs (CRSRA) is the US licensing authority for remote sensing activities in space. This is authorized by the National and Commercial Space Programs Act of 2010, Title 51 U.S.C., Subtitle VI (NCSPA), along with the special restriction on satellite imagery of Israel. This authority is implemented by 15 CFR Part 960, "Licensing of Private Land Remote Sensing Space Systems."

A remote sensing system is defined by 15 CFR Part 960, Section 960.3 as including "any device, instrument, or combination thereof, the space-borne platform upon which it is carried, and any related facilities capable of actively or passively sensing the Earth's surface, including bodies of water, from space by making use of the properties of the electromagnetic waves emitted, reflected, or diffracted by the sensed objects. A system consists of a finite number of satellites and associated facilities, including those for tasking, receiving, and storing data. Small, hand-held cameras shall not be considered remote sensing space systems."

From the CRSRA website: An operating license for a remote sensing system must be obtained by "any person subject to the jurisdiction or control of the USA who wishes to operate a private remote sensing space system, including but not limited to: an individual who is a citizen of the USA; a corporation, partnership, association, or other entity organized or existing under the laws of the USA or any state, territory, or possession thereof; or any other private space system operator having substantial connections with the USA or deriving substantial benefits from US law that supports its remote sensing satellite operations [59]."

Thus, ownership of the remote sensing system is the criterion for determining CRSRA jurisdiction over the system.

28.6.1 Licensing Requirements

As a first step, the applicant will file an initial contact form, to allow CRSRA to determine if a full application will be required. See the following for instructions and access to an online form for initial contact. Based on the initial contact information, CRSRA will either issue a waiver letter indicating that no license is necessary or indicate that a license must be obtained.

The application process is extensive, and the format of the application is not fixed. The process as of this writing consists of an interactive dialogue between the applicant and CRSRA. As with radio licensing, it is recommended that those without experience seek support from those who have experience with the system. Submittal both on paper and electronic is required.

Information required includes applicant identification, charter and representation, ownership, agreements with a foreign nation, launch data, orbit data, sensor resolution, communications, method for controlling data, and other factors regarding data security and availability. An ODAR is required with the application, as it is for radio transmission licensing. The federal government reserves the right to restrict remote sensing in specific areas at any time, aka "shutter control."

28.6.2 Fees, Timeline, and Post Issuance Obligations

The US government does not require a fee to obtain a remote sensing license. CRSRA indicates that 120 days should be allowed before determination for the issuance of a license, after the application is complete and accepted. This will allow for review by federal agencies, coordinated by CRSRA. At this time, CRSA is experiencing ever-increasing rates of application, and 120 days should be considered a goal rather than a guarantee. Licensees must provide a summary of system information that can be made public, within 30 days of issuance of a license. An annual audit submission is also required to identify any changes since issuance of license.

28.7 Export Control Laws

The US government, like other countries with high-technology capabilities, seeks to control the export of sensitive equipment, services, or technology to promote national security interests and foreign policy objectives, and prevent proliferation of weapons. Space technologies, or technologies utilized as part of a space program, are considered by the US government as inherently sensitive and are subject to US export control laws. The export status of a given space technology affects who can be a recipient of such export, including launch providers.

Therefore, US-based nanosatellite developers/operators must be cognizant of and comply with US export control laws [60].

The US export controls regime is a two-tiered framework addressing technology transfer concerns by classifying products, services, or technology, collectively termed "articles" or "items" as either dual use, meaning potential for civilian uses, as well as military uses, or as those items specifically designed or modified for military applications referred to as a defense articles or services.

The Export Administration Act (EAA) of 1969 established the Export Administration Regulations (EAR), which cover dual-use items. Dual-use items encompass hardware, software, and certain encryption algorithms. The Bureau of Industry and Security (BIS), an agency within the US Department of Commerce, enforces the EAR through the Commerce Control List (CCL). Items on the CCL require an export permit or qualify for exemption. To determine if an export permit is required, the BIS provides a procedure to determine if an item has an Export Control Classification Number (ECCN). If an item is subject to jurisdiction under the EAR and does not have an ECCN, then the item is designated as EAR99. Most commercial products are EAR99 and generally do not require an export permit.

For space technologies, the CCL relevant classifications are:

A.1: "600 series" refers to ECCNs in the "xY6zz" format on the CCL. Items controlled under the "600 series" were previously controlled on the United States Munitions List (USML) or are covered by the Wassenaar Arrangement Munitions List (WAML) and include certain items formerly classified under ECCNs ending in −018. The "6" indicates that the entry is a munitions entry on the CCL. The "x" represents the CCL category (0 through 9) and "Y" the CCL product group (A through E). In most cases, the "zz" represents the WAML category. The "600 series" constitutes the munitions ECCNs within the larger CCL.

Section 22 U.S.C. 2778 of the Arms Export Control Act (AECA) provided the general authority to control which defense articles and services can be exported out of the USA. The President has delegated authority to promulgate regulations to the United States Department of State, under the International Trade in Arms Regulations (ITAR). The defense-related articles and services subject to ITAR are commonly referred to as the USML. The primary categories of the USML related to space technologies are Categories IV and XV, but that does not imply checking those categories is sufficient. For example, certain materials with aerospace applications such as ablatives used in rockets nozzles fall under Category XIII. Certain high-energy substances such as hydrazine that fall under Category II are other examples.

One should refer to the USML (Part 121 of the CFR, Chapter I, Subchapter M) to determine if a given technology, data, or equipment is classified under ITAR. Hardcopy or electronic versions of the ITAR may be purchased, but a web formatted version is freely available at www.ecfr.gov.

28.7.1 General Principles, Requirements, and Common Misconceptions

A comprehensive treatment of US export controls is outside the scope of this chapter. Instead, a small sample of the more important general principles and requirements is presented, and common misconceptions refuted.

Aerospace technology fundamentally is a dual-use technology, having civil and military uses. Therefore, one should never assume that research, commercial activities, equipment or technology used, or the outputs involving aerospace technology are exempt from export controls.

A point of confusion is the definition of fundamental research. The ITAR regulations provide for exemption under what is commonly referred to as "fundamental research." A common misconception is that research, and the products derived thereof, are exempt from the ITAR. The fact is that research may or may not be subject to export controls. Universities often conduct research under contract to the US DOD. A hypothetical example is the case of research project involving space propulsion, which would fall under the USML Category XV spacecraft and related articles. Also, in this example and typically the research project is subject to an information disclosure clause. Although the project is not providing a defense service, the research is deemed to be export controlled under Category XV. Hence, any transfer of research-related information to a foreign person without the express approval of the DoD would constitute an export violation. Many universities have in-house export control representatives and retain services of law firms experienced in export controls to deal with these kinds of scenarios. The reader is invited to review the fundamental research definition from the ITAR regulations [61].

Any person engaged in the manufacturing or export of defense services or providing defense services is required to register with Directorate of Defense Trade Controls (DDTC). The ITAR CFR 121.1 provides specific registration requirements.

US-based nanosatellite developers/operators should understand the concept of export. Exports can occur as (i) sending or carrying items out of the US, (ii) transferring or disclosing data/technology, (iii) providing a defense service, and (iv) re-export, which means an item or information is transferred to a foreign person who will then transfer it to another country. An export license may not necessarily be required, but that determination is made by an authorized export control official and ultimately by the US government.

A foreign person is a term referring to any individual who is not a lawful US permanent resident. It also refers to any entity (e.g. company, trust, society, business associations, etc.) not incorporated or organized to do business or activities in the USA. The complete definition is available in the ITAR regulations.

Export controls must be considered for schedule planning. The process for determining the export classification of an article, applying for an export license (if required), and subsequent approval can take many months. One should also be aware that approval for export permit or license should not be assumed. Any export control request is subject to not only technical considerations but also a review in the context of current US national policy.

28.7.2 Export Control Reform

The EAA of 1979 provided legal authority to the President of the USA to control exports via various federal agencies for reasons of national security, foreign policy, or other national policy (e.g. trade sanctions) [62].

A brief history on the evolution of US export control is helpful for later discussion in this section. The EAA expired in 1990 and with exception of a brief period in 1994–1995 has not been renewed by Congress. In 1990, President George W. Bush extended the existing export regulations by an executive order under the International Emergency Economic Powers Act (IEEPA). Since that time, failure to achieve consensus on a cohesive regulatory framework between the executive branch, Congress, and industry has thwarted several attempts to renew the EAA or proposed variants. Subsequent administrations including, as of this writing, the Trump administration have extended authority of export regulations under the IEEPA [63–67].

The abovementioned brief history highlights how contentious the topic of export control is in the USA. This discussion purposely avoided the complexities. The reality is that over a dozen US agencies are involved in export control policy and/or enforcement, and the system is fraught with inefficiency and redundancy and requirements are often subjective. Complicating any attempts at reform are competing interests between the political parties, the branches of US government, and those of the private sector. Export controls remained status quo until a critical mass of support was achieved in favor of reform with support by senior administration leadership.

The most ambitious export control reform was initiated under the Obama administration. Following are the remarks in 2010 by then Secretary of Defense Robert Gates, noted during his tenure at the CIA as Deputy Director for Intelligence, "The length of the list of controlled technologies outstripped our finite intelligence monitoring capabilities and resources … We were wasting our time and resources tracking technologies you could buy at RadioShack … We need a system that dispenses with the 95 percent of 'easy' cases and lets us concentrate our resources on the remaining 5 percent [68]." Following an interagency review, the Obama administration proposed the following changes implemented in three phases:

- Transition from the two control lists, that is, the CCL and USML to a single-tiered control list.
- Establish an integrated enforcement center and single enforcement agency.
- Transition to a single Information Technology infrastructure.
- Transition to a single licensing system and agency.

By the end of 2016, Phases 1 and 2 were completed, with 18 of the 21 USML categories having been reviewed with proposals for reconciling the USML and CCL lists [69]. As of this writing, the CCL and USML are yet to be consolidated and enforcement duties are still shared among several agencies, notably the BIS and DDTC. The most visible changes are how the USML and CCL lists are defined. Key principles are: (i) lists are based on objective criteria, rather than design intent, and (ii) the principle of "positive" controls.

An example serves to illustrate this. Take USML Category XV(a)(2), which describes spacecraft that "autonomously detect and track moving ground, airborne, missile, or space objects other than celestial bodies, in real-time using imaging, infrared, radar, or laser systems." A commercial satellite that tracks ships via their automated information system (AIS) transmissions, but does not use imaging, infrared, radar, or laser systems, would not be captured by this paragraph.

Although a single export control list is yet to be implemented, both the CCL and USML have been updated to reflect a tiered approach, that is, the most sensitive technologies remain on the USML, while less-sensitive technologies were moved to the CCL. In 2014, a major outcome of the export reform initiative was how select space technologies are controlled. Prior to 2014, the majority of satellites and related components were controlled under very broad definitions as defense articles subject to ITAR (i.e. USML Category XV), this meant that anything part of a space system—even items like bolts were controlled as ITAR. As a result of the 2014 changes, communications satellites, remote sensing satellites with certain performance characteristics, spacecraft parts, accessories, and ancillary components have been moved from the USML to the CCL.

Another important reform outcome was simplification of the process for determining the export classification. A principle advertised by both the BIS and DDTC is "Order of Review." Essentially, a step-by-step process flow is used to determine the classification. Implementing the OOR concept has led to development of resources to aid users. For instance, the BIS has implemented an online CCL Order of Review decision tool [70].

As of this writing, with exception of an executive memorandum [71] implementing changes to the US Conventional Arms Transfer Policy (essentially addressing proposed changes to the USML not implemented during the Obama administration), the Trump administration has not put forth a position on export controls, but neither has it issued any changes or roll backs to reforms initiated in 2010.

Although the US government continues to implement export control reform [72], nevertheless US export control laws are complex, and penalties for noncompliance may be severe. Nanosatellite developers/operators are strongly advised to coordinate with their sponsoring organization, Export Control Authority, or retain professional legal counsel with experience in US export control law as appropriate.

28.8 Conclusion

Current regulatory practices for spectrum allocation have been deemed insufficient by many nanosatellite operators who are unable to meet the basic requirements. There is widespread feeling that meeting regulatory requirements, international and domestic, is among the greatest challenges for these operators. Along with this, there is a feeling that separate regulations for this class of satellite/operator would be appropriate.

28.8.1 International Efforts

The increased utilization of nanosatellites has led to increased international attention to spectrum coordination issues unique to these platforms. An outcome of the World Radiocommunication Conference 2012 (WRC-12) was Resolution 757, placed on the WRC-18 agenda "consideration of regulatory procedures for notifying satellite networks needed to facilitate the deployment and operation of nanosatellites and picosatellites." However, ITU working group documents presented at WRC-15 do not support modification of procedures in Articles 9 and 11 with respect to small satellites. The ITU position remains that spectrum coordination for satellite services is a necessary function, regardless of satellite class. The ITU does acknowledge the need for improved outreach and providing assistance for compliance to regulatory procedures [73].

International regulatory approaches toward nanosatellites are continually challenged, as new proposals are submitted to the World Radiocommunication Conferences. World Radiocommunication Conference 15 (WRC-15) included a proposal from the US delegation that set out to "examine the procedures for notifying space networks and consider possible modifications to enable the deployment and operation of nanosatellites and picosatellites." The proposal called for further study of Resolution 757, "Regulatory aspects for nanosatellites and picosatellites," which had been initiated at WRC-12. The resolution addresses the challenges, including short development times, typically of 1–2 years, along with short operational periods that typically are less than 5 years. It also addresses potential changes to current international regulatory standards. Another larger issue is placing nanosatellites within the correct frequency range. In many instances, nanosatellites and picosatellites have been licensed for frequencies on the 30–3000 MHz range that are not allocated for the service of the mission. Although such operation may be consistent on a noninterference basis with FCC Part 5, it remains inconsistent with allocations in the Red Book. The Resolution 757 invites WRC-18 to consider separate regulations for smaller satellites, possibly accompanied by modifications to the frequency allocation table that might resolve this inconsistency [74].

28.8.2 US Efforts

As this first edition goes to press, a US response to this challenge is the FCC Notice of Proposed Rulemaking, IB Docket No. 18-86. This proposes creation of a special class of small satellite networks in the Part 25 (Commercial) service. Those qualifying under these rules would have, compared to existing Part 25 rules, a reduced fee structure and reduced time for license application processing. In exchange they would meet a number of restrictions—10 satellites or less, launched into LEO, not requiring exclusive use of spectrum and meeting certain orbital debris requirements [75].

28.8.3 New Space

The future of space utilization is sometimes referred to as "New Space." This dynamic reflects the fact that significant venture capital is being invested in the space sector, whereas previously funding was dominated by government programs. Since 2006, over 400 emerging space companies have been funded by private equity totaling US$10 billion. About US$2.5 billion of that value has been directed toward small satellites technology and development [76]. The advent of significant venture capital, and commensurate expectation for return on investment, will drive companies to innovate, likely in ways that will continue to stress the US space regulatory framework.

As the American space enterprise enters the second decade of the twenty-first century, the authors foresee new drivers, economic activity, and policy changes for the nanosatellite phenomena. First, nanosatellite developers are integrating recent technological advancements into their designs, such as electric propulsion, advanced control algorithms, aerodynamic orbit control, and innovative form factors to name a few [77–81]. Also there are more launch opportunities, and at lower prices, with access to space enabled by the plethora of small launch vehicles under development by dozens of established and emerging providers [82–84].

On the horizon are new mission concepts for deploying at altitudes less than 300 km, or what the authors refer to as extremely low Earth orbit (ELEO). It is well understood that some conventional mission concepts (e.g. remote sensing) would benefit from the ability to operate at lower orbital altitudes [85]. Nanosatellites are particularly well suited to make exploitation of this volume of space in Earth orbit feasible. These low orbits and limited lifetimes also are suited to simple, low-cost experiments, and the limited lifetime and low initial orbit are expected to show lower risk of collisions and residual orbital debris.

The authors also see the trend continuing for increased numbers and diversity of nontraditional actors—including startups with consumer-oriented business cases and educational institutions ranging from tier 1 universities to elementary schools in developing nations [86–88].

Also, as satellites shrink ever smaller, designers are developing techniques to increase the radar signatures, and thereby trackability, of the smallest satellites. As examples, these techniques have involved reflectors, use of radio-frequency identification (RFID) technology, and use of fold out panels [89, 90].

A potential new trend is the convergence of nanosatellites and emergent space-based communications services in support for the Internet of Things (IoT). As of 2017, 10 startup firms at various stages of maturity and funding have announced plans to provide machine–to-machine data services [91].

Finally, as the number of nanosatellites operating in LEO continues to increase, there will be commensurate pressure for these platforms to conform to the broader regulatory framework applied to conventional satellites. As example, we can expect that nanosatellite operators will be required to adopt means for positive identification [92] and cyber-resiliency [93].

References

1 Bower, J.L. and Christensen, C.M. (1995). Disruptive Technologies: Catching the Wave. *Harvard Business Review* (January/February).

2 Woellert et al. (2011). *Advances in Space Research* 47: 663–684. https://doi.org/10.1016/j.asr.2010.10.009.

3 Erik Kulu, Nanosatellite and CubeSat Database (www.nanosats.eu), accessed August 19, 2018.

4 Williams, C., Doncaster, B., and Shulman, J. (2018). *Nano/Microsatellite Market Forecast*, 8e. SpaceWorks Enterprises, Inc.

5 NASA, "Small Satellites to Get Their Own Ride to Space" (http://www.nasa.gov/feature/small-satellites-to-get-their-own-ride-to-space), accessed July 12, 2018.

6 US DoD University Nanosat Program (universitynanosat.org), accessed August 19, 2018.

7 National Science Foundation CubeSat program, NSF18553 June 13, 2018.

8 SkyCube (http://www.kickstarter.com/projects/880837561/skycube-the-first-satellite-launched-by-you), accessed May 12, 2015.

9 STMSat-1 (https://www.nasa.gov/mission_pages/station/research/experiments/1885.html), accessed March 16, 2019.

10 NASA, State of the Art of Small Spacecraft Technology (sst-soa.arc.nasa.gov/01-introduction), accessed August 19, 2018.

11 Sa'id Mosteshar, (2014). "Authorization of Small Satellites Under National Space Legislation" (paper presented at Small Satellites: Chances and Challenges, Vienna, Austria, March 29, 2014).

12 UNOOSA.Org (http://www.unoosa.org/oosa/en/ourwork/spacelaw/treaties/introliability-convention.html), accessed August 18, 2018.

13 Schwartz, B. and Berlin, M.L. (1982). After the fall: an analysis of canadian legal claims for damage caused by cosmos 954, 27. *McGILL Law Journal* 676: 705–712.

14 Canada, Operation Morning Light (nuclearsafety.gc.ca/eng/resources/canadas-nuclear-history/past-presidents/alan-prince.cfm), accessed August 18, 2018.

15 US 14 CFR Part 440.7 and Appendix A to Part 440.

16 US State Department (http://www.state.gov/e/oes/sat), accessed August 18, 2018.

17 UN COPUOS, (1999). Technical Report on Space Debris, 99-82839-pr

18 NASA Orbital Debris Program Office (orbitaldebris.jsc.nasa.gov/faq.html), accessed August 19, 2018.

19 Kessler, D. A partial history of orbital debris: a personal view. *Orbital Debris Monitor* 6 (3): 16–20. 1 July 1993, and Vol. 6, No. 4, pp. 10–16, 1 October 1993, edited by Darren McKnight.

20 Weeden, B., The Evolution of U.S. National Policy for Addressing the Threat of Space Debris, IAC-16.A6.8.3.

21 U.S. Government Orbital Debris Mitigation Standard Practices, (www.orbitaldebris.jsc .nasa.gov/library/usg_od_standard_practices.pdf), accessed March 16, 2019.

22 IADC, 34th session, IADC-1997-1 (www.iadc-online.org/Documents), accessed 8/19/18.

23 IADC, Terms of Reference (www.iadc-online.org), accessed 8/19/18.

24 Liou, J.-C. and Johnson, N.L. (2010). Controlling the growth of future LEO debris populations with active debris removal. *Acta Astronautica* 66: 648–653. https://doi.org/10 .1016/j.actaastro.2009.08.005.

25 Johnson, N.L. (1993). *Advances in Space Research* 13 (8): (8)5–(8)20, Kaman Sciences Corporation.

26 Mueller, J., Hofer, R., and Ziemer, J. (2010). *Survey of Propulsion Technologies Applicable to Cubesats*. Jet Propulsion Laboratory, NASA available at hdl.handle.net/2014/41627.

27 Supra note, 2,Section 5: "Economics of Cubesat Technology," pg. 20.

28 Liou, J.-C. and Johnson, N.L. (2006). Risks in space from orbiting debris. *Science* 311: 340–341.

29 Liou, J.-C. and Johnson, N.L. (2008). Instability of the present LEO satellite populations. *Advances in Space Research* 41: 1046–1053.

30 National Space Policy of the United States of America, pg. 4, 8/28/10.

31 US Space Policy Directive-3, National Space Traffic Management Policy, 8/18/18.

32 NASA Procedural Requirements for Limiting Orbital Debris, NPR 8715.6A.

33 NASA Technical Standard 8719.14A, Process for Limiting Orbital Debris.

34 Ibid, 33.

35 SpaceNews, Shelton Orders Shutdown of Space Fence (spacenews.com/36655shelton-orders-shutdown-of-space-fence/), accessed 7/21/18.

36 SpaceNews, Lockheed Martin lands $914M Space Fence contract (spacenews.com/ 40776lockheed-martin-lands-914m-space-fence-contract/), accessed 7/21/18.

37 GAO, Space Surveillance Network: New Way Proposed to Support Commercial and Foreign Entities, 6/7/02.

38 2004 National Defense Authorization Act P.L. 108–136, Section 913.

39 Supra note 37.

40 National Research Council (2011). *Limiting Future Collision Risk to Spacecraft: An Assessment of NASA's Meteoroid and Orbital Debris Programs*. Washington, DC: The National Academies Press https://doi.org/10.17226/13244.

41 (2012). *Global Issues, Selections from Congressional Quarterly Research*, 126. CQ Press.

42 NASA Orbital Debris Program Office (2009). Satellite Collision Leaves Significant Debris Clouds. *Orbital Debris Quarterly News* 13 (2): 1–2. April 2009. Archived from the original (PDF) on 5/27/10, accessed 7/21/18.

43 Space News, JSpOC Conjunction Alerts Could Be Improved, Group Says (spacenews .com/jspoc-conjunction-alerts-could-be-improved-group-says/), accessed 7/22/18.

44 Combined Space Operations Center established at Vandenberg AFB (stratcom.mil/ Media/News/News-Article-View/Article/1579497/combined-space-operations-center-established-at-vandenberg-afb/), accessed 7/22/18.

45 FCC Memo 40011, Subject: File No. 0305-EX-CN-2017, 12/12/17.

46 US Space Policy Directive-3, National Space Traffic Management Policy, 6/18/18.

47 SpaceNews, DoD has deep expertise in "space situational awareness," or SSA, whereas Commerce faces a steep learning curve (spacenews.com/defense-department-turning-over-space-traffic-management-to-commerce-but-details-still-unclear/), accessed 7/22/18.

48 CFR, Title 47, Part 97, subpart A–97.1, Basis and purpose.

49 CFR, Title 47, Part 5, section 5.1, Basis and purpose.

50 CFR, Title 47, Part 25, section 25.1 (presageinc.com/contents/experience/satellitereform/ contents/briefingbook/technology/1962act.pdf) section 201C; section 501C-6 of Maritime; Sections I–III of Communications Act.

51 FCC, Online Table of Frequency Allocations, (https://transition.fcc.gov/oet/spectrum/ table/fcctable.pdf), accessed 3/21/19.

52 FCC, Office of Engineering and Technology Fee Filing Guide (https://docs.fcc.gov/ public/attachments/DOC-353917A1.pdf), accessed 3/21/19.

53 FCC, Consumer Guides (fcc.gov/research-reports/guides/frequently-asked-questions-faq-processing-earth-station-applications), accessed 8/19/18.

54 SpaceNews (spacenews.com/39459planet-labs-cubesats-deployed-from-iss-with-many-more-to-follow/), accessed 8/27/18.

55 CFR, Title 47, Chapter 1, Subchapter A, Part 1. Subpart G §1.1107 (www.ecfr.gov), accessed 3/21/19.

56 CFR47, Chapter 1, Subchapter A, Part 1, Subpart G §1.1156 (www.ecfr.gov), accessed 3/21/19.

57 CFR47, Chapter 1, Subchapter B, Part 25, Subpart B, §25.165 (www.ecfr.gov), accessed 3/21/19.

58 FCC IB Docket No. 12-267, "In the Matter of Comprehensive Review of Licensing and Operating Rules for Satellite Services," Comments of Planet Labs Inc. and Comments of Spire Global Inc.

59 NOAA, CRSA FAQ (https://www.nesdis.noaa.gov/CRSRA/generalFAQ.html), accessed 3/21/19.

60 US State Department, Overview of U.S. Export Control System (state.gov/strategictrade/overview/), accessed 8/19/18.

61 US State Department, DDTC (www.pmddtc.state.gov/regulations_laws/itar.html), accessed 8/19/18.

62 EAA (legcounsel.house.gov/Comps/eaa79.pdf), accessed 7/15/18.

63 Bush George H.W., (1990). Executive Order 12730 (Sep 1990), continuation of export controls.

64 Clinton William J., (1994). Executive Order 12923 (Jun 1994), continuation of export controls.

65 Bush George W., (2001). Executive Order 13222 (Aug 2001), continuation of export controls.

66 Obama Barack H., (2009). Presidential Notice August 13, 2009, continuation of export controls.

67 Trump, Donald J., Presidential Notice 82 FR 39005.

68 Robert Gates, SECDEF (archive.defense.gov/Speeches/Speech.aspx?SpeechID=1453), accessed 7/16/18.

69 U.S Department of State Archive, (2009–2017). Export Control Reform Update (state.gov/r/pa/prs/ps/2016/02/252355.htm), accessed 7/17/18.

70 BIS (www.bis.doc.gov/index.php/export-control-classification-interactive-tool), accessed 7/16/18.

71 Trump, D., US Conventional Arms Transfer Policy (whitehouse.gov), accessed 7/17/18.

72 U.S. State Department, Export Control Reform (pmddtc.state.gov/ECR/index.html), accessed 8/19/18.

73 ITU, (2015). ITU Symposium and Workshop on small satellite regulation and communication systems, Prague, Czech Republic, 2–4 March 2015, Status of ITU-R studies related to small satellites, Ubbels, Innovative Solutions In Space BV and Buscher, Technische Universität Berlin (itu.int/en/ITU-R/space/workshops/2015-prague-small-sat/Presentations/WP7B.pdf), accessed 8/19/18.

74 ITU Resolution 757 (http://www.itu.int/dms_pub/itu-r/oth/0c/0a/R0C0A00000A0025PDFE.pdf), accessed 8/19/18.

75 FCC Proposes To Streamline The Application Process For Small Satellites (https://www.fcc.gov/document/fcc-proposes-streamline-application-process-small-satellites), accessed 3/21/19.

76 Sweeting, M. (2018). Modern small satellites—changing the economics of space. *Proceedings of the IEEE* 106 (3).

77 Reducing spacecraft drag in very low Earth orbit through shape optimization, DOI: 10.13009/EUCASS2017-449.

78 Omar, S. R., and Wersinger, J. M. (2015). Satellite Formation Control Using Differential Drag. In 53rd AIAA Aerospace Sciences Meeting (p. 0002).

79 Tummala, A.R. and Dutta, A. (2017). An overview of cube-satellite propulsion technologies and trends. *Aerospace* 4 (4): 58.

80 Haiping, C., & Zhaokui, W. (2011). Ultra-low Earth orbit formation flying control using aerodynamic forces. In IAA Conference on Dynamics and Control of Space Systems.

81 SpaceNews, "Will Thinsats inspire the next generation of engineers and scientists? (spacenews.com/will-thinsats-inspire-the-next-generation-of-engineers-and-scientists/), accessed 8/10/18.

82 Forbes.com (forbes.com/sites/saadiampekkanen/2018/02/26/small-rocket-industry-heats-up-across-apac-as-space-is-declared-open-for-business/#5879574a9bbd), accessed 8/10/18.

83 SpaceNewsMag (spacenewsmag.com/feature/how-big-is-the-market-for-small-launch-vehicles/), accessed 8/10/18.

84 Wekerle, T., Pessoa Filho, J.B., da Costa, L.E.V.L., and Trabasso, L.G. (2017). Status and trends of smallsats and their launch vehicles—an up-to-date review. *Journal of Aerospace Technology and Management* 9 (3): 269–286. doi.org/10.5028/jatm.v9i3.853.

85 Virgili-Llop, Josep, Roberts, Peter, Hao, Zhou, Ramio, Laia, Beauplet, Valentin. (2014). Very low Earth orbit mission concepts for Earth observation. Benefits and challenges.

86 NASA (nasa.gov/feature/goddard/2018/after-launch-two-nasa-educational-cubesats), accessed 8/10/18.

87 The Teaching Channel.org (http://www.teachingchannel.org/cubesat-engineering-unit-boeing), accessed 8/10/18.

88 National Academies of Sciences, Engineering, and Medicine (2016). *Achieving Science with CubeSats: Thinking Inside the Box*, Chapter: 3, CubeSats as a tool for education and hands-on training. Washington, DC: The National Academies Press https://doi.org/10.17226/23503.

89 SRI, mstl.atl.calpoly.edu/~bklofas/Presentations/DevelopersWorkshop2009/5_Cousins-SRI_Tag.pdf, accessed 8/14/18.

90 SecureWorld (http://swfound.org/media/205410/bw_nrc_cubesats_space_debris_22june2015.pdf), accessed 8/14/18.

91 SpaceNews, Fleet details 100 nanosat constellation for Internet of Things connectivity (http://spacenews.com/fleet-details-100-nanosat-constellation-for-internet-of-things-connectivity), accessed 8/10/18.

92 SpaceNews (http://spacenews.com/laser-license-plate-could-improve-identification-of-cubesats), accessed 8/10/18.

93 SpaceNews (http://spacenews.com/no-encryption-no-fly-rule-proposed-for-smallsats), accessed 8/10/18.

29 III-2c

International Organizations and International Cooperation

Jean-Francois Mayence

Belgian Federal Science Policy Office (BELSPO), Brussels, Belgium

29.1   Introduction

Although international intergovernmental organizations have played a historic role in international cooperation since the end of the Second World War, they have remained second row actors in the field of outer space. While Europe had opted for an institutional cooperation among member states active in the exploration and use of outer space, first with the European Space Conference, then with ESRO[1] and ELDO,[2] eventually moving forward with European Space Agency (ESA), other spacefaring nations have kept their energy focused on national effort or bilateral cooperation. Considering the inherent international dimension of outer space, one could have expected a more active involvement of international institutions.

With organizations such as Intelsat and Inmarsat, a new role for intergovernmental bodies was considered in the field of satellite applications. However, the privatization of their operational activities marked a stop in that direction. At global multilateral level, international organizations do not seem to be a solution supported by states to achieve space cooperation. The case of Europe is clearly an exception: with the European Space Conference, in 1962, followed by ESRO and ELDO and eventually the establishment of ESA in 1975 and European Organisation for the Exploitation of Meteorological Satellites (EUMETSAT) in 1986, European states have continuously pursued their cooperation in outer space through institutional frameworks, allowing the pooling of funds, an integrated management, and a sharing of responsibilities in the implementation of projects. No doubt that this option has allowed Europe to gain its place in the "top 3" of the world space powers. With the growing involvement of the European Union as an institutional actor, developing its own programs and projects under its constitutional competence, space has become a more political area. Time will tell whether the union will be seen as a suitable framework for complementing or replacing ESA, or will remain a partner, sometimes in competition with its own member states.

---

1 European Organization for Space Research.
2 European Organization for Launchers Development.

Nanosatellites: Space and Ground Technologies, Operations and Economics, First Edition.
Edited by Rogerio Atem de Carvalho, Jaime Estela, and Martin Langer.

This chapter, however, is dedicated to global international organizations and cooperation, starting with the first of all: The United Nations. It will also cover other nonregional organizations, as well as the main national agencies and administrations, as far as their policies and programs in connection with nanosatellites are concerned.

29.2 The United Nations and Affiliated Organizations

29.2.1 General Considerations

Nothing in the 1945 Charter establishing the United Nations Organization and its principles as a constituent treaty explicitly deals with outer space. At most, Chapter IX on International Economic and Social Cooperation can be seen as the general framework in which subsequent institutions and bodies of the United Nations have been established, including United Nations Educational, Scientific and Cultural Organization (UNESCO) and United Nations Committee on the Peaceful Uses of Outer Space (UNCOPUOS).

A connection with outer space is rather to be found in Article I of the Charter, stating the main purposes of the United Nations, which are:

1. To maintain international peace and security, to take effective collective measures for the prevention and removal of threats to the peace, and for the suppression of acts of aggression or other breaches of the peace, and to bring about by peaceful means, and in conformity with the principles of justice and international law, adjustment or settlement of international disputes or situations which might lead to a breach of the peace.
2. To develop friendly relations among nations based on respect for the principle of equal rights and self-determination of peoples, and to take other appropriate measures to strengthen universal peace.
3. To achieve international cooperation in solving international problems of an economic, social, cultural, or humanitarian character, and in promoting and encouraging respect for human rights and for fundamental freedoms for all without distinction as to race, sex, language, or religion.
4. To be a center for harmonizing the actions of nations in the attainment of these common ends.

The interest for outer space as an area of technologies dedicated to supporting the achievement of those goals is clear, particularly when it comes to items 3 and 4.

Quite naturally, outer space has become an area of cooperation or, by default, a field of peaceful rivalry exempted from war and conflict between nations. Ironically enough, it was an instrument of destruction invented during the war that helped mankind achieving its ultimate dream: reaching beyond the sky. Von Braun and Korolev, both engineers appointed to the development of war weapons, were tasked by their respective governments with the conception of launchers able to send objects, and ultimately, human beings, in orbit or beyond.

The Cold War has deeply marked the United Nations. The "Lady" smells like the old time, with its corridors covered with red carpets and its deep seats plastered with moleskin, designed to keep the secrets of diplomatic backstage unheard. The atmosphere contrasts with the frantic business of space advancement. Indeed, space is a risky business, two words that do not really belong to UN vocabulary. Therefore, the adequacy between UN and the space world has not always proven optimal.

Nevertheless, the United Nations has shown some proactivity in adopting a corpus iuris dedicated to the space domain. Today, some provisions of the five UN treaties on outer space may seem a bit obsolete, while others have not yet been applied due to their anticipative character (i.e. exploitation of celestial bodies). International space law is the main achievement of the United Nations in the field of outer space. Operational involvement of the United Nations in outer space has later developed through the use of applications serving their purposes and programs (UNDP,[3] UNEP,[4] UNOSAT,[5] etc.), but so far, the United Nations has not provided a global institutional approach for the management of outer space activities, that is, through the establishment of a dedicated world organization.[6]

With the Space Applications Programme, the United Nations Office for Outer Space Affairs (UNOOSA) provides support to educational and capacity-building initiatives in the field of space-related sciences and technologies.

The role of the United Nations in outer space is thus threefold, acting:

- As an international policymaker, lawmaker, and regulator.
- As a space applications' user.
- As a capacity builder.

In each of those three aspects, the importance of nanosatellite technology is acknowledged.

29.2.2 UNCOPUOS and Space Law

Established in 1959 as a subsidiary body of the General Assembly, reporting to its 5th Committee on Decolonization and Special Political Matters, the Committee for the Peaceful Uses of Outer Space is the cradle of the five United Nations treaties on outer space and of the subsequent resolutions adopted by the General Assembly on various matters concerning space activities.

The composition of the committee has evolved through time, starting with 24 "founding" states[7] not necessarily representative of the contemporary super space powers, but rather of the geopolitical blocks.

Fifty five years later, the committee counts more than 70 member states representing a large sample of nations variably involved in outer space. Some are big spacefaring nations

---

3 United Nations Development Programme.
4 United Nations Environment Programme.
5 UNOSAT is the operational satellite program of UNITAR (United Nations Institute for Training and Research). This program aims at collecting and exploiting satellite data for humanitarian and environmental causes and from various sources, notably through partnership and support from other (inter)governmental or nongovernmental organizations.
6 Contrary to what exists in maritime activities (International Maritime Organization, IMO), or weather forecast and climate studies (World Meteorological Organization, WMO), for instance.
7 Albania, Argentina, Australia, Austria, Belgium, Brazil, Bulgaria, Canada, Czechoslovakia, France, Hungary, India, Iran, Italy, Japan, Lebanon, Mexico, Poland, Romania, Sweden, the Union of Soviet Socialist Republics, the United Arab Republic, the United Kingdom of Great Britain and Northern Ireland, and the United States of America.

with large operating capacities, including launch facilities, others are developing capacities in relation with their national needs and others do not have national capacities, but are users of space products and applications, notably for sustainable development policies, disaster management, and service infrastructure. Every state represented at UNCOPUOS has an interest in outer space and the number of spacefaring nations is constantly rising. Small satellite technologies are one of the good explanations for this phenomenon: more affordable, less complex systems benefiting from more competition on the launchers market make it possible for developing countries to consider entering the space game. Those so-called emerging spacefaring nations are sometimes newcomers in this area, but their application as member of UNCOPUOS does not necessarily imply their accession to the United Nations outer space treaties. At least, such a formal condition is not imposed to candidate countries.

Enhancing the participation of states in the corpus iuris spatialis remains however a key factor in the development of international cooperation: cooperating states or institutions seek to refer to common rules, standards, and practices generally accepted at international level. The question remains whether those standards and practices are sufficiently developed and to which extent they can be subject to consensus between states. That question is particularly relevant when it comes to nanosatellites.

Nothing in the United Nations outer space treaties or in the United Nations General Assembly resolutions specifically refers to satellites with regard to their type, their function, their size, or their mass. The treaties deal with the concept of "space objects." The 1975 UN Convention on the Registration of Objects launched into outer space explicitly refers to objects launched in an earth orbit or beyond. That is all. The word "satellite" does not appear anywhere in the United Nations outer space treaties.[8] The notion of "space object" includes any man-made object present in outer space, as well as any component of that object. The question whether space debris qualifies as space object is disputed, but we do not see any legal ground according to which a piece of a space object, even though resulting from an accidental destruction, should be considered as something else than a space object itself. The comparison between nanosatellites and space debris is not innocent: actually, from legal and operational standpoints, satellites which cannot be maneuvered in orbit would be handled the same way as space debris for collision risk management.

If the legal specificity of nanosatellites must be demonstrated, it should actually be with regard to their technical and operational characteristics. A large proportion of nanosatellites so far are nonmaneuverable spacecraft. In terms of liability, this has important consequences. Liability for a damage caused in orbit is based on the evidence of a fault having caused the damage. Apart from the moment of its positioning at the end of the launch phase, a nonmaneuverable satellite is not subject to operation during its life in orbit. It is therefore very difficult to consider that its trajectory could be affected by negligence or misconduct. The pending issue is whether launching a nanosatellite with any means to maneuver it once positioned in outer space could be, considering the overpopulation of objects in orbit, assimilated as a wrongful act. This would imply that the launching state (the state having launched or procured the launch or provided its territory or its facilities to the launch)

---

8 Except in the 1979 UN Agreement Governing the Activities of States on the Moon and other Celestial Bodies, where the term is used in reference to natural satellites.

could not ignore that the positioning of the nanosatellite would generate a situation where the probability of collision with other space objects was beyond what is reasonable.

The particular case of nonmaneuverable satellites was addressed in Belgium by a revision of the provisions of the law dedicated to the operation of space objects.[9] Before that revision, the operator as defined by the law was identified as he who has the actual control of the object. In the case of nonmaneuverable objects, that definition led to identifying the launch service provider as the operator (since he was in charge of performing the only phase of control on the object). To avoid this situation, which would have resulted in putting the liabilities associated to the object on the head of the launch service provider, the law has been modified. It now explicitly provides for the situation where the object cannot be maneuvered once positioned into orbit. In such a case, the law identifies the operator as he who has exercised final authority over the positioning phase, that is, he who has procured the launch. This example from the Belgian space law illustrates the discussions held at international level on the notion of "space activities" as provided for in Article VI of the 1967 UN Treaty on Principles Governing the Activities of States in the Exploration and Use of Outer Space, including the Moon and Other Celestial Bodies, and to whom the responsibility for such activities must be attributed.

Nanosatellites may nevertheless be equipped with propulsion or orientation means. The affordability of such systems with regard to the economic value of the mission, but also with regard to the risk incurred (collision and cost of tracking), is constantly improving, leading to a new generation of (very) small satellites that could be deorbited "on demand." Now, this helps solving one issue, but, in the meantime, it raises another one: who will be in charge of monitoring and operating the flight of the satellite and what will be the cost thereof?

Those questions lead to addressing the phenomenon of emerging new actors in the space operation business. Newcomers are different from the traditional satellite operators providing services to the final users. They actually are the conceptor, the manufacturer, and the final user at the same time. Universities, research centers, small industries now have the (financial and technical) capacity to design, develop, manufacture, and launch (very) small satellites. Now the question is whether they have the expertise and skills to operate spacecraft in full compliance with applicable rules and standards. This question is often considered as not relevant, since the nanosatellite is "only" meant to be launched and left alone in orbit, only for the purpose of generating a signal or transmitting data. What happens in orbit after the launch, beyond the calculation of an appropriate orbit, is not seen as an issue. Therefore, no requirement is integrated in the definition and development phases to manage in-orbit operation. It is likely that a better guidance from states, through their national regulations, would be valuable to that extent, raising awareness of nanosatellite designers and developers about the legal and technical requirements implied by the launch of the satellite. Most of the time, the national space law procedures (whenever they exist) take place already too late in the process: license is required prior to the launch, but not before the development is already completed.

In their wisdom, the drafters of the United Nations outer space treaties have opted for the broad notion of "space object." Who could blame them for that when golf balls are swung

---

9 Law of 17 September 2005, as revised on 1 December 2013 (see http://www.belspo.be/belspo/space/beLaw_en.stm).

out of the International Space Station (ISS) for promotional purposes, or body ashes are dispersed as an ultimate tribute to cosmic dreamers? Objects of all sorts have been launched in outer space. Technological progress makes it even possible to dramatically reduce the size and the mass of spacecraft and to launch smaller but much more complex objects than before.

The story of the very first artificial satellite of the Earth shows how size matters sometimes, even though not as a legal factor. In prevision of the International Geophysical Year 1957/1958, the Soviets were planning the launch of "Object D," a spacecraft weighing more than 1 ton embarking scientific payloads. Realizing they were behind schedule and fearing to be bypassed by the Americans, the USSR Council of Ministers postponed the launch of "Object D" and replaced it by a much smaller spacecraft. The first Sputnik weighted less than 100 kg and was 58 cm large. Its payload consisted in a simple radio transmitter. Actually, the size of Sputnik 1 allowed it to be optically tracked from the ground. In that respect as well, things have considerably evolved. All in all, we can affirm that the Space Age started with a small satellite.

The current trend to reduction of the spacecraft's size can be explained technically, by the miniaturization of data and processing storage capacity, and economically, by the cost of launch. But once in outer space, small size is not necessarily an advantage. New technologies against the multiplication of space debris, such as active debris removal (ARD) or in-orbit servicing (IOS), are not facilitated by this trend. Conjunction assessment for collision avoidance also requires enhanced tracking capabilities to monitor small objects or debris able to neutralize satellite systems.

Another legal issue concerns CubeSats and their particular single or multiple unit(s) configuration (1U, 3U, 6U, etc.). Registration of objects launched into outer space is foreseen by Article VIII of the 1967 UN Outer Space Treaty, at least when it comes to its legal effects (jurisdiction and control of the state of registry over the object) and stated as an obligation under the provisions of the 1975 UN Registration Convention. Now, this latter text defines the "space object" as including component parts thereof, as well as its launch vehicle and parts thereof. In the case of multiple unit CubeSats, each unit may be considered as an autonomous element, even though all units put together form a single spacecraft. Units integrated in a single spacecraft may come from different manufacturers, as well as from different countries.[10] It is not certain, especially in this latter case, that states will develop a continuous practice in registering the spacecraft as a whole and not each unit separately. This might raise issues when it comes to exercising jurisdiction and control over the spacecraft from which each unit would be subject to a different state's authority.

UNCOPUOS is aware of small satellites regulatory issues: they have been addressed at several occasions, in particular during the 53rd session of the UNCOPUOS Legal Subcommittee, in March 2014, by the ECSL/IISL opening Symposium.[11] This symposium

---

10 The QB50 project illustrates the situation where a single mission could be placed under the authority of several states; each of them entitled to consider the mission as performed under its jurisdiction and national legislation. This could notably lead, in such a case, to multiple states of registry for each cubesat, integrated or not as a unit in the spacecraft.

11 Presentations made at the symposium can be found on the following webpage: http://www.unoosa.org/oosa/COPUOS/lsc/2014/symposium.html.

was usefully complemented by a conference organized by University of Vienna, in the margin of the subcommittee's session [1].

Again, the process of democratization of outer space was clearly assessed: we may anticipate a future where every service provider would have its own constellation in orbit. In the longer term, why not imagining individuals operating their own CubeSat from their cell phone? This would raise questions with regard to the sustainable use of outer space as a natural resource and, in the meantime, as an environment [2], which is certainly a major part of the terms of reference of UNCOPUOS. A coordinated approach at international level is required in order to implement a coherent policy with regard to the use of orbital resources.

Besides the legal and regulatory aspects under its mandate, UNCOPUOS also reviews the implementation, by the UNOOSA of the UN Space Applications Programme. Since its launch in 1971, this program serves as a framework for coherent educational initiatives and projects toward capacity-building in the field of space sciences, space technologies, and space applications. Considering the purpose of the UN Space Applications Programme, in particular with regard to developing countries, it is obvious that the development of nanosatellite technologies has been followed with great interest.

More generally, the scientific and technological potential associated to small satellites has been reviewed by UNCOPUOS at several occasions since mid-1990s [3]. The idea that developing countries could be more than "users" of space applications, that is, buy or build their own space systems, have it launched, and operate it, was not obvious until that time. The evolution of mind can be spotted between the lines of the UNISPACE Conferences[12] reports. Of course, the world dramatically changed between UNISPACE II (1982) and UNISPACE III (1999) [4], starting with the computer and telecommunications exponential development (microprocessor, telematics, web, etc.), then moving from the collapse of the Soviet world to a post-Cold War Era guided by the motto of globalization. Satellite technologies were at the same time more complex and more accessible: more complex because of the growing sophistication of payloads and systems, more accessible because of the cheaper and smaller components available to a larger public and the "opening" [13] of the launchers market.

All in all, the question of small satellites, for example, nanosatellites, within UNCOPUOS illustrates their double face: on the one hand, they offer promising, affordable solutions for small applications systems, demonstration projects, and educational initiatives, while on the other hand, they represent a risk in terms of orbital population and "debris" monitoring, as long as they are not subject to a clear and global regulatory policy.

29.3 International Telecommunications Union

The International Telecommunication Union (ITU) is the competent UN specialized agency to deal with the assignment of radio frequencies and their use at international level. Its establishment and its missions have been provided for by international treaties,

---

12 The United Nations has organized three world conferences on outer space: UNISPACE I in 1968, UNISPACE II in 1982, and UNISPACE III in 1999.
13 By "opening," we mean here the multiplication of launch service providers in a competitive environment independent from the discretionary control of governments.

and compliance with the principles and rules of the ITU system is therefore mandatory. With the number of nanosatellites rising, ITU has faced new issues.

Basically, the purpose of the ITU system is double:

1) Allowing a fair and equitable sharing of orbital resources (radio frequencies + slots) among all nations and for different kinds of utilizations.
2) Avoiding harmful interference at international level.

Considering these double purpose nanosatellites must once again be regarded from different perspectives. They are often used in the frame of educational projects or scientific missions whose results benefit a large community of users, notably from developing countries or countries without independent space capacity. To that extent, they should be considered as a tool for the development of space technologies to the benefit of all. In the meantime, nanosatellite projects seem to have difficulty to cope with the whole (and rather heavy) administrative road through ITU and national instances. In fact, the responsibility of national authorities may be pointed out, considering that ITU Radio Regulations (no. 11.2) provide for the obligation to notify the Radiocommunication Bureau.

The use of amateur radio frequencies by a large number of scientific or educational nanosatellites is also at stake. That part of the spectrum was originally dedicated to allow amateur applications and experiments in a restrictive manner. It is doubtful that the increasing use (and, eventually, saturation) of such frequency bands corresponds to the intention of the international regulator. Other bands dedicated to special purpose (e.g. experimental stations) are also used for nanosatellite missions, raising similar issues [5].

From the nanosatellite community, issues are raised too, especially with respect to the heavy and long procedure to obtain international recognition of the frequency assignment that does not always fit the calendar of the projects. ITU Radio Regulations (no. 9.1) stipulate that a publication of the frequencies considered must happen prior to the assignment by national authorities in a time lapse of 5 years: no more than 7 years and no less than 2 years before the entry in function of the satellite. Integrating ITU constraints into small satellite projects is a tough challenge, but after all, it is part of the learning process. That being said, a path of reflection for the future could be for ITU to elaborate procedure and requirements dedicated to nanosatellites, starting with a legal definition of the term. Ideally, this reflection should be led in close consultation with UNCOPUOS in shaping the "clear and global regulatory policy" we were referring to in the preceding text.

29.4 Other United Nations Agencies and Bodies

By nature, an organization such as the United Nations is called upon to play an active role in the exploration and use of outer space. From huge missions to small field projects, the United Nations has integrated the space dimension in its institutional jobs: preventing collision with asteroids, foreseeing human presence and activities on extraterrestrial ground, deterring weaponization, predicting natural disasters and mitigating their consequences, helping better communication among people, and improving health, safety, and awareness are some of the tasks that the organization carries out in outer space.

Besides the two main organizations active in the space area (UNCOPUOS and ITU), a large number of agencies, bodies, or programs have integrated space solutions and technologies in their usual business. The relation with outer space and satellite applications varies from one organization or body to another.

In 1980, an interagency consultation mechanism was set up within the United Nations. The meetings are annual and serve as an area of exchange of information and consultation between various organizations and bodies of the UN family involved in the use of space technologies and applications. Today, some 26 organizations, bodies, and program executives are participating in this mechanism.

United Nations' various missions in peacekeeping, disaster management, sustainable development, and natural resource management are supported by the use of geospatial data, including satellite data. However, due to the fact that such data are provided through partnership networks with satellite system operators and service providers, it is difficult to assess the role of small satellite and nanosat technologies in connection with United Nations' needs.

29.4.1 UNITAR/UNOSAT

The operational Satellite Applications Programme (UNOSAT) set up by the United Nations Institute for Training and Research (UNITAR) aims at collecting satellite data from various sources in order to provide a quick response to crisis or natural disaster management. To that extent, small satellite systems may provide solutions.

UNOSAT works in close partnership with data providers as well as field actors, such as nongovernmental organizations with humanitarian purpose. Like it is the case with the International Charter for Space and Major Disasters,[14] which was initiated by ESA and Centre national d'études spatiales (CNES, French government space agency) with the support of other space agencies and the involvement of regional coordinators, small satellites resources can be seen as a valuable complement to large satellites. The idea is to establish a responsive global network through which accurate data can be provided to end users in a timely and flexible, nonbureaucratic manner.

29.4.2 UNESCO

Similarly to what was achieved 10 years later within UNCOPUOS with the adoption of the United Nations General Assembly Resolution on Principles Governing the Use by States of Artificial Earth Satellites for International Direct Television Broadcasting,[15] UNESCO has adopted recommendations on the use of satellites for broadcasting of cultural or educational content.[16] Since the late 1960s, satellites have been considered by UNESCO as a topic for education and economic development.[17]

---

14 See Section 29.4.5.
15 UNGA Resolution 37/92 of 10 December 1982.
16 Declaration of Guiding Principles on the Use of Satellite Broadcasting for the Free Flow of Information, the Spread of Education and Greater Cultural Exchange, done on 15 November 1972.
17 See also a concrete application with UNESCO Intergovernmental Oceanographic Commission's Sea Level Station Monitoring Facility (http://www.ioc-sealevelmonitoring.org).

In 2010, UNESCO initiated a project of development of a dedicated microsatellite (100 kg) with scientific payloads (30 kg) [6]. The spacecraft was procured by UNESCO and industrial partners from Russia and USA were entrusted with the development and the launch of the satellite. The project was meant to promote science and education, as well as public awareness. It was remarkable to the extent that the satellite would be the first "United Nations spacecraft," unlike satellites used so far in partnership with national operators and data providers.

Interestingly enough, the United Nations Organization has not deployed satellite systems of its own so far. This situation can be easily explained by the cost of such projects and their redundancy with existing systems. But the propagation of nanosatellites for operational purposes may offer opportunities in the future for UN agencies to acquire their own assets and reduce their dependency with regard to external supply of data and products.

29.4.3   UNDP

The United Nations Development Programme is a frequent user of satellite services and products. One example among others is its involvement in the "SatElections" partnership project, together with the European Space Agency, the European Union, and the Independent Electoral Commission of the Democratic Republic of Congo. The project was financially supported by the governments of Belgium, Luxembourg, and Italy.

Once again, it can be noticed that UN programs are basically in need of user-oriented applications developed through small targeted projects carried on in the frame of "intelligent" partnerships involving all actors of the value chain. This economic and technical environment seems particularly favorable to the use of nanosatellite systems.

29.4.4   UNEP

Very much like UNDP, the United Nations Environment Programme has integrated satellite data as a resource in the implementation of its programs and the achievement of its goals. Here again, the supply in space products goes through partner institutions or commercial channels. But small technologies are considered as an important link between global observation systems and local reporting. For instance, the use of air drones is part of the UNEP Global Environmental Alert Service and allows the acquisition of focused imagery at local scale.[18]

29.4.5   Other UN Agencies and Bodies

Other specialized agencies of the United Nations have recourse to satellite services and products in implementing their mission: FAO (Food and Agriculture Organization), IMO (International Maritime Organization), WMO (World Meteorological Organization), WHO (World Health Organization), IAEA (International Atomic Energy Agency), and ICAO (International Civil Aviation Organization). What has been said for one is true for the others: those world institutional users might benefit from flexible and affordable solutions

---

18 See http://www.unep.org/pdf/UNEP-GEAS_MAY_2013.pdf.

provided by nanosatellites, today as well as in the future, considering the technological progress that makes such spacecraft more and more responsive to users' needs.

Before leaving the big family of the United Nations where examples of uses of outer space are countless, two noninstitutional initiatives deserve to be mentioned here: UN-SPIDER and the International Charter on Space and Major Disasters.

UN-SPIDER is a platform for space-based information for disaster management and emergency response established by UNOOSA (see the preceding text). It serves as a focal point for the access and transmission of data based on parametric situations (type of data, type of hazards, coverage, etc.).

The International Charter on Space and Major Disasters is an operational mechanism through which national civil protection authorities may call for the support of live satellite data acquired after a dedicated (re)programing of satellites. This mechanism allows providing field actors with accurate data in a minimum of time after the occurrence of the disaster, through the activation of a network of satellite operators. The relevance of nanosatellites in disaster prevention or mitigation has been well demonstrated over the past years.[19] Institutionalization of response mechanisms at international level shows how small satellites may serve as tools for intergovernmental cooperation in this field.

29.5 Non-UN Organizations

It would be unrealistic to cover all international intergovernmental organizations in the world in this chapter, even though space technologies have become an indispensable resource to governmental policies. Considering the topic of nanosatellites, we propose to focus on two intergovernmental organizations whose activities have or may have an impact on this type of space technologies.

29.5.1 UNIDROIT

The International Institute for the Unification of Private Law, headquartered in Rome, Italy, has developed a set of conventions and agreements meant to harmonize private law rules among state parties. In the field of financial guarantees on investment in mobile assets, such as railway rolling stock, aircraft, or space systems, UNIDROIT has adopted a convention allowing the creditor to recover its money through uniformed procedure in all state parties. A specific protocol dealing with space assets was adopted in Berlin in 2012.[20] Some criticisms from governments, as well as from the industrial sector, have been expressed about this instrument. Until now, four countries have ratified the protocol.[21]

---

19 Cf. projects such as Quakesat, developed by QuakeFinder (Stellar Solutions). See also the 2010 UN/Austria/ESA Symposium on Small Satellite Programmes for Sustainable Development: Payloads for Small Satellite Programmes, notably the presentation by Prof. Sir Martin Sweeting: Microsatellites: Moving from Research to Operational Missions.
20 Protocol to the Convention on International Interests in Mobile Equipment on Matters Specific to Space Assets, held in Berlin on 9 March 2012.
21 Burkina Faso, Germany, Saudi Arabia, and Zimbabwe (see www.unidroit.org).

A remarkable item defined under the protocol is the notion of "space asset."[22] Under that term, the object of the (international) guarantee may be the satellite as such or any part thereof that can be separately registered. As previously explained, registration of nanosatellites made of separate units (CubeSats) cannot be excluded in the absence of regular practice in that respect. But in the context of the UNIDROIT Protocol on Space Assets, the question may take another dimension: such a practice would allow submitting each unit to its own legal regime, including financial guarantees applying to it. This may result in a complex situation where several elements of the spacecraft would fall under different rules and be subject to different rights or interests.

The UNIDROIT Protocol may be seen as an instrument to facilitate the funding of space projects, in particular in developing nations. However, it must be used with caution, especially to the extent that it may end up with a legal fractioning of space assets, which may have consequences to third parties.

29.5.2 NATO and Military Nanosatellites

Military and defense-related purposes contrast with the usual perception of scientific and educational (even ludic) vocation of nanosatellites. However, the interest of states in developing very light satellite systems as a tactical support for their defense operations provides another illustration of the operational capacity of nanosatellites.

As we have seen it, United Nations peacekeeping missions are already relying upon the use of geospatial data, including satellite imagery, as well as of satellite telecommunication and positioning. Apart from that, national military programs have integrated nanosatellites in their space strategy. For instance, the US Department of Defense has identified several areas of missions where CubeSats may provide an appropriate solution (such as space situational awareness, offensive and defensive counterspace operations [7], or weather forecast [8]).

The growing involvement of nanosatellites in the future of defense missions depends on several factors related to the development of technology, such as:

- The accuracy of data.
- The reliability of propulsion.
- The on-board autonomy.

The North Atlantic Treaty Organization (NATO) is certainly the most representative of all defense-related intergovernmental institutions in the world. Outer space capacities have

---

22 Art. I, 2, (k), of the Protocol: "space asset" means any man-made uniquely identifiable asset in space or designed to be launched into space, and comprising

(i) a spacecraft, such as a satellite, space station, space module, space capsule, space vehicle, or reusable launch vehicle, whether or not including a space asset falling within (ii) or (iii) below;

(ii) a payload (whether telecommunications, navigation, observation, scientific, or otherwise) in respect of which a separate registration may be effected in accordance with the regulations; or

(iii) a part of a spacecraft or payload such as a transponder, in respect of which a separate registration may be effected in accordance with the regulations;

together with all installed, incorporated, or attached accessories, parts and equipment, and all data, manuals, and records relating thereto.

been integrated in NATO's strategy at an early stage, through the availability of national capacities developed by NATO member states, such as the USA and the United Kingdom (Skynet satellites). From 1991 to 1993, NATO had developed its own satellite communication capacity (NATO 4A and NATO 4B), on the basis of the Skynet 4 generation. Today, NATO 4 satellites have been decommissioned and NATO relies on the national capacities of its member states (i.e. France, United Kingdom, and Italy) to provide strategic access to satellite communication systems.

There are no real indications today that NATO considers direct investments in the development of CubeSats for its missions, although what has been said in the preceding text about military nanosatellites remains basically valid for what concerns an intergovernmental organization. The role of NATO in the establishment of the Von Karman Institute, in Sint-Genesius-Rode, in the suburbs of Brussels, must be acknowledged considering that the institute has acquired a level of excellence in the development of CubeSats, notably with the QB50 project. This (civilian) project, funded by the European Commission and consisting in the simultaneous launch and sequential deployment of roughly 50 CubeSats for scientific purpose, has been regarded as highly significant in the way that CubeSat missions should be implemented and monitored not only from a technical but also from a regulatory standpoint. Even though the Von Karman Institute is now an independent and autonomous nonprofit academic institution, it was established in 1956 on a proposal by the famous physicist Theodore Von Kármán, then chair of NATO Advisory Group for Aerospace Research and Development. Indirectly through its historical involvement, NATO has thus contributed to the development of a European capacity in CubeSat technology.

29.5.3 Intergovernmental Agreement on the International Space Station

Obviously, this international agreement, which qualifies as a treaty but does not establish any legal entity, has no connection with nanosatellite development or operation. Nonetheless, the use of the ISS as a launch platform for nanosatellites, picosatellites, or femtosatellites has been considered in several occasions. The in-orbit positioning of very small objects from the ISS happened with various objects, such as a golf ball or a spacesuit. So-called "hand launches" have been performed since 1982, using different orbital platforms (especially by the Russians). There is no need to say that nanosatellites are natural candidates for this kind of operation. Several project teams in the world have studied this technique [9, 10].

The legal framework of the ISS Intergovernmental Agreement 1998 does not provide for any rule on the use of the station as an orbital launch platform. However, several provisions of this agreement are relevant to the use of the station by the partners.[23] Article 9 of the agreement stipulates that the details and modalities of use of the infrastructure and equipment of ISS are agreed among the representing agencies of the partners (NASA, ROSKOSMOS, JAXA, CSA, and ESA). Any use of the infrastructure or equipment by a non-partner or by a private entity under a partner's jurisdiction must receive prior approval

---

23 The "partners" in the ISS are the USA, the Russian Federation, Japan, Canada, and the member states of the European Space Agency.

from the other partners. The use of ISS is limited to peaceful purposes. Another requirement is the fact that the use of ISS by one partner must not seriously affect the use by other partners or interfere with it.

Apart from those principles, the ISS Intergovernmental Agreement leaves open the legal issues raised by the launches performed from ISS. Those issues are:

- The liabilities for the damage caused by the object launched.
- The registration of the object launched.

Article 16 of the ISS Intergovernmental Agreement excludes any action in liability from one partner to another in the context of activities covered by the agreement. This waiver of liability is obtained through the definition of "Protected Space Operations," which reads as follows:

"The term 'Protected Space Operations' means all launch vehicle activities, Space Station activities, and payload activities on Earth, in outer space, or in transit between Earth and outer space in implementation of this agreement, the MoUs, and implementing arrangements. It includes, but is not limited to:

(1) Research, design, development, test, manufacturing, assembly, integration, operation, or use of launch or transfer vehicles, the Space Station, or a payload, as well as related support equipment and facilities and services.

(2) All activities related to ground support, test, training, simulation, or guidance, and control equipment and related facilities or services.

"Protected Space Operations" also include all activities related to the evolution of the Space Station, as provided for in Article 14 [Evolution].

"Protected Space Operations" exclude activities on Earth that are conducted on return from the Space Station to develop further a payload's product or process for use other than for Space Station related activities in implementation of this agreement."

At first sight, this definition seems wide enough to cover the case of orbital launch from ISS. But the question remains whether such use actually falls under the scope of the agreement, as required by the definition itself. For instance, would the commercial launch of a nanosatellite procured by a non-partner entity and performed from an element of ISS qualify as a "Protected Space Operation." We tend to bring a positive answer to that question. Indeed, even though the launch of commercial nanosatellites from ISS is not, per se, an activity related to the development, the deployment, the evolution, or the operation of ISS, it participates in its utilization. Furthermore, such a launch would need to be approved by all partners, making it an authorized activity and, thereby, fall under the scope of the ISS Intergovernmental Agreement. Liability issues, such as the damage caused subsequently by the object to ISS, for example, by an orbital collision, can be settled through the agreement or implementing arrangements between the states and parties concerned.

When it comes to liability with regard to third parties outside the scope of the ISS Intergovernmental Agreement, things become more complex. The object (e.g. nanosatellite) launched from ISS may cause a damage in outer space or on the ground. In the former case, if there is evidence that the damage is due to a fault (e.g. a false maneuver at the moment of the launch resulting in the object acquiring a wrong orbit), the "launching state(s)" may be held liable. A "launching state" is defined by the United Nations treaties on outer space as any state having either performed the launch, procured the launch, or provided its territory or its facilities to the launch of the object having caused the damage. According to this definition, the launch of a nanosatellite from ISS may involve several states:

a) The state having registered the ISS element from which the object is launched.
b) The other state partners if ISS is considered as a single, integrated facility used for the launch.
c) The state having procured the launch of the object.

Moreover, the question remains whether the launch of the object should be considered as being performed from Earth to ISS, or from ISS to outer space, or even as a multiphase operation consisting in two successive launches. According to those interpretations, the state(s) involved in the primary launch (Earth to ISS) may or not be held liable as "launching state." This question is particularly relevant in the case of a damage caused on the ground (or to a flying aircraft) where absolute[24] liability is applicable. In the case of damage in orbit, the fault will likely be attributed to the person having performed the maneuver from ISS.[25]

The second issue is the registration of the object launched from ISS and is directly connected to the identification of the "launching state." Only (one of) the "launching state(s)" of the object (i.e. nanosatellite) may register the object according to the United Nations treaties on outer space. By doing this, the state exercises jurisdiction and control over the object and on-board the object. It also comes in frontline as far as the liability is concerned, being formally identified as (one of) the "launching state(s)" of the object on the United Nations Secretary General's register of space objects.

Arrangements between states involved may provide solutions to this situation: the launch procurement should foresee either the registration by the state of origin of the nanosatellite as a payload carried from Earth to ISS (this constitutes the most likely arrangement), or the registration by the state having provided the launch, either from Earth to ISS or from ISS to outer space.

Issues over liabilities and registration are very intricate and complex. The importance of prior arrangements between states involved must not be underestimated. In the specific case of nanosatellites, practice has shown that such projects being carried out by nontraditional space business actors often lead to ignore many legal requirements that can have an economical or technical impact on the project. In the case of ISS activities, it can be expected that governmental agencies in charge of the utilization policy of the station will take due care of legal and regulatory assessments, but many nanosatellite projects do not integrate such procedures within their protocols and roadmaps.

29.6 Main Non-European Spacefaring Nations

Our review of major international intergovernmental organizations' space policies or activities shows a close connection with national policies and programs of their member states.

---

24 "Absolute liability" means objective (without fault) and unlimited.
25 Furthermore, the final destination ("positioning") of the nanosatellite is only attained once the launch from the ISS is performed, not before.

We can even affirm that, apart from organizations whose core business is the development of space technologies and applications, none of them have launched a consistent program aiming at acquiring an autonomous space capacity for the purpose of supporting their institutional mission. As previously explained, this is due to cost-efficient policies, avoiding redundancy, and giving priority to existing assets.

With the advent of nanosatellites and their growing operational capabilities, it is not unreasonable to foresee an evolution in the institutional space services. Organizations may be recognized by states—and, at the first place, their own member states—as independent entities, pursuing their own goals and needing their own tools, independently from those of the governments.

This review must be complemented by that of the national policies, which might bring interesting clue on what we should expect as far as the development of nanosatellites is concerned. The purpose of this section is certainly not to provide a comprehensive description of national policies with regard to nanosatellites (if such policies exist at all), but rather to identify some trends through projects implemented at national level.

29.6.1 USA

NASA has already a strong experience in nanosatellite development. Through a dedicated Small Spacecraft Technology Program, NASA collaborates with universities in some 13 projects (2013) in various nanosatellite technology areas, such as communications, navigation, propulsion, power, science instrument capabilities, and advanced manufacturing.

In the same program, NASA developed a Nanosatellite Launch Adapter System, designed to "increase access to space while simplifying the integration process of miniature satellites, called nanosats or CubeSats, onto launch vehicles."

Besides, the US Department of Defense is also investigating how the further technological development of nanosatellites can provide sufficient autonomy and capacity to involve them in strategic missions (cf. the preceding text).

29.6.2 Russia

In 2011, Russia and Israel concluded an agreement on the setting up of a joint center for nanosatellites ("RINI"). The interest of Russia as space veteran for the industry of a relatively newcomer—Israel—shows that nanosatellite technology is quite specific and does not always flow from traditional archetypes.

Besides, the Russian Space Agency, ROSKOSMOS, has also supported R&D in the field of propulsion means though a new type of engine without mass reaction emission, as well as of satellite orientation.

As a launch provider, Russia is also keen to adapt its offerings to new customers with smaller payloads but heavy requirements [11].

29.6.3 India

India launched its first nationally built nanosatellite in 2011. Named "Jugnu," the small device had been developed by a team of students. The project illustrates the difficulties that academic teams may have in coping with high standards (i.e. those of ISRO, the Indian Space Research Organization), while working with limited technical and financial means [12].

29.6.4   Canada

In April 2012, the first Canadian Nanosatellite Workshop was held in Quebec City. That workshop resulted in a set of findings and recommendations for a future national policy promoting the development and the use of nanosatellites in a both competitive and cooperative context [13].

In particular, the cost-effectiveness of nanosatellites was acknowledged with regard to environmental and fiscal constraints. The need to foster international cooperation and partnership was highlighted, as well as a need for a dedicated programmatic framework.

29.6.5   Japan

In 2014, Kagawa University conducted the STARS-II mission, using a two-piece nanosatellite in a "mother–daughter" configuration, one of the satellites being used to video-monitor the deployment of a tether in orbit [14]. The STARS-II spacecraft re-entered the atmosphere 2 months after its launch.

This mission illustrates the multiple solutions that combined nanosatellites may offer in orbit, forming one single system made of several devices. This feature might also be considered as relevant from a legal point of view when it comes to defining a registration policy for this kind of systems.

In 2002, Japan also launched the UNISEC initiative [15] as a nongovernmental organization dedicated to facilitating and promoting space development projects at academic level. Such projects include the design, the manufacturing, and the launch of small satellites and hybrid rockets. The ambition of UNISEC is to establish, by 2020, an international cooperative network connecting universities of more than 100 countries.

29.6.6   China

China has always been at the forefront of production systematization. The capability to manufacture nanosatellites according to an industrial pattern is no real challenge for the Chinese space sector. The effort concentrates on developing launching and positioning systems that would allow fair price and technically efficient services. Nanosatellites are viewed as a large-scale potential market.

Beihang University is working on a project of orbital injection of cluster flight of nanosatellites, taking into account their cooperative character as well as the need to optimize fuel consumption during the deployment phase [16].

29.6.7   Developing Countries

Following the previous considerations on the interest of sustainable development initiatives for small satellites, be it for their well-established value in capacity building or for their growing operational capacities, the attractiveness of nanosatellites for developing countries may seem obvious. However, we call for prudence with such an assumption.

First of all, the concept of "developing country" remains quite vague. Global economy does not provide for consensual criteria to circumscribe it. Poor countries with regard to GDP per capita might know a high rate of growth. Furthermore, in the space economy, traditional parameters are not necessarily applicable: countries such as Nigeria or Vietnam have established space policies and space programs, including the procurement of satellite system operations, while remaining with a very low Human Development Index. Promising cooperative projects are also undertaken at regional level, from which a multiplicative effect can be expected: the New Partnership for Africa's Development serves as an incentive framework for initiatives such as the African Resource Management Satellite Constellation (ARMC) in which Algeria, Kenya, Nigeria, and South Africa are participating.

Second, the needs of so-called developing countries in terms of space services and applications do not necessarily correspond to those provided through nanosatellites that have a narrower scope.

Third, any satellite system not only requires space infrastructure, but also ground segment with the means to operate the satellite and to collect and process data. This requires recurring financial resources as well as expertise and training.

We advise against a simplistic vision according to which nanosatellites would be the panacea to the "space divide" between industrial nations and developing nations. Instead, we would encourage a much better sharing of existing resources and cooperative programs for capacity-building, dedicated application development, and transfer of technologies.

That being said, it is true that some small satellite models have already been "commercialized": the PROBA class provides an example of microsatellites affordable to developing countries and providing solutions for their geospatial data needs. The big question is whether nanosatellites, in their various uses (science, observation, defense, etc.), will, alone, offer valuable resources to countries that do not enjoy corresponding infrastructure, network, knowledge, or operational capacities. Even nanosatellites require supporting infrastructure and operational expertise that is not (yet) available to all states throughout the world.

The temptation to see nanosatellites as a way to achieve an autonomous status in outer space is deceptive in itself. International cooperation remains the best way to enhance global sustainable progress and a more equitable benefit of space resources.

29.7 Conclusions

(1) We have observed a substantial evolution of nanosatellite missions from scientific and engineering education toward operational and applicative systems (remote sensing, in-orbit demonstration, integrated application, etc.). This evolution bodes well for a technological development of nanosatellite technologies and applications. The nanosatellite community extends now from scientists and students to service providers and institutional users.

(2) The overpopulation in orbit calls nevertheless for a careful assessment of regulatory and policy aspects: launch pace, orbital parameters, sharing of resources, tracking, etc.

Nanosatellites may share characteristics with space debris: regulating their operation is the governments' business and duty. This requires an in-depth reflection over the concept of rideshare and piggyback launches, over the development of dedicated launchers, including air-to-space launches (from an aircraft), as well as over end-of-life disposal. This also requires integrating a new type of actors in the space sector: small or medium emerging companies offering access to outer space at very affordable conditions.

So far, the mass of the satellite has not been considered as a relevant legal criterion. This might change in the future by identifying the spacecraft according to specific requirements with regard to its positioning, its tracking, or its operation (including the use of radio frequencies).

(3) To this end, international cooperation remains a key element in defining coherent national policies that would not allow forum shopping: small operators selecting the launch country with respect to the (absence of) applicable regulations. We must be cautious that smooth dissemination of and easy access to nanosatellite technology do not allow nanosatellite operators to escape the legal and regulatory framework applicable to their activity. International cooperation remains the best manner to impose technical standards integrating safety and security concerns.

(4) In the famous book, Small Satellites: Past, Present, and Future, Henry Helvajian (The Aerospace Corporation, California) offers prospective considerations on "The Generation After Next: Satellites as an Assembly of Mass-Producible Functionalized Modules." In conclusion to the same chapter, Helvajian proposes the idea of "defining the production unit not as the satellite but as an element of the satellite—the multifunctional module," as the "goal of the economies of scale" [17].

This vision opens a new landscape for the future: as we have witnessed the rise of the personal computer together with microprocessor technologies in the 1980s, should we expect PCs (personal CubeSats) to establish their reign over social economy through this century? This poses the basic question of our tendency to individualism versus our capacity to mutualize resources as a form of progress.

Planning a more equitable access to space is only half the way to sustainable progress. The second element lies in our ability to share benefits in a reasonable exploitation of wealth. Nanosatellites may play a positive role in this endeavor, to the extent they are considered for their actual efficiency, which implies assessing a cost/added value ratio.

References

1 Mayence, J.F. (2016). QB50: legal aspects of a multinational small satellite initiative. In: *Small Satellites: Regulatory Challenges and Chances* (ed. Marboe). Brill Nijhoff.

2 J.F. Mayence, 2010. "Toward a Space Environment Law?," Fifth Eilene Galloway Symposium, Washington D.C., (http://www.spacelaw.olemiss.edu/events/pdfs/2010/galloway-mayence-paper.pdf).

3 UNCOPUOS documents A/AC/105/611 of 2 November 1995 (Microsatellites and Small Satellites: Current Projects and Future Perspectives for International Cooperation), A/AC.105/638 of 12–13 February 1996 (Symposium on Utilization of Micro- and Small Satellites for the Expansion of Low-Cost Space Activities) and A/AC.105/645 of 9–13

September 1996 (Report on the United Nations/Instituto Nacional de Técnica Aeroespacial/European Space Agency International Conference on Small Satellites: Missions and Technology).

4 A/CONF. 184/L.13 and A/CONF. 184/C.2/L.7 on the conclusions from the UNISPACE III Technical Forum Workshop on Small Satellites at the Service of Developing Countries (https://cms.unov.org/llsulinkbase/contenttree.aspx?nodeID=243).

5 Jakhu, R. and Pelton, J. (2014). *Small Satellites and their Regulations*, 58. Springer.

6 D. Platt, 2010. "UNESCOsat, The United Nations' First Satellite," in Proceedings of the 24th AIAA/USU Conference on Small Satellites, SSC10-XII-9.

7 C. Galliand, 2010. "Study of the Small: Potential for Operational Military Use of Cubesats," in Proceedings of the 24th AIAA/USU Conference on Small Satellites, SSC10-III-2.

8 Jackhu, R. and Pelton, J. (2014). Small satellites and their regulations. In: *SpringerBriefs in Space Development*, 16. New York: Springer.

9 Moses, R. (2002). HAND—the throwaway satellite. In: *Proceedings of the Bath Royal Literary and Scientific Institution*, 46–47. Bath Royal Literary and Scientific Institution.

10 P. Anderson, C. Swensson, C. Fish, J. Martineau, 2011. "The ISS as a Launch Platform for Phenomena of Interest," in Proceedings of the 25th AIAA/USU Conference on Small Satellites, SSC11-VI-2.

11 Webb, G.M. (2002). Overview of the Russian launch possibilities for small satellites. In: *Smaller Satellites, Bigger Business?* (eds. Rycroft and Crosby), 257. Kluwer.

12 India Today, 31 October 2011: ITT-Kanpur Launches Nano Satellite Jugnu (http://indiatoday.intoday.in/story/iit-kanpur-launches-nano-satellite-jugnu/1/158074.html).

13 Small is Beautiful—Report of the First Canadian Nanosatellite Workshop, 23 April 2012, Quebec City.

14 Kramer, H.J. (2002). *Observation of the Earth and its Environment: Survey of Missions and Sensors*. Springer-Verlag.

15 Cf. technical presentation made by the Japanese delegation on 10 February 2015, at the 52nd session of the Scientific and Technical Sub-Committee of UNCOPUOS (http://www.unoosa.org/pdf/pres/stsc2015/tech-45E.pdf), as well as www.unisec.jp.

16 Wen, C., Hao, Z., and Gurfil, P. (2014). Orbit injection considerations for cluster flight of nanosatellites. *Journal of Spacecraft and Rocket*: 1–13.

17 Helvajian, H. and Janson, S.W. (2008). *Small Satellites: Past, Present and Future*, 853. AIAA.

30 III-3a

Economy of Small Satellites

Richard Joye

KCHK—Key Capital Hong Kong Limited, Hong Kong

30.1 Introduction

The increasing role played by small satellites[1] is fueling the growth of the overall economy of the space industry. Small satellites are at the center of major themes attracting interest and funding: affordable access to space, low-cost broadband communication, asteroid mining, mission to Mars, life in space, and other future trends.[2] Small satellites enable the emergence of new players in the space value chain, allowing private, usually small companies to play a greater role alongside traditional industrial conglomerates.

The different segments of the space value chain are however not equally being disrupted by the economy of small satellites, and the economic impact significantly differs among actors of the space industry.

In this chapter, a closer look at the space industry framework will be taken to understand how the different segments benefit from the growing importance of small satellites and what the real, tangible economic impact is.

30.2 Rethinking the Value Chain

Nanosatellites, microsatellites, and small satellites unarguably altogether represent a central element of the NewSpace value chain. They are the trigger for change, enabling existing actors to evolve and new entrants to propose innovative products and services to play a role in this growing market, representing a multibillion-dollar opportunity.

The space industry value chain has been and continues to be disrupted by new, nimble, small space companies that have replaced (or are aiming at replacing) as well as complement well-established industry leaders and, in some countries or regions, government

---

1 In this chapter, the term small satellites, unless specified otherwise, includes the broad range of spacecraft from nano or pico size (sub-1–1 kg) up to the 250–500 kg series. Unless discussion specificities of larger spacecraft, the arguments presented herein are valid for nanosatellites, which is the focus of this whole book.

2 Please refer to the chapter on "Future Trends" to learn more about that and how small satellites are playing a central role.

Nanosatellites: Space and Ground Technologies, Operations and Economics, First Edition.
Edited by Rogerio Atem de Carvalho, Jaime Estela, and Martin Langer.

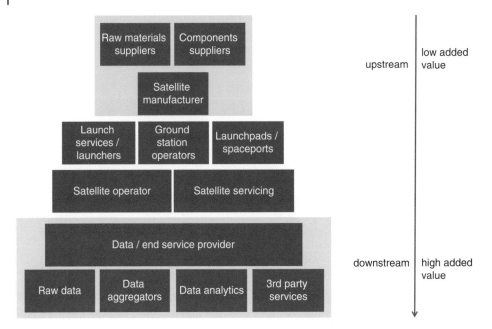

Figure 30.1 The space industry value chain.

institutions. This is to the extent that new players have now become integral components of the overall space framework, such as SpaceX, servicing the full spectrum of the industry.

But today, the overall economy of small satellites is still built on the interactions and the collaboration between actors of different types and sizes, combining traditional public and private companies, tier 1 space conglomerates, governments, academic and research centers, as well as, of course, NewSpace companies (Figure 30.1).

The utopic assumption praised by a young community of startup founders that a parallel value chain could be created and eventually would supersede the existing, traditional chain did not materialize yet. If it does, it will take time. Therefore, I am advocating the idea that NewSpace goes well beyond its current, widely accepted, definition, provided by NewSpace Global as "[…] a global industry of private companies and entrepreneurs who primarily target commercial customers, are backed by risk capital seeking a return, and seek to profit from innovative products or services developed in or for space." [1].

Certainly, the NewSpace category includes private enterprises backed by venture capital funding. But NewSpace represents a broader ecosystem within which companies of all kinds benefit from change and evolution from the traction generated by small satellites. This generation is witnessing the creation of a greater ecosystem where interests converge; a new, fully integrated and expanded value chain that no other industry has witnessed before.

30.3 A Hybrid Small Satellite Value Chain

A stand-alone, fully integrated NewSpace value chain, independent from traditional industry participants, is not achievable at the moment, and most probably for decades to come.

Small satellites are enabling rapid change, but some barriers prevent complete disruption, thus only creating a realignment of industry participants.

Barriers are in place because of the history of the current system and its participants and of 70+ years of building the space industry, still predominantly dominated, on a global scale, by governments and large conglomerates.

The following analysis of the elements that create a resistance to the development of an end-to-end small satellite economy is not fully applicable to all markets. It is highly relevant to North America, Europe, and Japan, but some regions and countries, for instance China and Russia, are following different paths due to the nature of their own space industries.

30.3.1 Irreplaceability of Key Players

Not all participants[3] of the space value chain can easily be replaced. Some segments of the ecosystem are not easily disrupted. Many sub-segments and their actors can benefit from the economics of small satellites, even though they are not focusing on the smallest types of satellites. But they are not building a specific economy from it, rather integrating the specificities of smaller spacecraft to capture the market demand, in ways that this does not impact their core, more profitable operations.

The impossibility to see players dedicated to small satellites in all segments of the value chain creates some bottlenecks and obstacles to faster growth. The most obvious bottleneck and segment to challenge is the one of the launch service providers.[4] There exists a lack of availability of affordable launches (or dedicated launch systems) to give easy, fast, frequent, and cheap access to space. To date, no NewSpace company has been able to consistently place nanosatellites to small satellites in orbit and make it a viable business model; SpaceX vehicles, for instance, have become heavier and their core business is far from being centered around nanosatellites to small satellites. Existing vehicles marketed as dedicated to the small satellite market are either too expensive, lack of launch opportunities, or are not truly dedicated to this segment of the market.[5] Other dedicated vehicles are still in development phase[6] or have failed.[7] We see some vehicles getting closer to commercialization at a larger scale, such as Rocket Lab, however they still have to build a solid track record and confirm the viability of their business model.

---

3 Not in the sense of a particular company or entity, but rather a group representing specific interests and realizing a specific role in the overall value chain.
4 With focus on manufacturers, operators, and their systems (i.e. Arianespace, United Launch Alliance (ULA), Soyuz, MHI, etc.).
5 For instance, European Space Agency's (ESA's) Vega is expensive and addresses the largest small satellite segment and the mid-market. Epsilon in Japan is also not very well suited for the nanosatellites to small satellites segment. Pegasus is not offering launch windows and is not affordable. Other systems are mainly ICMB/missiles reconversion into civil rockets, such as Minotaur I, Taurus, Super Strypi, Unha in South Korea, and Shavit in Israel.
6 Still in development or pre/early-commercial phase: Virgin Orbit LauncherOne, Rocket Lab Electron, LandSpace (China) LS-1, Generation Orbit GoLauncher 1, Interorbital Systems NEPTUNE, and several others.
7 The following are defunct companies and/or systems: XCOR Lynx, Swiss Space Systems SOAR, Firefly Space Systems (even though all its assets were recently acquired), Garvey Spacecraft Corporation (even though the company has recently been acquired), as well as many projects initiated or developed by tier 1 industry players (Lockheed Martin, Boeing, Khrunichev, CALT, etc.).

This specific segment of the industry will be discussed later in this chapter and how the economy of small satellites is impacting launch service providers, but at this stage it is worth mentioning that the lack of NewSpace launch system manufacturers and operators is a moderate threat to the rapid growth of the small satellite market.

This segment is presently being serviced by existing, heavy launch system. Launch opportunities are available as piggyback options on rockets and systems, such as Falcon, Soyuz, Delta, Atlas, Polar Satellite Launch Vehicle (PSLV), and even Ariane 5, H-IIB, or Long March. PSLV, a large rocket, has for instance the record in respect to the number of nanosatellites launched in 2017, with 101 nanosatellites launched [2].

Many of my fellow space industry professionals following the small satellite market would argue that this is just a matter of time before a greater disruption appears. That several NewSpace companies have tried to build and operate launch systems dedicated to the small satellite world, and more will follow, but failure overshadows success. Companies such as XCOR, Swiss Space Systems, or Firefly are now defunct. Promising companies are yet to demonstrate launches and a successful commercial activity with a strong track record. From New Zealand to Sweden, from China to the USA, progress is ongoing.

Some would argue that because space is tough and expensive, to demonstrate and then successfully operate a dedicated rocket is just a matter of time and money. But the economics of dedicated launch systems may simply not work.

In the short to medium term, the only viable option for the small satellite industry is to work, directly or indirectly, with existing, traditional rockets. Mainstream launch system operators are happy to outsource the handling of nanosatellites and microsatellites to space brokers and to some extent this applies to small satellites as well. While brokers and integrators have always been active in the space industry, new types of launch-related service providers have emerged. Companies are proposing turnkey solutions, telling end-users to solely focus on their core businesses. CubeSat-based platforms are offered from one unit to multiple units, with or without payloads, inclusive of the launch itself or not. NewSpace brokers are servicing end customers with an end-to-end service and are able to offer rideshare opportunities, often prebooking rides well in advance.[8]

New players will emerge, completing the existing offering in respect to launch service providers. NewSpace brokers, focusing on the smaller segment, are addressing a real demand, but there is a growing need for dedicated launch vehicles.

30.3.2 Interdependencies Between Small Satellite Industry Players

In any ecosystem in which the industrial chain depends on purchase orders and proactive customers, dependencies can be seen. But in the emerging NewSpace framework, interdependency between companies is a driver for inefficiencies and slower growth. Certainly, NewSpace is a nascent industry that is at the embryonic stage of its development. But the close relationships between groups of companies, all at the same stage of their R&D and funding process, is not often seen in other industries.

---

8 NewSpace brokers, such as Clyde Space and Innovative Solutions in Space (ISIS).

30.3.2.1 Startups Doing Business with Startups[9]

Many promising launch service companies have signed MOUs[10] and contracts with new satellite manufacturing ventures or end users. Often with the anticipation that such agreements, associated with public announcements, would boost investors' confidence and interest, meaning funding. But, in many cases as it has been demonstrated, such agreements are cashless, and one of the two parties becomes insolvent or ceases to exist before the contract generates a positive outcome. A lot of noise among NewSpace startups has been seen, with little tangible and credible execution. I have witnessed over a dozen NewSpace companies, in rather advanced stage, going out of business and impacting this fragile ecosystem.

To summarize the value chain and illustrate the problem, let us consider only three actors: the satellite manufacturer, the launch service provider, and the customer using the data or the service. It will be assumed that the manufacturer of the satellite is also the operator of the spacecraft. And, as it is often the case with the smallest satellites, it is also the company servicing the customer, getting and processing raw data received from the satellites to sell a semi-finished product to the customer, who will package a final product for its market. The satellite manufacturer has a concept or prototype (at best) and needs funding to develop a working platform and test it in space. Whether it manufactures the whole spacecraft or acquires a platform and develops the payload, it does not matter much. To secure investors, the manufacturer needs to demonstrate that it has a viable business model. That the hardware and the service it intends to offer are attracting customers. For this, the manufacturer needs to enter into an agreement (usually a MOU or a precommercial contract) with a customer. The customer is at the end of the value chain (before selling services or products to the end users) and can only sign a MOU because (i) as a NewSpace venture it has no cash to affect a commercial contract, and (ii) it needs to make sure the satellite will be placed in orbit and will work as planned before engaging hypothetical financial resources. To mitigate this risk of nonperforming, the manufacturer needs to sign a launch contract to confirm to investors and to the customer that it has a window to place the satellite in orbit and perform the contract. But lacking cash and operating in the small satellite segment, the manufacturer cannot easily sign a launch agreement with a traditional launch company. It may enter into an agreement with a space broker or, as seen on multiple occasions, with a company developing a dedicated launch system. Such company is usually at pretesting phase and requires additional funding to develop, manufacture, test, and operate its launch system. To show investors that they have "sold" launches can add some value.

But this is a very fragile chain. If one player gets delayed or stops its program, the whole chain is worthless.

Most of small satellite ventures (from manufacturers to launch vehicles and data companies) are not fully funded and require cash inflows to deliver their program, especially in non-US markets where funding for NewSpace companies is very difficult to secure.

Considering the last 5 years of the space industry, we can illustrate, in simplified terms, the probability of success or failure for the different types of players, considering whether

---

9 In the real sense: undercapitalized, prerevenue companies with, at best, a proof of concept.
10 Memorandum of Understanding.

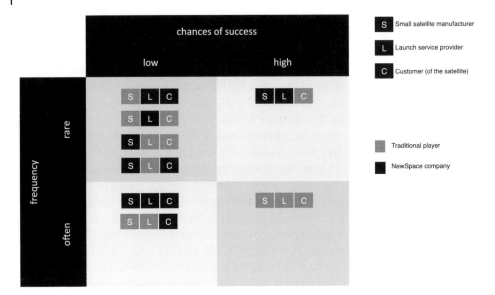

Figure 30.2 Mapping of the three main categories of players and the outcome of their agreements.

they are traditional actors or NewSpace companies. Agreements signed or announced (represented as frequency) will be showed in the following chart[11] (Figure 30.2).

New ventures that are cash-rich[12] and backed by large investment funds[13] or wealthy individuals or that are quickly able to obtain robust contracts (especially with the likes of NASA, Defense Advanced Research Projects Agency (DARPA), or the Department of Defense (DoD) in the USA) can position themselves as credible players in this ecosystem. It is worth mentioning companies such as Planet, O2b (now acquired), and SpaceX in its earlier years.

Some encouraging achievements outside of the USA have been seen, with companies in more advanced stage signing agreements with each other while receiving funding from investors. But, because of the complexity of the space value chain, some alliances and partnerships are unnatural. The space value chain has only been partially challenged and disrupted by the economics of small satellites.

30.3.3 Some Segments are Passive or Only Planting Seeds

The overall value chain of the space industry benefits from the growing attractiveness of small satellites. The previous section illustrates that launch service providers could see the lower end of the segment[14] as a distraction, nevertheless they benefit from it, that is, additional revenues "filling space" in the rocket where capacity exists beyond the primary payload(s). The operators of Atlas, Soyuz, or Mitsubishi H2 rockets do not need to embark

---

11 This illustrative chart is simplified to give readers a general perspective on what the space industry has achieved with NewSpace players. It does not have the granularity required to offer a full perspective.
12 Usually operating in the US.
13 Predominantly venture capital.
14 Mainly picosatellites to microsatellites.

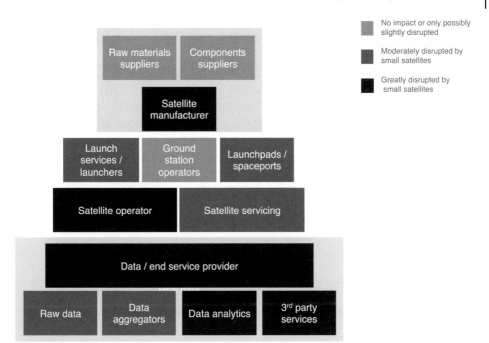

Figure 30.3 Measuring the disruption on the industry segments.

six or twelve CubeSats to survive. A launch shared with nanosatellites (even if in some cases was talking about a ride to the International Space Station (ISS) from where nanosatellites are deployed) adds risks, complexity, weight, and the risk–reward ratio is not necessarily positive. Yet, traditional space companies carry such spacecraft to space, for little money. Other factors influence them, such as market dominance, politics, national space agencies' agendas, technological evolution, or public funding.

The overall value chain can be looked again and the impact that small satellites have on each segment can be assessed (Figure 30.3).

Some kinds of players are being replaced or will be replaced by NewSpace companies, since some participants will not remain competitive against small satellites or will miss the business opportunity that is created.

There are also critical elements of the value chain that require substantial amount of money to exist with no real profitable business model. The economy of small satellites is benefiting them, but has, at the moment, very limited direct economic impact on them.

Ground monitoring stations are a good example of a full segment that benefits from the overall growth in the space industry, but do not directly and actively participate in any particular shift toward smaller spacecraft. They simply exist in the value chain and service new customers. Certainly, new monitoring services have been developed for smaller platforms but, overall, this segment has not changed much during the last decade.[15]

Launch facilities are critical actors of the space industry, even if their role is often overlooked while discussing the economics of space, because they are only recognized

---

15 Some companies are addressing this segment, on the service side (Leaf Space, BridgeSat, KSAT Light).

as infrastructure. But without launch pads, no launch can happen. And, at the moment, no horizontal take-off system exists in operation or shows a credible path to commercial success during this decade.

Space launch complexes are, in most cases, government-owned or, at least, government-funded infrastructure. In most of countries, they interact with the military and with civilian agencies and, in few countries, with private launch companies. They do or will benefit from the increasing business driven by the growth in the small satellite segment, indirectly through the increase of launch frequency and directly through the development of new concepts to attract small satellite companies and upcoming launch service providers, with various incentives.[16] From military or government complexes, launch pads can evolve to become commercial spaceports, but the transition is not smooth. Space launch facilities with existing launch pads and recurring businesses have tremendous advantages over brand new developments that are, for most of them, stuck at the concept phase with no funding and very limited business.

Three groups of players were identified competing to become spaceports for NewSpace companies:

1) Existing space launch complexes with launch services and operations (existing or past).
2) Conventional airports (military and/or civil)—brownfield initiatives.
3) Greenfield projects.

There are countless projects of spaceports, at different stages of development. In the USA, some commercial spaceports have already obtained Federal Aviation Administration (FAA) approval [3], adding more possible launch options on top of existing, government-owned and operated launch complexes, such as Cape Canaveral and Vandenberg. Outside of the USA, it was noticed with several commercial spaceport initiatives, all aiming at capturing the business of upcoming private launch companies, and small satellite dedicated vehicles and ancillary businesses. Commercial spaceport projects are burgeoning in Europe (United Kingdom, Germany, Switzerland, France, Spain, and others), in Asia Pacific (Malaysia, China, Australia, and others), as well as in emerging space regions (Africa, South/Central America).

Spaceport projects falling in category 1 described earlier have strong advantages over brownfield and greenfield initiatives. But the business model of the commercial spaceport goes way beyond launch operations. Spaceports want to offer integrated hubs for NewSpace operations, including launch services but also small satellite assembly facilities, R&D and innovation centers, as well as education and entertainment programs.

Some initiatives propose realistic plans; others are very ambitious real estate developments. Some have secured funding; some are still at the PowerPoint stage. Most are moving very slowly. Spaceports are often, themselves, startups, doing business with startups. Several projects have signed emerging launch service providers as anchor tenants, to only see them out of business after months or years with no tangible achievement in between and money lost.

Spaceports are a core element of the NewSpace value chain centered on small satellites.[17] But the slow emergence of credible business models and the slow pace at which some

---

16 This is the case, for instance, of the Space Florida initiative.
17 And other business opportunities such as space tourism.

elements of the NewSpace industry are being developed (for instance, dedicated launch systems) make it difficult for spaceports to become reality.

30.4 Evolution, Not Revolution?

The growing importance of small satellite platforms is often seen as an obvious illustration of the application of Moore's law to the space industry, that is, exponential capacities and performance with a significant reduction in costs, weight, and size, as well as time to market. As such one can argue that small satellites are merely an evolution and not a revolution. The fact that large industry players are being challenged and new companies emerge to eventually replace traditional actors is the law of business. Faster, more nimble companies work to offer new products and services that get adopted by the market. Technology advances and companies pioneering change will lead.

To some extent, small satellites are not different than laptop computers. But what makes the situation unique is that space is a difficult industry to disrupt and what has been happening for a decade now is unseen in this industry, unprecedented and somehow not anticipated by most of traditional players. Launch service providers did not see SpaceX coming. They have learnt their lessons, and no one is overlooking small satellite manufacturers, upcoming launch service providers, and data service companies.

Nanosatellites and microsatellites have enabled the emergence of new players who would not have been able to penetrate this market otherwise because of technical or funding requirements, and because of political concerns or access to suppliers, launch opportunities, and talents. This, overall, creates a huge boost to the economy of space.

Most of the tier 1 companies have overlooked the small satellite market, but their ability to catch up with the market must not be underestimated. They maintain dominant positions and, contrary to other industries, the industry strongly relies on government contracts. Or, more specifically, defense contracts that may not easily be granted to NewSpace companies.

Traditional space companies can reinvent themselves. They have the money and they can integrate other companies or technologies, can acquire the competition, or replicate. Airbus now owns Surrey Satellite Technology Limited (SSTL).[18] SES now fully owns O3b.[19] MDA bought DigitalGlobe.[20]

Large players are attacked but can survive. Not all launch service providers have to be SpaceX.

Manufacturers of heavy satellites, usually part of large space and/or defense conglomerates, will survive.

Players at the lower end of the value chain will also survive (suppliers of components or raw materials).

At the highest end of the value chain, the most significant impacts will appear. Interestingly, it is where the economic opportunity is the highest that the level of threat put on traditional actors is also the highest (more on this in the next section).

---

18 EADS Astrium, now part of Airbus Defence and Space, acquired the majority in Surrey Satellite Technology Limited in April 2008.
19 In May 2016.
20 In February 2017, completed in October 2017.

Where there are opportunities, there are threats of status quo. Companies that are not able to adapt, to their advantage, change is still slow, and the bottleneck created by scarce launch opportunities is helping them to progress.

Google, Facebook, and others have also entered the space race through acquisitions and internal R&D. But they have also sold their assets to companies that will be better fitted to offer them products and services. Not everyone can penetrate this market and be successful.

30.5 The Economics at Play

Not all segments of the space framework are disrupted equally by the growth in small satellite business. And within each segment, companies, depending on their size, nature, and geographical reach, are differently impacted. I tried to outline general trends.

The space value chain is being redesigned or at least complemented and beefed up. Let us look at where the economy of small satellites makes and will make the highest impact and how.

The following categories have been purposely omitted in the following analysis:

- *Material suppliers*: They benefit from the overall growth of the industry, but the size of spacecraft does not matter or have a direct impact on them.
- *Components suppliers*: New entrants are focusing on supplying the nanosatellites and microsatellites segment, for instance, suppliers of components for CubeSat platforms; but, overall, the economic impact of small satellites on them is marginal; there is growth but some limited long-term economic benefits.
- *Ground stations*: This segment was covered in the previous sections; there is some degree of disruption, but no specific business model is really emerging.
- *Spaceports*: The overall economic impact on space launch facilities in relation to small satellites may only become clear in years from now. Spaceports are not viable businesses if only considering launching nanosatellites few times a year. Their business model is really about hospitality and space tourism, not small satellites.

30.6 Satellite Manufacturers

Satellite manufacturers are obviously one of the most important categories to benefit from the small satellite industry. With close to 1200 nanosatellites and microsatellites launched in the last 6 years and close to 300 spacecraft in the 50–250 kg range [4, 5], this is with no doubt a growing segment of the market. Projects of constellations,[21] new applications forcing an increasing demand from end users, and emerging space nations getting access to space, all these factors, with many others, are here to stimulate this economy. But when it comes to revenues, this segment is not the one that will profit the most. The real value is not in hardware platforms.

---

21 Read Chapter 32 III-3c for more insights and thoughts on this.

The commercial nanosatellite[22] to microsatellite segment is predominantly dominated by new entrants. In the academic and research world, space agencies, universities, and research centers have greatly adopted such satellites, since very small platforms are affordable. From CanSats to microsatellites, everyone can build a spacecraft for few thousand dollars. Rideshare opportunities offer such manufacturers access to space at low cost. The popularity of the CubeSat platform confirms that the hardware, when it comes to the platform itself, has very little value in this segment of the market. NewSpace companies can order semi-finished satellites or work with space brokers who offer them the CubeSat-based platform in groups of 1–27 units.

Nanosatellite companies are burgeoning all around the world. Over 200[23] NewSpace companies, agencies, and research centers have been identified that offer hardware, from semi-finished spacecraft to ready-to-launch satellites, globally; and over a thousand providing services to nanosatellite companies and platforms. Many will not survive or evolve.

It was estimated that over 1000 nanosatellites[24] will be launched in the next 3 to 5 years [4, 5], or, at least, manufactured and be ready for launch. This number seems extremely high, since there are "only" around 2000 active satellites in orbit now [6], from around 9000 launched since the 1950s. And this does not come with some risks.[25] But 2000 nanosatellites brought to market by 200 or even 100 or 25 players means small business. The value is in the data and the service (see later) that those satellites will gather through new channels that do not use the existing infrastructure, and the new business models that they will enable.

In the on-boarded technologies and payload segment, the business is also quite small.[26] It is however worth noticing that some nanosatellite companies are focusing on on-boarded technologies specific to the smallest satellites, such as telecommunication systems, propulsion, batteries, optical or earth observation sensors, or microcomponents. These companies are looking toward this segment with the ambition to evolve into full-scale satellite manufacturing. Their business is hardware sale or servicing. With on-boarded labs, they may offer data retrieval and analytics to end users.

Considering platforms from around 50 kg and up to around 125 kg, a mix of NewSpace manufacturers and traditional companies has been observed. From individual spacecraft to constellations, the revenues are attractive enough to allow NewSpace companies to focus on the hardware with all-inclusive solutions or to develop specific products on existing platforms.[27] With around 200[28] [4, 5] satellites to be manufactured and launched in this segment in the next 3 to 4 years, with market value of over US$100 billion, this is an attractive area.

With the upper range, consisting of satellites of up to 500 kg,[29] large well-established satellite manufacturers are playing the greater role. They are not equipped to offer cost-efficient

---

22 And even smaller spacecraft, such as picosatellites of less than 1 kg and femtosatellites of less than 100 g.
23 Author's own analysis.
24 1–50 kg range.
25 Risks of collisions, space debris, and quality control for nanosatellites.
26 This is why the author omitted the components supplier segment in this analysis.
27 For instance, the now acquired Skybox Imaging with SSL Loral.
28 This is a conservative number that may be exclusive to some large constellations.
29 Where the industry agrees the high limit is for small satellites.

nanoplatforms or microplatforms. Volumes are too low to consider investing into new production chains. Even though it was noticed that tier 1 space companies are trying to bring innovation and offer low-cost, flexible platforms to tap this growing market, their focus will remain within the 250–500 kg+ range. NewSpace companies are disrupting this segment, but large conglomerates are capturing the largest share of the market.

Also, while there is business to be done in developing hardware for this weight range, the investment to be made remains substantial. A company can manufacture a nanosatellite platform, test few spacecraft in space, and become commercial for less than a million dollars. For a 250 kg platform, the investment remains massive with a very long time to market. The dominance of existing tier 1 manufacturers will prevail. They may not have the NewSpace mindset, but will certainly accommodate NewSpace partners and customers.

30.7 Launch Service Providers

This is the segment of the space industry that attracted most of media coverage, because a rocket launch remains something spectacular to the eyes of the general public. Also, because players such as SpaceX and Blue Origin have demonstrated vertical take-off and landing, which represents a significant milestone toward rocket reusability. Furthermore, some space companies have actively been promoting space tourism[30] and point-to-point suborbital flights, also with colonization of Mars or other "out of this world" ideas that make people dream of our future in space.

More realistically, disruption is slow because, apart from SpaceX, NewSpace launch service providers are yet to demonstrate successful commercial launches. But when it comes to the economics and the opportunity for upcoming launch service providers to challenge existing players, it is a tough call.

There have been 114 launches in 2018, 90 launches[31] in 2017, 85 in 2016, 87 in 2015, and 92 in 2014 [7, 8]. In 2014, 285 spacecraft were launched, 245 in 2015, 225 in 2016, and 366 in 2017 [9]. Interestingly, the number of launches has declined since 2014, until 2018, but 2018 sets a record in terms of the number of satellites placed into orbit. Mainly because of the many CubeSats that were launched. This chapter does not discus satellite production and launch forecast, addressed elsewhere in this book. But it is worth noticing that the overall trend is toward launches of smaller satellites. Because of the lack of dedicated launch vehicle for nanosatellites to small satellites, this actually translates into fewer launches until constellations of small satellites (50–250 kg) are ready to be deployed. It can be anticipated that fewer launches will occur in 2019, with an increasing number of launches from 2020 onward. At this time, dedicated launch vehicles shall hit the market and be fully commercial.

Launch service providers are looking for secured, large launch contracts, from credible and reliable customers. Order books of launch companies are full for the years to come, with some switching happening in case of delays or failure. Most of the launch service providers

---

30 Companies such as Virgin Galactic, XCOR, Swiss Space Systems, or SpaceX, even if with a high degree of failure and significant delays.
31 Orbital only, not considering test flights.

are aiming at domestic government launch contracts (military and/or civil) and contracts of institutional or supranational agencies. Mainly because they offer guarantees and prepayments, and also because they tend not to shop around for the best price (i.e. the DoD will always launch on an American rocket, the Japanese government will always favor a domestic launcher, etc.). On top of that, some contracts offer recurring business, which commercial customers, until larger constellations are being launched, usually cannot guarantee.

To carry nanosatellites, for those launch service providers, is a distraction. It represents a small to marginal business bearing some risks. They are capturing this spectrum of the market because there is no other option. They want to secure the constellations, even with 50 kg spacecraft, but the nanosatellite business will be rerouted to dedicated launch vehicles, as long as they can show competitiveness versus piggybacking on larger rockets. This is where affordable rockets and reusable solutions will be most needed. They will have to have a viable business model.

30.8   Satellite Operators

The world of satellite operators is being disrupted and civil satellite operators are now being heavily challenged. Telcos and content providers[32] have already faced significant competition on Earth, in an ultraconnected world. They are now facing increasing competition in space. Small satellites are enabling companies that have no history of satellite operations to propose similar services, often at a cheaper price, since their infrastructure cost (small satellites versus large geostationary spacecraft) is much smaller.

Moore's law is here in action. Small satellites can now match the performance of heavy geostationary platforms at a fraction of the cost, and with a shorter time to market.

Small satellites also have constraints and are limited in their performance. If a 250 kg spacecraft can today match the optical quality of a geostationary satellite of 5 tons, its low orbit prevents it from offering the same 24/7 live coverage. This means that, until constellations are fully deployed, operators have time to adapt. Competition will be fierce and coming from nontraditional players, forcing satellite operators to evolve and also change their business models, similarly to what telcos and cable operators have done on the ground.

It is basically the confrontation of two worlds: large satellites operated by traditional operators versus small satellites operated by new companies. But traditional operators are also venturing into small satellites (not down to the micro or nano size though), trying to expand their offering and adapt to smaller spacecraft orbiting in low Earth orbit (LEO) or medium Earth orbit (MEO).

Companies such as JSAT, SES, and Inmarsat are all taking necessary steps to upgrade their offering and become data-driven companies, offering value-added services and not services that end users could get at a lower price. Their business models are evolving toward the higher spectrum of the value chain.[33]

Competition emerges from different horizons. NewSpace ventures focusing on providing specific products or services through the operation of nanosatellites to small satellites,

---

32 Mainly TV.
33 See Section 30.10 and Chapter 32 III-3c for more insights on data-driven business models.

operating their own satellite(s) or of third parties. But it is possible to see large Internet companies moving into this segment of the industry, Google and Facebook for instance. Even if after some major announcements and investments, they decided to put their initiatives on hold or, at least, to sell the low value-added assets (manufacturing, satellites) to focus on operations, data, and services.

The economics of small satellites for this segment is still quite small. But there is tremendous potential, especially when operators are focusing not only on content and communication but also on smart data and analytics. Operators can control the most valuable part of the value chain. The chain is controlled by the one who owns, channelizes, and analyzes the data.

30.9 Satellite Servicing Providers

On-orbit satellite servicing has always been seen as the less attractive segment of the space value chain.

The typical image people have about this is astronauts repairing the Hubble Space Telescope. With this image comes the assumption that satellite servicing is only about repairing spacecraft, and that it can only be costly (even if spacecraft are used instead of humans) and certainly not become a profitable business. Indeed, until now, only on-demand activities or demonstration operations have been done, but nothing driven by a strong commercial business model.

But as the industry is changing, this segment is the one that shows tremendous potential. Several NewSpace companies are solely focusing on servicing solutions and established players are now actively looking into this from a commercial angle.[34]

Activities and services will go well beyond spacecraft servicing (refueling and repair). Another option is spacecraft manufacturing and assembly in space, about upgrade and expansion, and about in-orbit infrastructure building. With projects such as life in space and mission to Mars, satellite servicing will gain momentum. It will be discussed in detail in the next chapter.

30.10 Data and Solution Providers

Earlier, I highlighted that there is very limited value in nanohardware and microhardware components. Even larger spacecraft have only little intrinsic value. Instead of value, it is referred to barriers to entry, since building a satellite manufacturing facility and going commercial requires large financial resources and time.

Until recently, those barriers created a perceived value. To be able to manufacture, and to own or to operate large, heavy satellites could lead to a certain degree of monopoly. Leadership positions have been created in sectors such as telecommunication, satellite TV, geopositioning, Earth observations, and military solutions.

---

34 Instead of government contracts (military or civil) for on orbit repair or refueling.

Other segments of the value chain also developed a perceived economic value, for instance, the launch vehicle providers and space launch complexes, simply because there is no viable alternative to them, at least in the short to middle term.

Small satellites are enabling the democratization of space. They allow small private companies to offer services and products that are more affordable and, sometimes, better than the ones offered by larger companies.

New business models are emerging, leveraging LEO orbit spacecraft and disrupting all the sectors, from telecommunication to the military. The next chapter will detail and analyze them.

In a nutshell, small satellites are removing or at least significantly lowering those barriers. To obtain live pictures of its own territory from space, a nation would need to launch its own spacecraft. An emerging space nation may not be able to afford its own surveillance satellite that would traditionally cost more than US$250 million. A decade ago, the only option would have been to contract an operator or image company to buy the images, or buy a service for live or delayed monitoring. Today, the same nation discovers a range of options. By engaging its newly created space agency or promotion research with its top universities, the country could easily co-develop a CubeSat-based image satellite. This has been the case for several nations recently. Initiatives such as QB50 played a strong role in promoting affordable and easy access to space. For around US$150 000 in total costs, a nation or any end user can build a small spacecraft, and have it delivered to orbit. For some low multiples of this amount, the customer, who could also be manufacturer and the operator, can have a high-quality satellite, or a small constellation.

Small satellites support a rapid integration of manufacturers and operators, as well as end users. Simply because the value is lost with the hardware itself, everyone is aiming at getting closer to where the value is: the data.

This nation wants images of its own territory for monitoring purpose. Affordable solutions and easy access to space are key for that, offering a full new spectrum of possibilities. Certainly, there is some level of national price to be part of the club of space nations, but the satellite is just the medium. Apart from the high degree of independence from other nations or satellite operators, the true value in this case lies in the fact that the country has access to data that it fully controls, from acquisition to processing. Traditional customers themselves become space companies and data companies.

The value is in the data, but not just raw data. Data are a commodity. The value lies in a new integrated value chain that is built around the data, from collection or generation (remote sensing, telco capture, etc.) to transmission, and from receiving raw data to delivering value-added services with processed information.

30.11   A Shift Toward New Models

Spacecraft ranging from few grams to 500 kg may be just a normal evolution, Moore's law applied to space, but they are making a long-lasting impact by inducing change in a 70-year-old industry predominantly dominated by large industrial conglomerates and governments.

The overall value chain is not easy to fully replicate or replace with NewSpace companies, because some of its segments, such as launch system providers or space launch complexes, still need to exist until they are challenged by fully funded companies.

Space is a very large industry where trillions of dollars are at play. It is not always a profitable sector, but a tough one, especially outside government contracts.

Small satellites have started to change this landscape, allowing NewSpace companies to introduce innovative business models and deliver services that until recently were the monopoly of some of the world's largest companies. With a value chain evolving toward data and leaving the hardware space, the economics is shifting toward a new space industry.

To conclude, one can state that 2018 has seen some consolidation but also the successful demonstration of a commercial path for some NewSpace companies. Small satellites are only here to play a growing and greater role to disrupt, over time, this industry. Larger, well-established players may be challenged, and they will be pushed to evolve or will simply acquire smaller, more nimble companies to ensure they secure their future and play a role across all segments of the industry.

References

1 NEWSPACEGLOBAL. [Online], available: http://www.newspaceglobal.com/home [accessed 21 March 2019].

2 Samantha Mathewson, *India Launches Record-Breaking 104 Satellites on Single Rocket*, Space.com [online], available: https://www.space.com/35709-india-rocket-launches-record-104-satellites.html. 15 February 2017.

3 FAA—Federal Aviation Administration, [online], available: https://www.faa.gov/about/office_org/headquarters_offices/ast/about/faq/#ls1 [accessed 21 March 2019].

4 Satellite Industry Association, (2017). *State of the Satellite Industry Report* [online], available: https://www.sia.org/wp-content/uploads/2017/07/SIA-SSIR-2017.pdf, June 2017.

5 SpaceWorks, (2017). *2017 Nano/Microsatellite Market Forecast* [online], available: https://www.spaceworks.aero/nano-microsatellite-forecast-7th-edition-2017.

6 UCSUSA. [Online], available: https://www.ucsusa.org/nuclear-weapons/space-weapons/satellite-database [accessed 21 March 2019].

7 Spaceflight101. [Online], available: https://spaceflight101.com/calendar [accessed 21 March 2019].

8 Spacelaunchreport. [Online], available: http://spacelaunchreport.com [accessed 21 March 2019].

9 Lafleur Claude, (2019). *Spacecraft encyclopedia* [online], available: http://claudelafleur.qc.ca/Spacecrafts-index.html [accessed 21 March 2019].

Further Reading

Aerospace Industries Association, (2017). *Engine for Growth: Analysis and Recommendations for U.S.* Space Industry Competitiveness, May 2017.

Giovanni, F., Nicola, S., Michael, D., and Giovanni, C. (2016). *Small Satellites—Economic Trends*. Milano: Università Commerciale Luigi Bocconi.

Kalik Thomas, (2016). article, *Investing Big in Small Satellites*, the White House Office of Science and Technology, Policy, December 2016, retrieved through https://obamawhitehouse.archives.gov.

University of South Australia, Adelaide, the International Space University (ISU), (2017). SMALL SATS BIG SHIFT—RECOMMENDATIONS FOR THE GLOBAL SOUTH.

Sweeting, M. (2016). *Small Satellites—Changing the Economics of Space*. Surrey Space Centre.

Lal, B., Balakrishnan, A., Picard, A. et al. (2017). *Trends in Small Satellite Technology and the Role of the NASA Small Spacecraft Technology Program, IDA for NASA*. IDA Science and Technology Policy Institute.

NASA, (2010). On-Orbit Satellite Servicing Study Project Report, NASA GSFC,.

Matthias, L., article (2007). *Potential of Small Satellites,* GIM Interviews Dr. Rainer Sandau, Chairman. International Academy of Astronautics (IAA).

Klotz, Irene, *July* (2017). article, *Small satellites driving space industry growth—report,* retrieved through http://cnbc.com.

Dillay, Clow, (2015), article, Here's why small satellites are so big right now, retrieved through http://fortune.com.

31 III-3b

Economics and the Future

Richard Joye

KCHK – Key Capital Hong Kong Limited, Hong Kong

31.1 Introduction

Small satellites[1] are expected to play a central role in the future of the economy of space. They represent much more than a simple technology evolution. Small satellites are not only replacing or complementing larger, more expensive, sometimes obsolete geostationary platforms but they are also enabling new business models. They allow NewSpace companies, independent from large industrial conglomerates, sometimes backed by venture capital money, to emerge and offer disruptive products and services. Overall, they make the access to space easier and more affordable and open a full spectrum of unforeseen opportunities.

Small satellites come with constraints. They have limitations because of their mass, impacting their orbit height[2] and life in space, and also because of other factors, some linked to the economics of the NewSpace framework.[3] But the majority of small spacecraft have enough advantages to overcome those constraints. Affordable constellations of nanosatellites to small satellites are the answer to expensive fleets of geostationary satellites. They significantly reduce the problem of low orbit. Electric propulsion could be one answer to the refueling issue, especially when considering deep space exploration. New types of batteries will increase the lifetime of nanosatellites.

New platforms, more nimble, cheaper, and with shorter time to market, will pioneer new areas of the space framework. It is more likely that the first asteroid-mining activities will be carried by small spacecraft than vehicles weighting several tons. Also, debris removal will be carried by robots with mass below 100 kg and not by heavy platforms. The future of high-speed data transmission will be carried by constellations of low Earth orbit (LEO) payloads and not geostationary satellites.

Small satellites are at the center of megatrends. They initiate and open billion-dollar markets and accelerate existing trends.

1 In this chapter, the term small satellites, unless specified otherwise, includes the broad range of spacecraft from nano or pico size (sub-1 kg to 1 kg) up to the 250–500 kg series. However, the focus is really in the 10–250 kg range.
2 Usually restricting them to low Earth orbit (LEO).
3 For instance, a NewSpace company may have limited access to funds to maintain or repair a satellite, or to allow a platform to evolve.

Nanosatellites: Space and Ground Technologies, Operations and Economics, First Edition.
Edited by Rogerio Atem de Carvalho, Jaime Estela, and Martin Langer.
© 2020 John Wiley & Sons Ltd. Published 2020 by John Wiley & Sons Ltd.

31.2 Themes Shaping the Space Industry

This chapter will be focused on trends that are directly being set or accelerated by the growing importance of small satellites, and on trends that are reshaping the space industry or disrupting sectors. The chapter will also highlight sustainable trends that are long-lasting and will have global impacts.

It is therefore important to differentiate industrial, technological, and commercial trends set by small satellites from broader trends that benefit small platforms. One could argue that the miniaturization of spacecraft is a trend on its own, but it should not be considered as a trend. Miniaturization is part of the overall evolution of space technologies led by technological breakthroughs in components and, more generally speaking, Moore's law.

The emergence of NewSpace companies is not a trend by itself. It is the result of multiple factors creating an ecosystem favorable for small private companies to flourish. Venture capital investing into space as the new frontier is not a trend. It is the combination of the growing attractiveness of the space sector as an investment destination, the good chances of seeing high rates of return with the accessibility to deals that used to be the monopoly of large industrial groups and governments.

The mediatization of the space industry is not a trend that belongs to the economics of small satellites. Certainly, companies such as SpaceX, Blue Origin, and Virgin Galactic have helped reviving the general public interest in rocket launches and space travel, because of mass marketing and the charisma of some of their leaders. Movies such as Gravity, Interstellar, and The Martian are blockbusters, but small satellites have limited impact on Hollywood. They are just part of the global picture.

Some segments of the space value chain,[4] such as spaceports, traditional launch pads, and ground monitoring stations, are evolving but they are not creating new trends because of small satellites. Their natural evolution to cater for the needs of the small satellite industry is not opening new commercial markets. The same applies to component suppliers. Sub-trends or segments, such as 3D printing in space, will be omitted. These can benefit from small satellites but are not directly related to them.

The major themes that are shaping the future of the space industry will be described first and after that the trends created by the growing presence of small satellites will be enumerated and analyzed. These themes are impacting every segment of the space framework and make huge economic impacts.

31.2.1 Privatization of Space Activities

If small satellite manufacturers, new satellite operators, and dedicated launch system developers can exist, it is because the space industry has opened itself to a brand-new type of players. The overall space industry has traditionally been dominated by large industrial groups, usually aerospace and defense conglomerates[5] that are often publicly listed,

---

4 Please read the previous chapter for more information about the space value chain and the impact small satellites are making on the participants and segments.
5 Lockheed Martin, Boeing, or Airbus (under their respective group names and legal entities).

state-owned enterprises,[6] national or supranational space agencies, as well as the military. The privatization of space goes beyond space agencies or government entities contracting privately owned companies or large companies doing business together (for instance, a television network purchasing a satellite service). The privatization of space is a very recent phenomenon, motivated by different factors. On the legal side, private commercial space travel and exploration was illegal until very recently. Nations safeguarded space-related activities for national security reasons to control their access to space. For the last 10 years, the space industry, globally in a similar way to Russia and China, has experienced a shift toward the acceptance of private companies at all levels of the space value chain, from launch service providers to private commercial space travel operations, and from temporary or permanent, privately owned infrastructure in space to commercial experiments in the International Space Station. Funding concerns on decreasing budgets allocated to space agencies and research centers, as well as state-owned space companies in some countries, have pushed governments to look at the private sector for some of its space-related activities. This created a whole new dynamic and a modified economic framework. The old space industry is open for business and now integrates private companies to accelerate technology development and space activities at a pace unseen since the end of the cold war.

This chapter will not discuss the details and the pros and cons of the privatization of space. This is only an evaluation about the impact of the privatization of the small satellite economics and how the privatization of space activities is fueling the emergence of new trends.

The opening of the market has many benefits in terms of funding made available, efficiency, time to market, and innovation. Of course, this transition comes with challenges, from policy and legal concerns to quality control and risk management in space.

The impact private companies nowadays have on the economics of small satellites is crucial and unprecedented. Private companies will still need large financial resources to conduct deep space exploration or asteroid mining. They will most probably need to work with space agencies and large conglomerates. But the part played by private, small satellite actors is getting a huge opportunity boost.

31.2.2 Making Space Accessible and Affordable

The smallest satellite platforms are very affordable. A CubeSat-based platform equipped with high-resolution lenses can be acquired for not more than US$15 000. By pricing some R&D work, the service, and the launch, a nation, a company, or an individual can place a satellite in LEO for around US$150 000.

Because small spacecraft are affordable, they create a strong pressure on prices at all levels of the space value chain. Emerging space nations, private NewSpace companies, and research and academic centers, all can easily develop, engineer, and build satellites, or order ready-to-launch platforms, inclusive of the payload and the software. The 1000+ nanosatellites that are going to be built every year do not, by themselves, have a strong influence on launchers. But, when including microsatellites and small satellites, they do bring some

---

6 With different degree of government ownership. For instance, Progress Rocket Space Centre and RKK Energiya in Russia or CASC in China.

weight in the discussions with launch service providers. Small satellites, because they are cheap, require cheap launch services. They require easy access to space. The theme of making space accessible and affordable is a broad topic that impacts all segments of the market. Inexpensive nanosatellites, low-cost launch opportunities, constellations cheaper than a geostationary satellite, and cost-effective infrastructure will flourish in space. Space tourism comes at a huge premium,[7] but space tourism companies, if they succeed, aim at offering very affordable space flights.

Small satellites, since they are becoming a viable and very economical alternative to large, expensive platforms, are fitting perfectly in this global theme.

31.3 Megatrends

The privatization of segments of the space industry and the growing affordability of products and services are igniting many strong trends that will make long-lasting impacts on the economics of space.

When it comes to small satellites, four megatrends can be identified. These trends are not only opening new commercial opportunities around small satellites but also they are significantly impacting the whole space value chain, challenging existing players and models. They are reshaping, over time, the whole industry and will accelerate the transformation of the market (Figure 31.1).

Small satellites, because of their affordability, accessibility, and technology potential, are becoming more and more the platform of choice to established companies and NewSpace players. This creates a strong momentum that accelerates the trends they have started.

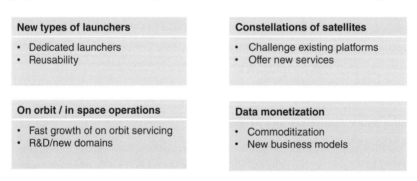

Figure 31.1 The four megatrends started by small satellites.

31.3.1 Launchers

In the previous chapter (Economy of Small Satellites), it was showed that the lack of launch opportunities and the unavailability of dedicated launchers represent a potential threat to the growth of the small satellite sector. Today, there is no rocket dedicated to the nanosatellite to small satellite sectors that can offer frequent launches and demonstrate a solid track record of commercial activities. The existing, operational smallest rockets offering dedicated launches are targeting the higher range of the

---

7 A ride to the ISS aboard a Soyuz vehicle typically costs between US$20 and 40 million.

spectrum,[8] such as Pegasus or Epsilon, and their launch opportunities are not frequent. On top of being more expensive than the option of piggybacking on a larger rocket, dedicated launchers targeting the smaller platforms are all still in development or have been abandoned[9] or are only starting to build their track record.

It was noted in the previous chapter that the lack of dedicated launchers is not really a bottleneck today, but only a threat to faster growth. Traditional launch service providers sometimes with the help of space brokers are launching nanosatellites to small satellites as secondary or tertiary payloads. There are some economic benefits of doing so, but, more often than not, the risk is not always worth the revenue. With rockets becoming heavier and even if prices are under pressure, traditional launch service providers can be expected to decline future opportunities, especially in the nanosegment and microsegment. Launch system providers will still be interested in providing services to the heavier small satellites or to fleets of satellites, usually part of upcoming constellations. Launches of three 120 kg+ payloads ignited strong competition between existing traditional players and upcoming dedicated launch systems (Figure 31.2).

Figure 31.2 Launcher megatrends and underlying trends.

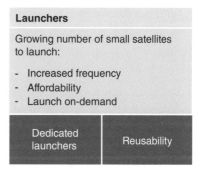

Small satellite manufacturers and customers are not only looking for launch opportunities but also they are changing the nature of their demand because of their needs:

- *Increased frequency*: The impact of the growing number of nanosatellites to small satellites on the overall number of rocket launches has only been limited. CubeSat-based satellites and other nanoplatforms and microplatforms have been launched in pools,[10] with no dedicated launch for them because the economics did not make sense. The increasing demand for launches in the nanosegment and microsegment will add pressure on the launcher segment. But because of the lack of interest for the smaller platforms, dedicated launchers are most required. The upcoming constellations of small satellites, with spacecraft in the 50–500 kg range, with an estimate of thousands of payloads to deliver to LEO, will require a number of launches that the current industry cannot deliver. Independently of budget, space launch companies cannot manufacture and deliver two, three, or five times their current volumes. Reusability is for instance seen as a solution by some of the players to absorb the growing demand.

---

8 250–500 kg.
9 Please read the previous chapter for more details on this, as well as the specific section of this book dedicated to the launch service providers.
10 Some of them, such as Planet's nanosatellites, have been deployed from the ISS and have been carried to the ISS as cargo in one launch.

- *Affordability*: Since small satellites are relatively inexpensive, manufacturers and customers are looking for the lowest possible price-per-kilogram. For the last few years, there is some pressure on price, with players such as SpaceX beating the competition on price or non-US launch companies offering sponsored launches. For small satellites as secondary or tertiary payloads, this is possible and prices below US$1000 per kilogram are still seen. Rocket operators maintain a strong position. They control the market. But once new dedicated launch systems come to market, the launch industry will quickly be democratized, and customers will have more choices. Low-cost launch systems are anticipated in China, in Russia, in Europe, and in the USA.
- *On-demand launches*: The maintenance of constellations of hundreds of satellites on orbit will require the deployment of servicing spacecraft within a very short time, or the launch of replacement satellites. On-demand launches certainly bear significant challenges, but the market demand will be very strong to motivate launch service providers to look into this. Reusable rockets and in-air deployment or horizontal take-off systems will try to address this demand. Logistic constraints are important, but economical aspects are also big obstacles. To maintain ready-to-launch rockets nearby launch pads and where payloads can quickly be mounted is very expensive.

The market pressure for more frequent, affordable, and on-demand rides to space is creating two main trends within the launch segment that manufacturers, existing or upcoming, cannot ignore:

- *Dedicated launch systems*: In less than two decades from now,[11] there will be several dedicated launch systems to service the small satellite segment. Small, expendable, low-cost rockets will focus on the smaller spacecraft,[12] while larger rockets, expendable or reusable, as well as air-launched and potentially horizontal take-off vehicles will service the microsatellite and the heavier satellite customers. Competition will drive prices down, up to a balance between a growing demand and a broadened offering.[13]
- *Reusable launch systems*: For large rockets and upcoming space launch systems targeting the 250 kg+ segment, reusability is a promising path to bring operating costs down and ensure sustainable competitiveness. To date,[14] SpaceX has successfully demonstrated the recovery of rockets and their reuse with main launch systems.[15] Other launch system providers are divided when it comes to reusability. Some have ruled this concept out, citing the heavy investment that needs to be made to develop this technology. Others challenge the long-term cost efficiency of reusable rockets or want to look at recycling components and not the whole first stage. Some are targeting low-cost expandable systems instead. There is no doubt that reusable launch systems will not be offered by all existing launch service providers. However, more interesting are the technology advances that this concept brings, and the future reusability will pioneer. Reusable launch systems will be critical when considering the eventual colonization of another planet. Reusability is also a main element when considering making the space industry eco-friendlier.

---

11 Around the year 2030.

12 For instance, CubeSat-based satellites, small formations, and constellations.

13 Please refer to the section of this book that covers upcoming launch systems for more details on existing and planned launchers.

14 2018.

15 And soon as boosters for the Falcon Heavy.

31.3.2 Constellations

Much has been written about satellite constellations. This is an area that benefits from large media coverage, through specialized publications and general articles for the masses. This is for several reasons: Because they represent massive projects with state-of-the-art technologies and thus interest a broad audience. Because of the promises they offer, in respect to low cost or free Internet coverage, telecommunications, and access to data. Because they are sponsored by US corporations and private companies that enjoy public recognition, such as SpaceX. And also because of the risks and challenges they bear. Overall, constellations of satellites are a game changer for mankind. Because they are so important, we will analyze them in more details in the next chapter.[16]

In this chapter, the global trends set by small satellites will be described. Constellations are one of the four identified megatrends (see Figure 31.3). But they play a very central role, if not *the* central role. Constellations already represent a driving force that stimulates the growth of the whole space industry. The OneWeb constellation, for instance, already provides jobs to hundreds of engineers and technical staff working on the project.[17] OneWeb, by itself, stimulates the manufacturing sector, since over 600 spacecraft are intended to be built. But this project also stimulates the launch segment, with bookings already taken on launch service providers' manifests, as well as many other related segments of the space industry, including often overlooked professional sectors, such as space lawyers, telecom specialists, among many others.[18] The Iridium NEXT constellation is not only enabling multiple launches and stimulating spacecraft manufacturing but it will also pioneer on-orbit servicing and on-orbit activities not seen until now.[19]

Constellations are not a new concept. The idea of multiple spacecraft placed in orbit to cover most of the Earth goes back to the 1940s. Theoretical models such as the one developed by John Walker in the early 1980s set frameworks that have been used by existing

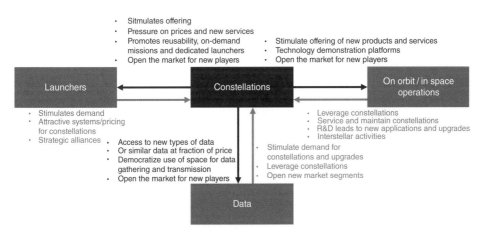

Figure 31.3 Interdependencies between constellations and the three other megatrends.

---

16 Even if the focus will be on networks of nanosatellites and not on all constellations coming to market.
17 Not only at OneWeb corporate level, but inclusive of all contractors, suppliers, and service providers.
18 Please refer to the next chapter for more details and numbers.
19 More on this in the section dedicated to in-space activities.

formations and are still being used as the basis for upcoming constellations. The size of announced constellations, such as OneWeb, will have unprecedented massive impacts. At the center of the small satellite ecosystem, constellations are exponentially magnifying the economics of the small satellite industry.

The four underlying benefits brought by constellations are (see Figure 31.4):

- Low cost connectivity.
- Advanced positioning systems.
- Various Earth observation (EO) activities.
- The emergence of a new type of operators.

Most of what constellations bring could be seen as an extension or an improvement of existing products and services. While this is true from a pure technology standpoint, it goes way beyond that can consider the impacts of this evolution on the whole space industry. Constellations are not only, for instance, offering an evolution of the existing geostationary orbit (GEO)-based platforms with data transmission (Internet, TV, etc.) but they are also drastically disrupting the sector, enabling low-cost products and services[20] and challenging well-established players who will have to reinvest themselves to survive.

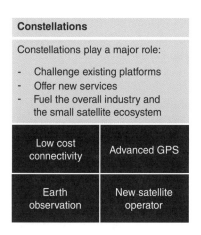

Figure 31.4 Constellation megatrend and underlying trends.

31.3.3   On-orbit and In-space Operations

On-orbit[21] and in-space operations are gaining tremendous momentum, mainly because of the growing impact small satellite constellations are anticipated to make, and already making in some areas, such as R&D.

From a pure statistical standpoint, more satellites in orbit means more in-space operations and servicing activities. But there is no linear growth, rather a dissymmetrical relationship between the number of satellites in orbit and space activities. This is due to the fact that the lower end of the spectrum, nanosatellites and even smaller spacecraft, up to 50 kg+, or more, will not need servicing at all. It is cheaper to let a CubeSat die than repairing it in space. To extend the life of a 50 kg satellite may be more expensive and more complex than

---

20 Compared to existing offerings.
21 In this chapter, on-orbit and in-orbit are used with the same meaning.

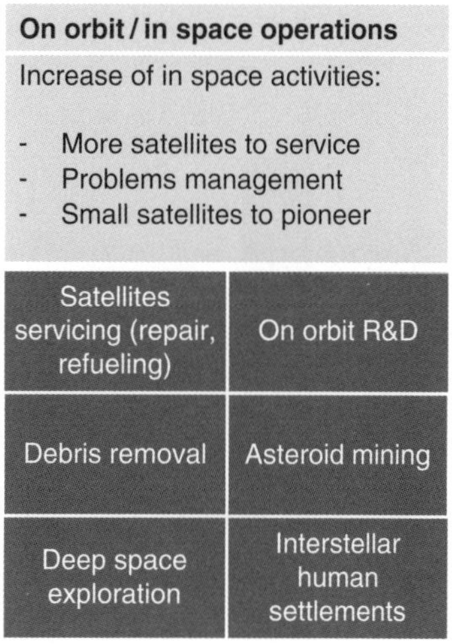

Figure 31.5 On-orbit/in-space operations megatrend and underlying trends.

On orbit / in space operations
Increase of in space activities:
- More satellites to service - Problems management - Small satellites to pioneer

Satellites servicing (repair, refueling)	On orbit R&D
Debris removal	Asteroid mining
Deep space exploration	Interstellar human settlements

replacing it with a newer, upgraded platform, but CubeSat platforms are ideal for technology demonstrations and R&D activities in space. Space is becoming an affordable live lab for experimentation (Figure 31.5).

The different types of services and the in-space activities will be mapped to understand their impact and how small satellites are supporting them[22] (see Figure 31.6):

1) *On-orbit servicing*: Any type of activity that involves observing or interacting with a satellite to anticipate, observe, or fix a problem or enhance capabilities or lifetime in space.
2) *On-orbit R&D*: Technology demonstration in space that will not be otherwise possible without the help of nanosatellites or micro satellites (mainly).
3) *Debris removal*: Use of microsatellites (mainly) to remove space debris, faulted spacecraft, and dead satellites of all sizes. By extension, clean up space.
4) *Asteroid mining*: Mining, extraction, and/or collection of minerals and other natural resources from asteroids, comets, and planets.
5) *Deep space exploration*: Small satellites will play a greater role in exploring deep space for, mainly, research purpose.
6) *Interstellar/human settlements*: In our quest for a second home in space, small satellites will play a central role for all purposes (infrastructure, cargo, manufacturing, energy, communications, food, etc.).

There exist interdependencies between the six sectors. For instance, R&D in space will be used to further advance technologies for asteroid mining or support human settlements. Servicing will mainly be performed by small satellites, which, in due course, will rely on in-orbit demonstration first.

In all of them, small satellites are playing a very central role. Affordable platforms and shorter development cycles will accelerate this megatrend. R&D in space, for instance, would not be considered as economical if technologies had to be demonstrated on GEO satellites.

---

22 For detailed analysis on each of the topic, please review the references provided and refer to specialized literature.

Activity	Type	Sub-category	nano	micro	small	Emergency	Scheduled	(Semi-) permanent	Economics	Timeframe
On orbit servicing	Servicing	Inspection Manipulation Maintenance Upgrade/ assembly							$$	Now
On orbit R&D	Demonstration	Propulsion and Power Upgrade Assembly Manufacturing							$$$	Now
Debris removal	Demonstration Operation	Inspection Removal							$	5-10 years
Asteroid mining	Demonstration Operation	Technology demonstration Mining Retrieval							$$$	5-20+ years depending on technical and non-technical challenges
Deep space exploration	Operation	Exploration							$	10+ years
Interstellar / human settlements	Demonstration Operation	All purposes (similar to LEO satellites) New purposes							$$$	15+ years

Figure 31.6 On-orbit and in-space activities impact on each class of satellites, economics, and timeframe. Note: This chart is meant to remain generic. In some cases, emergency servicing of critical nanosatellites or nanosatellites engaged into mining activities will be needed. But this chart maps where the 80% of the activities are supposed to happen.

In the short term, this megatrend will be dominated by on-orbit servicing, repair, and refueling. Debris removal is a very important topic, critical for the evolution of space activities. Recent laws and policies help but many challenges remain to make it happen. In any case and whatever happens, this will help demonstrating technologies and will support the small satellite market. Even if the economics is not yet very clear. R&D in space will be a major trend, enabling on-orbit research, but also manufacturing and assembly. Intelligent spacecraft will play a central role in the development of our future activities in space (mining, exploration, settlement), and small satellites will be part of every breakthrough.

31.3.4 Data

Small satellites are to space what laptops[23] have been and are to the computer industry. They are a logical evolution. Not a revolution by themselves, but a trigger to the emergence of disruptive business models. Laptops and smartphones have accelerated the deployment of fast wireless Internet access. The Internet, laptops, and smartphones have changed the way of communication, study, work, and purchase of goods. They have made Amazon, Twitter, Alibaba, and Netflix possible.

Small satellites are playing a very similar role. They are acting as a catalyst to the emergence of new business models, of new ways of doing things. They will be instrumental in creating the next billion-dollar companies. Amazon and Alibaba allow customers to buy almost anything with one click, and new companies will allow end users to use space to do things not done before (Figure 31.7).

Certainly, the hardware is important (you cannot order a product online without a device connected to the Internet), and it has seen the advantages small satellites bring, such as affordability and time to market. They have strong value to whoever owns and operate them, especially when considering constellations that could create new monopolies and at the same time they democratize the use of space. But the true value lies in the data. Since manufacturing, deploying, and operating a satellite becomes more and more affordable and easy, long-term sustainable business models will evolve around the access to the data and the treatment to value them.

With their own small satellite(s), nations will get access to images. They will control their data. Private companies can launch fleets of satellites to test technologies, to gather data, to retrieve them and treat them to offer unique, value-added products and services. The

Figure 31.7 Data megatrend.

Data / services

- Open Source / free data
- New business models
- NewSpace companies

Data analytics

3rd party services

---

23 And to some extent smartphones.

emergence of new operators, different from government entities and large conglomerates, will also allow companies to choose from different data providers. Price will impact the quality of service among other factors, such as functionalities, application program interfaces (APIs), and user interfaces.

Nanosatellites have allowed many companies to launch new services based on the data retrieved by one or more spacecraft. Small satellites are already supporting the transformation of many industries, from weather forecasting to agriculture. Farms are becoming smart and data-driven, thanks to data captured from space. Companies such as Planet Labs are operating their own fleet of satellites, but the true value of a company lies in the analytics it offers to its customers, that is, from data acquisition through the lenses of their fleet of spacecraft to the delivery of insights to their customers.

Constellations aiming at offering ultrafast connectivity will be able to capture, channelize, and absorb an enormous amount of data. Connecting large regions in Africa will open new markets, enable new products and services, leverage the access to space, and data transmission through LEO satellites.

31.4   Conclusion: The Space Industry Is in Mutation

The privatization of commercial space is reshaping the space value chain, allowing new entrants to propose products and services built around small satellites. From manufacturing to data analytics, the value chain is growing and changing, supported by four megatrends that will significantly impact not just the space industry, but many existing and future industries and sectors.

Further Reading

Pelton, J. (2015). *New Solutions for the Space Debris Problem*. SpringerBriefs in Space Development.

K. Landzettel, A. Albu-Schäffer, C. Preusche, D. Reintsema, B. Rebele, G. Hirzinger, 2006. *Robotic On-Orbit Servicing—DLR's Experience and Perspective*, DLR (German Space Center).

Carillo Kevin, 2016. *In-Orbit maintenance: the future of the satellite industry*, Toulouse Business School, with A. Krishna, M. Skenjana, V. Do, R. Yoshida, Q. Wang, and F. Rizzo.

NASA, 2010. *On-Orbit Satellite Servicing Study—Project Report*.

Li Yan, Cai Yuanwen, Xu Guofeng, Zhao Zhengyu, 2015. *On-orbit Service System Based on Orbital Servicing Vehicle*, Equipment Academy, Beijing.

B. Sommer, 2012. *From OOS 2002 to Present Vision & Accomplishments*, DLR (German Space Center), 2012. Aerospace Industries Association, *Engine for Growth: Analysis and Recommendations for U.S. Space Industry Competitiveness*, May 2017.

Facchinetti Giovanni, Sasanelli Nicola, Davis Michael, Cucinella Giovanni, 2016. *SMALL SATELLITES—economic trends*, Università Commerciale Luigi Bocconi, Milano, December 2016.

Kalik Thomas, 2016. article, *Investing Big in Small Satellites*, the White House Office of Science and Technology, Policy, December 2016, retrieved through https://obamawhitehouse.archives.gov

University of South Australia, Adelaide, the International Space University (ISU), *SMALL SATS BIG SHIFT—RECOMMENDATIONS FOR THE GLOBAL SOUTH*, 2017.

Sweeting, Martin, 2016. *Small Satellites—Changing the Economics of Space*, Surrey Space Centre.

Bhavya Lal, Asha Balakrishnan, Alyssa Picard, Ben Corbin, Jonathan Behrens, Ellen Green, Roger Myers, 2017. *Trends in Small Satellite Technology and the Role of the NASA Small Spacecraft Technology Program*, IDA for NASA, March 2017.

Lemmens Matthias, 2007. article, *Potential of Small Satellites, GIM Interviews Dr. Rainer Sandau, Chairman, International Academy of Astronautics (IAA)*, 2007.

Klotz, Irene, 2017. article, *Small satellites driving space industry growth—report, July 2017*, retrieved through http://cnbc.com.

Dillay, Clow, 2015 article, Here's why small satellites are so big right now, 2015, retrieved through http://fortune.com.

32 III-3c

Networks of Nanosatellites

Richard Joye

KCHK—Key Capital Hong Kong Limited, Hong Kong

32.1 Introduction

In the world of satellites, nanosatellites with weights ranging from 1–10 kg have played a major role in opening up the space industry to new entrants. Universities, labs, and research centers from across the globe have embraced the nanosatellite platform because of its low cost and high degree of standardization. CubeSat-based satellites, from 1 unit to 27, have allowed NewSpace companies to build profitable, sometimes disruptive businesses leveraging this new, affordable, and easy access to space. Countries have gained their space nation title, thanks to the tiny platform, in only few years starting with nothing else than the willingness to own, launch, and operate a satellite.

Nanosatellites, because of their very small mass, are of course bearing limitations. They are often used as demonstrators of technologies, as a proof of know-how, or represent the final step in an academic project with no real commercial sense. But nanosatellites working in networks have tremendous potential. Alongside constellations of larger satellites, fleets of nanosatellites represent a significant segment of the space value chain of the next decades.

32.2 Why Networks?

The laws of physics are modeling and impacting the capabilities and specificities of our satellites. Because of gravity, the Earth rotation, and other laws and phenomena, there is no perfect satellite mass, orbit altitude, inclination, and position that can meet all our needs and requirements.

One spacecraft can fulfill several uses or can be dedicated to a specific mission, such as imagery or telecommunication. Specific missions have requirements that one satellite, alone, cannot fulfill. For instance, to obtain a quasi full coverage of the planet for navigation systems is impossible with one spacecraft. To capture images of any point of planet Earth in real time is also not possible with one low Earth orbit (LEO), medium Earth orbit (MEO), or geostationary orbit (GEO) satellite. For such uses, networks of satellites are the

Nanosatellites: Space and Ground Technologies, Operations and Economics, First Edition.
Edited by Rogerio Atem de Carvalho, Jaime Estela, and Martin Langer.
© 2020 John Wiley & Sons Ltd. Published 2020 by John Wiley & Sons Ltd.

only possible solution. Also, in some cases, the economics or the convenience of a network of smaller spacecraft will take over a single GEO satellite.

32.2.1 Background: Networks are Not New

Traditionally, geostationary satellites, orbiting at around 36 000 km with a fixed position over an equatorial point, have dominated the telecommunication,[1] some sub-segments of the imagery industry, and other segments. They also come with constraints: GEO satellites cannot cover the Earth's poles and usually have very high manufacturing and launch costs, long time to market, or the need to plan for repairs and maintenance.

In a space industry led by government agencies and industrial conglomerates, where massive investments are required to develop, test, and deploy new technologies, the GEO segment could enjoy its market dominance with very limited competition. But the space industry is in deep transformation. New entrants challenge the old value chain. The miniaturization of components and technological advances are now opening a new era, with small satellites helping to reshape the space framework. NewSpace companies are considering small satellites to challenge GEO platforms, offering similar products and services and putting strong pressure on pricing.

The idea of networks of satellites is not new and even precedes the first satellites. Arthur Clarke [1], in 1945, 12 years before the launch of Sputnik 1 by the USSR, is usually recognized as the first person to offer a description of what a formation of satellites could achieve. Based on the work done by Germany during World War II, his theoretical study presents what rocket-launched spacecraft[2] could achieve in respect to global television distribution.

Very early on it was made clear that the use of single satellites could not fulfill all commercial applications that space could offer. A geostationary satellite is a good solution for live or quasi-live imagery of a rather broad territory of our planet. But if a nation is blind with the other two-third part of the globe, it becomes a problem, for instance, for border monitoring for national security. Networks are needed for several reasons to give global coverage (for all orbits, from GEO to LEO), and also to allow spacecraft to effectively exchange data with each other, and, together, offer a unique set of services.

The first networks of satellites were planned and then deployed to address technical challenges and limitations carried by geostationary, single function spacecraft. The first LEO constellations, such as Iridium and Globalstar, were not designed as commercially more attractive alternatives to GEO satellites. They were built to offer global mobile telecommunication coverage where geostationary satellites failed due to their equatorial fixed positions. Surfing on the mobile phone revolution, companies such as Motorola betted on a growing interest for voice and data communication, anywhere, anytime. Coverage and latency were addressed, and the problems of GEO platforms resolved. But the high price of products and services, built on high manufacturing, deployment, operational, and maintenance costs, did not meet their market and forced the operators into bankruptcy.[3]

The failure of Iridium and Globalstar in their early days is not an isolated example.

---

1 Voice, data, television.
2 In his paper, Arthur Clarke illustrates a German V-2 radio-guided enhanced rocket that would place itself in orbit at 42 000 km.
3 Iridium filed for Chapter 11 in August 1999. Globalstar in February 2002.

ORBCOMM, which also developed a network of LEO satellites for voice and data communication, backed by industry-leading space companies, filed for Chapter 11 in September 2000. Many projects, some in advanced stage, were halted in the late 1990s and early 2000s. Teledesic, with a very ambitious proposal to establish a global network of 840 120-kg LEO communication satellites,[4] backed by prestigious institutional investors and influential private individuals such as Bill Gates, was liquidated in 2002, after a series of wrong decisions and increasing costs.

Those companies, alongside many other projects, pioneered the world of constellations of LEO spacecraft, but they came too early. In the 1980s, when they originated as concepts they made a lot of sense. In a world dominated by very expensive yet limited GEO solutions, with no alternative for global voice and data coverage and no affordable mobile solution, Iridium, Globalstar, ORBCOMM, and Teledesic came to fill this gap. But at the time of their deployment or demonstration, the fast-growing Internet and the increasing number of mobile handset manufacturers and telcos, together with cheaper solutions, made their businesses no longer commercially viable. Nonspace solutions were more effective and affordable, and also the consumer market had changed. The targeted market no longer existed and new customers emerged asking for different services and solutions.

32.2.2 LEO and MEO Networks

Today, with the economics of small satellites, commercially viable business models for the telecommunication sector are in trend. Iridium, Globalstar, and ORBCOMM under new management and with new shareholders experience a second life and plan for new constellations to service new customers and offer new services.

They are transitioning to more agile enterprises, becoming NewSpace companies offering disruptive products in sectors such as the maritime industry, connected objects, or low-cost broadband. Many of their upcoming customers are NewSpace companies themselves, leveraging this revitalized access to space to bring new products to the market and compete with well-established players.

Constellations of LEO small satellites are enabling new business models and can survive alongside GEO offerings.

Within this ecosystem, nanosatellites are not necessarily directly competing with larger spacecraft, but rather offer an alternative to GEO or MEO/LEO constellations or a way to complement upcoming network (Figure 32.1).

32.2.3 Constellations: One Type of Network

Very large projects, such as the ones of SpaceX and OneWeb, have popularized the concept of constellations of satellites and have raised the general public interest. A typical upcoming LEO constellation comprises of hundreds of small spacecraft ranging from nanosize to 500 kg or more, injected in polar orbit on multiple orbital planes. They are often illustrated with the Earth inside a grid formed by several planes of multiple spacecraft.

---

4 The original design of 1994 included 840 satellites. A revised proposal cut down the number of spacecraft to 288, at a higher orbit of 1400 km.

Type of network	Pros	Cons	Challenges	Risks	Trends	Main uses
GEO	• Small formation for a large coverage • High efficiency • Easy deployment/dedicated launch • Simple formations (3) • Quality • Long life • Coverage at poles • Track record • Can handle several customers • Offer hosted payload opportunities • Can cover different frequencies/beams	• High latency • High costs (manufacturing, launch) • Long time to market • No easy upgrade • Expensive servicing/repairs • Usually operator is the owner	• Competition from MEO/LEO • Technological disadvantages in fast-moving world • New credible players entering market	• Concentration in one or more satellites • Incident at launch (loss of payload)	• Technology upgrade in space • GEO as a base for future space exploration	• Constellations • Telecommunications (voice/data) • Satellite TV • Imagery • Global positioning systems • Weather
MEO/LEO micro-small satellites	• Cost (manufacturing) relative to GEO • Cost (launch) depending on the size • Lesser latency than GEO • Full range of capabilities • Easy upgrade • Short time to market • Offer hosted payload opportunities • Can cover different frequencies/beams	• Low coverage, requires large networks • Overlaps, low efficiency • Overall deployment cost of network could be costly • Time for full deployment, many launches required • Orbital drag • Replacement instead of repairs	• Business models to be demonstrated (many failed) • Long-term advantages over GEO • Too many players entering the market	• Space junk • Inactive constellations (failed companies)	• Low cost broadband • IoT • New players entering the market → replace GEO	• Constellations, trailing formations, clusters • Broadband communication • Imagery • GPS and GPS-enhancement • Disaster monitoring • Earth/air monitoring • Space-linked applications • Agriculture
LEO nanosatellites	• Most affordable solution (manufacturing and launch) including ISS deployment • Easy upgrade/replacement of spacecrafts • Very short time to market • Easily accessible to new players • Latency better than MEO/GEO or higher LEO orbits	• Limited capabilities and uses • Low efficiency • Strong orbital drag • Low quality • No real strong track record • Mono-frequency (usually) • No hosted payload options	• Regulations • Lack of capital for long-term maintenance • Launch opportunities bottleneck	• Space junk • Inactive constellations (failed companies) • Quality control	• Low cost imagery • Low cost broadband • New players entering the market	• Constellations, trailing formations, clusters • Imagery • Earth monitoring • Disaster monitoring

Figure 32.1 Comparison of GEO, MEO, and LEO networks of satellites.

Constellations represent one type of network of satellites. LEO constellations rely on a large number of spacecraft, synchronized and coordinated by master spacecraft or from the ground, with a certain degree of overlap to offer global coverage. Constellations are required to offer live, global telecommunication services. However, many other applications do not require full live coverage of the Earth. Satellites in formation flying can be grouped into clusters and trails. Trailing networks are built with several spacecraft following each other on the same path. Clusters are groups of satellites following a similar path, but not necessarily on the same orbit or plane, in a concentrated formation.

Usually constellations are made from similar spacecraft. All of them have the same mass, the same technical capabilities and functions, and are built by the same manufacturer. They are launched in groups. Some satellites within the constellation can play a master role and carry greater computing or telecommunication capabilities, but generally 99% of the spacecraft are exact copies.

Trailing networks and clusters differ from constellations, since they are not bound to a unique type of satellite. Their missions are usually not telecommunication services but Earth observation (EO) and monitoring, including HD imagery, science, and weather.

They are often built from satellites of different masses, coming from different manufacturers. Each satellite can play a very specific role within the network. A-Train is a typical example, with six LEO satellites in a trailing formation, weighting between 500 and 2000 kg each, coming from different manufacturers and operators, with different purposes. Individually, each satellite retrieves and transmits a significant amount of data gathered from its on-boarded instruments. Since the satellites in the formation follow the same path in a tight formation, the analysis of their combined sets of data offers a full observation of the atmosphere and the Earth. Together, this network is a powerful, integrated solution.

The Cluster II formation of the European Space Agency is a good illustration of a flexible formation of four 1200 kg spacecraft. The choice of a cluster was made for technological reasons. The mission objective was to measure the impact of solar rays onto our Earth's magnetosphere. To provide accurate measures and 3D data visualization, only a cluster of satellites could be used, since no unique spacecraft could achieve this alone. Operators also needed to modify the range between the four satellites in order to obtain different measures from a different formation.

32.2.4 The Raison d'être of Networks of Small Satellites

Satellites in LEO and MEO networks can achieve missions that one spacecraft alone cannot. They offer five major advantages:

1. *Overcome physical constraints*: As a constellation, a network of small satellites can ensure global coverage that a formation of geostationary satellites cannot. A LEO constellation of hundreds of spacecraft can guarantee 100% of the Earth coverage for data acquisition and broadcasting. At the moment, there is no other satellite-based technology that can offer the same, even with ground relays.
2. *Increase performance*: A network, being a constellation, a trailing formation, or a cluster, can be the base for the delivery of products and services similar to geostationary

satellites but with increased capabilities; for instance, in the voice and data space, a LEO constellation will significantly reduce latency; and because of shorter time to market, a constellation can be upgraded with new spacecraft, faster than for GEO satellites.

3. *Reduce costs*: This is not always the case, but a formation of LEO small satellites is supposed to be more cost-effective than a single or more GEO satellites; launch costs and lack of launch opportunities still represent a bottleneck but over time this issue shall be resolved. The total cost of manufacturing and deploying a constellation of 200+ LEO satellites may still surpass that of three GEO platforms, but the ROI is expected to be much higher over time. They also offer shorter time to market, low maintenance, and better coverage and service for a large customer base.

4. *Increase spectrum of capabilities and more flexibility*: Satellites often fulfill only one mission. They can carry different instruments onboard, for instance for science missions, but usually don not mix different mission types. A TV broadband GEO satellite will not do imagery or science missions. In order to mitigate risk, GEO platforms rarely offer hosted payload opportunities, compared to MEO and LEO smaller satellites that offer more flexibility and adaptability. A trailing formation or a cluster can include satellites fulfilling different roles, with different mass and payloads. A formation can meet the overall requirements of a mission, with each individual spacecraft carrying specific tasks.

5. *Service and in-space mission*: Constellations are deployed with redundant satellites that can easily be used as replacement spacecraft. This significantly reduces maintenance risks and costs. Future constellations are planned with servicing and repairing robots. The future of constellations in space (for instance orbiting around other planets) is showing tremendous potential over single satellites. Decentralized functions and computing capacity will reduce spacecraft mass and costs.

Because of lower manufacturing costs, technological breakthroughs, and the shift toward a new space value chain that integrates NewSpace companies, networks of small satellites are becoming increasingly attractive.

32.2.5 Existing Networks

The media, including the specialized press, like to focus on the upcoming megaconstellations of SpaceX and OneWeb, or original plans outlined by Facebook and Google. The general public perception is that constellations are a new, very promising concept. On the contrary, it was also observed in the previous sections that the idea of networks of satellites originated in the 1940s and that constellations have already successfully been deployed.

If networks of satellites are not new, why no one talked about them before? This is because several constellations failed initially, because their services were targeting a small group of customers, usually institutional clients and professionals, and because the old space industry did not attract the general public interest in the 1980s and the 1990s. NewSpace companies have changed this. Most of planned constellations do not originate from space conglomerates. They originate from new ventures that see an opportunity, now that small satellites are affordable, to develop products and services to compete with existing solutions.

This is because almost anyone can now hope to become a satellite operator. Affordability and access have changed the playground of networks of satellites. Existing and upcoming networks can be listed to identify trends, challenges, and issues.

Groups of satellites do not always represent a formation, or even a network. If satellites of the same nature, sometimes bearing the same name, do not have a technological relationship in a network configuration, they are not considered as a formation, but just a fleet. This is the case, for instance, of most of broadcasting geostationary satellites. If one satellite is down, the others are not impacted and only the services provided around the defaulting satellite are disrupted. In this analysis, fleets of satellites were omitted that are not considered as a formation or a constellation, for instance, Arabsat, Astra, among others. Furthermore, military networks were omitted due to a lack of publicly available data. Also not considered were multiple services based on the same hardware, to avoid overlaps.[5] The list also excludes programs or services based on multiple satellites if the spacecraft are not organized as a network. This is the case of Landsat imaging program, for instance. The likes of MSAT, SkyTerra dual satellite systems are also not considered, since they are not really formations. Similarly, satellites part of multiple programs or networks based on hosted payloads such as Aieron on Iridium NEXT are not relevant for this work (Figure 32.2).

32.3 Opportunities for Networks of Nanosatellites

32.3.1 Network Trends

Originally, networks were deployed in formations or constellations to solve the problem of the geographical coverage of GEO satellites. Formations and clusters were then designed and launched to offer decentralized multicapacities, especially in weather, Earth, or atmosphere monitoring and science. Today, the majority of upcoming commercial networks are focusing on telecommunication or Earth observation.

32.3.1.1 Telecommunication

Upcoming MEO and LEO constellations are aiming at offering global coverage (sometimes with a regional focus, such as South East Asia/Pacific or Africa) for, mainly, data connectivity.[6] Over 40 constellations were counted, totaling over 1500 satellites in LEO and MEO orbits.

These networks are based on different technologies, some using direct access to communication devices (mobile handsets, connected objects), and some relying on ground relays or dedicated modems and terminals.

---

5 Such as SES Broadband, Outernet.
6 While some constellations may also offer traditional voice services, the majority is focusing on fast broadband connectivity.

Name	Sponsor	Country	Use	Status	Orbit	Sat type	Nr of sats	Start date/1st launch	Completion date	Notes
Cluster II	ESA	European Union	Science	Operating	HEO	Large 1,200 kg	4	2000	2000	
MMS	NASA	USA	Science	Operating	HEO	Large 1,360 kg	4	2015	2015	
Molniya	Russian military	Russia	Telco Voice TV	Operating Some satellites have been decommissioned	Specific	Large 1,400 to 1,800 kg	16	1965	1994	Using unconventional elliptical orbits at high altitude to cover Russia's Northern territories
Sirius	SES Sirius (now part of SES)	Sweden	Telco Voice, data TV	Defunct as constellation	GEO	Large 1,400 to 4,400 kg	4 (1 active - so no more network)	1989	1998	
TDRSS	NASA US military	USA	Telco Voice, data Support to space launch	Operating Some satellites have been decommissioned	GEO	Large 2,200 kg	3 main +6 spare	1983	2002	
1worldspace	WorldSpace	USA	Broadcasting Radio	Defunct	GEO	Large 2,750 kg	2	1998	2000	
Radiosat	Sirius XM Radio, Inc.	USA	Broadcasting Radio	Operating	GEO	Large 3,800 kg	6 (4 launched)	2000	2000 (1st 3) 2013	
JSAT constellation	SKY Perfect JSAT Group	Japan	Telco Voice, data TV	Operating Some satellites have been decommissioned	GEO	Large Over 1,400 kg	15 active	1989	2019	
Spacebelt	Cloud Constellation Corporation	USA	Data storage network	Planned	LEO	Medium 400 kg	12	2019	n/a	
TripleSat	Earth-i	UK	Earth monitoring (imagery)	Operating	LEO	Medium 447 kg	3	2015	2015	
Double Star	ESA and CNSA	European Union and China	Science	Terminated	HEO	Medium 660 kg	2	2004	2007	
LeoSat	LeoSat Enterprises	USA	Telco Broadband	Planned	LEO	Medium 670 kg	108	2021	2022	
Galileo	European Space Agency	European Union	GPS	Operating Still in deployment	MEO	Medium 675 kg	24 +6 spare	2005 (demonstrator) 2011	2020	
Iridium	Iridium (bought by private equity funds)	USA	Telco Voice, data	Operating	LEO	Medium 680 kg	66 +spare	1997	1998 Followed by replacements until 2017	

Figure 32.2 Extract from the full list of satellite networks – please refer to Appendix I for the full report.[a]

a) Please note that this list is not exhaustive because of the exclusions mentioned above (no inclusion of military networks, etc.). This is also valid at the time of writing, January 2018, in a very fast changing environment, with some recent update made in March 2019. New players can be announced at any time or plans for future constellations may be dropped. There are also planned constellations that have not been made public. During my various assignments, I have met with space agencies, research organizations, and public and private companies that are planning to deploy networks of satellites but did not make any public announcement yet. Expect some new players from South East Asia, Central and South America, as well as Africa.

The existing and planned networks are classified in three categories:

1. *Broad data connectivity*: The most followed and anticipated networks offer the promise of very fast, cheap, and global data connectivity. The objective is to connect the other 50% of the Earth population that does not have Internet connectivity, as well as to offer an alternative to existing broadband access, driving prices down. Unconnected people represent a huge opportunity. By allowing another 4 billion people to connect to the Internet, satellite owners, operators, and service providers (sometimes the same company) will tap a virgin market. Low cost or free data connectivity will help to quickly build a strong community that can easily be monetized. Since the value is not in the volume of data any longer but in the analytics and the size of the connected audience, disruptive business models will emerge.

2. *Internet of Things (IoT) and machine-to-machine (M2M) connectivity*: Connected objects are representing a multibillion-dollar industry. The IoT and M2M connectivity is gaining strong momentum, with capacity issues that constellations of satellites are aiming at addressing. Earth-based operators are all offering or considering offering solutions for connected objects using different protocols and systems (Sigfox, LoRa, Zigbee, and others). Satellite-based solutions can be cost-effective and offer fast connectivity and services to connected objects. This can also be compatible with automatic identification system (AIS). For broad data connectivity, the value will be in the services and products offered around the data, and not on the volume.

3. *Specialized communication and data services*: Some networks are solely focusing on specific sectors, such as maritime communication and data services. Interestingly, such networks can also be upgraded to support some of the megatrends identified earlier, such as asteroid mining and deep space exploration. Unlike the two other categories listed earlier, companies operating such networks of satellites will be able to charge fees for base services and value-added products.

32.3.1.2 Earth Observation/Monitoring (EO)

The first Earth observation or Earth monitoring satellites were deployed for imagery (both commercial and for governmental use, including national security) and for science (study of the atmosphere, Earth condition, monitoring of natural resources). The heavy costs linked to the development, launch, and operation of a network of GEO satellites created a barrier to entry, allowing monopolies to maintain their strong position in sectors such as image collection and distribution, weather, and mapping. Today, networks of small satellites allow new entrants to either directly compete with existing players or offer new services and products to new customers, in the same and new sectors.

The main two categories are:

1. *Affordable imagery*: Companies such as Planet Labs are democratizing the access to Earth imagery. Other existing and upcoming constellations and formations will be offering high-definition, low-cost imagery and data analytics. Like with data transmission, there is little value in the images themselves. Some companies will be offering high-resolution images for free or at a minimal cost to customers. The business models of operators will be to offer value-added data analytics. Leveraging progresses made in deep machine learning and artificial intelligence, analytical layers will meet traditional and NewSpace customers' growing needs. Affordable imagery can serve as the basis for specialized services.

2. *Specialized earth monitoring*: Several networks of spacecraft are aiming at offering specialized Earth monitoring solutions. Some are based on optical imagery and offer software solutions for dedicated sectors such as agriculture, emergency services, asset tracking, national security, natural resources, and many others. Constellations of Synthetic Aperture Radar (SAR) or infrared satellites are also being planned, as well as commercial and scientific networks with specific on-boarded instruments for dedicated monitoring. Commercially viable business models are being designed around technologies that used to be too expensive or too complex and that small satellites can now offer to a broader group of customers.

Easy access, affordability, global coverage, and disruptive applications and business models are benefiting and also stimulating the growing presence of networks of satellites of all sizes.

32.3.2 Nanosatellites in This Framework

Existing and prospective networks of satellites include spacecraft of all sizes, from picosatellites to heavy platforms of several tons. Irrespective of age, the large satellites are usually dedicated to GPS systems, TV broadcasting, telecommunication (voice and data), as well as scientific uses.

Satellites ranging from around 80 kg to 1 ton can be used for most types of applications: imagery, remote sensing, broadband communication, weather, science, AIS tracking, IoT, among others.

Nanosatellites, and by extension microsatellites, up to 20 kg were considered and can be grouped as follows (see Figure 32.3):

- Low cost broadband.
- Earth monitoring, generic imagery, or specialized sensors.
- IoT and M2M.
- Weather monitoring and analysis.
- Science and technology demonstration.

The preceding chart is simplified and does not include dropped projects. The total number of satellites includes failed and decommissioned spacecraft and presents the total, maximum number of satellites planned for each project.

Nanosatellites, due to their limited size, cannot compete with heavier spacecraft in terms of performance. A LEO constellation of 150 kg communication or imagery satellites will outperform a constellation of 4 kg CubeSats, irrespective of the number of satellites the nanosatellite constellation would be based on. Because larger platforms can carry heavier and multiple payloads, they can embark better electronics, computing, and power systems. Nanosatellites do, however, bring significant advantages that make them very attractive to existing and new players.

Nanosatellites, especially when based on the widely adopted CubeSat format, beat any other platform on pricing from manufacturing to payload integration. Standardized buses allow easy mount as secondary payload on rockets to deployment from the International Space Station (ISS). When meeting with some industry leaders in the nanosatellite segment, I have even heard the idea that the smallest platforms can be and should be seen as

Name	Use	Status	Orbit	Sat type	Target sats	1st launch
CICERO	Weather	Indeployment	LEO	Nano - 10kg	12	2017
GHGSat	EO	Indeployment	LEO	Micro - 15kg	20	2016
Landmapper	EO	Indeployment	LEO	Nano - 10kg	30	2014
Planet	EO	Indeployment	LEO	Nano - 5kg	140	2014
Sky and Space Global	Telco	Indeployment	LEO	Nano - 6kg	200	2017
Spire Lemur	Weather	Indeployment	LEO	Nano	125	2013
BRITE	Science	Operating	LEO	Nano - 7kg	6	2013
QB50	Science	Operating	LEO	Nano	50	2017
Adelis-SAMSON	EO	Planned	LEO	Nano	3	2018
Aerial & Maritime	AIS	Planned	LEO	Nano - 4kg	80	2018
AISTech	AIS	Planned	LEO	Nano	100	2018
Analytical Space	IoT	Planned	LEO	Nano		2018+
Astrocast	IoT	Planned	LEO	Nano	64	2018
Audacy	Telco	Planned	MEO	Nano - 4kg	3	2019
Blink Astro	IoT	Planned	LEO	Nano		2018+
Bluefield	EO	Planned	LEO	Micro - 12kg	20	2019
Capella Space	EO	Planned	LEO	Micro - 12kg	36	2018+
Fleet	IoT	Planned	LEO	Micro - 12kg	100	n/a
HawkEye 360	EO	Planned	LEO	Micro - 15kg	15	2018
Helios Wire	IoT	Planned	LEO	Micro - 20kg	30	2018
Hera Systems	EO	Planned	LEO	Micro - 12kg	9	2018+
Hiberband	IoT	Planned	LEO	Nano	48	2018+
Hypercubes	EO	Planned	LEO	Nano	12	n/a
ICEYE	EO	Planned	LEO	Nano	9	2018
KASKILO	IoT	Planned	LEO	Nano	300	2019
KEOSat	EO	Planned	LEO	Nano - 10kg	16	2018+
Kepler Communications	IoT	Planned	LEO	Nano	140	2018
Ocean-Scan	AIS	Planned	LEO	Nano	6	2019
OMS	Weather	Planned	LEO	Micro - 20kg	40	2018+
Planetary Resources	R&D	Planned	LEO	Micro - 15kg	10	2018
PlanetIQ	Weather	Planned	LEO	Micro - 20kg	18	2018+
Reaktor Space Lab	Science	Planned	Moon	Nano	28	2018+
Rupercorp	EO	Planned	LEO	Nano - 4kg		2019
SkiFy	Telco	Planned	LEO	Nano	60	2018+
Transcelestial	Telco	Planned	LEO	Nano		2018+
UnseenLabs	AIS	Planned	LEO	Nano - 5kg		2018+

Total	**1730**	
Planned	1494	(minimum)
In Deployment	200	
Operating	36	

Figure 32.3 Existing and upcoming networks of nanosatellites (base January 2018). Source: Courtesy of Nanosatelites Base.

disposable. Some may be lost during the injection and some will fail over time. All will deorbit and be replaced by a newer generation after 1 or 2 years. But, on top of price attractiveness and the concept of disposability, networks of nanosatellites offer tremendous advantages and promising opportunities.

The example of QB50, despite many challenges (to find a suitable launch vehicle, to coordinate the work done by 50 different players, and to finance the program), is a good demonstration of how a multinational constellation can be designed, using spacecraft manufactured by different actors, for different specific purposes. Altogether they serve a common mission, but individually each spacecraft plays a unique role.

Future networks of nanosatellites, on top of the one listed earlier, will include some degree of evolution. Networks can combine different clusters. Networks will also be based on combinations of different constellations, mixing sizes for different uses, such as a LEO constellation dedicated to imagery and another synchronized LEO network handling data services; both working together to offer one global, integrated product. Networks of nanosatellites are considered for Moon or Mars orbits.

32.4   Challenges and Issues

The growing number of networks of satellites is a confirmation of the importance played by small satellites.

They represent a huge opportunity, but constellations and formations also face challenges and bring issues.

32.4.1   Overcapacity

If all the planned constellations reach full deployment, overcapacity can be expected. If all of the OneWeb and SpaceX constellations are successful, they would, at best, create healthy competition. Indeed, it would take more than 2, 3, or 4 constellations to ensure everyone on Earth is connected to the Internet and at a good speed. This also applies to connected objects. In total, over 1000 spacecraft will be dedicated to broadband and IoT. This looks like a very high number, but in fact this may not meet consumers' requirements. To combine space constellations with ground infrastructure will be important and alliances between traditional telcos and space companies can be anticipated.

Overcapacity may rather come in the area of optical imagery, already well occupied by many players. NewSpace companies offering low-cost imagery will face direct competition from their customers on image acquisitions. Nations or companies could launch their own nanosatellite or constellations to get independent access to space and images of the Earth. Over time, this could be more economical than to purchase images or data from satellite operators or imagery companies. Several players aim at offering free images and charge for a service or an analytical solution. The key will be on the end services and value-added software.

32.4.2   Lack of Launch Opportunities

In the previous two chapters, the impact made by the growing number of small satellites on launch systems was discussed. At the present time, there are enough launch opportunities to absorb the demand, but soon demand will face a shortage of supply, unless dedicated launch systems come to market.

Some of the sponsors of the upcoming networks identified herein and in Appendix I have booked launches with established or prospective launch service providers. Some have switched from one operator to another due to expected delays in system delivery or because of funding concerns. New companies aiming at launching demonstrators or small constellations of nanosatellites are relying on space brokers or integrated service companies such as Innovative Solutions in Space (ISIS) and GomSpace, which are providing the hardware, sometimes the payload and the launch service.

Will SpaceX agree to launch satellites that may compete with their own constellation?

There could be two main problems:

1. Anticipated NewSpace launch systems do not come to market on time, or do not come to market at all. Some companies have already retired their systems or have gone bankrupt, and some others are experiencing significant delays of many years. There is a probability that between 2020 and 2022 constellations of nanosatellites face a real shortage of options to launch at a decent price.
2. Over the long run, constellations will require on-demand launch capabilities, which only dedicated low-cost systems can address. Spare satellites are launched alongside operating spacecraft, but in limited supply (between 1% and 3%), representing a serious concern for the continuity of the service.

The nanosatellite network industry is stimulating the demand for low-cost dedicated launch systems. But unless they come to market quickly, this segment may face a true setback.

32.4.3 Space Debris

Space debris is already a concern and active satellites from past or existing networks have caused problems; for instance, in 2009 when an Iridium satellite collided with a decommissioned Russian spacecraft.

The likelihood of the Kessler effect, whereby the density of satellite in LEO is such that spacecraft or space debris colliding will generate a cascade, is increasing because of the large number of small satellites that will be launched in the next few years. With over 2000 spacecraft to be launched in formations and constellations over the next 4–5 years, the level of risk increases significantly to a level unseen and unforeseen before.

The problem is not only the increasing quantity of spacecraft but also a combination of this with other factors. Nanosatellite constellations are built with spacecraft bearing a short lifetime in space. They have limited power because of their mass. They are subject to strong orbital drag because of their low orbit. Even if re-entry is planned, seeing 500 or more nanosatellites falling onto the Earth is a major risk. Some of these satellites will fail. The probability of seeing a nanosatellite of an emerging space nation or a university failing to power up in space or reach its orbit is much higher than with institutional, larger platforms. Some NewSpace companies are considered as hobbyists by the industry. Some companies want to mass-produce cheap platforms knowing the failing rate can be up to 30%. The risks of seeing uncontrollable spacecraft in space is significantly increasing.

R&D in space will become important and more activities will take place in orbit including the demonstration of space debris removal technologies, which, by itself, represents a high degree of risk.

Policy and quality control are key to ensure that spacecraft launched will behave as planned. Risk mitigation is a major component of each mission, and upcoming dedicated launch system providers, space brokers, and NewSpace satellite companies will have to address this very seriously.

32.4.4 Regulatory

This section of the book addresses the economics of small satellites, and this chapter focuses on the opportunity networks of nanosatellites bring to grow it. Regulatory and legal matters are nevertheless very important, as they bring constraints but also necessary guidelines and policies to mitigate some risks, such as space debris.

International and national space laws such as the Outer Space Treaty exist to partially regulate the access and the use of space. While, in practice, no regulatory body can truly prevent a foreign entity to launch a nanosatellite, and rules and guidelines are well followed enough to guarantee that the upcoming networks of nanosatellites cause minimal threat to other, existing or upcoming spacecraft or network. Space debris mitigation is partially addressed along with communication protocols and frequency attribution.

There are, however, some gray areas that policymakers of space nations need to clarify. Only few nations have a commercial space framework in place that truly regulates private commercial space companies, including dedicated launch systems and constellations. Servicing, replacement, sustainability, and decommissioning of future very large networks and activities, such as private deep space exploration and natural resources exploitation, are yet to be included in global policies, recognized and followed by all nations.

32.4.4.1 Conclusion: Networks of Nanosatellites are a Game-changing Technology

Networks of nanosatellites are disrupting this industry and the industries they target, such as Earth imagery, Earth monitoring, and broadband communication. They are enabling new business models and support the emergence of new companies that will compete with traditional players, operating fleets of large satellites. Networks of nanosatellites also come with challenges and risks that the industry needs to address, sooner than later.

Reference

1 Clarke, A. (1945). Extra-terrestrial relays: can rocket stations give world-wide radio coverage? In: *Wireless World*.

Further Reading

Abbany, Z. (2017). Inter-agency meeting to 'improve' space law on megaconstellations of satellites. In: *Deutsche Welle*.

Brodkin, J. (2017). SpaceX and OneWeb broadband satellites raise fears about space debris. In: *Ars Technica*.

Clarke, A.C. (1945). Extra-terrestrial relays—can rocket stations give world-wide radio coverage ? In: *Wireless World*.

Cookson, C. (2016). *Nano-satellites dominate space and spread spies in the skies. The Financial Times* July 2016.

Crisp, N.H., Smith, K., and Hollingsworth, P. (2015). Launch and deployment of distributed small satellite systems. *Acta Astronautica* 114: 65–78.

LeoSat (2017). *Faster Than Fiber – Espace & Exploration Magazine looks at LeoSat's Unique Constellation for Enterprise Data, February 2017. Le magazine de l'Aventure Spatiale N° 37. France.*

Fei, L. (2014). Satellite network constellation design. In: *Satellite Network Robust QoS-aware Routing*, 21–40.

Gill, E., Sundaramoorthy, P., Bouwmeester, J., Zandbergen, B., Reinhard, R., (2010). Formation flying within a constellation of nano-satellites: the QB50 mission, 6th International Workshop on Satellite Constellation and Formation Flying, Taipei.

Guillemin, T., (2015). LEO Constellations: What You Need to Know.

Halt, T., (2014). A Table of Commercial Satellite Constellations.

Henri, Y., (2017). Regulatory Challenges for Small Satellites and New Satellite Constellations, ITU.

Johnson, C. D., (2014). Legal and Regulatory Considerations of Small Satellite Projects.

'Mega-constellations' of satellites are causing 'space junk' in Earth's orbit, The Telegraph, April (2017).

Pérez Cano, J. S., (2015). Challenges and opportunities with small satellites, NexComm/SPACOMM Conference, Barcelona, April 2015.

Planet—Flock 1 Imaging Constellation, Through eoPortal, November (2013).

Planet Labs' remote sensing satellite system, Planet marketing material, (2013). CubeSat developers workshop, Utah.

Raja, P. (2015). *The Art of Satellite Constellation Design: What You Need to Know*. July: Astrome Technologies.

Sellitto, P., Dauphin, P., Dufour, G., Eremenko, M., Coman, A., Forêt, G., Beekmann, M., Gaubert, B, Flaud, J-M., (2011). Performance Assessment of Future LEO and GEO Satellite Infrared Instrument to Monitor Air Quality.

GPS, (2016). *Space segment: Constellation Arrangement*, through http://GPS.gov, June 2016.

Staedter, T., (2017). Constellations of Internet Satellites Will Beam Broadband Everywhere.

Suber, R. (2017). *Next Generation Satellites: The Path for the Pacific Islands*. Intelsat.

Vance, A., (2017). The Tiny Satellites Ushering in the New Space Revolution, through Bloomberg.

Velivela, V. (2015). Small satellite constellations: the promise of 'Internet for All'. In: *ORF Issue Brief*.

32.A List of Existing and Upcoming Networks of Satellites—January 2018, Updated March 2019ª

Please refer to Chapter III-3c for background and explanations on certain exclusions. Sources: Multiple, author's own research and interviews.

Name (A to Z)	Sponsors	Country	Use	Status	Orbit	Sat type	No. of sats	Start date/ first launch	Completion date
1worldspace	WorldSpace	USA	Broadcasting Radio	Defunct	GEO	Large 2750 kg	2	1998	2000
A-train	NASA Other space agencies	USA International	Earth monitoring Science, research	Operating	LEO	Small–Large 500–2000 kg	6	2002	2014
Adelis-SAMSON	Israel Technology Institute and the Israel Space Agency	Israel	Earth monitoring	Planned	LEO	Nano CubeSat-based	3	2018	2018+
Aerial & Maritime	GOMSpace	Denmark	AIS tracking	Planned	LEO	Nano CubeSat-based 4 kg	80	2018	2021
AISTech	AISTech Space	Spain	AIS tracking	Planned	LEO	Nano CubeSat-based	25, then 100	2018	2022
Analytical Space	Analytical Space	USA	Data IoT–M2M services	Planned	LEO	Nano CubeSat-based	n/a	2018+	n/a
Astranis	Astranis	USA	Telco Broadband	Planned	LEO	n/a	n/a	n/a	n/a
Astrocast	Astrocast SA (former ELSE)	Switzerland	Data IoT–M2M services	Planned	LEO	Nano CubeSat-based	64	2018	2019
Audacy	Audacy Space	USA	Telco Broadband	Planned	MEO	Nano CubeSat-based 4 kg	3	2019	2021
AxelGlobe – GRUS	Axelspace	Japan	Earth monitoring (imagery)	Planned	LEO	Micro 80–100 kg	50	2018	2022
BeiDou (ex-COMPASS)	BeiDou Navigation Satellite System (BDS)	China	GPS	Decommissioned	GEO	Large	3 + 1 spare	2000	2003

Name	Organization	Country	Application	Status	Orbit	Satellite size	Number	First launch	Completion
BeiDou-2	BeiDou Navigation Satellite System (BDS)	China	GPS	In deployment	GEO and MEO	Large	25 now in orbit out of 35 (5 GEO) (30 MEO and IGSO)	2007	2020
BGAN	Inmarsat	UK	Telco Voice, data	Operating	GEO	Large	3	2005	2013
BlackSky	BlackSky Global	USA	Earth monitoring (imagery)	In deployment	LEO	Micro 44–55 kg	60	2016 (demonstrator), 2018 constellation	n/a
Blink Astro	SpaceWorks Enterprises	USA	Data IoT–M2M services	Planned	LEO	Nano CubeSat-based	n/a	2018+	n/a
Bluefield	Bluefield	USA	Earth monitoring	Planned	LEO	Micro 12 kg	20	2019	2020
BRITE	Consortium of universities	Canada, Austria, Poland	Science	Operating	LEO	Nano 7 kg	6	2013	2014
Capella Space	Capella Space	USA	Earth monitoring (SAR)	Planned	LEO	Micro CubeSat-based 12 kg	36	2018+	n/a
CASIC Hongyun	CASIC	China	Telco Voice, data	In deployment	LEO	Small 600 kg	156	2018	2022
CASC	Hongyan	China	Telco Voice, data	In deployment	LEO	Small 147 kg	320	2018	2023

(Continued)

Name (A to Z)	Sponsors	Country	Use	Status	Orbit	Sat type	No. of sats	Start date/ first launch	Completion date
CICERO	GeoOptics	USA	Weather	In deployment	LEO	Nano CubeSat-based 10 kg	12 4 launched (only 1 active)	2017	n/a
Cluster	ESA	European Union	Science	Failed	HEO	Large 1200 kg	4	1996	n/a
Cluster II	ESA	European Union	Science	Operating	HEO	Large 1200 kg	4	2000	2000
DMCii	DMC International Imaging	For five governments	Earth monitoring for disaster monitoring	Operating Some satellites have been decommissioned	LEO	Micro and small 80–130 kg platforms	4 (1st Gen) 4 (2nd Gen) Presently 3 working	2002	2011
Double Star	ESA and CNSA	European Union and China	Science	Terminated	HEO	Medium 660 kg	2	2004	2007
Earth-i	Earth-i	UK	Earth monitoring (imagery)	Planned	LEO	Small 100 kg (+ or −)	15	2018	2019
Earthcube	Earthcube	France	Earth monitoring (thermal IR imagery)	Dropped	LEO	Nano	10	2018	2019
Facebook	Facebook	USA	Telco Broadband	Dropped	GEO	Large (possibly existing GEO sats)	n/a	n/a	n/a
Fleet	Fleet Space	Australia	Data IoT–M2M services	Planned	LEO	Micro CubeSat-based 12 kg	100	n/a	n/a

Galileo	European Space Agency	European Union	GPS	Operating / Still in deployment	MEO	Medium 675 kg	24 + 6 spare	2005 (demonstrator), 2011	2020
GEOS	NASA	USA	Weather	Operating / Still in deployment for update / Some satellites have been decommissioned	GEO	Large	2 in operation	1975	2020+
GHGSat	GHGSat Inc.	Canada	Earth monitoring	Operating / Still in deployment	LEO	Micro 15 kg	20 (1 in orbit)	2016	n/a
Global Xpress	Inmarsat	UK	Telco Broadband	Operating	GEO	Large	4	2013	2017
Globalstar	Globalstar (bought by private equity funds)	USA	Telco Voice, data	Operating / Some satellites decommissioned	LEO	Small 500 kg	48 + spare	1998	2000
Globalstar Second Gen	Globalstar (bought by private equity funds)	USA	Telco Broadband	Operating	LEO	Small/medium 650 kg	24 + more to follow	2010	2013
GLONASS	USSR/Russian governments	Russia	GPS	Operating / Still in deployment for K update / Some satellites have been decommissioned	MEO	Small/medium/large 250–1450 kg	24 (base configuration, but varied over time)	1982	1995 (1st Gen), continuous updates with M and K (2003, 2011, continuous)

(Continued)

Name (A to Z)	Sponsors	Country	Use	Status	Orbit	Sat type	No. of sats	Start date/ first launch	Completion date
Gonets	ISS Reshetnev (from Russian space agency)	Russia	Telco Voice, data	Operating Upgraded	LEO	Small 280 kg (2nd Gen)	13 (at present time, 12 2nd Gen and 1 1st Gen)	1996	2015 (2nd Gen)
Google	Google	USA	Telco Broadband	Unclear (Google sold sat imagery business but could plan for broadband)	LEO	n/a	>1000	n/a	n/a
GPS	US Government	USA	GPS	Operating Still in deployment for update Some satellites have been decommissioned	MEO	Medium/large 750–1600 kg	72 launched, 31 operational	1978	2016 (last), planned until 2034
HALO	Laser Light Com.	USA	Telco Broadband	Planned	MEO	n/a	12	n/a	n/a
HawkEye 360	HawkEye 360	USA	RF monitoring	Planned	LEO	Micro 15 kg	15	2018	2019
Helios Wire	Helios Wire	Canada	Data IoT–M2M services	Planned	LEO	Micro CubeSat-based 20–25 kg	30	2018	n/a
Hera Systems	Hera Systems	USA	Earth monitoring (imagery)	Planned	LEO	Micro CubeSat-based 12 kg	9	2018+	n/a
Hiberband	Hiber	The Netherlands	Data IoT–M2M services	Planned	LEO	Nano CubeSat-based	48	2018+	n/a

	Operator	Country	Application	Status	Orbit	Size	Number		
Hypercubes	Hypercubes	USA	Earth monitoring (imagery)	Planned	LEO	Nano	12	n/a	n/a
ICEYE	ICEYE Oy	Finland	Earth monitoring (SAR)	Planned	LEO	Nano	3 (1st Gen) 6+ (2nd Gen)	2018	2019
ICO	Pendrell Corporation (ex-ICO)	USA	Telco Voice, data	Defunct	MEO	n/a	10 + 2 spare	n/a	n/a
Infante	Tekever	Portugal	Earth monitoring Telecom (data)	Planned	LEO	Micro 25 kg	12	2020	2020
Intelsat	Intelsat	USA	Telco Broadband	Operating Still in deployment for update Some satellites have been decommissioned	GEO	Large (large fleet)	>30—not all working in the network	1965	2018+
Iridium	Iridium (bought by private equity funds)	USA	Telco Voice, data	Operating	LEO	Medium 680 kg	66 + spare	1997	1998, followed by replacements until 2017
Iridium NEXT	Iridium (bought by private equity funds)	USA	Telco Broadband	In deployment	LEO	Medium 860 kg	66 + spare (total 81)	2017	2018

(Continued)

Name (A to Z)	Sponsors	Country	Use	Status	Orbit	Sat type	No. of sats	Start date/first launch	Completion date
IRNSS	Indian Space Agency	India	GPS	Operating Still in deployment	GEO/GSO	Medium 614 kg	9 (7 launched; 6 working; 4 more planned) 4 more planned)	2013	2017 (1st Gen)
JSAT constellation	SKY Perfect JSAT Group	Japan	Telco Voice, data TV	Operating Some satellites have been decommissioned	GEO	Large Over 1400 kg	15 active	1989	2019
Kacific	Kacific	Singapore	Telco Broadband	MEO constellation project dropped GEO project planned	GEO (originally MEO)	Large >4000 kg	1	2019	2019
KASKILO	EightyLEO	Germany	Data IoT-M2M services	Planned	LEO	Nano CubeSat-based	300	2019	n/a
KEOSat	Karten Space	Spain	Earth monitoring (imagery) AIS tracking	Planned	LEO	Nano CubeSat-based 10 kg	14–16	2018+	n/a
Kepler Commu- nications	Kepler Communications	Canada	Data IoT-M2M services	Planned	LEO	Nano CubeSat-based	140	2018	2019+
Kleos	Kleos Space	Luxembourg	Geolocalization AIS tracking	Planned	LEO	n/a	20	2019	n/a
Koolock	Koolock	USA	Weather (IR imagery)	Planned	LEO	n/a	n/a	n/a	n/a

Name	Operator	Country	Application	Status	Orbit	Size	Number	Year	Year
Landmapper	Astro Digital (former Aquila)	USA	Earth monitoring for agriculture and disaster monitoring	In deployment	LEO	Nano CubeSat-based 10 and 20 kg	10 + 20	2014	2018
LeoSat	LeoSat Enterprises	USA	Telco Broadband	Planned	LEO	Medium 670 kg	108	2021	2022
MMS	NASA	USA	Science	Operating	HEO	Large 1360 kg	4	2015	2015
Molniya	Russian military	Russia	Telco Voice TV	Operating Some satellites have been decommissioned	Specific	Large 1400–1800 kg	16	1965	1994
Northstar	NorStar Space Data	Canada	Earth monitoring Space monitoring	Planned	LEO	n/a	40	2019	n/a
O3b	O3b Networks Ltd. now part of SES	USA (Luxembourg under SES)	Telco Voice, data Broadband	Operating Upgrade planned for 2021	MEO	Medium 700 kg	20 + 3 spare	2013	2019 2021 for mPower update
Ocean-Scan	SRT Marine		AIS Tracking	Planned	LEO	Nano CubeSat-based	6	2019	2019
OMS	Orbital Micro Systems	USA, UK	Weather	Planned	LEO	Micro CubeSat-based 20 kg	32–40	2018+	n/a
OneWeb	OneWeb	USA	Telco Broadband	Planned	LEO	Small 175–200 kg	648 in 1st Gen 1972 additional in 2nd Gen	2018	2020+

(Continued)

Name (A to Z)	Sponsors	Country	Use	Status	Orbit	Sat type	No. of sats	Start date/ first launch	Completion date
OQ Technology	OQ Technology	Luxembourg	Data IoT–M2M services	Planned	LEO	n/a	n/a	n/a	n/a
ORBCOMM	ORBCOMM (originally part of Orbital Sciences)	USA	Data IoT–M2M services	Operating Gen 1 partially replaced by Gen 2 Currently problems with 6 sats	LEO	Micro/small 42 kg (Gen 1) then 80 115 kg (Gen 2)	35 (Gen 1), 24 in operation today, 18 (Gen 2) more to come	1992 (Gen 1) 2014 (Gen 2)	1999 (Gen 1) 2015 (Gen 2)
Planet	Planet Labs	USA	Earth monitoring	Operating Still in deployment	LEO	Nano 5 kg	140 (more to be launched)	2014	>2018
Planetary Resources	Planetary Resources	USA	R&D Space observation	Planned	LEO	Micro CubeSat-based 15 kg	10, 2 demonstrators in orbit	2018	2019+
PlanetIQ	PlanetIQ	USA	Weather	Planned	LEO	Micro CubeSat-based 20 kg	12, then 18	2018+	2020
Pléiades	CNES	France and Italy	Earth monitoring	Operating	LEO	Medium 970 kg	2	2011	2012
QB50	Von Karman Institute, European Union	European Union, Global	Science, distributed sensors	Operating Some satellites have been decommissioned	LEO	Nano CubeSat-based	50	2017	2017
Quasi-Zenith	JAXA – Japan Space Agency	Japan	GPS	Operating Still in deployment	GEO/GSO	Large >4000 kg	4	2010	2018

Name	Company	Country	Application	Status	Orbit	Size/Mass	Number	Year	Year
Radiosat	Sirius XM Radio, Inc.	USA	Broadcasting Radio	Operating	GEO	Large 3800 kg	6 (4 launched)	2000	2000 (first three) 2013
RapidEye	RapidEye, now 100% Planet Labs ownership	Germany, now USA	Earth monitoring	Operating	LEO	Small 150 kg	5	2008	2008
Reaktor Space Lab	Reaktor Space Lab	Finland	Moon observation	Planned	Moon orbit	Nano	20–28	2018+	n/a
RUNNER	Terran Orbital and ImageSat Intl	USA, Israel	Earth monitoring (imagery)	Planned	LEO	n/a	n/a	n/a	n/a
Rupercorp	Rupercorp	Argentina	Earth monitoring	Planned	LEO	Nano CubeSat-based 4 kg	n/a	2019	n/a
Samsung	Samsung	Korea	Telco Broadband	Dropped	LEO	n/a	>4000	n/a	n/a
Satellogic	Satellogic	USA	Earth monitoring	Operating Still in deployment	LEO	Micro 37 kg	Min 25—Up to 300, 3 in orbit,	2016	n/a
Sirius	SES Sirius (now part of SES)	Sweden	Telco Voice, data TV	Defunct as constellation	GEO	Large 1400–4400 kg	4 (1 active, so no more network)	1989	1998
Siwei Star SuperView	Siwei Star Co., CASC	China	Earth monitoring	Operating Still in deployment	LEO	Small 560 kg	16 (4 in orbit)	2016	n/a
SkyFi	SkyFi	Israel	Telco Broadband	Planned	LEO	Nano	60	2018+	n/a

(Continued)

Name (A to Z)	Sponsors	Country	Use	Status	Orbit	Sat type	No. of sats	Start date/ first launch	Completion date
Sky and Space Global	Sky and Space Global	UK, Australia	Telco Voice, data Broadband	In deployment	LEO	Nano CubeSat-based 6 kg	200 (3 in orbit)	2017	2018+
Spacebelt	Cloud Constellation Corporation	USA	Data storage network	Planned	LEO	Medium 400 kg	12	2019	n/a
SpaceX	SpaceX	USA	Telco Broadband	Planned	LEO	Small <400 kg	400+	2018	2020
Spire Lemur	Spire Global	USA	Weather AIS tracking	In deployment	LEO	Nano CubeSat-based	125 (planned), 1 in orbit	2013	2018+
SPOT	Spot Image (previously CNES)	France	Earth monitoring	Operating Some satellites have been decommissioned	LEO	Medium/large 712 kg to >3000 kg	3 (from total 7 launched)	1986	2014
SpymeSat									
TDRSS	NASA US military	USA	Telco Voice, data Support to space launch	Operating Some satellites have been decommissioned	GEO	Large 2200 kg	3 main + 6 spare	1983	2002
Teledesic	Teledesic	USA	Telco Broadband	Defunct	LEO	Small 120 kg (+ or −)	840 + spare, then scaled down to 288 (1997)	1998 (demonstrator)	n/a
Telesat	Telesat	Canada	Telco Broadband	Planned	LEO	Micro/small 50–100 kg	117 + 3 spare	2020	2021

Terra Bella (ex-SkySat)	Skybox Imaging, later Google and now 100% Planet Labs ownership	USA	Earth monitoring	Operating	LEO	Micro/small 80–120 kg	24 (13 launched)	2016	2019
Thuraya	Thuraya	UAE	Telco Voice, data	Operating	GEO	Large >4000 kg	3 (1 in orbit)	2000	N/A 3rd sat failed
Transcelestial	Transcelestial	Singapore	Telco Broadband	Planned	LEO	Nano CubeSat-based	n/a	2018+	n/a
Trident	Trident Space	USA	Earth monitoring (SAR)	Planned	LEO	n/a	6–12	2019	n/a
TripleSat	Earth-i	UK	Earth monitoring (imagery)	Operating	LEO	Medium 447 kg	3	2015	2015
UnseenLabs	UnseenLabs	France	AIS tracking	Planned	LEO	Nano CubeSat-based 5 kg	n/a	2018+	n/a
Ursa Space Systems	Ursa Space Systems	USA	Earth monitoring (SAR)	Dropped	LEO	Nano	32	n/a	n/a
UrtheCast	UrtheCast	Canada	Earth monitoring (imagery and SAR)	Planned	LEO	Medium/large 700–1800 kg	16	2019	2020
VEOWARE	Veoware	UK	Earth monitoring (imagery)	Planned	LEO	n/a	n/a	2020	2020
Westar	Western Union	USA	Telco Voice, data	Defunct	GEO	Medium/large 574–1400 kg	7 (total launched)	1974	1993
XpressSAR	XpressSAR	USA	Earth monitoring (SAR)	Planned	LEO	n/a	4	2020	2020

Index

Nanosatellites: Space and Ground Technologies, Operations and Economics, First Edition.
Edited by Rogerio Atem de Carvalho, Jaime Estela, and Martin Langer.
© 2020 John Wiley & Sons Ltd. Published 2020 by John Wiley & Sons Ltd.